C.D. PEMBLETON ASSOCIATES
HWY. CRASH RECONSTRUCTION SPECIALISTS
2711 SPRAGUE DRIVE
WALDORF, MD 20601-3022

7/96

Roadway Defects and Tort Liability

By
John C. Glennon, D. Engr., P.E.

Lawyers & Judges
Publishing Company, Inc.

This publication is designed to provide accurate and authoritative information in regard to the subject matter covered. It is sold with the understanding that the publisher is not engaged in rendering legal, accounting, or other professional service. If legal advice or other expert assistance is required, the services of a competent professional person should be sought.

–From a *Declaration of Principles* **jointly adopted by a Committee of the American Bar Association and a Committee of Publishers and Associations.**

Copyright © 1996 by Lawyers & Judges Publishing Co. All rights reserved. All chapters are the product of the Authors and do not reflect the opinions of the Publisher, or of any other person, entity, or company. No part of this book may be reproduced in any form or by any means, including photocopying, without permission from the Publisher.

Lawyers & Judges Publishing Company, Inc.

P.O. Box 30040
Tucson, AZ 85751-0040
(520) 323-1500
(602) 323-1500
FAX (520) 323-0055

Glennon, John C.,
    Roadway defects and tort liability / by John C. Glennon.
      p. cm.
    Includes bibliographical references and index.
    ISBN 0-913875-17-1
    1. Liability for traffic accidents—United States. 2. Highway law—United States. 3. Roads—Design and construction—Safety regulations—United States. 4. Roads —United States—Maintenance and repair—Safety measures. I. Title
KF1325.T7G58   1996
343.7309'42—dc20
[347.303942]   96-15339
                                                          CIP

ISBN 0-913875-17-1
Printed in the United States of America
10 9 8 7 6 5 4 3 2 1

Table of Contents

Chapter 1
Roadway Safety and Tort Liability Overview 1
  1.1 Diagnosing the Traffic Safety Problem 2
  1.2 Driver Performance 4
      Information Handling 5
      Driver Perception-Reaction Time 6
  1.3 Driver Expectancy, Roadway Consistency, and Positive Guidance 10
      Driver Expectancy 12
      Roadway Consistency 13
      Positive Guidance 14
  1.4 Nature of Roadway Defects Today 15
  1.5 Basic Legal Concepts 17
      Negligence 17
      Contributory vs Comparative Negligence 18
      Joint and Several Liability 18
      Standard of Care 18
      Notice of a Defect 19
      Sovereign Immunity 19
      Governmental vs. Proprietary Functions 19
      Discretionary vs. Ministerial Acts 20
      Design Immunity 20
  1.6 Authoritative Sources of Information 21
      Roadway Design 21
      Roadway Maintenance 22
      Traffic Control Devices 22
      Work Zone Safety 23
      Roadside Safety 23
      Rail-Highway Crossings 23
  1.7 The Intended Focus of This Book 24

Chapter 2
Roadside Safety 27
  2.1 Single-Vehicle Collision Characteristics 30
  2.2 Characteristics of Roadside Encroachments 32
  2.3 Roadside Hazard Model 34
  2.4 Clear Zone Standards 38
  2.5 Functional Roadside Elements 40
      Roadside Slopes 40
      Bridges 43
      Drainage Facilities 43
      Light Poles 45
      Sign Supports 48
      Traffic Signal Supports 53
      Curbs 53

2.6 Non-Functional Roadside Elements 56
    Trees 56
    Utility Poles 56
    Traffic Barriers 58
    Other Fixed Objects 58
2.7 Technical Aspects of Roadside Hazard Cases 59
    Roadway Widening Projects that Sacrifice Roadside Safety 59
    Steep Side Slopes 60
    Removable Objects Within the Clear Zone 61
    Moveable Objects Within the Clear Zone 61
    Non-Breakaway Functional Objects Within the Clear Zone 63
    Unprotected Functional Objects Within the Clear Zone 63
    Defective Traffic Barriers 63
    Pavement Edge Drops 64
2.8 Typical Defense Arguments 64

Chapter 3

Traffic Barriers 67
3.1 Roadside Barriers 68
    Performance 69
    Warrants 69
    Designs 70
    End Treatments 74
    Placement 78
3.2 Bridge Rails 81
    Performance 82
    Designs 83
    Transitions 85
    Retrofitting Substandard Bridge Rails 85
3.3 Median Barriers 86
    Warrants 87
    Designs 87
    End Treatments 92
    Transitions 95
    Placement 95
3.4 Crash Cushions 95
    Concept 95
    Designs 97
    Placement 101
3.5 Technical Aspects of Traffic Barrier Defect Cases 101
    Guardrail Defects 102
    Bridge Rail Defects 113
    Median Barrier Defects 118
    Crash Cushion Defects 118
3.6 Typical Defense Arguments 121

Table of Contents

Chapter 4

Stopping Sight Distance 125
    4.1 Historical Perspective 125
    4.2 Stopping Sight Distance Design Standards 127
    4.3 Accident Experience with Restricted Stopping Sight Distance 129
    4.4 Functional Analysis of Stopping Sight Distance Requirements 132
    4.5 Technical Aspects of Stopping Sight Distance Defects Cases 135
        Hillcrests with Deficient SSD on Narrow Roadways 135
        Hillcrests with Deficient SSD that Hide Nearby Intersections 137
        Sharp Roadway Curves Hidden by SSD-restricted Hillcrests 138
        Hidden STOP Signs 138
        A Project to Flatten a SSD-restricted Hillcrest Can Produce No Safety
        Benefit 138
        Sharp Roadway Curves with Restricted SSD 140
        Sharp Roadway Curves on Freeways with a Concrete
        Median Barrier 141
    4.6 Accident Reconstruction Aspects 142
    4.7 Typical Defense Arguments 142

Chapter 5

Intersection Sight Distance 145
    5.1 Sight Distance Requirements for Uncontrolled Intersections 145
    5.2 Sight Distance Requirements for Yield-Controlled Intersections 146
    5.3 Sight Distance Requirements for Stop - Controlled Intersections 147
    5.4 Rail-Highway Grade Crossing Sight Distance 148
    5.5 Technical Aspects of Intersection Sight Distance Defect Cases 149
        Uncontrolled Intersections with Deficient Sight Distance 150
        Stop-Controlled Intersections with Sight Restrictions 153
        Sight Obstructions at Freeway Interchanges 154
        Sight Obstructions at Driveways 154
    5.6 Accident Reconstruction Aspects 156
    5.7 Typical Defense Arguments 156

Chapter 6

Slippery Pavements and Hydroplaning Sections 159
    6.1 Design and Construction of Pavements 161
    6.2 Mechanics of the Tire-Pavement Interface 162
    6.3 Frictional Requirements of Traffic 164
    6.4 Measurement of Pavement Skid Resistance 169
    6.5 Accidents on Wet Pavements 172
    6.6 Standards for Skid Resistance 173
    6.7 Hydroplaning 178
        Mechanics of Hydroplaning 179
        Pavement Texture 180
        Pavement Drainage 181
        Pavement Wheel Ruts 182
        Vehicle Tires 183

6.8 Technical Aspects of Slippery Pavement or Hydroplaning Cases 184
6.9 Accident Reconstruction Aspects 187
6.10 Typical Defense Arguments 188

Chapter 7

Pavement Edge Drops 193
   7.1 Characteristics of a Pavement Edge Drop Traversal 194
   7.2 Writings and Research on Pavement Edge Drops 197
      Ivey and Griffin (1976) 197
      Nordlin, Parks, Stoughton, and Stoker (1976) 198
      Klein, Johnson, and Szostak (1977) 198
      Zimmer and Ivey (1982) 200
      Graham and Glennon (1984) 203
      Glennon (1985) 205
      Olson, Zimmer, and Pezoldt (1986) 208
      Ivey and Sicking (1986) 209
      Ivey, Mak, Cooner, and Marek (1988) 211
      Humphreys and Parham (1994) 214
   7.3 Recommended Standards for Pavement Edge Drops 215
      Recommended Practices for Normal Maintenance 216
      Recommended Contract Provisions for Resurfacing Roadways 216
      Recommended Practices for Construction Zones 216
   7.4 Technical Aspects of Pavement Edge Drop Cases 219
   7.5 Typical Defense Arguments 221

Chapter 8

Rail-Highway Grade Crossings 225
   8.1 Accident Statistics 225
   8.2 Legal Responsibilities of Railroads and Roadway Agencies 226
   8.3 Legal Requirements of Drivers at Crossings 227
   8.4 Stages of the Rail-Highway Grade Crossing Maneuver 228
   8.5 Vehicle-Train Conflict Circumstances 230
   8.6 Driver Needs 231
   8.7 Traffic Control Devices for Rail-Highway Grade Crossings 235
      Crossbuck Signs 237
      Advance Warning Signs 238
      Pavement Markings 238
      Turn Prohibition Signs 239
      Stop or Yield Signs 239
      Flashing Light Signals 239
      Automatic Gates 241
      Train Detection 242
      Traffic Signals Near Grade Crossings 243
   8.8 Sight Distance Requirements 243
      Moving Roadway Vehicle 244
      Stopped Roadway Vehicle 247

Table of Contents

8.9 Rail-Highway Grade Crossing Improvements	248
Eliminate Crossings	248
Remove Devices at Abandoned Crossings	250
Improve Sight-Restricted Crossings	251
Bring Crossings Up to Standard	253
Install Flashing Light Signals	253
Install Automatic Gates	254
Add Cantilever Signals	254
Install Active Advance Warning Signs	255
Install DO NOT STOP ON TRACKS Sign	255
Install Turn Prohibition Signs	255
Install Warning Bells	255
Install Crossing Illumination	255
Improve Roadway Horizontal Alignment	256
Improve Roadway Vertical Alignment	256
Improve Roadway Cross-Section	257
Improve Crossing Surfaces	257
Improve Train Conspicuity	258
Reduce Roadway Speed Limit	258
Reduce Track Speed Limit	259
Install Combinations of Improvements	259
8.10 Technical Aspects of Rail-Highway Grade Crossing Defect Cases	259
Sight Triangle Obstructions	261
Nearby Intersections	261
Sharp Angle Crossing Approach	262
Poor Maintenance of Signs	262
Lack of Advance Warnings	264
Poor Flashing Signal Visibility	264
Steep Crossing Grade	265
Rough Crossings	266
8.11 Accident Reconstruction Aspects	266
Train Braking Performance	266
Crush Damage Method	267
8.12 Typical Defense Arguments	269

Chapter 9

Roadway Curves	**273**
9.1 Vehicle Cornering	273
9.2 Tire Pavement Skid Resistance	274
9.3 Roadway Curve Design Standards	278
Safe Side Friction Factors	278
Maximum Superelevation Rates	279
Minimum Curve Radius	279
9.4 Roadway Curve Accident Characteristics	280
9.5 Signing for Roadway Curves	283
Curve and Turn Signs and Advisory Speed Plates	283
Combination Curves	286
Supplementary Curves Signs	287
9.6 Delineation for Roadway Curves	288

9.7 Technical Aspects of Roadway Curve Defect Cases 288
    Narrow Roadway 290
    Tangent Roadway With False Visual Cues 291
    Low or Negative Superelevation 292
    Rough or Uneven Pavements 292
    Slippery Pavements 292
    Other Elements 293
9.8 Accident Reconstruction Aspects 293
9.8 Typical Defense Arguments 293

Chapter 10

Construction and Maintenance Zones 297
    10.1 Fundamental Principles 298
    10.2 Traffic Operational Elements 299
        Tapers 300
        Detours and Diversions 302
        One-Way Traffic Control 302
        Flagging 302
        Mobile Operations 303
        Pedestrian Safety 303
        Worker Safety 306
        Typical Applications 306
    10.3 Traffic Control Devices 306
        Regulatory Signs 312
        Road Closure Signs 312
        Warning Signs 313
        Detour Signs 314
        Channelizing Devices 315
        Warning Lights 317
        Advance Warning Arrow Displays 317
        Pavement Markings 319
        Temporary Traffic Signals 319
    10.5 Technical Aspects of Construction or Maintenance Zone Defect Cases 320
    10.6 Typical Defense Arguments 322

Chapter 11

Traffic Control Devices 325
    11.1 Traffic Control Device Requirements 326
        Design 326
        Placement 327
        Operation or Application 327
        Maintenance 327
        Uniformity 327
    11.2 Driver Information Needs 327

11.3 Traffic Signs	328
STOP Signs	328
YIELD Signs	330
Warning Signs	330
Road Closure Signs	333
Wrong Way Traffic Control	333
11.4 Traffic Markings	334
Longitudinal Lines	335
Transverse Pavement Markings	337
Object Markers	338
Roadway Delineators	340
11.5 Traffic Signals	341
Traffic Signal Warrants	341
Number and Arrangement of Lenses per Signal Face	346
Flashing Signals	348
Size of Signal Lenses	349
Locations and Visibility of Signal Faces	350
Height of Signal Faces	350
Shielding of Signal Faces	352
Signal Out of Operation	352
Yellow Phase-Change Intervals	352
Pedestrian's Signal Needs	352
Hazard Identification Beacon	353
Intersection Control Beacon	354
Left-Turn Signal Displays	354
11.6 The MUTCD at Court	355
MUTCD Word Meanings	355
MUTCD Contradictions	357
Lack of Continuity	358
Plaintiff vs. Defendant	360
Summary	361
11.7 Technical Aspects of Traffic Control Device Defect Cases	362
Hidden Traffic Signs	362
Illusive Traffic Signs	364
Non-Reflective Signs	364
Failure to Warn	365
Wrong Sign for Location	366
Unclear Messages	366
Non-Standard Traffic Signs	366
Damaged Traffic Signs	368
Warning Sign in Only One Direction	368
Partially Obscured Signs	369
Signs that Directly Indicate Negligence	369
Missing End of Road Markers	369
Obstructions in the Roadway	372
Missing or Faded Pavement Markings	372
Isolated Traffic Signals	372
Signal Visibility Defects	372
11.8 Typical Defense Arguments	372

Chapter 12

Roadway Maintenance 375
   12.1 Unpaved Surfaces Maintenance 375
        Corrugations 377
        Wheel Ruts 378
        Soft Spots 378
        Dust 378
        Potholes 378
   12.2 Asphalt Pavement Maintenance 378
        Corrugations and Shoving 379
        Potholes 379
        Upheavals and Settlements 379
        Flushing 381
        Cracks 381
        Pavement Edge Drops 382
        Wheel Ruts 382
        Aggregate Polishing 382
        Debris on the Pavement 382
   12.3 Concrete Pavement Maintenance 383
        Surface Texture Distress 383
        Scaling 383
        Crack and Joint Distress 383
        Spalling 385
        Blowups 385
        Frost Heaves 386
        Pumping 386
        Pavement Edge Drops 386
        Debris on the Pavement 386
   12.4 Shoulder Maintenance 387
        Earth Shoulders 387
        Sod Shoulders 388
        Aggregate Surfaced Shoulders 388
        Surface Treated Shoulders 389
        Paved Shoulders 389
        Maintenance of Pavement Edge Drops 389
   12.5 Roadside Drainage and Vegetation Maintenance 389
        Drainage Facilities 390
        Vegetation Control 390
        Roadside Hazard Control 390
   12.6 Traffic Barrier Maintenance 390
        Guardrails 391
        Bridge Rails 392
        Median Barriers 392
        Crash Cushions 392
   12.7 Traffic Control Device Maintenance 394
        Traffic Signs 394
        Traffic Markings 395
        Traffic Signals 395
        Roadway Lighting 395

12.8 Snow and Ice Control	396
Winter Maintenance Policies and Plans	396
Storm Watch	398
Training	398
General Maintenance Procedures	399
Snow Plowing	399
Chemical and Abrasive Application	401
12.9 Technical Aspects of Maintenance Defect Cases	402
Washboard, Wheel Ruts, and Holes on Unpaved Roadways	403
Windrows of Gravel on Unpaved Roadways	403
Dust on Unpaved Roadways	403
Roughness, Potholes, and Bumps on Paved Roadways	403
Gravel on the Pavement	405
Flooded Pavements	406
Soft Shoulders	406
Icy Bridges	406
Sight Distance Obstruction Created by Snow Plowing	406
Snow Plow Cloud	407
Hydroplaning Sections in the Winter	407
12.10 Typical Defense Arguments	407

Appendix
Parts of Title 23 U.S. Code Related to Highway Safety April 1, 1995 411

About the Author 511

Index 513

Acknowledgments

I am indebted to several people for their contribution to my learning and well-being. My learning was best fostered by my mother, Virginia, and my father, Bill, whose constant attention to teaching me the correct ways to deal with people and situations has branded my style of endeavor. I owe them everything.

My great mentors in the fields of roadway design, traffic engineering, and roadway safety were John W. Hutchinson, John E. Baerwald, and Jack Keese. These three men gave me a keen sensitivity to and knowledge of the complexities of roadway safety.

A special thanks goes to Bonnie, my best friend and loving wife, who picks me up when I stumble, gives me continuing good advice across wide areas of endeavor, and provides me with constant companionship.

Thanks also to Cookie Wohlschlaeger, who typed the original manuscript for this book, including numerous edits. Your patience is appreciated!

Roadway Defects and Tort Liability

By
John C. Glennon, D. Engr., P.E.

Chapter 1

Roadway Safety and Tort Liability Overview

Motor-vehicle safety is a comprehensive pursuit that requires knowledge of the interactions between the driver, the vehicle, and the roadway environment. The needs for safety must be met promptly and continually.

Some officials plead, "We have been fighting motor-vehicle accidents for years and there is no real solution. With more vehicles and more drivers, there are bound to be more accidents." This statement may be true if we confine our efforts to the popular *solutions* of the past. A common approach has always been to search for a single panacea. More recently, however, the systems approach suggests that reducing the *frequency* and *severity* of motor-vehicle accidents requires a set of solutions aimed at maximizing the safety payoff from available funds.

The evolution of motor-vehicle safety efforts can be characterized into three distinct periods, which are described below as the Campaign Era, the Action Era, and the Priority Era.

In the *Campaign Era*, before 1966, well-meaning public information campaigns abounded with the popular notion being to make the driver drive better. He was told "Speed Kills," "The Life You Save May Be Your Own," and "Stop, Look, and Listen." Although these slogans may have a temporary effect on some drivers, their long-term effect is highly questionable.

Also during this era were the major post-war road-building campaigns. Although these new roads were generally built to minimally acceptable standards for improved travel and safety, roadway officials preached that new roadways meant greater safety, while at the same time they ignored the dire needs of the older existing roadways. With all of the attention to cleaning up hazardous roadsides since the late 1960's, it is hard to believe that a prevailing attitude in the 1950's was that any driver who inadvertently ran off the roadway into a large rigid signpost or tree *deserved* the consequences of his own imprudence.

The Campaign Era also saw large enforcement campaigns designed with the assumption that more enforcement equals fewer accidents. Patrols were stepped up in those areas that had high-accident rates. Although a decline in accidents was usually reported for the selected road section, a closer examination usually indicated no accident reductions for the total jurisdiction. This kind of selective enforcement is still popular today.

During the 1960's, motor-vehicle accidents increased at an alarming rate. A response to this trend was the 1966 National Highway Safety Act, which ushered in the *Action Era*. This act, with its 16 highway safety standards (see Appendix for the 1995 version), promoted the idea that the only way to achieve motor-vehicle safety was to do something about everything.

The Federal and State governments began pouring money out to attack all aspects of the problem at once. Money was spent on promoting legislation, improving vehicle safety standards, improving program coordination, conducting research, building more comprehensive accident reporting and filing systems, improving vehicle registration and driver license record systems, improving training for police officers, driver education instructors, ambulance personnel, and bus drivers, providing more equipment such as squad cars, ambulances, radar units, breathalyzers, and driving simulators, ad infinitum. But the money was spread to thin and no one aspect has yet been adequately solved.

The Action Era quickly evolved into a new era: the *Priority Era*. Officials began to diagnose high-potential target areas and to concentrate more money and effort toward those countermeasures with higher demonstrated safety payoffs. For example, in the roadway safety effort, both high-accident locations and roadside objects that create high average occupant injury severity's have been better addressed in recent years.

1.1 Diagnosing the Traffic Safety Problem

An effective traffic safety program must aim at two particular aspects of the problem. It should look toward the reduction of those kinds of accidents that occur most frequently and also those kinds that most often result in severe injuries or fatalities.

The National Safety Council[1] publishes very detailed accident statistics from which some focus can be gained. Table 1.1 compiled from 1993 data, points out some high-potential target areas.

Among the notable observations are that 83% of urban accidents involve three kinds of collisions, with the two-vehicle angle accident (intersection) heading the list. Single-vehicle and pedestrian accidents account for a majority of urban fatal accidents.

For rural accidents, again a few kinds of collisions make up the majority for both total accidents and fatal accidents. But here, both the most frequent kind of accident and also the most frequent kind of fatal accident involves a single vehicle, with head-on and angle accidents being other significant contributors to fatalities.

Table 1.1 U.S. Accident Statistics[1]

U.S. Urban Accidents (1993)

Type of Accident	Number of Accidents	% of All Urban Accidents
2-Vehicle (angle collision)	2,330,000	28.8
2-Vehicle (other)	2,220,000	27.4
2-Vehicle (rear-end collision)	2,150,000	26.5
1-Vehicle	1,060,000	13.1
Other	340,000	4.2
TOTAL	8,100,000	100.0

U.S. Urban Fatalities (1993)

Type of Accident Accidents	Number of Accidents	% of All Urban Fatal
1-Vehicle	4,900	35.8
Pedestrian	3,400	24.8
2-Vehicle (angle collision)	1,900	13.9
2-Vehicle (other)	1,500	10.9
2-Vehicle (head-on)	900	6.6
Other	600	4.4
2-Vehicle (rear-end)	500	3.6
TOTAL	13,700	100.0

U.S. Rural Accidents (1993)

Type of Accident	Number of Accidents	% of All Rural Accidents
1-Vehicle	1,350,000	35.5
2-Vehicle (other)	720.000	18.9
2-Vehicle (angle collision)	680,000	17.9
2-Vehicle (rear-end collision)	600,000	15.8
Animal	390,000	10.3
Other	60,000	1.6
TOTAL	3,800,000	100.0

U.S. Rural Fatal Accidents (1993)

Type of Accident Accidents	Number of Accidents	% of All Rural Fatal
1-Vehicle	9,800	41.7
2-Vehicle (head-on collision)	3,800	16.2
2-Vehicle (other)	3,600	15.3
2-Vehicle (angle collision)	3,500	14.9
Pedestrian	2,000	8.5
Other	800	3.4
TOTAL	23,500	100.0

From these statistics, four target areas that include about two-thirds of all accidents and 81% of all fatal accidents can be identified as follows:
1. Pedestrian accidents
2. Single-vehicle accidents
3. Two-vehicle angle accidents
4. Two-vehicle head-on accidents

Other significant factors to consider in selecting target areas are:
- Although more accidents occur in urban areas, 63% of all fatal accidents occur in rural areas;
- The age group of 15-24 is involved in more than 25% of all fatal accidents even though this group has less than 15% of all drivers;
- Trucks are over one-third of all the vehicles involved in fatal accidents even though they make up less than one fourth of all registered vehicles.

1.2 Driver Performance

An understanding of driver performance is important to safe roadway design and traffic control. According to Alexander and Lunenfeld,[2] driving requires several simultaneous performance activities that can be classified into three levels: control, guidance, and navigation.

The *control level* of driver performance includes the activities and information needed for normal physical manipulation of the vehicle. The driver maintains lateral and longitudinal control using the steering wheel, accelerator, and brake. Information about how well the driver has controlled the vehicle is transmitted by dynamic vehicle feedback and by vehicle displays.

The *guidance level* of driver performance relates to the task of selecting a safe speed and path. The driver evaluates the temporal situation, makes speed and path decisions, and translates these decisions into control actions. Typical guidance activities are lane positioning, car following, overtaking, passing, cornering, response to traffic control devices, and accident avoidance. Most of these activities require complex decisions, judgments, and predictions. Information at this level is primarily transmitted by the roadway design, the traffic control devices, and traffic events.

The *navigational level* includes the driver's ability to plan a trip and to execute that plan from start to finish. Information at this level is taken from maps, guide signs, and landmarks.

These three driver performance levels form a hierarchy of information-handling complexity (see Figure 1-1). At the control level, performance is relatively

Figure 1-1 Levels of the Driving Task.

simple and is mostly subconscious for most drivers. At the guidance and navigation levels, information handling is more complex, and drivers need more processing time to make decisions and respond safely to information inputs.

Drivers use many of their senses to gather information. They feel changes in vehicle dynamics through forces generated by acceleration. They feel road surface texture through vibration of the steering wheel. They hear car horns, truck engines, sirens, and train whistles. And, they see other traffic, roadway alignment, and traffic control devices. Most information is received visually.

Drivers normally perform several tasks at the same time. They look for information sources, make several decisions, and perform needed control actions. Because several sources of information compete for a driver's attention, that information most crucial to safe driving must be in the field of view, must be available when needed, must be conveyed in a clear, simple message, and must command the driver's attention and respect.

Information Handling

Effective and efficient information handling is crucial to safe driving. This task can be made more difficult when drivers are presented with a multitude of information, transmitted from a variety of sources, and received through several human sensory channels. Drivers may sometimes need to receive and process several sources of information to determine their relative importance, to make accurate interpretations, to decide on appropriate actions, and to commit these actions in a very limited time frame.

When drivers are required to process competing information sources under a time constraint, they need to assign relative priorities as a basis for making decisions. Similarly, roadway agencies need a basis for deciding on what information is important to the driver. The concept of *primacy* is useful for this process.

Primacy refers to the relative safety contribution by each level of driver performance and its related information. Control has the highest primacy and navigation has the lowest. Because a control or guidance failure can result in a collision, those levels have a higher primacy than the navigation level where failure is much less likely to result in a collision.

Primacy also relates to gradients within a performance level. At the control level for example, locking the wheels through braking is more serious than a casual inadvertent encroachment onto a flush paved shoulder. At the guidance level, failing to avoid an immediate hazard can be more crucial than momentarily exceeding the speed limit.

Driver Perception-Reaction Time

The time that drivers use to process information and respond is called their perception-reaction time. Like visual acuity, color vision, and driver attitude, perception-reaction time varies by individual. Perception-reaction time also varies with decision complexity, information content, and driver expectancy (see Section 1-3). The more complex the decision, the more information required to make the decision, the longer the perception-reaction time, and the greater the chance for failure.

Figure 1-2 shows that the driver perception-reaction time needed for collision avoidance is made up of detection, recognition, decision, and action. The driver must first be able to see the hazard. Here the hazard's visibility, contrast, and target value, must attract the driver's attention away from competing visual stimuli. Second, the driver, mainly through experience, must be able to recognize the hazard. Third, the driver must decide whether to brake or steer to avoid collision, when the proper choice among those alternatives may not be clear. And, fourth, the driver must take the decided action. This evasive action may be right or wrong depending on the quality of the information, the driver's experience and temporal state, and the time available. Also, under the extreme conditions of competing information and minimal time to collision, indecision can prevent the driver from ever responding.

The relationship between information content and perception-reaction time depends on the amount of information (bits) needed to resolve uncertainty. Whether a decision is expected also affects perception-reaction time. Johannson

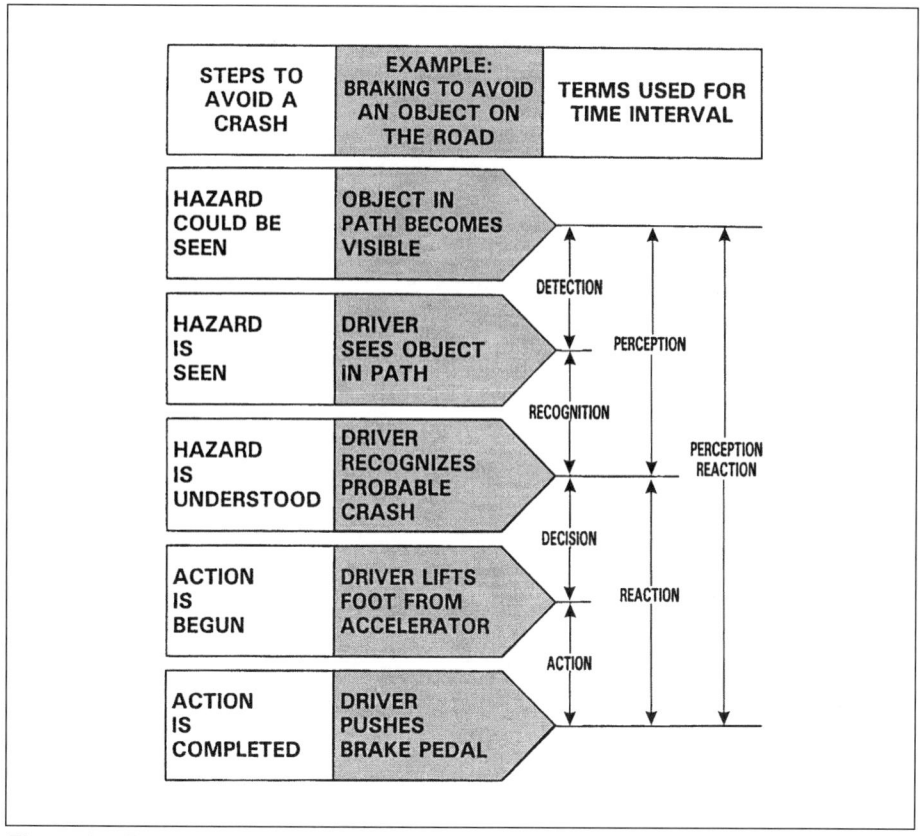

STEPS TO AVOID A CRASH	EXAMPLE: BRAKING TO AVOID AN OBJECT ON THE ROAD	TERMS USED FOR TIME INTERVAL
HAZARD COULD BE SEEN	OBJECT IN PATH BECOMES VISIBLE	DETECTION
HAZARD IS SEEN	DRIVER SEES OBJECT IN PATH	PERCEPTION / RECOGNITION / PERCEPTION REACTION
HAZARD IS UNDERSTOOD	DRIVER RECOGNIZES PROBABLE CRASH	DECISION
ACTION IS BEGUN	DRIVER LIFTS FOOT FROM ACCELERATOR	REACTION / ACTION
ACTION IS COMPLETED	DRIVER PUSHES BRAKE PEDAL	

Figure 1-2 Elements of Driver Perception-Reaction Time (Courtesy of Criterion Press, Inc.)

and Rumar[3] measured perception-reaction time for expected and unexpected signals. When information was expected, perception-reaction time ranged from 0.6 second to more than two seconds. When the signal was unexpected, perception-reaction time varied from one second to over 2.7 seconds. Figure 1-3, shows this relationship for the median driver, while Figure 1-4 shows this relationship for the 85th-percentile driver. These figures relate to *bits* of information, a term used to quantify the smallest amount of information needed to decide between two alternatives. A complex decision (say 4 bits or more) may exceed the driver's capacity to respond, resulting in either a very long perception-reaction time or complete indecision.

Figure 1-5 shows a distribution of driver perception-reaction times, measured by Olson, et. al.,[4] for an unexpected object in the driver's path under actual roadway conditions. The 50th-percentile measure from this study was 1.0 second, and the 85th-percentile measure was 1.5 seconds. Also shown in Figure 1-5 is

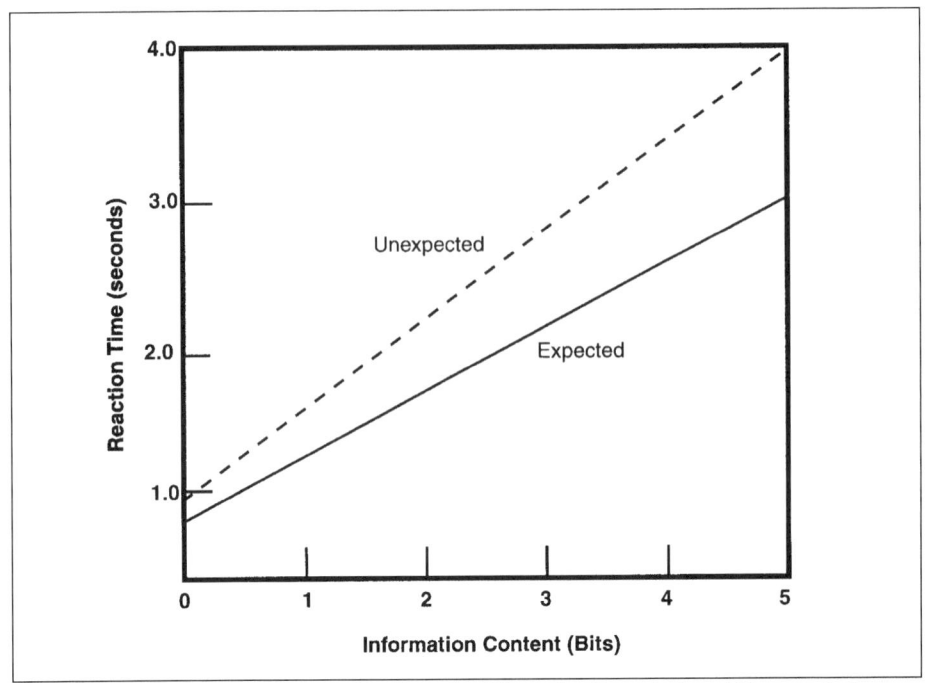

Figure 1-3 Median Driver Perception-Reaction Times for Expected and Unexpected Information.[3]

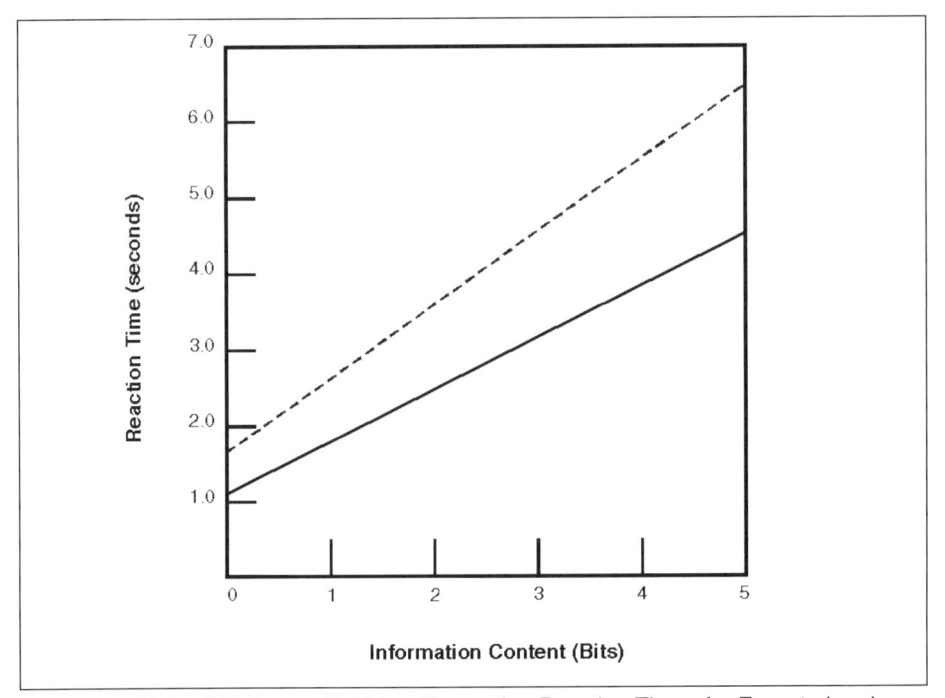

Figure 1-4 Eighty-fifth Percentile Driver Perception-Reaction Times for Expected and Unexpected Information.[3]

Roadway Safety and Tort Liability Overview 9

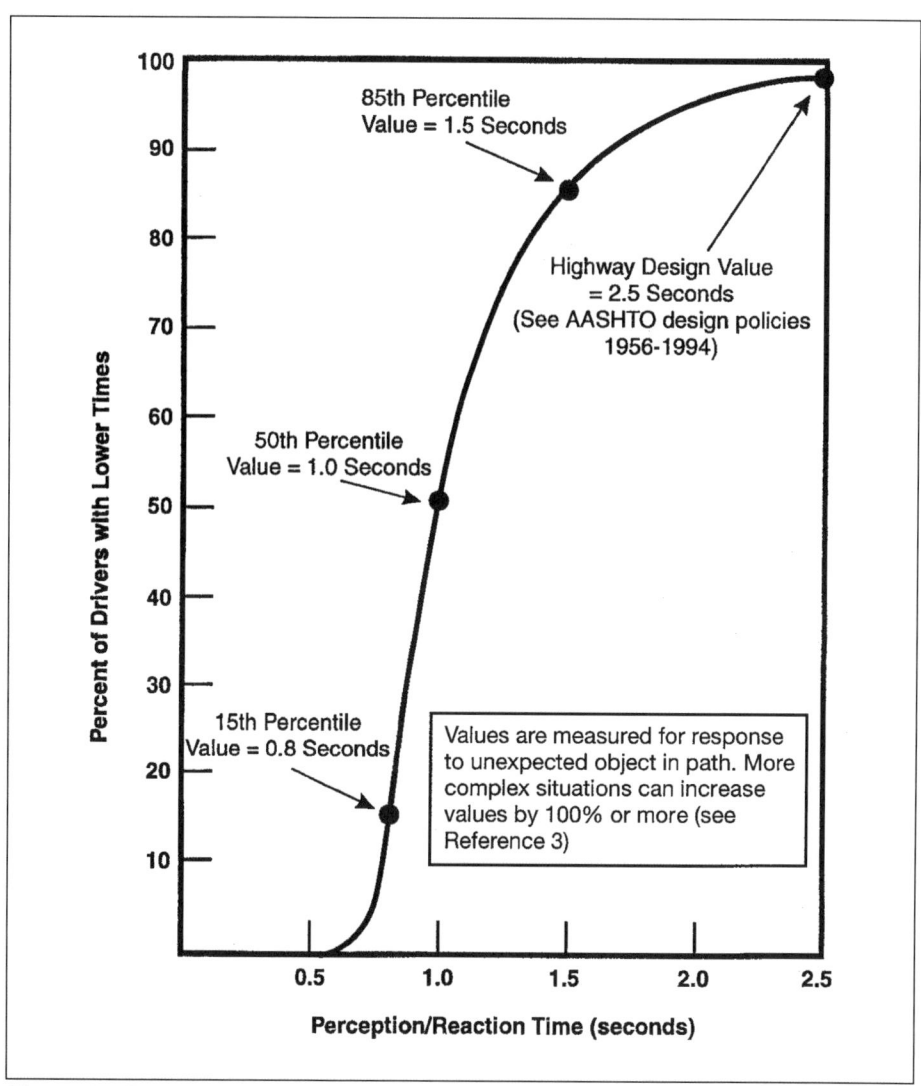

Figure 1-5 Distribution of Driver Perception-Reaction Times.[4] (Courtesy of Criterion Press, Inc.)

the perception-reaction time used in all the design policies of the American Association of State Highway and Transportation Officials (AASHTO) since 1954 (listed in Figure 1-10 toward the end of this chapter).

Figure 1-6 is a conceptual comparison between: (1) the average simple *reaction time* value of 0.75 second, measured in controlled laboratory tests; (2) the median perception-reaction time value of 1.00 second, measured by Olson, et. al.[4]; and (3) the design perception-reaction time value of 2.50 seconds used by AASHTO.

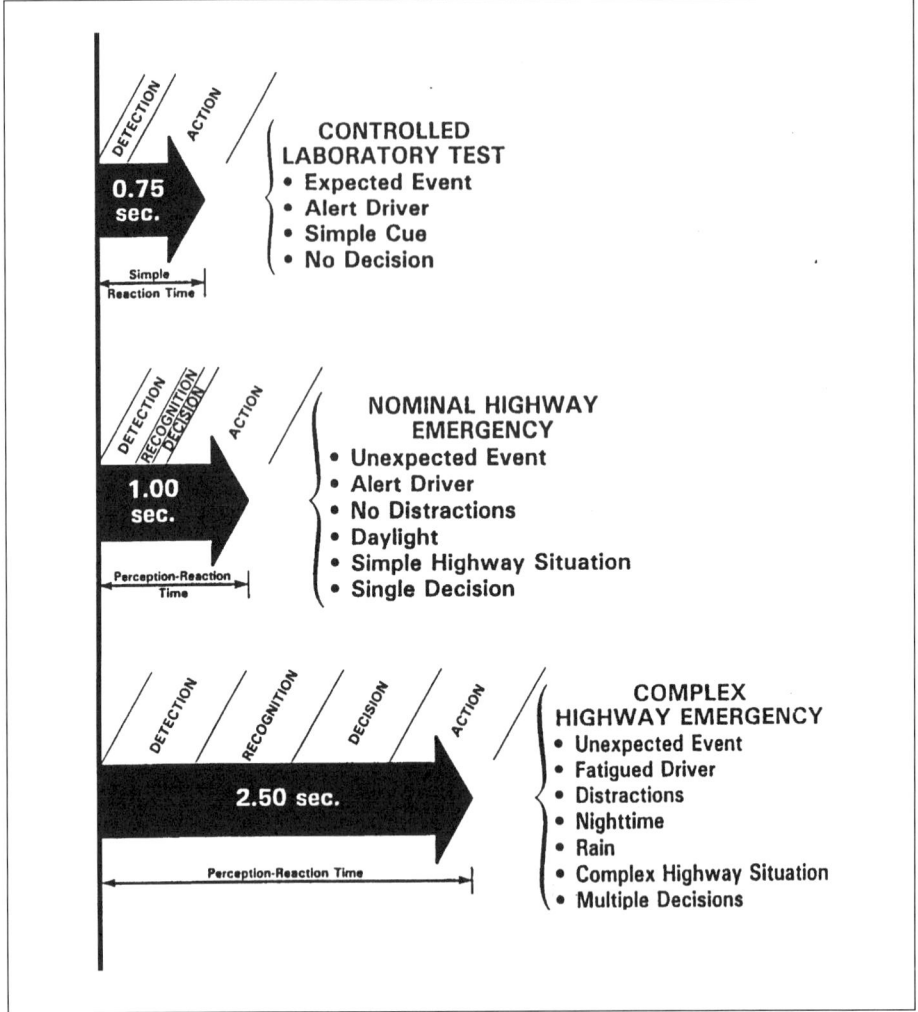

Figure 1-6 *Comparison of Driver Perception-Reaction Times with Simple Reaction Time. (Courtesy of Criterion Press, Inc.)*

Roadway design features and traffic control devices should account for driver perception-reaction time by recognizing that drivers vary in their responses, taking longer to respond when roadway conditions are complex or unexpected.

1.3 Driver Expectancy, Roadway Consistency, and Positive Guidance

Traffic accidents reflect an inefficiency of the roadway system. They occur when the performance of one system component — the driver — falls short of the performance required by the other system components. The demands placed on the

Roadway Safety and Tort Liability Overview

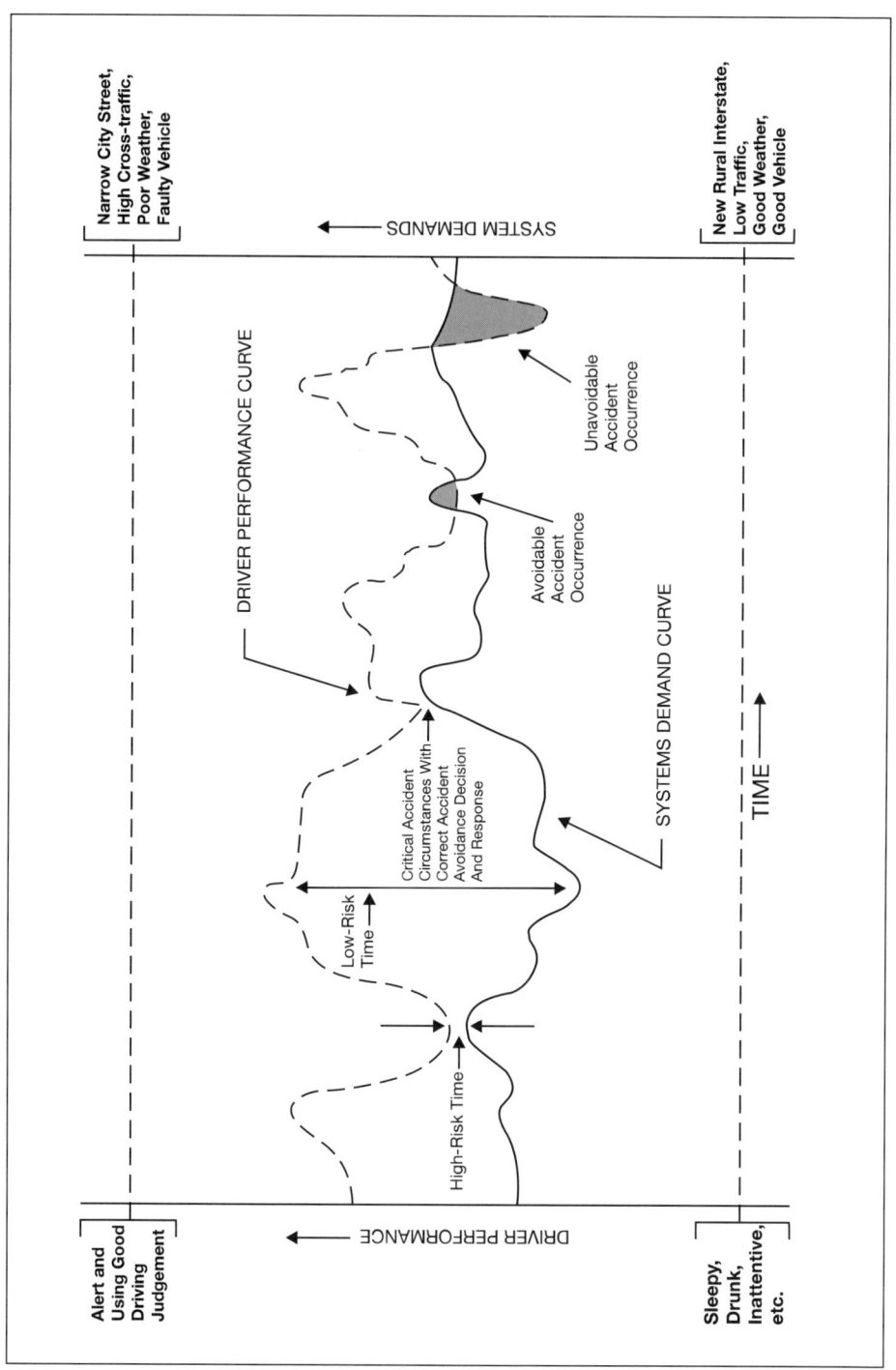

Figure 1-7 *Interaction of Driver Performance and System Demands that Create Accident Circumstances.*

driver by the system depend on the vehicle characteristics and the state of the driving environment (roadway design, pavement friction, traffic controls, visual aspects, climatic factors, etc.)

Both driver performance and system demands fluctuate with time, as shown conceptually in Figure 1-7, which suggests the probabilistic nature of motor-vehicle accidents. The performance of most drivers is usually adequate to meet the demands of the system at most locations during most of the time. Accidents happen at those places or during those times when driver performance is inadequate to meet the demands of the system. These failures can occur either because of low driver performance, because of very high system demands, or because of a combination of lower driver performance and higher system demands. A level of driver performance low enough to result in an accident at one place could very well be more than adequate at other places where the system demands are not so great.

The real solution in motor vehicle safety, therefore, is one of matching the limited sensory and motor capabilities of the driver to the requirements of the driving task. This means that the design, maintenance, and traffic control of the roadway must be compatible with the limitations of most drivers. When they are not, system failures (motor-vehicle accidents) can be expected.

Driver Expectancy

Roadway defects that cause motor-vehicle accidents are often an inconsistent feature of the roadway that violates the expectancy of the driver. Driver expectancy[2,4,5,6] is a human factors term used to describe the readiness of the driver to respond in predictable and successful ways to events, situations, or the presentation of information. It is primarily a function of the driver's experience both with his total driving exposure and with his most recent driving exposure. If the driver's expectancy is met, performance tends to be adequate for the demands of the roadway. When the driver's expectancy is violated by an inconsistent roadway feature, longer response times and incorrect behavior can result.

People expect certain things to operate in ways they have commonly experienced before. When people enter an unfamiliar dark room, they expect to find a light switch on the wall near the door. They also expect to move the switch up to turn the lights on or move it down to turn the lights off. When the switch is somewhere else in the room, when the person has to move the switch down to turn the lights on, or when the person must turn a rheostat knob, they will take longer to respond to these unexpected circumstances.

When a driver's expectancy is violated, the driver will either take longer to respond properly or, even worse, the driver may respond wrong or not at all.

If, for example, a unsigned roadway curve with a 30-mph safe speed is hidden over a sharp hillcrest on a 55-mph roadway, drivers may have great difficulty steering the curve, particularly at night and when they are strangers to the area. This may seem to be an extreme example, but it is all too common on local rural roads in the U.S.

What the driver expects for an upcoming roadway section is largely influenced by what was experienced over the previous roadway sections. Therefore, a small stop sign partially hidden by brush encountered after five miles on a 55-mph roadway with no stops would be a serious violation of driver expectancy.

Driver expectancy is conditioned not only by very recent experiences, but also by those things learned through past experience. For example, stop signs are red, warning signs are yellow, all signs are reflectorized, railroad crossings are marked with a crossbuck, yellow pavement markings are placed on the left side, and roadway dimensions such as grade, alignment, width, and number of lanes do not normally change dramatically without warning signs. When these uniform practices are violated at a roadway location, erratic driver behavior will be manifest, particularly at night or under adverse weather.

Roadway Consistency

This term relates to the sameness of the roadway from one section to the next. Roadway inconsistencies are sudden changes in the nature of the roadway. Because inconsistencies violate a driver's expectancy, either the roadway should be upgraded, which is not always feasible, or remedial measures should be applied to correct false driver expectancies. For example, on the hidden sharp roadway curve, both placing a curve warning sign with an advisory speed plate in advance of the curve and also placing a large arrow sign on the curve, will usually provide the positive guidance needed to restructure the driver's expectancy.

Other examples of inconsistency are:
- A two-lane roadway that suddenly narrows to a one-lane roadway
- A paved roadway that suddenly changes to a gravel roadway
- A construction zone with a substantial pavement edge drop that has no advance warning
- A blind, uncontrolled intersection that is driven through in an area where most intersections have either clear sight triangles or STOP signs
- An intersection with frequent right-angle conflicts that is hidden over a hillcrest
- A bridge that is narrower than the approach roadway
- A T-intersection or dead-end roadway that has no traffic control (at night)

Positive Guidance

This is a popular traffic engineering concept, which means using traffic control devices to overcome the violations in driver expectancy created by inconsistent roadway features (see Figure 1-8). Positive guidance information is that which increases the driver's probability of selecting both the speed and the path most appropriate for the operating conditions of the roadway. Positive guidance is based on the premise that a driver can be given sufficient information about an inconsistent roadway feature at the time he needs it and in a form that he can use to avoid an accident. This concept involves an analysis of the information the driver needs to perform the driving task safely. It includes selecting appropriate traffic control devices, their location, and the format of information presented.

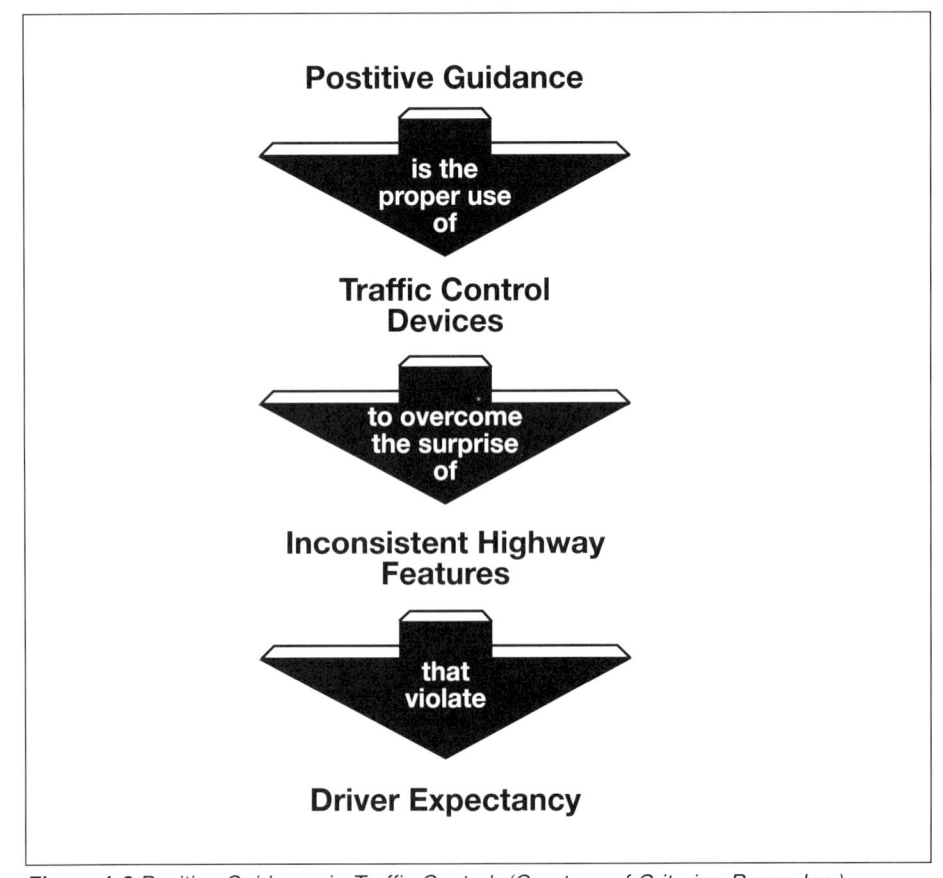

Figure 1-8 Positive Guidance in Traffic Control. (Courtesy of Criterion Press, Inc.)

> **HIGHWAY DEFECTS**
> **Rule of Thumb**
> *Although the State of the Art in Roadway Design and Traffic Control Has Advanced Greatly in the Last 40 Years, Faulty Practices and Applications Continue to Persist on U.S. Roadways, Particularly at the Local Jurisdiction.*

1.4 Nature of Roadway Defects Today

Although the annual number of motor-vehicle deaths in the U.S. hit an all-time high of 56,000 in 1972, it has generally declined since then to a 31-year low of 40,800 in 1992, even though the vehicle-miles of travel have increased over 40%. This general decline is not only attributed to the nationwide reduction of speed limits in 1973 but also to other collision severity reduction measures such as upgrading roadsides and improving the crashworthiness of the vehicle.

Despite this exceptional progress made in reducing fatal accidents, roadway defects are still a common feature in the U.S. Most of the progress has been made by improving State roadways. On the other hand, most local jurisdictions and some States have not progressed at all, usually because of the lack of commitment to having well-trained employees who know the current technology for safe roadways. Still common today are new intersections built at hazardously sharp angles, new guardrails that are too short to develop beam strength on impact, unsigned T intersections, unsigned and unmarked sharp roadway curves, newly paved roadways with hazardous pavement edge drops, etc. Then too, many roadway defects such as sight distance obstructions, pavement edge drops, hazardous roadsides, etc. happen because of the lack of proper maintenance.

This book is intended to highlight the roadway defects that are most commonly associated with tort claims involving serious injuries or fatalities. The book focuses on the key technical aspects of these roadway defects. The topics of the book chapters generally align with the key roadway defect categories. Example roadway defects claimed in tort actions are listed in Figure 1-9.

1. **Guardrails** that lack crashworthy end treatments where vehicles and passengers are impaled upon impact.
2. Very **steep embankment slopes** that are close to the edge of the traveled way and do not have guardrail protection.
3. **Guardrails** with too few posts to develop adequate strength to prevent vehicle penetration.
4. Uncrashworthy **pipe or wooden bridgerails** that will not redirect impacting vehicles at speeds above 5 mph.
5. **Guardrail flares** to bridgerail ends that are not attached and, therefore, redirect colliding vehicles into the end of the bridgerail.
6. Large unprotected **rigid objects** close to the edge of the traveled way.
7. Substandard **sight distance** over hillcrests or around sharp roadway curves.
8. Substandard **sight distance** at uncontrolled intersections.
9. Substandard **sight distance** at stop-controlled intersections.
10. **Sharp roadway curves** with no signing or delineation.
11. **Sharp roadway curves** hidden over hillcrests with substandard sight distance and inadequate signing.
12. Repaved roadways where the unpaved shoulder was not treated, leaving a 2-inch or greater **pavement edge drop** between the pavement and shoulder.
13. Unpaved shoulders that become eroded or rutted leaving a 2-inch or greater **pavement edge drop** between the pavement and shoulder.
14. Pavements with worn wheel tracks where **hydroplaning** is a likely occurrence during wet weather.
15. **Traffic signals** with yellow clearance phases that are too short for the speed of traffic and the width of the intersection.
16. **Traffic signals** with 8-inch lens, which have inadequate target value for high-speed traffic.
17. Small **stop signs** hidden around a roadway curve or over a hillcrest on a high-speed rural roadway where drivers do not expect to see a stop sign because of having traveled several miles without a stop.
18. **Rail-highway grade crossings** without train-activated warning devices where sight obstructions limit the driver's view of an oncoming train to less than 2 seconds.
19. **Rail-highway grade crossing** approaches that are steep, narrow, rough, or at a severe angle to the track and consequently not only limit the driver's time to see trains but also distract the driver from searching for trains.
20. Lane-closure tapers in **construction zones** that are much too short for the speed of traffic.
21. Inadequately marked **road closures**.
22. **Slow-moving maintenance vehicles** with inadequate advance warning devices.

Figure 1-9 Common Highway Defects.

Roadway Safety and Tort Liability Overview 17

1.5 Basic Legal Concepts

The following discussion is intended for the non-legal reader. It is not intended as a discourse of legal opinions, but is simply an engineer's perspective of the basic concepts applied in tort liability actions involving roadway agencies. Readers wanting more precise information about these and other legal doctrines and concepts should consult their local law libraries. One publication that may be particularly helpful for roadway engineers is *Practical Guidelines for Minimizing Tort Liability*.[7]

Roadway agencies have a common law and court-imposed duty to operate public roadways in a safe condition, so as to not expose motorists to undue hazard. Where existing hazards cannot be eliminated, roadway agencies have a duty to warn drivers of these hazards.

Roadway agencies are held liable when they fail to exercise reasonable care in the design, construction, and maintenance of public roadways. Unless sovereign immunity applies, States, Counties, and Cities are subject to common-law rules of negligence and may be liable for failure to exercise ordinary care. Statutes may also impose duties for the safe operation of roadways. In some States, violation of these statutes is negligence, per se.

Roadway tort claims plead for repayment of damages for personal injury and property loss suffered in motor vehicle collisions. To have a valid tort action, four conditions must exist:

1. The plaintiff suffered **damages** in a motor-vehicle collision.
2. The defendant owed a **legal duty** to the plaintiff. Roadway agencies have a duty to the motoring public to provide reasonably safe travel.
3. The defendant was negligent because of a **breach of duty**. The roadway agency failed to design, construct, or maintain a safe roadway.
4. The negligence of the roadway agency was a **proximate cause** of the collision. A proximate cause is an element in a natural or continuous sequence that produced the collision, without which the collision would not have occurred.

The following paragraphs attempt to further explain some of the more important concepts and doctrines associated with tort liability actions related to roadway defects.

Negligence

Negligence is the breach of a legal duty. It is the failure to exercise the reasonable care that a prudent person would use under similar circumstances. Stated in the most popular vernacular, negligence can either be an act of omission or an act of commission. It is either not doing something that a reasonable man guided by ordinary circumstances would do, or it is doing something that a reasonable man would not do.

Contributory vs Comparative Negligence

Under the old classic concept of negligence, a basic tort action implied that the plaintiff must not be guilty of contributory negligence to be successful in his suit. Over the last 20 years, however, many States have passed statutes replacing the doctrine of contributory negligence with the doctrine of comparative negligence. The doctrine of comparative negligence says that even though the plaintiff shares a portion of the total negligence, he is able to recover for those portions assigned to the defendants. For example, if the defendants are found to be 60% negligent and the plaintiff is found to be 40% negligent, then the plaintiff can recover 60% of the awarded judgment from the defendants. In some States, however, recovery by the plaintiff is barred under comparative negligence when the plaintiff's portion of negligence equals or exceeds the defendant's portion (50% rule).

Joint and Several Liability

The laws in some States hold that a defendant, who is liable for any portion of a plaintiff's damages, is jointly of severally liable for all damages along with any other defendant who is held liable. What this doctrine means is that if one or more defendants is unable to pay their portion of an award, the plaintiff can recover the entire award from any remaining defendant(s), even if they are only found 1% negligent.

Standard of Care

The basic standard of care for roadway agencies is *reasonable safety for all motorists*. Where a hazardous condition exists, the reasonableness of the risk of harm is measured by: (a) the likelihood of harm; (b) the gravity of harm posed by the condition; (c) the usefulness of the condition for other purposes; (d) the availability of methods to correct the condition; and (e) the burden of removing the condition.

In a tort action, the plaintiff must clearly prove negligence. The most effective way to offer this proof is to show how the agency failed to construct, design, or maintain the roadway according to their own standards, policies, or guidelines. In the absence of these agency documents, the court test normally is the commonly accepted good practices promulgated by authoritative national bodies in their standards, policies, or guidelines. Two of the most common examples used to establish a standard of care in roadway defect cases are the *Policy on*

Geometric Design of Highways and Streets[8] and the *Manual on Uniform Traffic Control Devices*[9] (see Section 1.6 for more discussion).

Notice of a Defect

The duty of a roadway agency to correct hazardous conditions arises when it has either actual or constructive notice of the roadway defect. Most courts hold that the roadway agency must have sufficient advance notice of the defect to have had reasonable opportunity to either correct the roadway defect or to warn of its hazard.

Under the concept of constructive notice, however, the duty to act arises when the hazardous condition is a direct result of the roadway agency's own negligence. Thus, proof of actual notice may not be necessary if the agency should have known of the hazard. For example, constructive notice may arise when a roadway defect has existed for such a time and is of such a nature, that the roadway agency should have discovered the defect by reasonable diligence.

Sovereign Immunity

The sovereign immunity doctrine emerged early in the United States as the Supreme Court held that Federal and State governments were immune from suit. The rule was based on a misconception of English common law, which immunized the King as sovereign for wrongs committed by him or his agents because "the King could do no wrong."

In 1946, the United States passed the Federal Tort Claims Act, which waived sovereign immunity for the Federal government but not the States. The law allowed litigation of tort claims against the United State government in the Federal court. In the last thirty years, the doctrine of sovereign immunity has either been completely waived or modified in most of the States. Modifications have come in the form of tort claim acts which prescribe certain limits on litigation. In some States, relief is made available through a special tribunal or a legislative claims board.

Governmental vs. Proprietary Functions

In 1842, the governmental/proprietary distinction was first developed by the courts of New York as a test of municipal tort liability. This distinction is now widely applied across the U.S. in those States where some immunity passes on to municipalities for governmental functions, but not for proprietary functions. Government functions are those that can only be performed adequately by a government unit such as police, fire protection, or courts. Proprietary functions are those that could be supplied by private concerns. Simply, proprietary functions are those services that derive revenue, such as water, gas, and electric supplies.

Discretionary vs. Ministerial Acts

The distinction between discretionary acts, which are usually exempt from liability, and ministerial acts, which may create liability, was developed by the courts and still prevails.

Ministerial functions usually involve clearly-defined tasks performed with minimum leeway on personal judgment and do not require any comparison of alternatives before undertaking the duty to be performed. Routine roadway maintenance may be an example of a ministerial function.

The concept of discretionary function, as employed by the courts, is the duty to make a choice among valid alternatives. Discretion requires comparing alternatives and exercising independent judgment in choosing a course of action. The exemption from liability for discretionary acts is rooted in common law. The defense, in many cases, resorts to a plea of discretionary immunity, which may be extended to many functions performed by roadway agencies. Discretion is a tricky concept and becomes a term of art, because its definition often depends on degree. The courts consider many factors to reach their decisions about the function involved, including: (1) the importance to the public; (2) the extent which governmental liability might impair the exercise of the function; (3) the availability of remedies other than tort suits to the individuals affected; (4) and an intrinsic belief, stemming from the *separation of powers doctrine*, that certain government decisions should not be subject to judicial review.

When considering discretionary functions, the courts have generally held that roadway agencies are not insurers against accidents. The courts have often ruled against liability when roadway agencies have carried out a reasonable plan of roadway improvements. The adoption of improvement plans, the designation of funds, and the setting of priorities for improvement are often held out as legitimate discretionary functions.

The most clear-cut example of the application of the exemption for discretionary functions is found in design immunity statutes, as described below.

Design Immunity

Several States have enacted design immunity statutes to provide legal protection to government agencies for the design of public projects. These statutes generally immunized agencies and employees from liability when the roadway design was in general accord with standards prevailing when the roadway was built. The courts have also noted exceptions to design immunity: (a) where the approval of a design was arbitrary or unreasonable; (b) where the plans were prepared without adequate care; (c) where the constructed roadway had an inherently dangerous defect from the beginning of actual use; and (d) where the agency had notice of a design defect that became obviously or manifestly dangerous following its construction.

1.6 Authoritative Sources of Information

For anyone interested in the safety aspects of roadway design, construction, maintenance, and traffic control, the most authoritative sources are found in the publications of the following organizations:

1. **FHWA.** The Federal Highway Administration, U.S. Department of Transportation, 400 Seventh St., S.W., Washington D. C., 20590, TEL. 202/366-4000 (formerly the Bureau of Public Roads).
2. **AASHTO.** The American Association of State Highway and Transportation Officials, Suite 225, 444 N. Capitol St., N.W., Washington, DC, 20001, TEL. 202/624-5800 (formerly the American Association of State Highway Officials).
3. **TRB.** The Transportation Research Board, National Academy of Sciences, 2101 Constitution Ave., N.W., Washington, DC, 20418, TEL 202/334-3214 (formerly the Highway Research Board).
4. **ITE.** The Institute of Transportation Engineers, 525 School St., S.W., Washington, DC, 20024, TEL. 202/554-8050 (formerly the Institute of Traffic Engineers).
5. **States.** Most States have roadway design manuals, maintenance manuals, traffic control device manuals, and a publication with a title such as *Standard Specifications for Constructing Roadways and Bridges.*

Of these organizations, the AASHTO publications are the most important when discussing design issues, the FHWA publications are most important when dealing with traffic control issues, and the ITE, TRB, and State publications cover a broad range of subjects that relate to roadway safety.

Roadway Design

AASHTO, founded in 1914, has promulgated formal roadway design standards since 1940. Generally accepted roadway design standards prior to 1940 can be found both in the textbooks of the time and some of the early annual proceedings of AASHO. Although the more current AASHTO design policies have disclaimers that say the publications do not represent standards, these AASHTO publications not only continue to reflect current thinking about both minimally acceptable and desirable roadway design dimensions, but also they are recognized worldwide as the best definition of good roadway design

- *A Policy on Highway Types (Geometric), 1940*
- *A Policy on Intersections at Grade, 1940*
- *A Policy on Sight Distance for Highways, 1940*
- *A Policy on Criteria for Marking and Signing No-Passing Zones on Two-and Three-Lane Roads, 1940*
- *A Policy on Design Standards, 1945*
- *A Policy on Geometric Design of Rural Highways, 1954*
- *A Policy on Arterial Streets in Urban Areas, 1957*
- *A Policy on Geometric Design of Rural Highways, 1965*
- *A Policy on Design of Urban Highways and Arterial Streets, 1973*
- *A Policy on Geometric Design of Highways and Streets, 1984*
- *A Policy on Geometric Design of Highways and Streets, 1990*
- *A Policy on Geometric Design of Highways and Streets, 1994*

Figure 1-10 AASHTO Design Policies.

practice. Figure 1-10 lists the major AASHTO geometric design policies that are of interest in roadway defect cases.

Roadway Maintenance

Defined as the care of existing roadbeds, pavements, signs, markings, signals, and other existing safety devices, roadway maintenance is discussed in most major roadway references. This subject is most clearly addressed in the 1987 *AASHTO Maintenance Manual*[10] and the 1985 *Street and Highway Maintenance Manual*[11] of the American Public Works Association (APWA).

Traffic Control Devices

The FHWA is the caretaker of standards for traffic control devices including signs, signals, and markings. The *Manual on Uniform Traffic Control Devices*[9] *(MUTCD)* is part of *Title 23 of the US Code* and requires all States to adopt its provisions to insure the uniform application of traffic control devices on all public roadways. Some States adopt the national *MUTCD* by legislative reference; other States prepare their own MUTCD's, usually replicating the national manual but with added specificity.

Roadway Safety and Tort Liability Overview   23

The most recent full edition of the *MUTCD* was published in 1988, with a revision of Part VI, on construction and maintenance work zones, published in 1993. The previous edition of the *MUTCD* was in 1978. A future edition is in preparation. In addition to specifying the uniform application of traffic signs, signals, and markings on all public roads, the *MUTCD* covers the special location topics of work zones, rail-highway grade crossings, pedestrian and school crossings, and bikeways.

The *Traffic Control Devices Handbook (TCDH)*,[12] published in 1983 by FHWA, is a special adjunct to the *MUTCD*. Its intent is to lend clarity to the application of the *MUTCD* standards, with particular focus on the engineering studies required for the proper application of traffic control devices.

The *Traffic Engineering Handbook*,[13] published by ITE, is also a major source of information about the application of traffic control devices as well as roadway design principles.

Work Zone Safety

The two major sources dealing with the operation of construction, maintenance, and utility work zones are the FHWA publications *Manual on Uniform Traffic Control Devices*[9] and the *Traffic Control Devices Handbook*.[12]

Roadside Safety

This important safety concern is covered widely by many publications of TRB. With regard to standard applications, two particular AASHTO publications are important. The 1987 *Roadside Design Guide*[14] is the most recent attempt to codify the definition of roadside *clear zones* and the application of roadway guardrail, bridgerails, median barriers, crash cushions, and breakaway signs and light posts. The predecessor of this document was the 1977 *Guide for Selecting, Locating, and Designing Traffic Barriers*.[15] Another historically important document by AASHTO was the 1967 publication *Highway Design and Operating Practices Related to Highway Safety*.[16] FHWA released follow-up publications in 1968, 1973, and 1978 with the title, *Handbook of Highway Safety Design and Operating Practices*.[17]

Rail-Highway Crossings

Four documents are particularly useful when analyzing the safety of rail-highway crossings. The *Manual on Uniform Traffic Control Devices (MUTCD)*[9] defines the use and application of advance warning signs, crossbucks, flashing signals, and crossing gates. The *Traffic Control Devices Handbook, (TCDH)*[12] adds considerable specificity to the *MUTCD*. The FHWA *Railroad-Highway Grade Crossing Handbook*[18] is the most comprehensive treatment of grade crossing safety, including the recommendations for minimum safe sight triangles. The

TCDH and the AASHTO *Policy on Geometric Design of Streets and Highways*[8] also give dimensions for minimum safe sight triangles at grade crossings.

1.7 The Intended Focus of This Book

This book recognizes that not only do plaintiffs sometimes file frivolous suits against roadway agencies, but also that roadway agencies occasionally offer irrational arguments in an attempt to defend well-founded suits. However, the intention here is not to dwell on these extremes, but to recognize that most States have tort laws that allow for the negligence of all parties to be compared. And, as such, the acts or omissions of the roadway agency are often legally weighed against the actions or lack of actions by the drivers involved in a roadway collision.

Although this book should be of help to both plaintiff and defense lawyers and to expert witnesses in the litigation of roadway defects cases, the author's hope is that it will serve best as a safety toolbox for roadway agencies. From an ongoing dialogue about the major issues surrounding the most prevalent roadway defect categories, a clear focus on prevalent roadway safety problems and their best candidate solutions should emerge.

This book has the following major topics:

1. Roadside Hazards
2. Sight Distance Obstructions
3. Skid Resistance and Hydroplaning
4. Pavement Edge Drops
5. Rail-Highway Grade Crossings
6. Roadway Curves
7. Roadway Construction and Maintenance Zones
8. Traffic Control Devices
9. Roadway Maintenance

Although each chapter is written to stand alone, all roadway defect chapters generally have common topics, including: (a) prevailing and historical standards; (b) accident circumstances; (c) elements of hazard; (d) technical aspects of roadway defects cases; (e) typical defense arguments; (f) rules of thumb; and (g) key references. Other topics sometimes discussed are prevailing research knowledge and accident reconstruction aspects.

Because future editions of this book are anticipated as the knowledge base broadens, readers are encouraged to submit their ideas, corrections, and arguments regarding any and all of the material either included or not included in this book.

Endnotes

1. National Safety Council, *Accident Facts*, 1994

2. Alexander, Gerson J., and Lunenfeld, Harold, *Positive Guidance in Traffic Control*, Federal Highway Administration, 1975.

3. Johannson, C., and Rumar, K., *Driver Brake Reaction Time*, Human Factors, Volume 13, No. 1, 1971

4. Olson, P.L., et. al., *Parameters Affecting Stopping Sight Distance*, Transportation Research Board, NCHRP Report 270, 1984.

5. Alexander, Gerson J., and Lunenfeld, Harold, *Driver Expectancy in Highway Design and Traffic Engineering*, Federal Highway Administration, 1983.

6. Lunenfeld, H., and Alexander, G.J., *A User's Guide to Positive Guidance*, Federal Highway Administration, 1990.

7. Lewis, Russell M., *Practical Guidelines for Minimizing Tort Liability*, Transportation Research Board, NCHRP Synthesis 106, 1983.

8. American Association of State Highway and Transportation Officials, *A Policy on Geometric Design of Highways and Streets*, 1990.

9. Federal Highway Administration, *Manual on Uniform Traffic Control Devices*, 1988. (available through Lawyers and Judges Publishing Co.)

10. American Associated of State Highway and Transportation Officials, *AASHTO Maintenance Manual*, 1987.

11. American Public Works Association, *Street and Highway Maintenance Manual*, 1985.

12. Federal Highway Administration, *Traffic Control Devices Handbook*, 1983.

13. Institute of Transportation Engineers, *Traffic Engineering Handbook*, 4th Edition, 1992.

14. American Association of State Highway and Transportation Officials, *Roadside Design Guide*, 1987.

15. American Association of State Highway and Transportation Officials, *Guide for Selecting, Locating, and Designing Traffic Barriers*, 1977.

16. American Association of State Highway Officials, *Highway Design and Operating Practices Related to Highway Safety*, 1967.

17. Federal Highway Administration, *Handbook of Highway Safety Design and Operating Practices*, 1968, 1973, and 1978 Editions.

18. Federal Highway Administration, *Railroad-Highway Grade Crossing Handbook*, 1986

Other Sources

Glennon, John C., *The MUTCD at Court*, 1991 Proceedings, Institute of Transportation Engineers.

Glennon, John C., and Harwood, Douglas W., *Highway Design Consistency and Systematic Design Related to Highway Safety*, Transportation Research Record No. 693, 1978.

Glennon, John C., *Priorities for Highway Safety*, Better Roads, September, 1976.

Glennon, John C., *The Desirability of Highway Design for Operating Speeds Above 70 mph*, 1970 Proceedings, Institute of Traffic Engineers.

Glennon, John C., *Major Areas of Highway Defect Tort Claims*, Proceedings, AIRIL 95, Inaugural International Conference on Accident Investigation and the Law, Queensland University of Technology, Brisbane, Australia, 1995.

Chapter 2

Roadside Safety

In the past, only limited efforts were spent by roadway agencies on improving roadsides for safety. Many roadway design engineers felt that drivers, who allowed their vehicle to encroach on the roadside, deserved the consequences of their imprudence. As the evidence mounted in the 1960's of an alarming increase in single-vehicle accidents and fatalities, however, the philosophy began to change and consideration of the injuries incurred by ran-off-road motorists started to become an important aspect of roadway design.

As an overreaction to the early recognition of the need for improved roadside safety, guardrail was thought to be a panacea. Large quantities of guardrail were placed either where none was needed, or where other roadway improvements could have eliminated that need. The tragic side of what was a promising trend in roadside safety was the alarming increase of spectacular collisions with guardrails that were not adequate for their intended purposes.

During the late 1960's, major improvements in roadside safety design began to evolve. Embankments and cut-slopes were flattened to increase the chance of recovery for errant vehicles. Full shoulder widths and areas free of fixed objects were provided along the roadside. Better guardrail, bridgerail, and median barrier designs were developed, breakaway sign and light posts were conceived, and effective impact attenuators (crash cushions) were discovered.

The most significant change in roadside safety efforts came in 1967 when AASHTO published *Highway Design and Operational Practices Related to Highway Safety,*[1] which strongly urged cleaning up U.S. roadsides. This report (referred to as the Yellow Book) was supported by the proceedings of the Congressional Blatnik Committee,[2] and the Federal Highway Administration, which initiated a comprehensive attack on roadside safety. This Federal effort was first documented by a letter from F.C. Turner, Director of the Bureau of Public Roads and followed by the Federal Highway Administration's Instructional Memorandum 21-11-67. Some excerpts, shown in Figures 1-1, and 1-2 from these documents demonstrate the strong commitment of the Federal Highway Administration to roadside safety.

(To all State Highway Department Administrators)

Dear:

By means of this letter, I call to your attention the recently issued American Association of State Highway Officials report entitled "Highway Design and Operational Practices Related to Highway Safety". The Bureau of Public roads concurs fully in the report's recommendations and conclusions and considers it to be one of the most important documents ever developed by the joint efforts of the Bureau of Public Roads and AASHO. We wish to assist every State highway department in applying its findings beginning as soon as possible in 1967 and continuing on a large scale for as long as is necessary to provide the highest possible level of roadway safety on the Federal-aid highway systems. We pledge to you such assistance as is necessary to allow the State highway departments to plan and program the use of Federal-aid funds to expedite the accomplishment of this objective.

The Bureau is placing its full support and resources behind a concentrated major effort to implement the recommendations of the AASHO report and I earnestly solicit your own support in this joint effort. I therefore urge you as the Chief Administrative Officer of your State highway department to examine fully the recommendations of the AASHO report to determine from a safety viewpoint those features of existing highways which constitute hazards to highway users; and to establish an active corrective program along the lines which the report suggests.

This is the first communication to you under my new title in the infant Department of Transportation. I know how busy you are and how many urgent priority items there are to claim your attention; I shall not be imposing on your time very often in this manner; but the overwhelming importance of this subject impels me to do so in this instance. We shall be sending through the normal channels such memoranda as appear necessary to aid in implementing this program; but I want you to know that our purpose will be to remove every possible hindrance to your being able to cooperate effectively in this important endeavor. We expect to provide you with the most liberalized procedural tools that we can devise under the law. I solicit your own personal support and I know that you have the same dedicated interest in this matter that we have.

Figure 2-1 Excerpts from a May 8, 1967 Letter from F.C. Turner, Director, Bureau of Public Roads, to State Highway Department Administrators.

SUBJECT: Safety Provisions for Roadside Features and Appurtenances

The February 1967 Report of the Special AASHO Traffic Safety Committee — HIGHWAY DESIGN AND OPERATIONAL PRACTICES RELATED TO HIGHWAY SAFETY — is approved by the Bureau of Public Roads for use on Federal-aid highways.

Enclosed is a copy of a letter I have sent to the top administrative officials of each State highway department offering our full cooperation and assistance in applying the findings of the report to the existing Federal-aid systems beginning as soon as possible in 1967 and continuing on a large scale for as long as is necessary to provide the highest possible level of roadway safety.

On completed Federal-aid highways each State highway department is asked to establish an active corrective program to apply the findings of the February 1967 Report. Public Roads requests that all features of geometric, structure dimension and roadside element design that can effect safety of the motorist who strays from the roadway be given careful consideration by the State. Each State should evaluate the seriousness of the existing condition as measured by the more safe conditions recommended by AASHO in the new Report and prepare its program for corrective work on previously constructed highways on the several Federal-aid systems. The most serious existing conditions should be assigned highest priority for correction. Corrections should provide the safer condition to the degree as outlined by the AASHO Report. . . .

. . . .Public Roads Division Engineers are to take a broad and liberal viewpoint with regard to approving programs proposed by the State highway department for work of the types described in the February 1967 Report. . . . Projects can cover sizable lengths of highway and may cover several or all types of roadside features. For example, a project might include relocation or adjustment of signs, installation or modification of guardrail, removal of and/or protection from the varied hazardous roadside elements, etc., on as long a section of highway as may be proposed by the State.

Figure 2-2 Excerpts from Federal Highway Administration Instructional Memorandum 21-11-67, May 19, 1967.

Starting in 1967, the Federal Highway Administration and some States funded specific programs to reduce the roadside hazard on existing roadways. These programs all followed the same general strategy, which, as paraphrased from the Yellow Book, simply says:

1. **Remove** unnecessary roadside hazards close to the traveled way.
2. **Move** those roadside hazards that cannot be removed to less exposed places.
3. **Reduce** the impact severity of those roadside hazards that cannot be moved.
4. **Protect** motorists with energy attenuation or deflection devices at those roadside hazards that cannot otherwise be improved.

Over the years, roadside hazards have been the subject of many tort claims. Claims relate to steep embankments, faulty barriers, large trees, rigid utility poles, light poles, and other fixed objects placed too close to the traveled way, curbs that cause vehicles to vault, etc. Although considerable effort has been expended by State roadway agencies to improve roadsides since 1967, most of the 1968 concerns expressed by Congressman John A. Blatnik[2] still ring true today, particularly for local streets and highways:

> *Our Subcommittee has made an in-depth study of still another potentially dangerous component of the road, the roadside. This study proves beyond a shadow of a doubt that the roadside in too many cases has been designed in a manner so fraught with danger as to be comparable, without overtaxing the imagination, to tank traps. Testimony at our hearings, abundantly documented by hundreds of photographic slides, pointed out some of the traps openly laid. The unyielding light posts, the massive signposts, the sturdy trees, the exposed bridge abutments are all more than a match for any car or any truck, for that matter.*

Many of these concerns were reiterated in 1974 by the consumer advocate tome, *The Yellow Book Road*,[3] published by the Center for Auto Safety.

2.1 Single-Vehicle Collision Characteristics

From 1976 to 1986, about 450,000 people died on U.S. roadways. Of these, about 190,000 people died in single-vehicle collisions (ran-off-road, hit fixed objects, or overturned — a frightening 42% of all roadway fatalities. In a collision, the chances of death are three times greater in single-vehicle collisions than all other motor-vehicle collisions. Clearly, positive actions taken to reduce both the frequency and severity of single-vehicle collisions could help change this grim testament of roadway death and injury.

Another reason for paying special attention to this kind of collision is that it represents the most visible form of failure in the driver-vehicle-roadway system. Circumstances caused by other drivers and other vehicles are less important in single-vehicle accidents than in multi-vehicle accidents. Therefore, roadside improvements aimed at reducing the frequency and severity of single-vehicle accidents offer one of the most clear-cut opportunities for roadway safety.

The hazards associated with roadsides are obvious to anyone who drives with his/her eyes open. Even with roadway alignment that is adequate, great danger often lies in what happens to an errant vehicle after it leaves the traveled way. Roadside hazards do not always give drivers a chance to avoid, or at least survive, a collision. As shown in Table 2.1, the roadside elements that contribute heavily to the consequences of single-vehicle collision are trees, utility poles, guardrails, embankments, culverts, curbs, bridge abutments and piers, bridge rails, and various other poles.[4]

Quantitatively, the degree of collision hazard associated with a roadside hazard can be defined in several ways. The collision hazard is the potential for a particular roadside location to experience a given time rate of collisions with some average injury consequence. At any roadside location, then, the degree of collision hazard is a function of two variables: collision frequency and collision

Table 2.1 Annual Fixed Object Fatalities by Object Type

Fixed Object	1983	1984	1985	1986	1987
Tree/shrub	2841	3021	2989	3444	3299
Utility Pole	1377	1426	1298	1494	1406
Guardrail	1310	1446	1258	1374	1326
Embankment	1288	1264	1211	1332	1396
Culvert/ditch	1259	1198	1337	1472	1393
Curb/wall	865	899	982	960	861
Bridge/overpass	803	738	628	577	571
Concrete barrier	263	240	225	197	203
Sign or light support	488	480	508	551	538
Other pole/support	495	434	481	518	495
Fence	434	455	431	478	484
Building	110	105	101	100	108
Impact attenuator	16	10	14	9	18
Other fixed object	565	629	630	699	729
TOTALS	12114	12345	12093	13206	12827

severity. When comparing two roadside locations, if both have the same collision frequency, the location with the lower collision severity is less hazardous. On the other hand, if two locations have the same collision severity, the location with the lower collision frequency is less hazardous.

2.2 Characteristics of Roadside Encroachments

One of the first formal and well-documented recognitions of the roadside safety problem came from outside the roadway engineering community. Researchers[5] at General Motors (GM) in the late 1950's became alarmed and studied crashes on their proving grounds. They found that of 236 crashes over a 6-year period, 80% involved a collision off the roadway, resulting in 64 lost days of work from injured drivers. After GM cleared its roadside environment of fixed objects and steep slopes, they experienced a subsequent 6-year period with no lost work days.

In 1966, Hutchinson and Kennedy[6] added further definition to the roadside encroachment event for freeways. Through field observation of physical evidence left by roadside vehicular encroachments on medians, they developed roadside encroachment rates as a function of traffic volume (Figure 2-3). They also defined exceedance distributions for the lateral displacement (Figure 2-4) and the encroachment angle (Figure 2-5) for roadside encroachments.

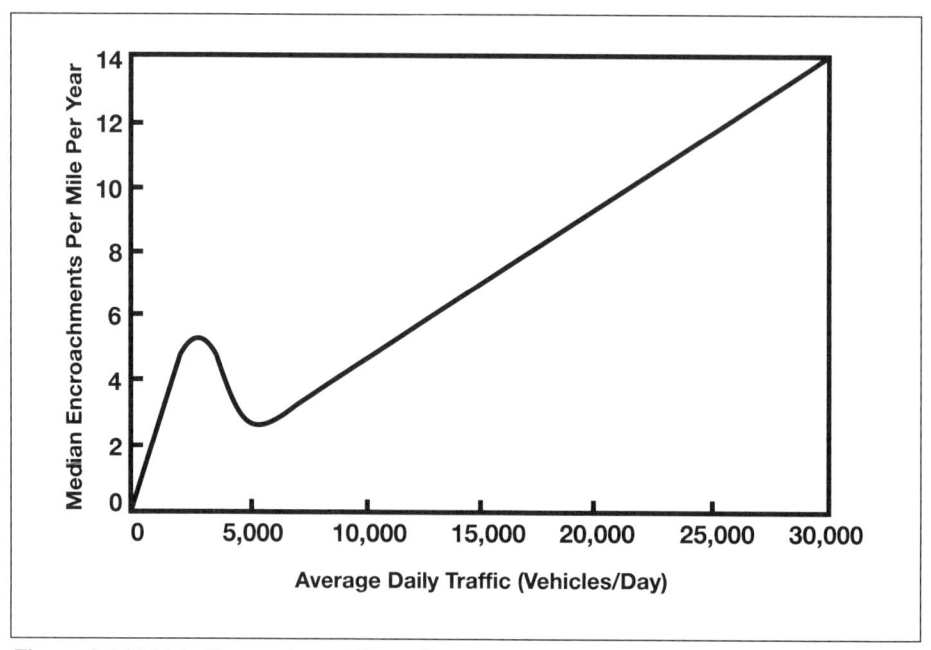

Figure 2-3 Vehicle Encroachment Rates.[6]

Roadside Safety

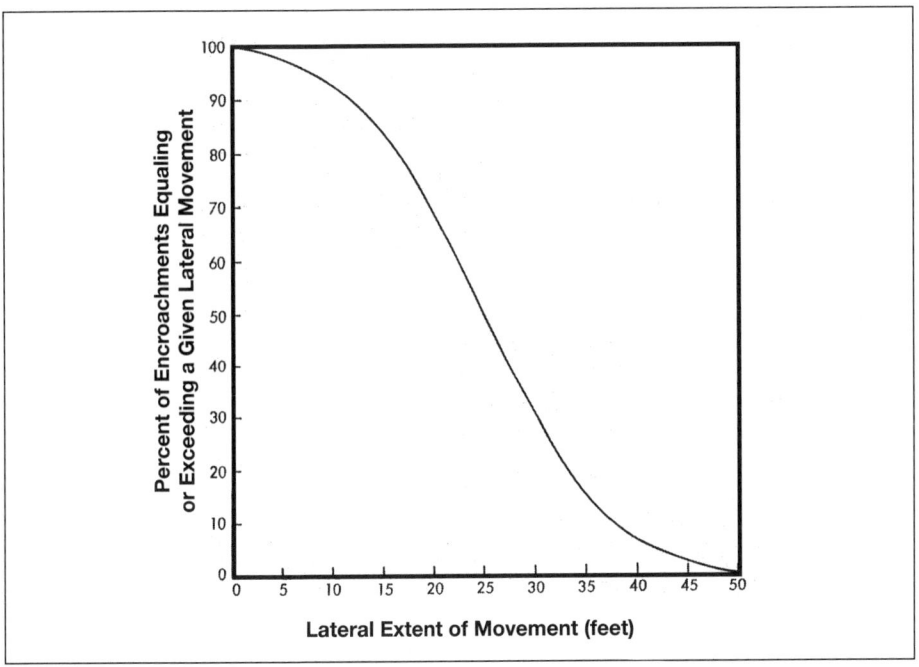

Figure 2-4 Excedance Distribution of Lateral Displacements for Vehicles Encroaching on the Roadside.[6]

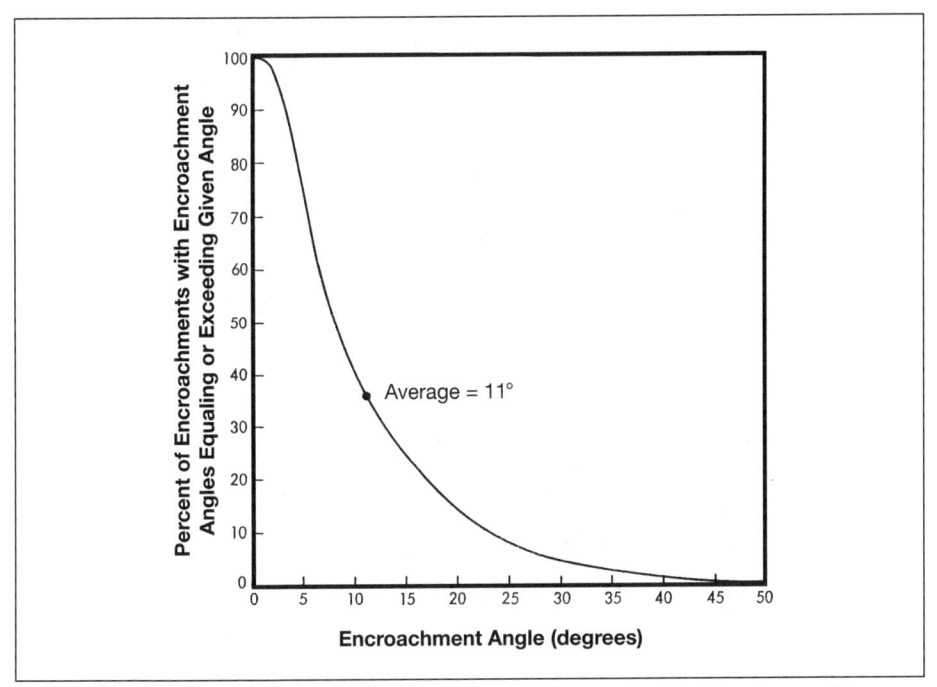

Figure 2-5 Excedance Distribution of Encroachment Angles for Vehicles Encroaching on the Roadside.[6]

In 1975, Glennon and Wilton[7] expanded the knowledge base on roadside encroachments by studying urban arterial streets and rural two-lane roadways and developing exceedance relationships similar to those of Hutchinson and Kennedy.

2.3 Roadside Hazard Model

A roadside hazard model developed by Glennon[8] in *NCHRP Report 148*, provides a basic analysis technique for comparing roadside environments. This model, adapted into a slightly less objective form, is also presented in the *Roadside Design Guide*[4] and the *Guide for Selecting, Locating and Designing Traffic Barriers.*[9]

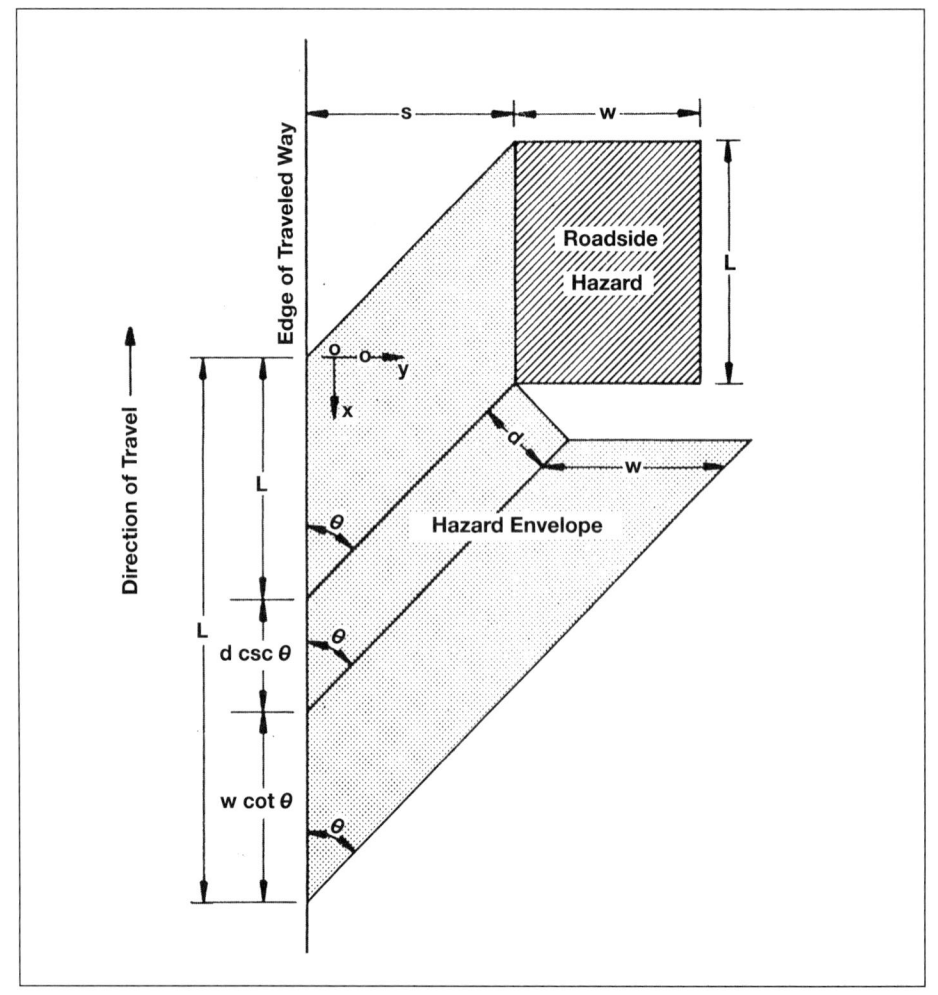

Figure 2-6 Schematic of Roadway Length Where Vehicles are Exposed to Impact with a Roadside Hazard.[8]

Table 2.2 Roadside Obstacle Severity Indices[7,8]

Roadside Obstacle	Severity Index (Fatal plus Injury Accidents Divided by Total Accidents.)		
	Freeway	Rural Surface Roadway	Urban Street
Utility Poles	0.55	0.45	0.40
Large Trees	0.60	0.50	0.45
Signposts			
Large Rigid Post	0.60	0.50	0.45
Small Rigid Post	0.35	0.30	0.25
Breakaway Post	0.20	0.20	0.20
Light, Traffic Signal, and Railroad Signal Poles			
Rigid	0.50	0.40	0.35
Breakaway	0.20	0.20	0.20
Curbs	0.40	0.35	0.30
Guardrails			
Short (less than 100 ft)	0.40	0.35	0.30
Long (greater than 100 ft)	0.35	0.30	0.25
Roadside Slopes			
Fill slopes			
2:1 or steeper	0.70	0.60	0.50
3:1	0.55	0.45	0.40
4:1	0.40	0.35	0.30
5:1	0.30	0.25	0.20
6:1 or flatter	0.20	0.15	0.15
Cut slopes			
1:1 or steeper	0.70	0.60	0.50
1-1/2:1	0.55	0.45	0.40
2:1	0.40	0.35	0.30
3:1	0.30	0.25	0.20
4:1 or flatter	0.20	0.15	0.15
Culverts	0.55	0.45	0.40
Bridge Abutments and Piers	0.70	0.60	0.50
Bridgerails			
Bridgerail—smooth	0.40	0.35	0.30
Parapet-type bridgerail	0.60	0.40	0.45
Bridgerail end	0.60	0.50	0.45
Retaining Walls and Fences	0.40	0.35	0.30
Fireplugs	–	0.30	0.25

The Glennon model depends on the concept that an injury-producing roadside collision occurs as a chain of four conditional events. First, the vehicle must be within the length of roadway, L, where a collision with a particular roadside hazard is possible. Second, the vehicle must encroach on the roadside. Third, the vehicle must travel far enough off the edge of the roadway to collide with the roadside hazard. And, finally, the collision must be of sufficient force to produce an occupant injury. This chain of events suggests a conceptual approach for evaluating the degree of hazard for roadside situations. This approach considers the vehicular exposure, the expected vehicular encroachment rate, the expected distribution of encroachment angles, the expected distribution of lateral displacements of encroaching vehicles, and the severity, size, and lateral placement of the roadside obstacle. Figure 2-6 is a schematic illustration of the roadway length, L, associated with a particular roadside obstacle. The hazard envelope is defined by the locus of the right front corner of the colliding vehicle as follows:

$$H = V[P(E)][P(C/E)][P(I/C)]$$

where:

- H = hazard index; expected number of fatal plus nonfatal injury accidents per year;
- V = vehicle exposure; number of vehicles per year passing through section L;
- P(E) = probability that a vehicle will encroach on the roadside within section L; encroachments per vehicle. This probability is a function of the length of exposure, L, and other environmental variables such as the geometric design of the roadway;
- P(C/E) = probability of a collision, given that an encroachment has occurred; accidents per encroachment. This probability is a function of the angle of encroachment, θ; the vehicle's lateral displacement (measured from the right-front corner of the vehicle), y; the lateral placement of the roadside obstacle, s; and the size of the obstacle, l and w; and
- P(I/C) = probability of an injury accident, given a collision; fatal plus nonfatal injury accidents per total accidents.

The description of the variables above suggests that a more explicit mathematical relationship is required to truly evaluate the hazard index of a particular roadside situation. For the average encroachment angle of 11° and a vehicle width of six feet, the more explicit equation is as follows:

Roadside Safety

$$H = \frac{E_f S}{10,560} \left\{ l \cdot P[y \geq s] + 31.4 \cdot P[y \geq s + 3] \right.$$

$$\left. + \frac{5.14w}{n} \sum_{j=1}^{n} P\left[y \geq s + 6 + \frac{w(2j-1)}{2n} \right] \right\}$$

where

H = hazard index; number of fatal plus nonfatal injury accidents per year, associated with a one-directional roadway. For median analysis, the hazard index is computed for each roadway separately and added;

E_f = encroachment frequency, number of roadside encroachments per mile per year. This parameter is taken from Figure 2-3 for various ADT rates. Values for urban arterials and rural two-lane roadways are found in Reference 7.

S = severity index, number of fatal plus nonfatal injury accidents per total accidents. Table 2.2 shows objective values developed by Glennon and Wilton[7] and by Glennon.[8] Different (subjective) values are given in Reference 4.

l = longitudinal length of the roadside obstacle, feet;

$P[y > ...]$ = probability of a vehicle lateral displacement greater than some value. These probabilities are taken from Figure 2-4. Values for urban arterials and rural two-lane roadways are found in Reference 7.

y = lateral displacement, in feet, of the encroaching vehicle; measured from the edge of the traveled lanes to the outside front corner of the vehicle at impact;

s = lateral placement, in feet, of the roadside obstacle; measured from the edge of the travel lanes to the longitudinal face of the roadside obstacle.

w = lateral width of the roadside obstacle, feet;

n = number of analysis increments associated with the obstacle width. A reasonable subdivision is 2.0 feet; and

j = the number of the obstacle-width increment under consideration, counting from 1 to n.

2.4 Clear Zone Standards

As evidence mounted in the early 1960's that roadside collisions were not only one of the most prevalent kinds of collisions but also one of the most severe, the prevailing negative attitude of roadway agencies about driver's involved in roadside collisions began to change. From the late 1960's up to the present, the prevailing attitude recognizes that many different kinds of drivers run off the road for many different reasons and that they and their passengers should not necessarily be punished with life-threatening injuries. Hence the term, *forgiving roadside,* has gained wide use and recognition.

A forgiving roadside has a *clear zone* alongside the roadway that is reserved for emergencies and is designed to safely accommodate vehicles that inadvertently encroach on the roadside. The forgiving roadside philosophy foresees that even roadways with well-designed alignment should be maintained with a clear zone; one that has gentle slopes free of fixed objects so the driver of an errant vehicle can recover with relative safety. More specifically, the AASHTO *Roadside Design Guide*[4] defines *clear zone* as follows:

> Clear Zone—The total roadside border area, starting at the edge of the traveled way, available for safe use by errant vehicles. This area may consist of a shoulder, a recoverable slope, a nonrecoverable slope, and/or a clear run-out area. The desired width is dependent upon the traffic volumes and speeds, and on the roadside geometry.

Since 1967, the term *clear zone* has evolved to where more definitive standards have been recommended. The 1977 *AASHTO Guide for Selecting, Locating, and Designing Traffic Barriers,*[9] first specified clear-zone widths relative to roadway speed and roadside slope. The 1984 AASHTO *Policy on Geometric Design for Streets and Highways*[10] gave further definition for applying the 1977 AASHTO Barrier Guide principles to various classes of roadway. Then, in 1989, AASHTO published their current policy on clear zones in the *Roadside Design Guide.*[4] These values are shown in Figure 2-7, which shows clear-zone widths ranging from 7 feet for lower-volume, lower-speed roadways to 46 feet for higher-volume, higher-speed roadways. Clear zone widths in Figure 2-7 also vary by slope ratio, with steeper slopes commanding greater clear zone widths.

As shown by Glennon, Neuman, and Leisch[11] and also by Perchonak, et. al.,[12] roadside collisions are much more likely to occur on roadway curves. Not only are encroachments more frequent, but the maximum lateral displacement of encroaching vehicles increases as the degree of curve increases (degree of curve is

Roadside Safety

Design Speed	Design ADT (vehicles per day)	Clear Zone Widths (in feet from edge of driving lane)				
		FILL SLOPES		CUT SLOPES		
		6:1 or flatter	5:1 to 4:1	3:1	4:1 to 5:1	6:1 or flatter
40 mph or less	Under 750	7-10	7-10	7-10	7-10	7-10
	750-1500	10-12	12-14	10-12	10-12	10-12
	1500-6000	12-14	14-16	12-14	12-14	12-14
	Over 6000	14-16	16-18	14-16	14-16	14-16
45-50 mph	Under 750	10-12	12-14	8-10	8-10	10-12
	750-1500	12-14	16-20	10-12	12-14	14-16
	1500-6000	16-18	20-26	12-14	14-16	16-18
	Over 6000	18-20	24-28	14-16	18-20	20-22
55 mph	Under 750	12-14	14-18	8-10	10-12	10-12
	750-1500	16-18	20-24	10-12	14-16	16-18
	1500-6000	20-22	24-30	14-16	16-18	20-22
	Over 6000	22-24	26-32	16-18	20-22	22-24
60 mph	Under 750	16-18	20-24	10-12	12-14	14-16
	750-1500	20-24	26-32	12-14	16-18	20-22
	1500-6000	26-30	32-40	14-18	18-22	24-26
	Over 6000	30-32	36-44	20-22	24-26	26-28
65-70 mph	Under 750	18-20	20-26	10-12	14-16	14-16
	750-1500	24-26	28-36	12-16	18-20	20-22
	1500-6000	28-32	34-42	16-20	22-24	26-28
	Over 6000	30-34	38-46	22-24	26-30	28-30

Figure 2-7 Roadside Clear Zone Standards.[4] (Courtesy of Criterion Press, Inc.)

equal to 5,730 divided by the radius, in feet). Figure 2-8, taken from the *Roadside Design Guide*,[4] shows the adjustment to the normal clear zone standards recommended for roadway curves.

K_{cz} Curve Corrective Factor								
Degree	Design Speed						
of Curve	40	45	50	55	60	65	70
2.0	1.08	1.10	1.12	1.15	1.19	1.22	1.27
2.5	1.10	1.12	1.15	1.19	1.23	1.28	1.33
3.0	1.11	1.15	1.18	1.23	1.28	1.33	1.40
3.5	1.13	1.17	1.22	1.26	1.32	1.39	1.46
4.0	1.15	1.19	1.25	1.30	1.37	1.44	
4.5	1.17	1.22	1.28	1.34	1.41	1.49	
5.0	1.19	1.24	1.31	1.37	1.46		
6.0	1.23	1.29	1.36	1.45	1.54		
7.0	1.26	1.34	1.42	1.52			
8.0	1.30	1.38	1.48				
9.0	1.34	1.43	1.53				
10.0	1.37	1.47					
15.0	1.54						

$CZ_c = (L_c)(K_{cz})$
Where:
 CZ_c = clear zone on outside of curvature, ft.
 L_c = clear zone distance, ft.
 K_c = curve correction factor
Note: Clear zone correction factor is applied to outside of curves only. Curves flatter than 2.0° do not require an adjusted clear zone.

Figure 2-8 *Adjustments to Clear Zone Width for Roadway Curves.*

2.5 Functional Roadside Elements

For the purpose of the discussion here, roadside elements are divided into two separate categories: functional and non-functional. What is meant by *functional* are those roadside elements that are either a part of the basic geometric design of the roadway or serve to enhance its traffic movement. Under this category are roadside slopes, bridge structures, drainage facilities, roadway lighting, traffic control devices, and curbs. Under the *non-functional* category are elements such as trees, utility poles, traffic barriers, and other fixed objects.

Roadside Slopes

Roadside slopes on embankments and in cuts should provide a reasonable opportunity for recovery by an errant driver. Flat slopes and generous rounding of slope breaks are the features of a traversable terrain. The normal design features encountered in sequence by the errant driver are the side slope, the ditch, and the backslope.

Roadside Safety 41

Figure 2-9 shows various side slopes drawn to scale. Since 1940, AASHTO[10,13,14,15,16] has recommended a practical maximum steepness of 4:1 for side slopes. Starting in about 1967, AASHTO[1] began to recommend 6:1 side slopes as desirable, particularly on Interstate highways. In 1984, AASHTO[10] made a concession, in deference to research to the contrary (but perhaps in response to liability pressures), by suggesting that 3:1 slopes may be minimally acceptable under certain circumstances. For any side slope steeper than 3:1, AASHTO recommends that traffic barriers be considered.

The most significant research on the hazard of roadside terrain was published in 1975 by Weaver, Marquis, and Olson.[17] This research, which used computer simulation and controlled vehicle test runs on various terrains, clearly demonstrated lower levels of force as the side slope became flatter. The research not

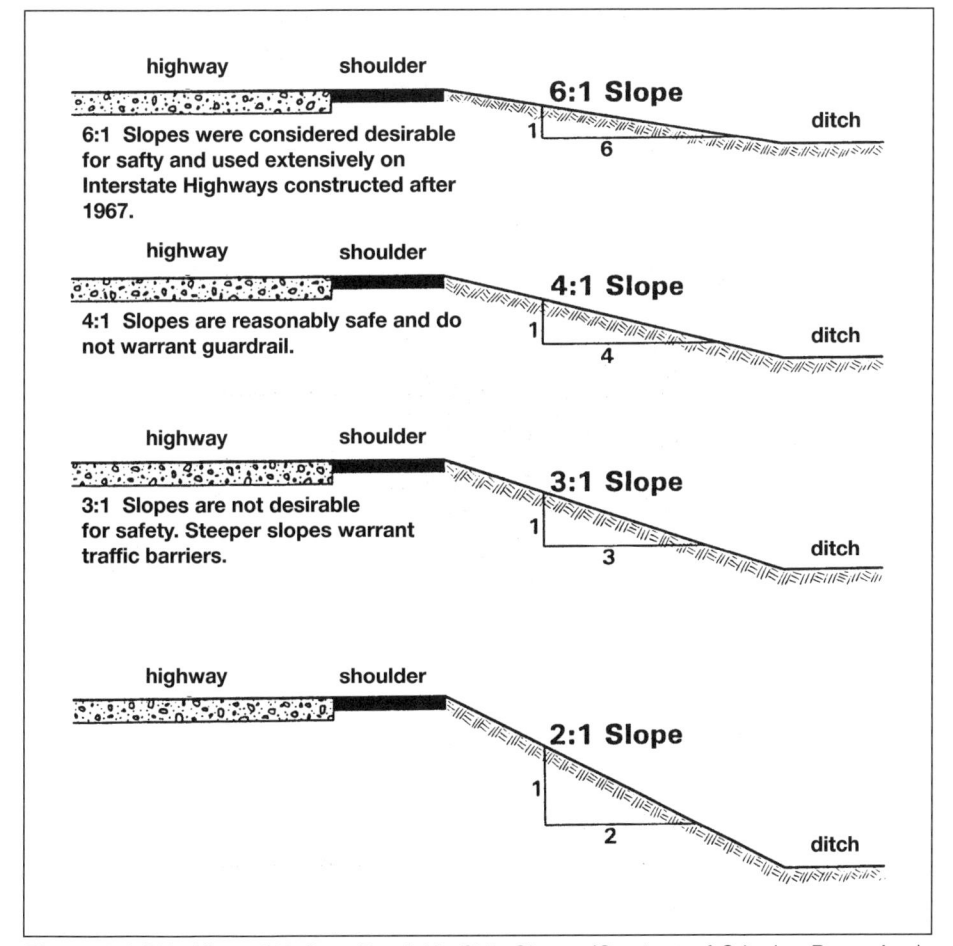

Figure 2-9 Side View of Various Roadside Side Slopes (Courtesy of Criterion Press, Inc.)

only validated the wisdom of the 4:1 minimum side slope recommended by AASHTO over the years, but also showed the benefits of rounding all of the slope breaks: between shoulder and side slope, between side slope and ditch, and between ditch and backslope. Figures 2-10 shows the recommendations of the *Roadside Design Guide*[4] based on the Weaver, Marquis, and Olson research.

Zeeger, et. al.,[18] developed relationships between single-vehicle accidents and field-measured side slopes along 1776 miles of roadway in four States. These relationships were used to estimate the accident reduction benefits of various side-slope flattening projects. The percent reductions are shown in Table 2.3. This table assumes that the side slopes to be flattened are relatively clear of rigid objects.

The use of flatter slopes not only reduces accident frequency, but also reduces accident severity. This relationship results because the likelihood of vehicle rollover decreases with flatter slopes. Average occupant severity from rollover accidents is exceeded only by the average severity of pedestrian and head-on accidents. Zeeger, et. al.,[18] found that side slopes of 5:1 or flatter were needed to significantly reduce the incidence of vehicle rollover.

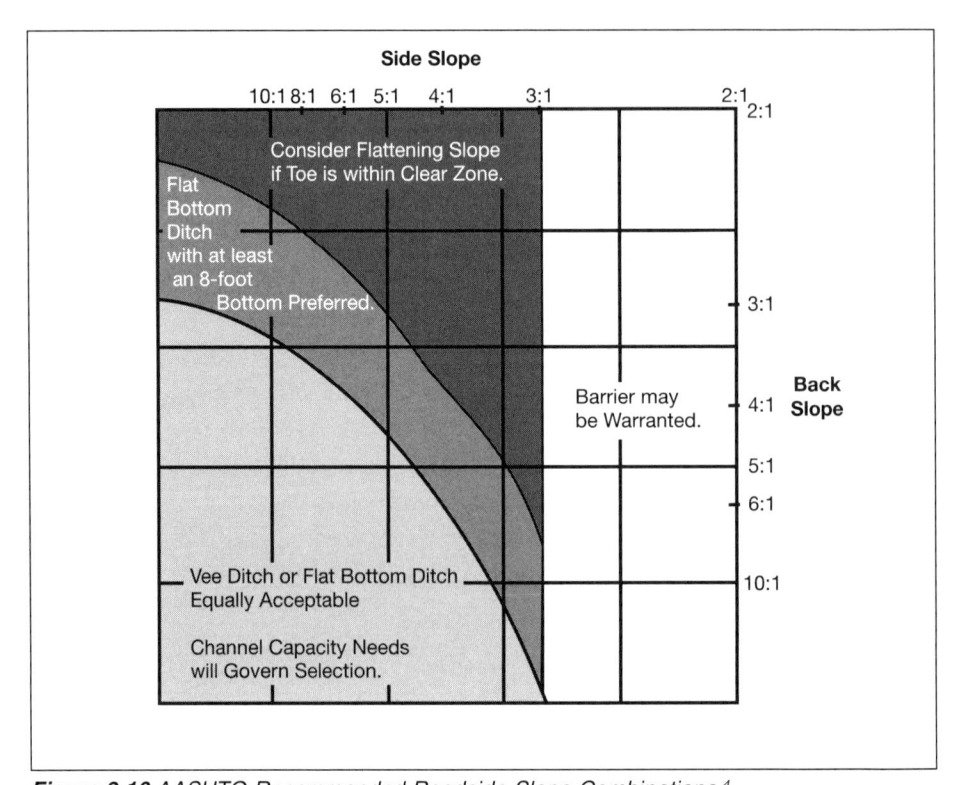

Figure 2-10 AASHTO Recommended Roadside Slope Combinations.[4]

Roadside Safety

Table 2.3 Percentage Reduction in Accidents for Various Side Slope Flattening Projects[18]

Existing Side Slope	\multicolumn{2}{c}{4:1}		\multicolumn{2}{c}{5:1}		\multicolumn{2}{c}{6:1}		\multicolumn{2}{c}{7:1 or Flatter}	
	Single Vehicle Accidents	Total Accidents	Single Vehicle Accidents	Total Accidents	Single Vehicle Accidents	Total Accidents	Single Vehicle Accidents	Total Accidents
2:1	10	6	15	9	21	12	27	15
3:1	8	5	14	8	19	11	26	15
4:1	—	—	6	3	12	7	19	11
5:1	—	—	—	—	6	3	14	8
6:1	—	—	—	—	—	—	8	5

Bridges

Two kinds of roadside hazard can be created by roadway bridges. The first of these includes both the steep approach side slopes and the bridge rail for traffic going over a bridge. Bridge rails need to be strong and smooth as do approach guardrails to both eliminate contact with the bridge rail end and to protect errant vehicles from the steep approach side slopes. The subject of bridge rails is covered more fully in Chapter 3.

The second kind of hazard is associated with bridges supports — abutments, piers, and columns — that can be exposed to traffic under a bridge. These supports, which are usually massive concrete, can snag errant vehicles causing excessive decelerations and significant injuries to vehicle occupants. Because of limiting costs, these bridge supports are often placed within the clear zone (Figure 2-11), requiring traffic barriers or crash cushions to minimize impact severity.

Drainage Facilities

To maintain the integrity of the roadway structure and to provide for natural drainage, an extensive drainage system is built into most roadways. However, the structures and waterways needed for this drainage system can present serious hazards to vehicles that run off the road. These structures includes dikes, headwalls, ditches, channels, culvert ends, and inlets. As seen from Table 2:1, over 10% of all fixed object fatalities result from impacts with culverts and ditches.

Figure 2-11 Dangerous Bridge Supports.

When considering drainage structures, good practice requires consideration of both hydraulics and safety, in the following order of priority:
1. Unnecessary drainage structures should be eliminated.
2. Necessary drainage structure should be placed to minimize their hazard.
3. Structures that cannot be eliminated should be designed to minimize impact severity.
4. Where the first three objectives cannot be reasonably met, guardrails should be considered.

A culvert pipe parallel to the roadway (Figure 2-12), running under a crossroad or driveway, can be serious hazard when it is in the clear zone. Ran-off-road vehicles sliding along a ditch can snag, spin out, and/or overturn when colliding with the pipe end. To minimize the impact severity when these pipes are in the clear zone, the pipe should be flush with the fill slope of the crossroad. In addition, for pipes of 30-inch diameter or more, a grate should be placed to ramp errant vehicles over the pipe end.

Culvert pipes perpendicular to the roadway pose a similar hazard because they too can cause impacting vehicles to snag, spinout, and/or overturn. These pipes and/or their headwalls should be flush with the sideslope. Again, pipes with diameters greater than 30 inches should also have grates over the pipe opening.

Roadside Safety

Figure 2-12 Dangerous Drainage Pipe Parallel to Roadway.

Culvert headwalls are one of the more insidious hazards on rural roadways. They are often within 10 feet of the travel lanes and many times are closer than 4 feet, as shown in Figure 2-13. Headwalls, which not only are close to the travel lanes but also protrude six inches or more above the surface, are frequently involved in serious roadside collisions. Culvert headwalls are safest when they are outside of the clear zone and/or are flush with the side slope. If protruding culvert headwalls within the clear zone cannot be moved or redesigned, they should be shielded with traffic barriers.

Light Poles

Street lighting is normally concentrated along high-volume roadways (Figure 2-14), at isolated intersections, and at freeway interchanges. AASHTO publishes a special booklet[19] to encourage that light poles are either made to breakaway or are placed where they are less likely to be hit. Recommendations to minimize the hazard of light poles are:

1. Whenever possible, light poles should be placed behind guardrails, on retaining walls, or outside the likely path of a ran-off-road vehicle.
2. Breakaway slip bases or frangible transformer bases should be used whenever the light pole is exposed to moderate or high-speed traffic.

Figure 2-13 Dangerous Culvert Headwalls.

Figure 2-14 Continuous Roadway Lighting.

Roadside Safety

3. The clearance to a light pole shall be at least two feet beyond either the face of a barrier curb or at least two feet beyond the outside edge of a shoulder.

Breakaway light poles supports are designed to release or break in shear (Figure 2-15) when impacted at a typical bumper height of 20 inches. Care is needed to locate these devices where superelevation, side slope, or other geometry will not adversely influence the striking height of the typical bumper.

Generally, a breakaway light pole will fall near the path of an impacting vehicle. The mast arm usually rotates and falls pointing away from the roadway, so that the falling pole is not a hazard to other traffic. However, these falling poles may sometimes be a hazard to pedestrians or bicyclists on sidewalks or special paths. Also falling poles in narrow medians on divided roadways could be a hazard to opposing traffic.

The height of breakaway light poles should not exceed 55 feet. Also, to prevent serious consequences if a pole should fall on a vehicle, the weight of the pole should not exceed 1,000 pounds.

Figure 2-15 *Light Pole with Breakway Transformer Base.*

Roadside safety should be a prime consideration when designing roadway lighting. Higher mounting heights can significantly reduce the number of poles necessary. The preferred method for lighting major freeway interchanges is the tower lighting shown in Figure 2-16, which requires far fewer poles that can be located much further from the roadway.

Sign Supports

Although roadway signs must be placed so they effectively communicate messages to the driver, they should also be placed or designed to present the least possible impact hazard. When possible, signs should be placed behind existing guardrail, on bridges over the roadway, or a sufficient distance from the edge of the travel lanes to minimize exposure to ran-off-road vehicles. When sign supports cannot be placed out of harms way, breakaway supports or crash cushions should be used.

Figure 2-16 Tower Lighting.

Roadside Safety

Roadway signs are classified as overhead signs, large roadside signs, and small roadside signs. *Overhead signs*, including sign bridges (Figure 2-17) and cantilevered signs (Figure 2-18), generally require massive supports which

Figure 2-17 Overhead Sign Bridge.

Figure 2-18 Cantilevered Sign.

cannot be made breakaway. Where possible, overhead signs should be installed or relocated on nearby overpasses (Figure 2-19). Overhead sign supports within the clear zone should be placed as far from the travel lanes as feasible and shielded by crashworthy barriers.

Large roadside signs usually have areas of 50 square feet or more. They typically have two or more breakaway supports. The basic concept of the breakaway sign support is to have adequate bending strength, but minimal shear strength. Current designs provide a structure that will resist most wind and ice loads, yet fail in a safe and predictable manner when struck by a vehicle. Figure 2-20 shows, however, that breakaway signs sometime fail in very heavy winds.

Slip-base breakaway mechanisms will activate under impact as the two parallel plates slide apart when the connecting bolts are pushed out. Figure 2-21 is a typical design. The upper hinge design (see Figure 2-22) consists of a slotted fuse plate on the expected impact side and a saw cut through the front flange and the web of the post up to the rear flange. The rear flange then acts as a hinge when the post rotates upward. Figure 2-23 shows how the breakaway sign will stand on one or more supports as one support is impacted, that support slips and rotates upward, and the vehicle passes underneath the sign.

Figure 2-19 Large Sign Placed on an Overpass.

Roadside Safety

Figure 2-20 Breakaway Sign Toppled by High Winds.

Figure 2-21 Breakaway Sign Slip Base.

Figure 2-22 Slotted Fuse Plate on a Breakaway Sign.

Figure 2-23 Typical Impact with a Breakaway Sign.

Roadside Safety

Small roadside signs have areas less than 50 square feet with one or more supports. Typically small sign supports are either driven into the soil, set in drilled holes, or mounted on a separate base. The most widely used supports are steel U-posts, wood posts, standard steel pipe, and square steel tubing.

Breakaway features are used on a small percentage of the total small roadside signs. For breakaway wooden supports, the shear strength is reduced by drilling holes perpendicular to the direction of impact (Figure 2-24). For single steel pipe supports, a slip base (Figure 2-25) angled up away from the direction of impact is used not only to reduce the impact forces but also to flip the sign up and over the impacting vehicle.

Traffic Signal Supports

As with any roadside object, traffic signals supports and controller cabinets should be placed as far as possible from the travel lanes. Figure 2-26 shows the devastation caused when a vehicle impacts a traffic signal pole close to a 55-mph roadway. The *Manual on Uniform Traffic Control Devices*[20] has the following to say about traffic signal supports:

> *Supports at a street with curbs shall have a horizontal clearance not less than two feet from the face of the curb. Where there is no curb, supports shall have a horizontal clearance not less than two feet from the edge of a shoulder, within the limits of normal vertical clearance.*
>
> *No part of a concrete base for a signal support should extend more than four inches above the ground level at any point...*
>
> *On medians, the above minimum clearances for signal supports should be obtained where practicable. Any supports which cannot be located with the required clearances should be of the breakaway type or should be guarded if at all practical.*

Curbs

Curbs are used for drainage and for pavement edge support and delineation. AASHTO[10] classifies curbs as either barrier curbs or mountable curbs. Barrier curbs are usually six inches or higher with nearly vertical faces. Mountable curbs are usually four inches or less and have sloping faces that are easily traversed.

Neither barrier nor mountable curbs are desirable on high-speed roadways because an out-of-control vehicle sliding sideways is prone to rollover when impacting the curb. Another problem with curbs on high-speed roadways is that, when they are placed in front of traffic barriers, they may cause vehicles to mount the barrier.

54 Roadway Defects and Tort Liability

Figure 2-24 Breakaway Wooden Sign Posts.

Figure 2-25 Angled Breakaway Slip Base for Small Signs with Pipe Support.

Roadside Safety

Figure 2-26 Fatal Impact with a Rigid Traffic Signal Support.

2.6 Non-Functional Roadside Elements

What is meant by *non-functional* are those roadside elements that are neither a part of the basic geometric design of the roadway nor serve to enhance traffic movement. Under this category are trees, utility poles, traffic barriers, large rocks or boulders, advertising signs, subdivision signs, large mailbox structures, brick planter boxes, utility cabinets, etc.

Trees

Trees are not only the roadside object most often hit, but also cause more fatalities than any other object. Single-vehicle collisions with trees account for nearly 25% of all fixed-object fatal accidents, causing nearly 3,000 deaths each year.

Trees will abruptly stop a vehicle when their trunk diameter exceeds four inches. These larger trees must be considered as hazardous fixed objects when they are near the edge of the travel lanes. They are non-yielding and cause great damage to errant vehicles and great injury to vehicle occupants. Although smaller trees may be relatively safe, their danger will increase with each year of growth. Obviously, AASHTO clear zone standards should be followed whenever possible both in not planting trees and in cutting existing trees within the clear zone.

Removing hazardous trees can be difficult because of emotional community and/or environmental issues. These issues can often be minimized by mowing clear zones to prevent seedlings from becoming established. Then too, when cutting down trees, alternative plantings can be substituted that are not only aesthetically pleasing and environmentally beneficial but also less hazardous when hit by an errant vehicle. If a group of large trees is close to the travel lanes, a properly designed and installed traffic barrier might provide an alternative to cutting the trees.

Utility Poles

Utility poles are a major kind of roadside hazard accounting for about 1400 fatalities per year in the U.S. Utility poles occupy roadway rights-of-way pursuant to State and local laws under various degrees of regulation. Standards for locating utility poles vary depending on the type of roadway, the available space within the roadway right-of-way, and the degree of control exercised by the roadway agency. In rural areas, many existing utility poles are located within a few feet of the traveled way. In urban areas (Figure 2-27), they are often only inches from the travel lanes where, because of the front overhang, an automobile can collide with the pole without leaving the roadway.

Utility poles within the right-of-way should be placed as far as possible from the travel lanes. In urban areas where roadways have curbs, utility poles should

Roadside Safety

Figure 2-27 Utility Pole Placed Close to the Roadway.

be located as far as possible behind the curb face. In areas with sidewalks, poles should be placed behind the sidewalk if possible. Joint-use poles should be used where more than one utility is involved. To further minimize the number of poles, the longest feasible spans should be used. Poles should not be placed where they are vulnerable to ran-off-road vehicles, such as on the outside of roadway curves or within traffic islands. For new construction, utility lines should be buried underground whenever feasible.

Any effective program to relocate, rearrange, or convert existing overhead utility lines must overcome legal obstacles, such as who will pay. Under present laws, most roadway agencies do not have legal authority to pay for utility relocation other than on Interstate projects. Where legislative authority is required for action, this authority should be sought.

Traffic Barriers

The term *traffic barrier* is generally used to include roadside barriers (guardrails), bridgerails, median barriers, and crash cushions. These are also called roadside safety appurtenances to generally indicate their function to shield hazardous elements within the clear zone.

Traffic barriers can also be considered as roadside hazards in that, depending on their design, they have some expected severity of impact. Usually their designed impact severity is relatively low. When traffic barriers are improperly designed, placed, installed, or maintained, however, they can become a serious hazard.

Because the subject of traffic barriers is fairly broad, Chapter 3 is devoted entirely to traffic barriers.

Other Fixed Objects

Many other non-functional fixed objects dot the roadside landscape. Large rocks or boulders, such as shown in Figure 2-28, are sometimes arranged along the roadway. Other objects include advertising signs, subdivision signs, large mailbox structures, brick planter boxes, utility cabinets, etc.. These all represent fixed hazards, which should be treated within the clear zone concept.

Figure 2-28 Large Boulder Placed Close to the Roadway.

> **ROADSIDE HAZARDS**
> **Rule of Thumb**
> *Remove, Redesign, or Protect Roadside Hazards Within 30 Feet of the Traveled Way on Freeways or Within 20 Feet of the Traveled Way on Other Roads.*

2.7 Technical Aspects of Roadside Hazard Cases

Roadside hazards are a leading topic of roadway defect tort claims on all roadways, but most particularly on roadways with speed limits of 45 mph or higher. Unprotected steep side slopes and large fixed objects close to the edge of the traveled way are obvious to plaintiff lawyers and injured parties alike as contributors not only to a collision occurrence but also more directly to collision severity. When an errant driver had no chance for recovery either because of a steep slope, a large tree, an unprotected bridge pier or rail, a rigid pole or post, or a culvert too close to the travel lanes, negligence of the roadway agency may be claimed. Clearly, the most important technical aspect is whether the roadway agency adhered to some reasonable dimension of clear zone width (Figure 2-7). Other arguments are whether particular objects were even necessary, whether they could have been designed for lower impact severity, or whether they could have been placed in less vulnerable positions.

In considering the negligence of the roadway agency for various roadside hazards, the date of installation is often important. 1940 is not only when a minimum 4:1 side slope was first recommended but also when guardrail was recommended on slopes steeper than 4:1. Before 1967, recommendations for the offset of fixed objects were minimal. In 1967, the general recommendation was for a 30-foot clear zone. Later, after 1977 (Figure 2-7), more comprehensive clear zone values were promulgated.

Roadway Widening Projects that Sacrifice Roadside Safety

Over the last 25 years, roadway projects that improve existing facilities have outnumbered projects that build new facilities. One prevalent kind of improvement project has involved widening of the roadbed to upgrade narrow lanes, to upgrade narrow shoulders, or to provide shoulders where none previously existed.

Sometimes these widening projects, which are usually done in the name of safety, can degrade the total safety of the roadway because roadside dimensions are decreased as roadbed dimensions are increased. Steep side slopes and rigid fixed objects within the clear zone effected by these changes are clear candidates for roadway defect claims.

Steep Side Slopes

Side slopes (Figure 2-29) of 3:1 or steeper that are close to the travel lanes on high-speed roadways can be candidates for roadway defect claims. Many times, if the colliding vehicle overturned on a steep side slope, that slope may be found not only as a proximate cause of the accident but also as a primary contributor to the severity of the resulting injuries.

Figure 2-29 *Steep Side Slope.*

Roadside Safety

Removable Objects Within the Clear Zone

Clearly, non-functional (Figure 2-30) objects such as large trees, boulders, and advertising signs can be removed from the roadway right-of-way. Because motorist injuries resulting from impact are often devastating, these objects have no place in the clear zones of high-speed roadways. When, for example, a vehicle impacts a large tree (Figure 2-31) within the clear zone on a high-speed roadway, the tree can be judged as both a proximate cause and the major contributor to occupant injuries. If emotional, environmental, or other aesthetic issues had previously overruled removal of the object, then other options of protecting the motoring public should have been implemented.

Moveable Objects Within the Clear Zone

When rigid objects such as sign supports, light poles, utility poles, and culverts are impacted within the clear zones, the tort liability question might be why

Figure 2-30 Tree Hazard is Inconsistent with Placement of Guardrail.

Figure 2-31 Fatal Impact with a Large Tree.

they were not placed or moved either outside of the clear zone or in a less vulnerable location behind a traffic barrier. Figure 2-32 shows a multiple fatal accident involving a large sign pole within the clear zone.

Figure 3-32 Fatal Collision with an Unshielded Large Sign Post.

Non-Breakaway Functional Objects Within the Clear Zone

The technology for breakaway supports for large signs, small signs, lightpoles, etc. has existed for almost 30 years. Where non-breakaway supports are still in place within the clear zone, particularly on high-speed roadways, the roadway agency has exposure to tort litigation. When these objects exist and cannot be moved, the next best option is to install breakaway devices.

Unprotected Functional Objects Within the Clear Zone

Unprotected bridge abutments, piers, and columns, non-crashworthy bridge rails, bridge rail ends, and culvert headwalls close to the travel lanes constitute serious hazards (Figure 2-33) to the motoring public, particularly on high-speed roadways. When these objects are close to the travel lanes and cannot feasibly be moved, traffic barriers are the suitable protection. Without this protection, roadway agencies can be found negligent for not exercising their duty to provide safe travel for the motoring public.

Defective Traffic Barriers

Of all the traffic barriers in place today probably more than half have some defect that can substantially increase the occupant injury severity when impacted. Barriers that are too low, too high, too short, have too few posts, have too few bolts, or do not have crashworthy ends are prevalent on U.S. roadways. A complete chapter in this book, Chapter 3, is devoted to traffic barrier defects.

Figure 2-33 Fatal Impact with an Unshielded Median Bridge Support.

Pavement Edge Drops

As suggested in the 1974 AASHTO publication, *Highway Design and Operational Practices Related to Highway Safety*,[21] pavement edge drops can be considered as a roadway element that is contrary to the concept of a forgiving roadside. Because pavement edge drops are a prevalent roadway defect claim, an entire chapter in this book, Chapter 7, is devoted to this subject.

2.8 Typical Defense Arguments

When defending tort claims that allege a defective roadside, roadway agencies generally argue that there are no required standards for roadside design, and that the AASHTO *Roadside Design Guide*[4] and other similar documents are simply guidelines. In other words, if the roadway agency chose to implement a certain roadside improvement in one place but didn't use the same improvement at a similar location, the argument is that those decisions were entirely within their discretion.

Another major defense argument is the claim of design immunity. If, for example, the roadway was originally designed in 1954 before many significant roadside safety guidelines or standards were recognized, the roadway agency could be immune from liability. Likewise, for those roadside obstacles where the roadside safety technology has been constantly improving since about 1964, the roadway agency will argue that a 1970 design that is now known to be hazardous cannot be the subject of liability because of the design immunity statute.

At least one State has statutory protection against most barrier defects. In that State's tort claims act, the roadway agency is only liable for defects within the "improved" roadway, which only includes travel lanes and shoulders.

Beyond the available statutory arguments given above, roadway agencies often use other case-specific defenses that argue either that the driver was the sole negligent party or the condition did not warrant improvement. Some of these arguments are:

1. The driver was negligent for running off the road.
2. The driver was negligent for speeding.
3. The driver was negligent for driving while intoxicated.
4. The driver was negligent for falling asleep at the wheel.
5. The subject location had no prior history of accidents.
6. The roadway agency lacked adequate funds to improve existing roadside hazards.
7. The roadway agency used a priority scheme for programming improvements, and this location was low on that priority list.

Roadside Safety

8. Roadside safety improvements were being delayed because the entire section of roadway was planned for reconstruction.

Endnotes

1. American Association of State Highway Officials, *Highway Design and Operational Practices Related to Highway Safety*, 1967.
2. Blatnik, John A., *The Need for Highway Safety Consciousness*, Roadside Hazards, The Eno Foundation, 1968.
3. Center for Auto Safety, *The Yellow Book Road: The Failure of America's Roadside Safety Program*, 1974.
4. American Association of State Highway and Transportation Officials, *Roadside Design Guide*, 1989.
5. Stonex, K.A., *Roadside Design for Safety*, Highway Research Board Proceedings, 1960.
6. Hutchinson, J.W. and Kennedy, T.W., *Safety Consideration In Median Design*, Highway Research Record 162, 1966.
7. Glennon, John C., and Wilton, Cathy J., *Effectiveness of Roadside Safety Improvements*, Federal Highway Administration, Report No. FHWA-RD-75-23, 1974.
8. Glennon, John C., *Roadside Safety Improvement Programs on Freeways: A Cost-Effectiveness Priority Approach*, Transportation Research Board, NCHRP Report 148, 1974.
9. American Association of State Highway and Transportation Officials, *Guide for Selecting, Locating, and Designing Traffic Barriers*, 1977.
10. American Association of State Highway and Transportation Officials, *Policy on Geometric Design of Streets and Highways*, 1984.
11. Glennon, J.C., Neuman, T.R., and Leisch, J.E., *Safety and Operational Considerations for Design of Rural Highway Curves*, Federal Highway Administration, 1983.
12. Perchonok, K., et. al., *Hazardous Effects of Highway Features and Roadside Objects*, Federal Highway Administration, Report No. FHWA-RD-78-202, 1978.
13. American Association of State Highway Officials, *A Policy on Highway Types: Geometric*, 1940.
14. American Association of State Highway Officials, *A Policy on Geometric Design of Rural Highways*, 1954.
15. American Association of State Highway Officials, *A Policy on Geometric Design of Rural Highways*, 1965.
16. American Association of State Highway and Transportation Officials, *A Policy on Geometric Design of Streets and Highways*, 1990.
17. Weaver, Graeme D., Marquis, Eugene L., and Olson, Robert M., *Selection of Safe Roadside Cross Sections*, Transportation Research Board, NCHRP Report No. 158, 1975.
18. Zeeger, C.V., et. al., *Safety Effects of Cross-Section Design for Two-Lane Roads*, Federal Highway Administration, Report No. FHWA-RD-87-008, 1987.

19. American Association of State Highway and Transportation Officials, *Standard Specifications for Structural Supports for Highway Signs, Luminaries, and Traffic Signals*, 1990.
20. Federal Highway Administration, *Manual on Uniform Traffic Control Devices*, 1988. (available through Lawyers and Judges Publishing Co.)
21. American Association of State Highway and Transportation Officials, *Highway Design and Operational Practices Related to Highway Safety*, 1974.

References

Graham, J.L., and Harwood, D.W., Effectiveness of Clear Recovery Zones, Transportation Research Board, NCHRP Report No. 247, 1982.

Chapter 3

Traffic Barriers

Unlike the other fixed objects discussed in Chapter 2, both functional and non-functional, traffic barriers are safety devices that are used to compensate for an otherwise unsafe roadside. They are used to protect motorists from steep embankment slopes, from large rigid objects close to the travel lanes, and from close-by opposing high-speed traffic. Generally, roadways should be designed to minimize the need for traffic barriers.

As discussed here, *traffic barrier* is a broad term used to encompass all devices that are placed within the clear zone to shield roadside hazards. They include roadside barriers (guardrails), bridge rails, median barriers, and crash cushions.

The basic performance function of traffic barriers is to minimize occupant injuries by reducing vehicle impact decelerations. Properly designed roadside barriers, median barriers, and bridge rails reduce vehicle impact deceleration by preventing snagging and by redirecting the vehicle along the smooth face of the barrier. Properly designed crash cushions attenuate the energy of the impacting vehicle by crushing a sufficient length of the system to allow tolerable decelerations. Current performance criteria for traffic barriers are given in *NCHRP Report 350*.[1] The predecessors to this document were *NCHRP Report 260*,[2] and *NCHRP Report 153*.[3]

A variety of crashworthy traffic barriers are available for different applications. Each barrier design has specific components and dimensions necessary for the device to properly function. Where non-crashworthy barriers are used, where the wrong device is deployed, where necessary barrier components are missing, where barrier dimensions are incorrect, or where barriers are placed incorrectly, rather than being safety devices, traffic barriers become roadside hazards. When a "safety device" is a proximate cause or the primary contributor to occupant injuries in a roadside collision, the negligence of the roadway agency can become the subject of a tort claim.

In reference to traffic barriers, the AASHTO *Roadside Design Guide*[4] defines the term *crashworthy* as follows:

> **Crashworthy** - *A feature that has been proven acceptable for use under specific conditions either through crash testing or in-service performance.*

Prior to about 1965, the term crashworthy was not commonly used. Most traffic barriers were not designed to be crashworthy and those that were occurred only because of the application of good common sense. In 1968, *NCHRP Report 54*[5] by Michie and Calcote was the first attempt to formally define what were crashworthy roadside and median barriers. Then in 1970, *NCHRP Report 86*[6] by Olson, Post, and McFarland attempted to define what were crashworthy bridge rails. What followed these documents was an explosion of technology on traffic barriers.

Much knowledge has been gained on traffic barrier systems through research projects using analytical studies and full-scale crash tests. The downside of this otherwise beneficial evolution was the continual tinkering with standards as designs adopted since 1965 proved to be not as good as previously thought. This leaves the U.S. roadside with many not-so-old devices that are now unacceptable. On a positive note, the Federal Highway Administration and the Transportation Research Board have been moving away from this device-by-device look-and-see approach, moving toward a mission-oriented research program aimed at developing standards that will stand the test of time.

The following sections will discuss each of the different kinds of traffic barriers, including their performance requirements, warrants for placements, structural characteristics, and placement specifications.

3.1 Roadside Barriers

Prior to about 1967, not only did objective criteria for roadside barriers (guardrails) not exist, but also the structural designs varied greatly from State to State. Guardrails in place ranged from flimsy pipe or lumber rails, to a series of heavy timber guide posts spaced at about 10-foot intervals, to heavy steel W-section rail mounted on steel I-beam posts. Standards for minimum roadside side slopes that required roadside barriers varied not only from State to State but also from district to district within a State. Roadside barriers placed to shield fixed objects were often only 10-12 feet long with 2 support posts, a design that would not prevent penetration and subsequent collision with the roadside hazard for impacts at highway speeds.

Performance

Guardrails are normally placed to shield motorists from natural or man-made roadside hazards when these hazards cannot be removed, moved to a less vulnerable point, or made less severe. They are also occasionally used to protect pedestrians, cyclists, and school yards from errant vehicles. Guardrail should only be placed where the average consequence of striking the guardrail is less severe than the average consequence of traversing the unprotected roadside. To accomplish this goal, guardrails need to be long enough to shield the hazard, be strong enough and high enough to contain the vehicle, and be designed to allow a smooth redirection of the vehicle with tolerable decelerations. *NCHRP Report 350*[1] specifies the standard crash tests needed to assure that guardrails perform as needed.

Traditionally, most guardrails have been designed, tested, and installed with the intention of containing and redirecting automobiles, vans, and light trucks. With this intent went the understanding that these guardrails should not be expected to perform as well for large buses and trucks. More recently, however, several roadway agencies have developed and used guardrails capable of containing and redirecting both large and heavy vehicles. These barriers are usually applied on roadways with large volumes of heavy truck traffic and /or where the consequences of guardrail penetration by a large vehicle are extraordinarily severe.

Warrants

The height and slope of embankment are the basic factors for determining barrier need on steep side slopes. The warranting criteria, shown in Figure 3-1, were first presented as a national standard in the 1977 AASHTO publication *Guide for Selecting, Locating, and Designing Traffic Barriers*.[7] These same criteria are also contained in the 1989 AASHTO publication, *Roadside Design Guide*.[4] They are based on studies by Glennon and Tamburri[8] and by Ross, et. al.,[9] which compared the severity of impacts with embankments to the severity of impacts with traffic barriers. Referring to Figure 3-1, side slope and height combinations that plot above the curve will warrant guardrail, while combinations that plot below the curve do not.

Guardrail warrants for roadside obstacles are a function of the obstacle's impact severity, its size, and its lateral placement. The guardrail should only be installed when the results of a guardrail impact on average are clearly less severe than an impact with the roadside obstacle. The most clear-cut fixed object locations that require guardrail are approaches to bridge rails and around bridge columns and piers within the clear zone.

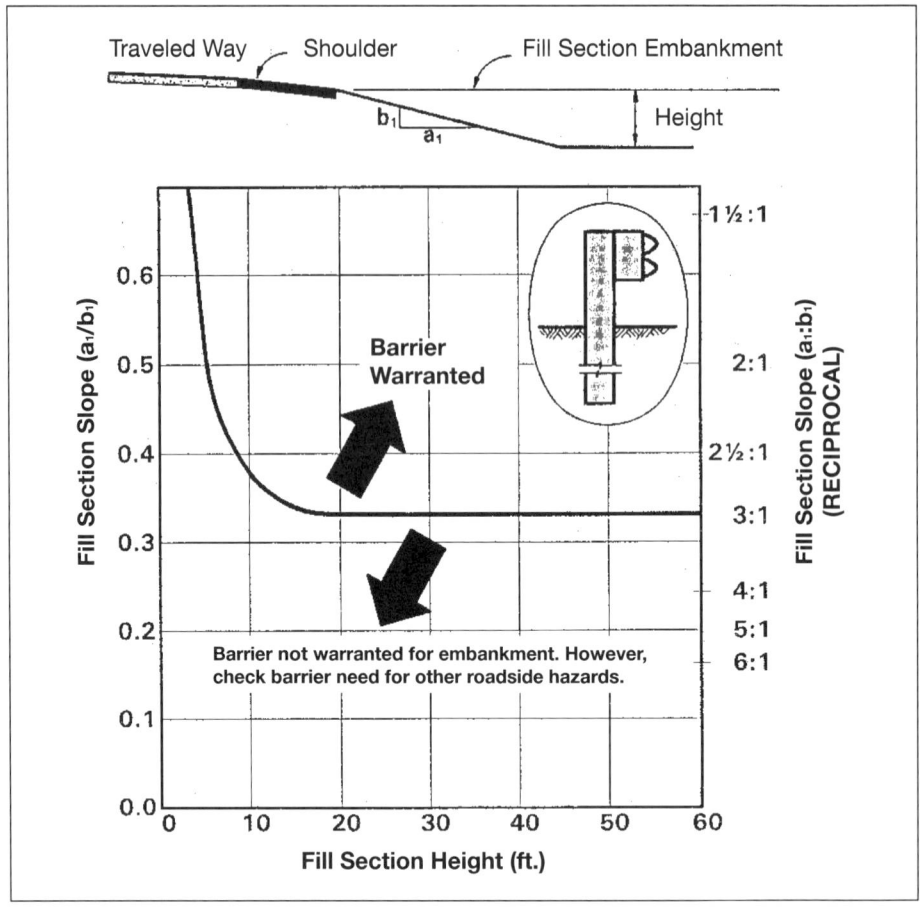

Figure 3-1 Warrants for Roadway Guardrail.[4]

Designs

Guardrail designs can be classified as flexible, semi-rigid, or rigid depending on their impact deflection characteristics. The most common guardrail systems used today are:

Flexible Guardrails
- 3-Strand Cable
- Weak-post W-beam

Semi-Rigid Guardrails
- Blocked-out W-beam
- Box Beam
- Blocked-out Thrie-beam

Rigid Guardrail
- Concrete Safety Shape

Traffic Barriers

Properly installed flexible guardrails are usually more forgiving than other guardrails because their greater impact deflection lowers the forces on the vehicle and its occupants. The most common designs are the 3-Strand Cable and the Weak-post W-Beam.

The *3-Strand Cable* guardrail, shown in Figure 3-2, has 3/4-inch steel cables spaced 3 to 4 inches apart and mounted 27 to 30 inches high on weak posts of steel or wood. Its primary advantages include low initial cost, effective performance over a wide range of vehicle sizes and installation conditions, low impact forces, and an open design that prevents snow drifting. Its primary disadvantages are the large clear area needed behind to allow up to 11 feet of deflection, and the relatively long sections that are rendered non-functional after an impact. The lateral deflection can be reduced by closer post spacing.

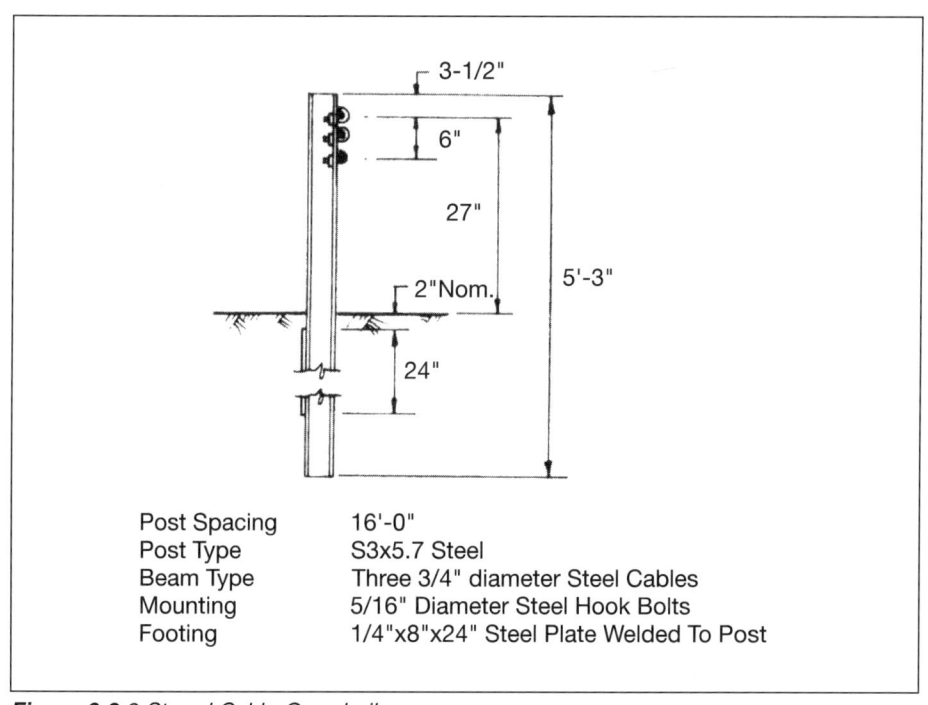

Post Spacing	16'-0"
Post Type	S3x5.7 Steel
Beam Type	Three 3/4" diameter Steel Cables
Mounting	5/16" Diameter Steel Hook Bolts
Footing	1/4"x8"x24" Steel Plate Welded To Post

Figure 3-2 *3-Strand Cable Guardrail.*

The *Weak-post W-beam* guardrail, shown in Figure 3-3, behaves much like the 3-Strand Cable. The system, however, is vulnerable to either vehicle vaulting or underride when placed on irregular terrain.

Semi-rigid guardrails are the most widely used in the U.S. The most common designs are the Blocked-out W-beam guardrail shown in Figure 3-4 and, to a much lesser extent, the Box Beam guardrail shown in Figure 3-5.

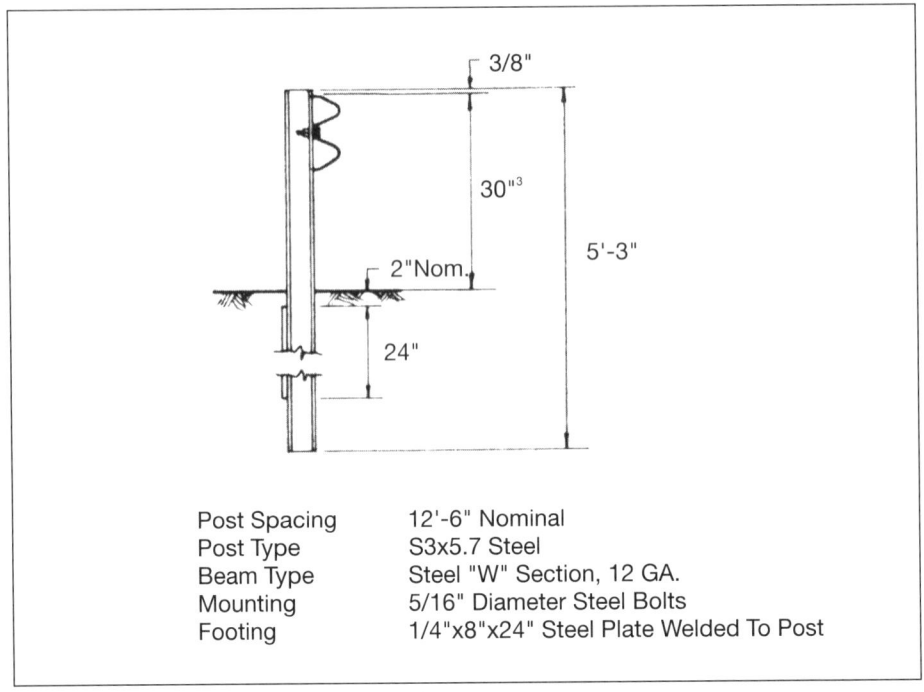

Post Spacing	12'-6" Nominal
Post Type	S3x5.7 Steel
Beam Type	Steel "W" Section, 12 GA.
Mounting	5/16" Diameter Steel Bolts
Footing	1/4"x8"x24" Steel Plate Welded To Post

Figure 3-3 Weak-post W-beam Guardrail.

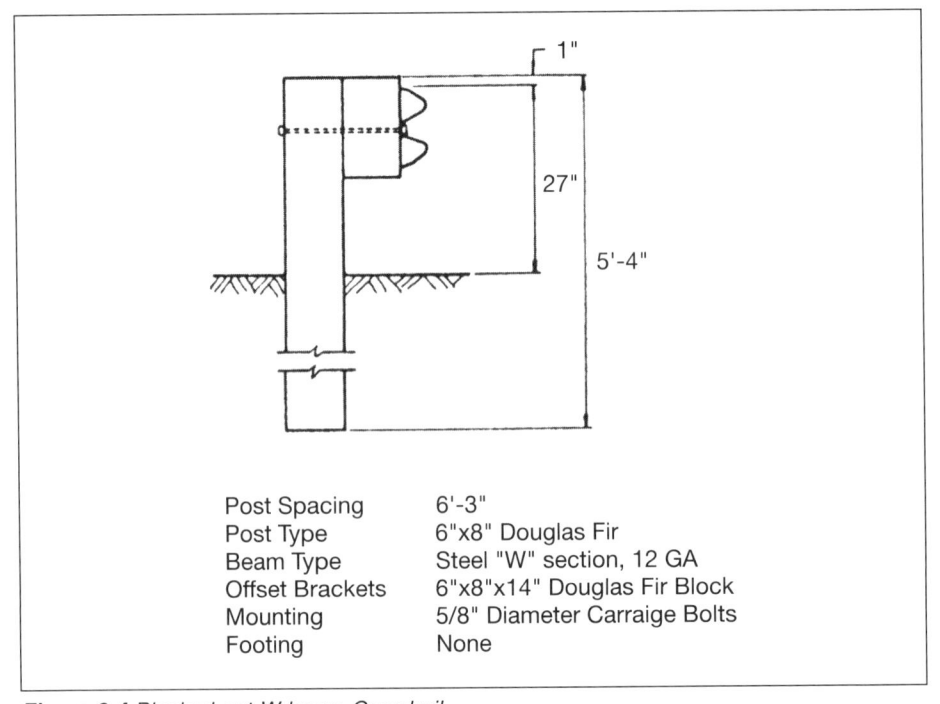

Post Spacing	6'-3"
Post Type	6"x8" Douglas Fir
Beam Type	Steel "W" section, 12 GA
Offset Brackets	6"x8"x14" Douglas Fir Block
Mounting	5/8" Diameter Carraige Bolts
Footing	None

Figure 3-4 Blocked-out W-beam Guardrail.

Traffic Barriers 73

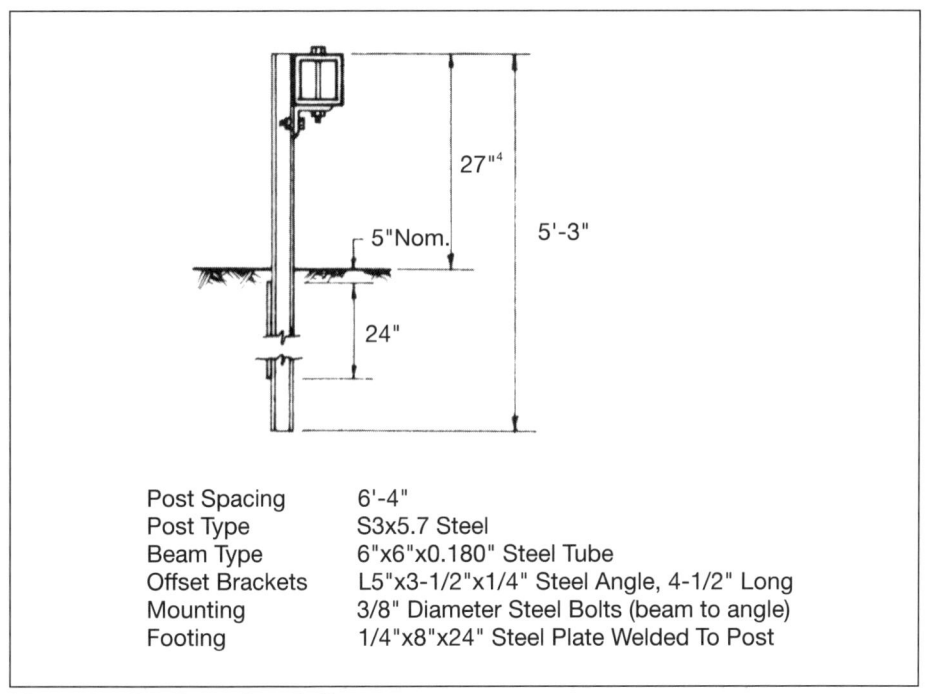

Figure 3-5 *Box Beam Guardrail.*

Post Spacing	6'-4"
Post Type	S3x5.7 Steel
Beam Type	6"x6"x0.180" Steel Tube
Offset Brackets	L5"x3-1/2"x1/4" Steel Angle, 4-1/2" Long
Mounting	3/8" Diameter Steel Bolts (beam to angle)
Footing	1/4"x8"x24" Steel Plate Welded To Post

Dimensions shown: 27" nominal, 5" Nom., 5'-3", 24".

Blocked-out W-beam guardrails using wood or steel posts are used extensively at bridge rail ends, bridge abutment and piers, and on steep side slopes, because they only deflect up to about 2 feet. The block-out has two functions: (1) to minimize wheel snagging on the posts; and (2) to maintain the height of the guardrail on impact. Several acceptable post designs are in use.

The *Box Beam* guardrail has weak posts that are designed to break or tear away, thereby distributing impact forces to adjacent posts. This system will only deflect about 3 feet but its effectiveness is very sensitive to mounting height and terrain irregularities.

The *Blocked-out Thrie-beam* shown in Figure 3-6 is a taller and stronger version of the Blocked-out W-beam. It is capable of restraining a wider range of vehicle sizes under a wider range of impact speeds and angles. Although it more effectively contains large trucks, occupants in impacting automobiles, vans, and light trucks will experience slightly higher impact decelerations than with Blocked-out W-beam guardrails.

The *Concrete Safety Shape* is a rigid guardrail with a sloping front face and a vertical back face. Except for the back face, its design details and performance are identical to the Concrete Median Barrier shown later in Figure 3-24. It is

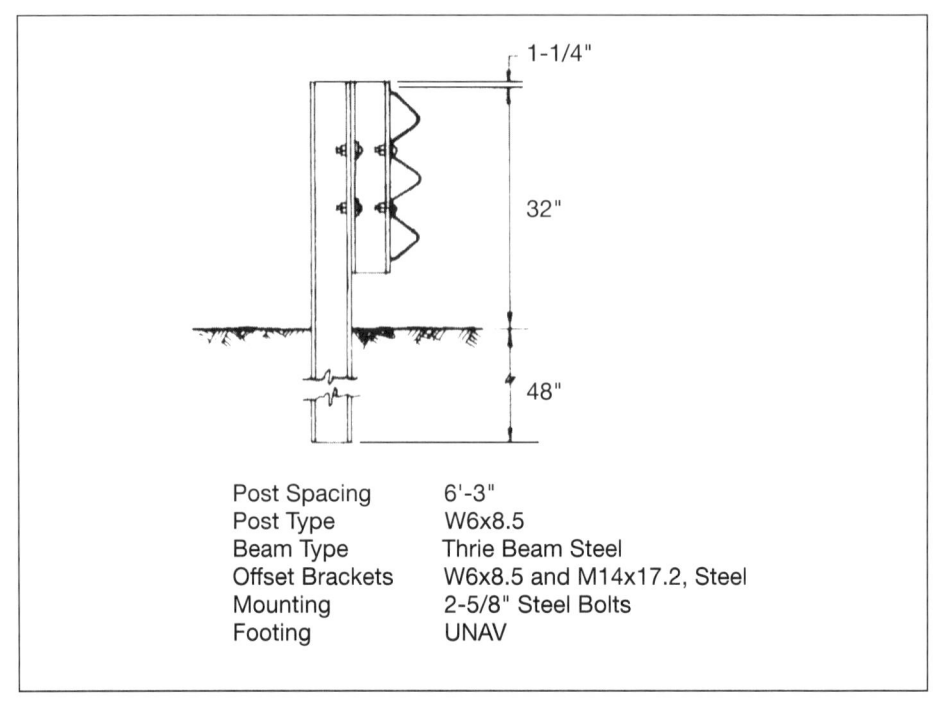

```
                    1-1/4"

                    32"

                    48"

Post Spacing      6'-3"
Post Type         W6x8.5
Beam Type         Thrie Beam Steel
Offset Brackets   W6x8.5 and M14x17.2, Steel
Mounting          2-5/8" Steel Bolts
Footing           UNAV
```

Figure 3-6 Blocked-out Thrie-beam Guardrail.

usually used where little room is available for guardrail deflection such as in front of a rock outcropping or in front of bridge abutments, piers, or columns.

End Treatments

An untreated end of a guardrail is extremely hazard. When a guardrail is hit by a vehicle directly on the end, the beam can penetrate the passenger compartment and impale the occupants. For this reason, a crashworthy end treatment is considered essential when the guardrail end is within the clear zone. To be crashworthy, the end treatment should not spear, vault, roll, or suddenly decelerate the vehicle during direct impacts. At the same time, the guardrail near the end should be capable of developing the full tensile strength of the standard rail element.

Crashworthy guardrail end treatments didn't exist prior to the 1960's. Early attempts to avoid the impalement of vehicles and their occupants by guardrail ends included flaring the unprotected end away from traffic or burying it in an upstream backslope. Not only did the flared end treatment not prevent spearing but it also created larger impact angles and resulting penetration for side hits near the end of the rail. Burying guardrail ends in an upstream backslope is always effective.

Traffic Barriers

In the late 1960's, many States adopted the *Turned-Down End* treatment developed in Texas. For this treatment, used on Blocked-out W-beam guardrails, the W-beam rail was twisted from the vertical section to a horizontal section and reduced from full height to ground level over a distance of 25 feet (see Figure 3-7). This design allows the impacting vehicle to pass over the end without impalement. However, several years of testing and field experience has shown the tendency of these end treatments to not only vault and roll vehicles but also to channel vehicles into the area of hazard. Although thousands of Turned-Down End treatments are currently in place, this treatment is no longer considered an acceptable design for installation.

Figure 3-7 Guardrail Turned-down End Treatment.

In the early 1970's, the *Breakaway Cable Terminal (BCT)*, shown in Figure 3-8, was developed to minimize the spearing and rollover tendencies of earlier guardrail end treatments. For end impacts, the first two posts are designed to break away allowing the rail to bend away from the vehicle. The cable which anchors the rail to the ground allows the beam to function in tension when a side impact happens near the end. To insure proper performance of the BCT, the guardrail needs to be extended such that the area behind the BCT is traversable.

76 Roadway Defects and Tort Liability

Figure 3-8 Guardrail BCT End Treatment

Figure 3-9 SENTRE™ Guardrail End Treatment (Courtesy of Energy Absorption Systems, Inc.)

The BCT has been the most popular end treatment for about 20 years. However, because of moderate to high decelerations on impact, the BCT has not only been modified to produce the slightly improved ELT and MELT variations but also is now giving way to several types of energy attenuating treatments. The SENTRE (Figure 3-9) and TREND (Figure 3-10) treatments are produced by Energy Absorption Systems, Inc. Other similar devices are available. The SENTRE consists of telescoping Thrie-beam panels, slip-base posts, and sand-filled plastic containers. The SENTRE can be directly attached to either a Thrie-beam guardrail or with a transition to a W-beam guardrail. It needs to be installed on a concrete pad. The TREND is similar to the SENTRE, except it has some different hardware because it is intended to shield the end of rigid barriers.

Figure 3-10 TREND™ Barrier End Treatment (Courtesy of Energy Absorption Systems, Inc.)

Perhaps one of the most cost-effective guardrail end treatment available today is the *ET-2000 Extruder Terminal*, shown in Figure 3-11 which functions much differently than other terminal designs.[10] A die on the end of the W-beam acts as an extruder when a vehicle impacts the end. The die bends the rail at 90°, flattens it, and projects it out away from the vehicle. The kinetic energy of the vehicle is dissipated by the work of bending and flattening the rail.

Figure 3-11 Guardrail ET-2000 Extruder End Treatment.

Placement

The following dimensions are pertinent to the placement of guardrail:
- Lateral Offset from Edge of Travel Lanes
- Clearance to Fixed Objects
- Length of Need

Lateral Offset - For optimum guardrail performance, the area between the travel lanes should be flat, free of fixed objects, and as wide as possible. Besides improving barrier performance, a wide, flat, unobstructed area gives the driver a better chance of avoiding impact with the guardrail. Also, because drivers on the nearest travel lane tend to shy away from objects that are too close to the edge of that lane, a lateral offset that minimizes this effect should be provided. The *Roadside Design Guide* recommends the lateral offsets shown in Table 3.1.

Traffic Barriers

Table 3.1 Recommended lateral Offsets for Guardrails[4,7]

Design Speed (mph)	Shy Line Offset (feet)
30	3.5
40	5.0
50	6.5
60	8.0
70	10.0
80	12.0

Terrain Effects - Because guardrails are designed to withstand impact at a normal vehicle bumper height, they should not be placed behind curbs or on side slopes steeper than 6:1. Also, if guardrail supports are less than 2 feet from the hinge point between the shoulder and the side slope, vehicle impact forces may push the posts out through the slope surface.

Clearance - The distance between a guardrail and the fixed object being shielded is an important factor in guardrail selection and placement. Obviously, if the guardrail deflects on impact more than this clearance distance, the vehicle will hit the object. The Blocked-out W-beam guardrail can generally be expected to have an impact deflection of about two feet on high-speed roadways. As the available distance from the guardrail to the fixed object decreases, the stiffness of the barrier must be increased. Where a guardrail shields a steep side slope, and impact deflection is not so critical, a clearance of two feet or more will usually be adequate.

Length of Need - The length of guardrail needed to shield a hazard depends on the size and shape of the hazard and its lateral distance from the travel lanes. Factors for determining this length of need are given in both The *Roadside Design Guide*[4] and The *Guide for Selecting, Locating, and Designing Traffic Barriers.*[7]

An effective guardrail needs to be extended far enough upstream to prevent most errant vehicles from getting behind the guardrail and striking the roadside hazard. The guardrail should also extend far enough downstream to prevent the vehicle from hitting the roadside hazard. If opposing traffic can be expected to crossover and hit the roadside hazard, the downstream end of the guardrail should be extended far enough to shield the hazard from opposing traffic.

Figures 3-12 and 3-13 shows the factors considered in determining the length of need both for adjacent traffic and for opposing traffic. The Runout Length, L_R, and the Lateral Extent of Hazard, L_H, are the primary factors. L_R, the theoretical distance that a vehicle encroaching on the roadside needs to come to a stop, is a function of speed and available friction. The values used by AASHTO

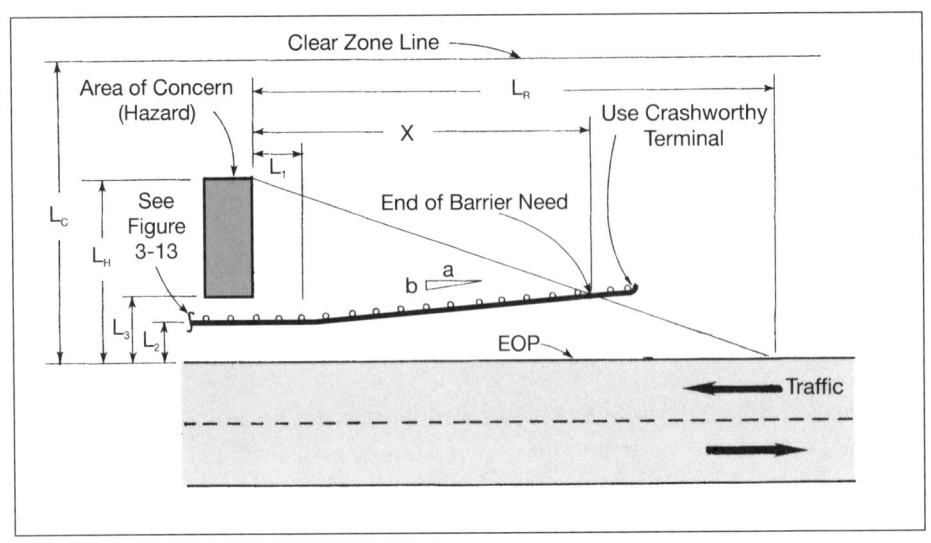

Figure 3-12 Approach Guardrail Layout.

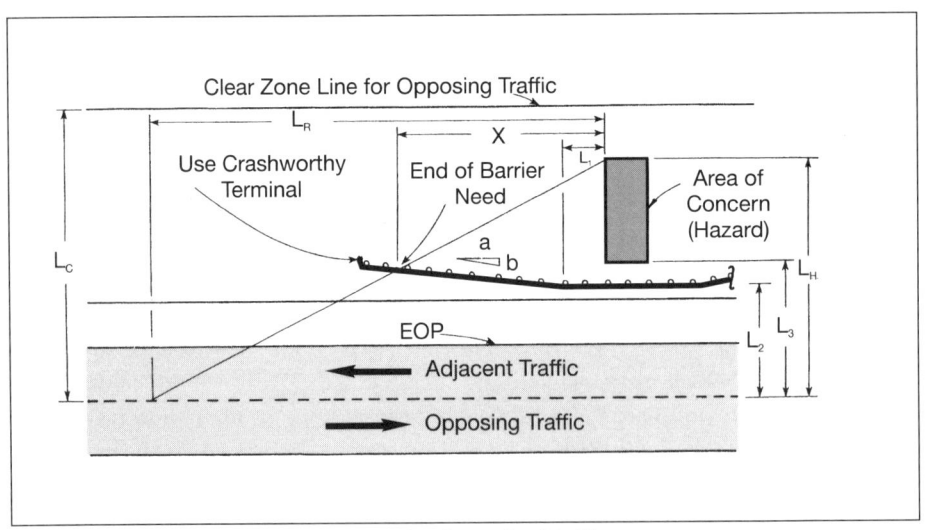

Figure 3-13 Guardrail Terminal Layout for Opposing Traffic.

are given in Table 3.2. Once L_R and L_H are known, the length of barrier depends the tangent length L_1, needed upstream of the hazard, its lateral distance from the travel lanes, L_2, and the flare rate, a:b, designed for the installation.

Guardrail Flare Rate - Flaring of guardrail has several purposes as follows:

1. Minimize the total length of guardrail needed (see Figure 3-12).
2. Locate the guardrail end farther from the travel lanes.
3. Transition the guardrail to a hazard near the travel lanes.

Traffic Barriers

Table 3.2 Suggested Runout Length for Guardrails[4,7]

Design Speed (mph)	Runout Length, L_R (ft) for Various Traffic Volumes			
	Under 800 veh/day	800-2000 veh/day	2000-6000 veh/day	Over 6000 veh/day
30	130	140	160	170
40	180	200	220	240
50	240	260	290	320
60	300	330	360	400
70	360	400	440	480

On the other hand, the greater the flare rate, the greater will be the impact angles. Also greater flare rates will increase the probability of an impacting vehicle being redirected into or across the travel lanes. Table 3.3 shows suggested flare rates for guardrails.

Avoidance of Gaps - Short gaps of less than 200 feet between adjacent sections of guardrails should be avoided. The adjacent guardrail should be joined to make one continuous section. Also short gaps between the upstream end of a guardrail and a structure or rock face should be closed by extending the guardrail and anchoring it to the upstream rigid object.

Table 3.3 Suggested Flare Rates for Guardrails[4,7]

Design Speed (mph)	Flare Rate		
	Guardrail Inside Shy Line	Guardrail Outside Shy Line	
		Right Guardrail	Semi-Rigid Guardrail
30	13:1	8:1	7:1
40	17:1	11:1	9:1
50	21:1	14:1	11:1
60	26:1	17:1	13:1
70	30:1	20:1	15:1

3.2 Bridge Rails

Bridge rails were placed first and foremost to keep vehicles from running off the edge of the bridge. AASHO standards[11] in 1944, however, were more concerned with the aesthetics of the rail than with its crashworthiness, as follows.

> *Substantial railings along each side of the bridge shall be provided for the protection of traffic. Consideration shall be given to the architectural features of the railing to obtain proper proportioning of its various members and harmony with the*

structure as a whole. Consideration should also be given to avoiding, as far as consistent with safety and appearance, obstruction of the view from passing vehicles.

By 1961, AASHO,[12] added a direct concern for the occupants in errant vehicles, as follows:

...Roadway railings preferably shall be designed in such a manner as to present smooth longitudinal surfaces on the traffic side, unbroken by vertical posts or any other projections.

In 1965, AASHO[13] began talking more directly about the crashworthiness of bridge rails, by saying:

While the primary purpose of traffic railing is to contain the average vehicle using the structure, consideration should also be given to protection of the occupants of a vehicle in collision with the railing...

In 1973, AASHTO[14] added concern for the impact consequences of the bridge rail end by saying:

Careful attention shall be given to the treatment of railings at the bridge ends. Exposed rail ends, posts, and sharp changes in the geometry of the railing shall be avoided. A smooth transition by means of a continuation of the bridge barrier, guardrail anchored to the bridge end, or other effective means shall be provided to protect the traffic from direct collision with the bridge rail ends.

Current bridge rails are normally constructed of either concrete posts and railings, a concrete safety shape, or a combination of metal and concrete. Bridge rails differ from most roadside barriers in that they are integrated into the bridge structure and are a rigid design that does not deflect on impact.

Performance

As with other traffic barriers, bridge rails are primarily designed to contain and redirect automobiles, light trucks, and vans. Recently, however, several roadway agencies have recognized the need for bridge rails that can redirect buses and large trucks at locations where the penetration of large vehicles could be particularly disastrous. At the other extreme, bridges on low-volume, low-speed roadways do not require the same bridge rail that is used on high-speed, high-volume freeways.

In 1989, AASHTO published *Guide Specification for Bridge Railings*,[15] which requires full-scale crash testing of all railings used on new construction. In this publication, they introduced the concept of multiple performance levels, providing a procedure for selecting bridge rails according to the needs of traffic.

Traffic Barriers

Designs

Figure 3-14 shows a variety of bridge rail cross sections that have previously been approved by AASHTO.[7] Many other bridge rail designs are in use today on U.S. roadways; some good, some bad, and some ugly (see Figure 3-15). All bridge rails that are within the clear zone should have approach guardrails to protect motorists from impacting the end of the bridgerail.

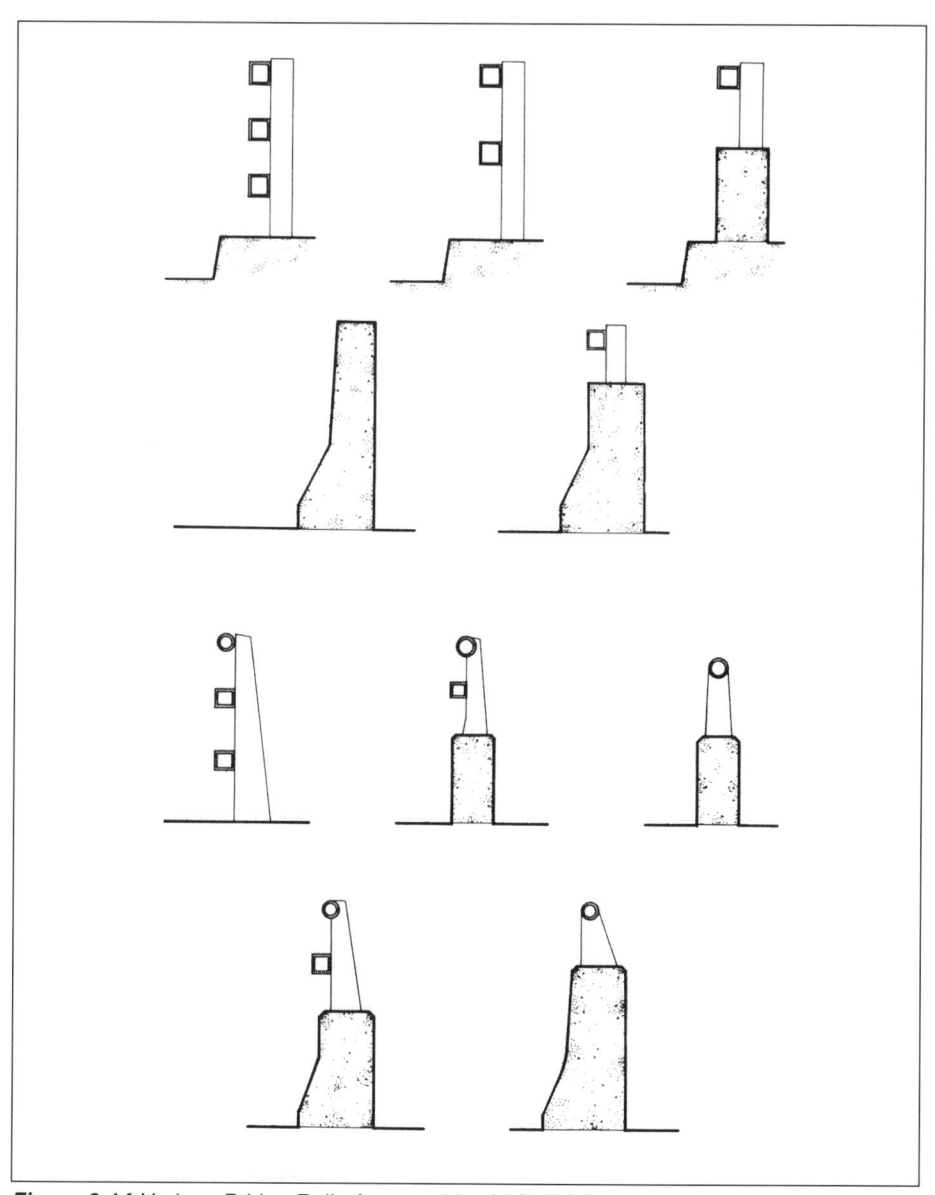

Figure 3-14 Various Bridge Rails Approved by AASHTO.[7]

84　　　　　　　　　　　　　　　　　　　　Roadway Defects and Tort Liability

The Good

The Bad

The Ugly

Figure 3-15 *Bridge Rails in Place.*

Traffic Barriers

Transitions

As shown in Figure 3-16, a transition section is needed where a semi-rigid approach guardrail meets the rigid bridge rail. The transition design should provide a gradual stiffening between the two elements so that vehicular pocketing, snagging, and penetration will not occur at this location.

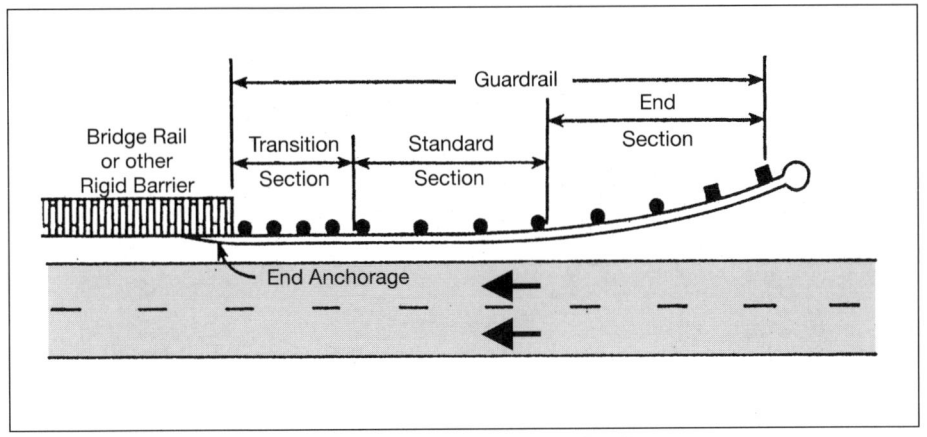

Figure 3-16 Transition Between a Guardrail and a Rigid Barrier.

Retrofitting Substandard Bridge Rails

Most bridges designed before 1964 will have rails that either: (1) have insufficient strength to prevent vehicle penetration; (2) have an open-faced design that can cause vehicle snagging and/or impalement; (3) have curbs or walkways that can cause vehicles to vault the rail; and/or (4) have missing or inadequate approach guardrails.

Rebuilding a bridge rail on an existing bridge often may not be feasible either economically or technically. For this reason, most bridge rails that have been improved for crashworthiness over the last 35 years have been retrofitted with elements that either improve the rail strength or provide a smoother impact surface. The *Roadside Design Guide*[4] shows several retrofit designs.

One of the most common retrofits for bridge rails is to rebuild the approach W-beam or Thrie-beam guardrail so that it continues across the bridge, bolting the new rail to the face of the old concrete, timber, or steel rail (see Figure 3-17). This relatively inexpensive treatment can add strength, reduce snagging potential, and provide continuity between the approach guardrail and the bridge rail. Also, if the bridge has a curb, the retrofit rail can be blocked out so the rail face is flush with the curb face.

Figure 3-17 Bridge Rail Retrofit.

Many older bridges have a fairly wide raised pedestrian walkway with a pedestrian handrail on the outside. Under these circumstance, there is often enough space to bolt a crashworthy metal post and beam barrier onto the walkway with the barrier face flush with the curb face.

Low volume rural roads often have short steel or wood deck bridges with flimsy timber or pipe rails. Depending on the structure, the old rail can sometimes be removed and replace with a thrie-beam mounted on steel posts that are bolted to the side of the bridge deck.

The concrete safety shape common to new bridges can often be added in front of a non-crashworthy bridge rail as an economical retrofit. The keys to this design are whether the structure can carry the additional dead load and whether the existing curb and railing can accommodate the anchorage requirements associated with the anticipated impact forces on the new barrier.

3.3 Median Barriers

Medians fulfill a variety of functions on divided roadways, the most important of which is to provide a recovery area for ran-off-road vehicles. Desirably, the median should be wide enough to allow the errant vehicle to recover before crossing into opposing lanes and causing a head-on collision. On rural roadways, medians should be 60 to 90 feet wide. On urban roadways, where right-of-way is limited and very expensive, the medians should be 10 to 30 feet.

On roadways where the traffic volume is low and the median is wide, the probability of a vehicle crossing the median is low. But, on roadways where the median is narrow and the traffic volumes are high, which typifies urban Interstate roadways, the probability of a vehicle crossing the median can be high.

Median barriers are designed to prevent ran-off-road vehicles from crossing the median and potentially causing serious head-on collisions with opposing traffic. Median barriers are designed symmetrically to redirect vehicles striking either side of the barrier. They are most often used to separate opposing traffic on divided roadways where medians are relatively narrow. They are also used as a positive divider along heavily-traveled roadways to separate through traffic from local traffic. As with other type of traffic barriers, median barriers should only be installed if striking the barrier is less severe than the consequences if the barrier was not there.

Warrants

Figure 3-18 shows the AASHTO warrants[4,7] for placing median barriers. They are generally warranted for higher traffic volumes and narrower medians. For traffic volumes below 5,000 vehicles per day, or for medians wider than 50 feet, the probability of cross-median encroachments is low. For these conditions, a barrier can only be warranted by a history of cross-median accidents.

Designs

Median barrier designs can be classified as flexible, semi-rigid, or rigid. Flexible barriers, such as the *3-Strand Cable* barrier and the *Weak-post W-beam* barrier, are used in wider medians, where larger barrier deflections can be accommodated. The 3-Strand Cable median barrier is essentially the same as the roadside barrier, shown in Figure 3-2, except one of the three cables is installed on the opposite side of each post from the other two. This barrier is relatively inexpensive, performs effectively on slopes up to 6:1, and can be used to advantage on moderately irregular terrain. Because the impact deflections of this design range up to 11 feet, it should only be installed in medians of 22 feet or wider. The Weak-post W-beam barrier shown in Figure 3-19 can deflect up to seven feet. Because its performance is very sensitive to terrain irregularities, the Weak-post W-beam barrier has very limited use.

Semi-rigid median barriers are used extensively in the U.S.. The most common designs are the Blocked-out W-beam median barrier shown in Figure 3-20 and, to a lesser extent, the Box Beam median barrier shown in Figure 3-21.

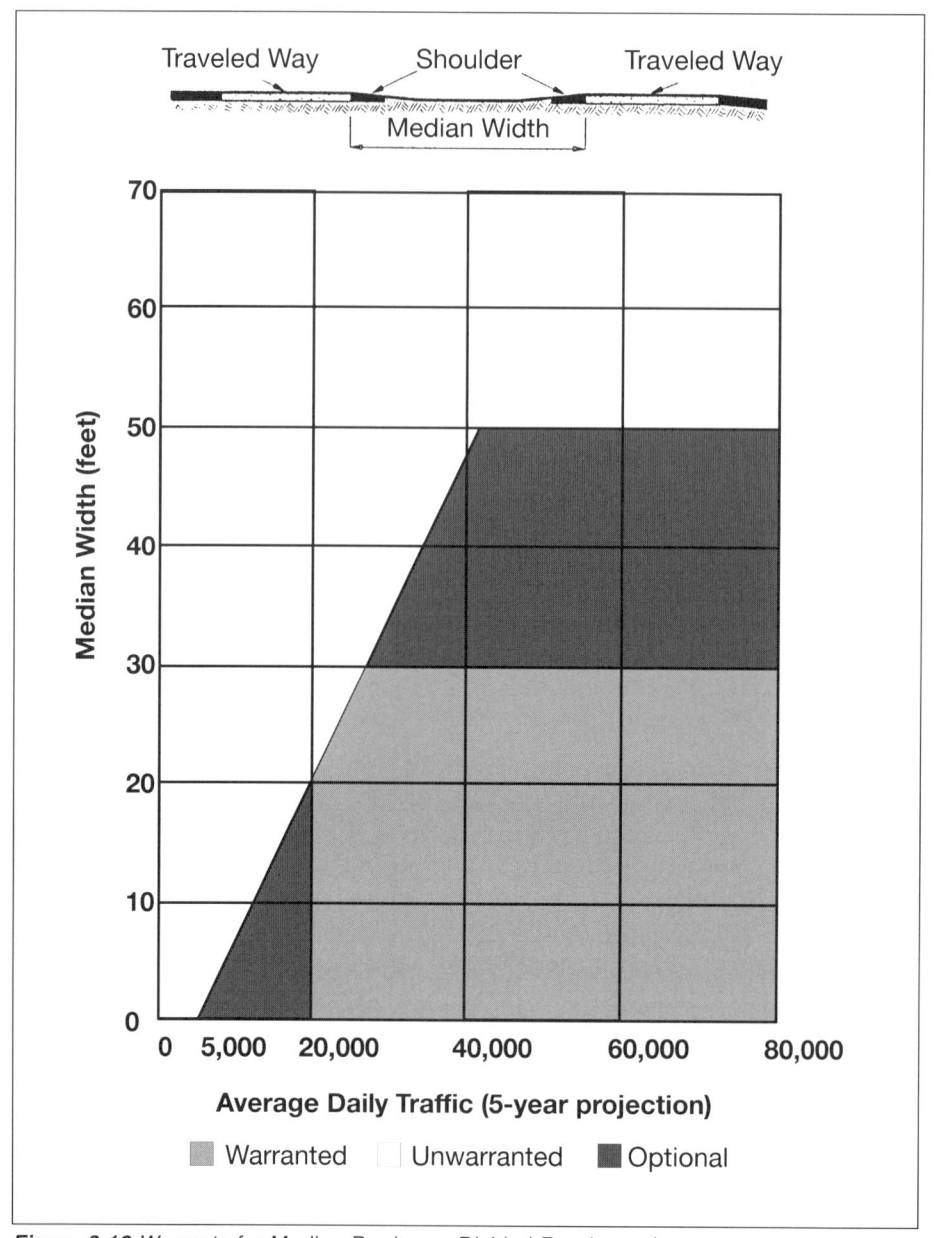

Figure 3-18 Warrants for Median Barrier on Divided Roadways.[4]

Traffic Barriers 89

Figure 3-19 Weak-post W-beam Median Barrier.

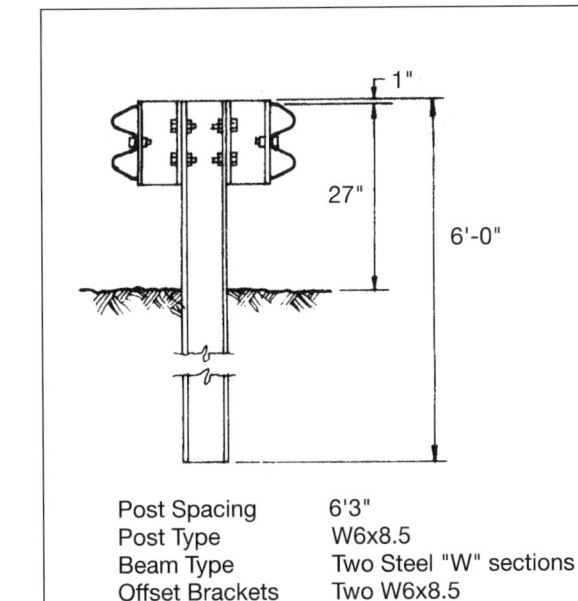

Post Spacing	12'6"
Post Type	S3x5.7
Beam Type	Two Steel "W" sections
Offset Brackets	None
Mounting	5/16" bolts
Footing	8"x1/4"x24" Steel Plate Welded to Post

Figure 3-20 Blocked-out W-beam Median Barrier.

Post Spacing	6'3"
Post Type	W6x8.5
Beam Type	Two Steel "W" sections
Offset Brackets	Two W6x8.5
Mounting	5/8" Diameter Steel Bolts

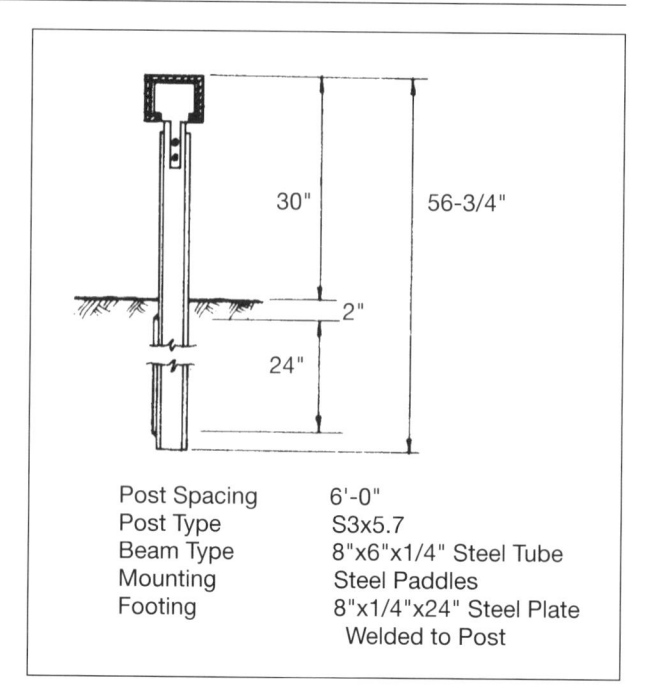

Figure 3-21 Box Beam Median Barrier.

Post Spacing	6'-0"
Post Type	S3x5.7
Beam Type	8"x6"x1/4" Steel Tube
Mounting	Steel Paddles
Footing	8"x1/4"x24" Steel Plate Welded to Post

    Blocked-out W-beam median barriers, using wood or steel posts, have been used to prevent crossover accidents in medians of 10-16 feet. Impact deflections for these semi-rigid systems are generally in the 2-foot range. Although Figure 3-20 shows a standard height of 27 inches, some roadway agencies use a 30-inch height as a greater insurance against crossover accidents, particularly for larger vehicles. To minimize wheels snagging on the posts with this higher mounting height, a separate lower rub rail (Figure 3-22) is often added to the design.

    The *Box-Beam* median barrier, shown in Figure 3-21, has a 6-foot post spacing, an 8 x 6 x 1/4 inch steel tube rail, and a height of 30 inches. Its expected impact deflection is about 3 feet. As with the Weak-post W-beam, this barrier is not suitable in medians with terrain irregularities.

    The *Blocked-out Thrie-beam* median barrier, shown in Figure 3-23, is similar to a W-beam barrier but, because of its larger size, is capable of restraining a wider range of vehicle sizes under a wider range of impact speeds and angles.

    Rigid median barriers are warranted in narrow medians where no barrier impact deflection can be tolerated. The *Concrete Median Barrier*, often referred to as a New Jersey barrier, is by far the most widely used median barrier in the U.S. (see Figure 3-24). It's popularity comes from its relatively low cost and its maintenance-free field record.

    With narrow medians, most impacts on a barrier will be at low angles. The Concrete Median Barrier has been proven to be very effective at minimizing

Traffic Barriers 91

Figure 3-22 Thirty-Inch High Blocked-out W-beam Median Barrier with a Rub Rail.

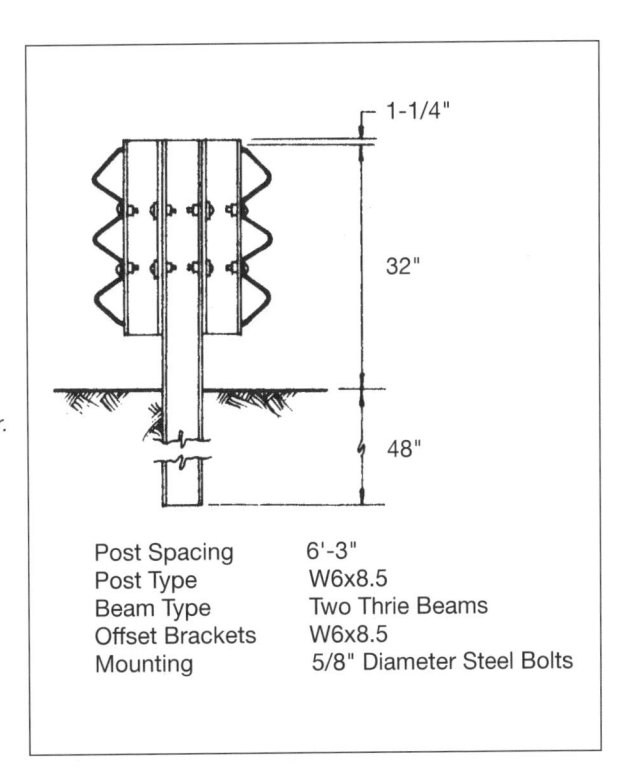

Figure 3-23 Blocked-out Thrie-beam Median Barrier.

Post Spacing	6'-3"
Post Type	W6x8.5
Beam Type	Two Thrie Beams
Offset Brackets	W6x8.5
Mounting	5/8" Diameter Steel Bolts

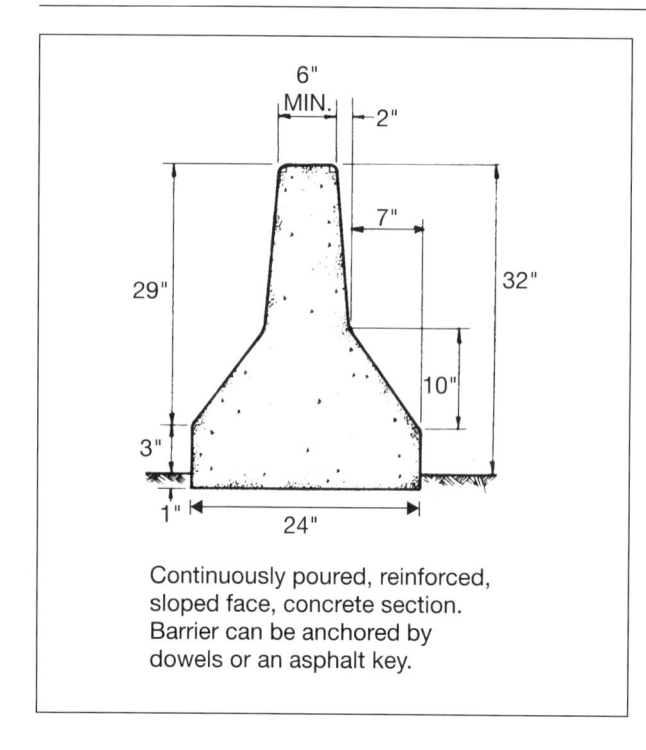

Figure 3-24 Concrete Median Barrier.

Continuously poured, reinforced, sloped face, concrete section. Barrier can be anchored by dowels or an asphalt key.

vehicle damage, barrier damage, and impact deceleration for low-angle impacts both in crash tests and through several years of field experience.

The Concrete Median Barrier has two troublesome performance characteristics. For high-angle, high-speed impacts, automobiles may become airborne and climb to the top of the barrier. Under this circumstance, fixed objects on top of the barrier, such as light poles or sign supports, may snag the vehicle. The second factor, is the tendency of large trucks to roll toward the barrier and contact either fixed objects on top of the barrier or bridge piers immediately behind the barrier. Also, large trucks will sometimes climb or roll over the barrier. For these reasons, many jurisdictions are now opting for the 42-inch *Tall Wall* concrete barrier shown in Figure 3-25. Not only is this barrier more effective in containing large trucks, but it also is an effective glare screen for shielding oncoming headlights at night.

End Treatments

A crashworthy end treatment is essential if a median barrier is started where it is exposed to high-speed impacts. Impact with the end of an untreated metal beam barrier can result in intolerable impact forces or impalement. Impact with the end of an untreated concrete barrier can also result in intolerable impact forces.

Traffic Barriers

Figure 3-25 Tall Wall Median Barrier.

To be crashworthy, the end treatment should not impale, vault, snag, or roll the vehicle. The following is a description of some of the more common end treatments seen:

1. *Crash Cushion* - In wider medians, sand barrel crash cushions (Figure 3-26) are effectively used to shield the ends of median barriers. For narrow medians, the 2-foot wide GREAT crash cushion (Figure 3-27) is often used.
2. *Tapered-Down Ends* - This type of treatment has been used both on metal beam and concrete median barriers to eliminate the spearing or high impact forces associated with untreated ends. This treatment, however, should be limited to places where end impacts are unlikely because of the tendency for vehicles to vault or overturn off of the taper.
3. *Flared Ends* - In variable-width medians, a median barrier can sometimes be introduced in a wider section where it can be flared away from traffic.
4. *Anchored in an Earth Beam or Backslope* - this technique is applied where existing terrain can be used to bury the start of the barrier.

94 Roadway Defects and Tort Liability

Figure 3-26 Sand Barrel Crash Cushion Shielding the End of a Rigid Barrier.

Figure 3-27 GREAT™ Crash Cushion Shielding the End of a Median Barrier.

Transitions

Proper transitioning between semi-rigid and rigid barriers is important to prevent snagging or excessive deflections. Although specific design details may vary, the concerns here are the same as those for transitioning from a guardrail to a rigid bridge rail as discussed in Section 3.2.

Placement

Other than factors already discussed, median barriers will function best when the approach terrain is flat and free of any irregularities. Curbs placed in front of barriers are discouraged. For sloped medians of 10:1 or flatter, the barrier can be placed in the center of the median. For sloped medians steeper than 10:1, various barrier arrangements, including staggered barriers separated for each side, may be warranted depending on the cross section.

3.4 Crash Cushions

Crash cushions (also called impact attenuators) are barrier devices that prevent ran-off-road automobiles, vans, and light trucks from impacting rigid objects that cannot be removed, relocated, or made breakaway. They are primarily designed to smoothly decelerate vehicles that impact the device head-on. For glancing impacts, they should redirect the vehicle away from the fixed object.

The most common application of a crash cushion is at a freeway exit ramp gore on an elevated or depressed structure where a bridge rail end or a pier requires shielding. Other frequent applications include, the ends of median barriers, the ends of retaining walls, at median bridge columns, and at large sign, signal, or utility poles that are within the clear zone.

Concept

Commonly-used crash cushions are designed to either absorb the energy or transfer the momentum of an impacting vehicle. An energy absorbing (compression) crash cushion will smoothly decelerate an impacting vehicle usually by crushing a deformable material contained in the crash cushion. Another compression design transfers energy from the vehicle by discharging water from the top of plastic tubes. These compression crash cushions need a rigid backup to hold them in place so they can resist the impact forces. Figure 3-28 illustrates how a compression crash cushion functions.

An inertial crash cushions is designed to transfer the momentum of the impacting vehicle to an expendable mass of material, such as sand, within the crash cushion. These crash cushions do not need a backup because the kinetic energy of the vehicle is not absorbed but is transferred to other masses. Figure 3-29 illustrates how an inertial crash cushion functions. Basically, the vehicle imparts

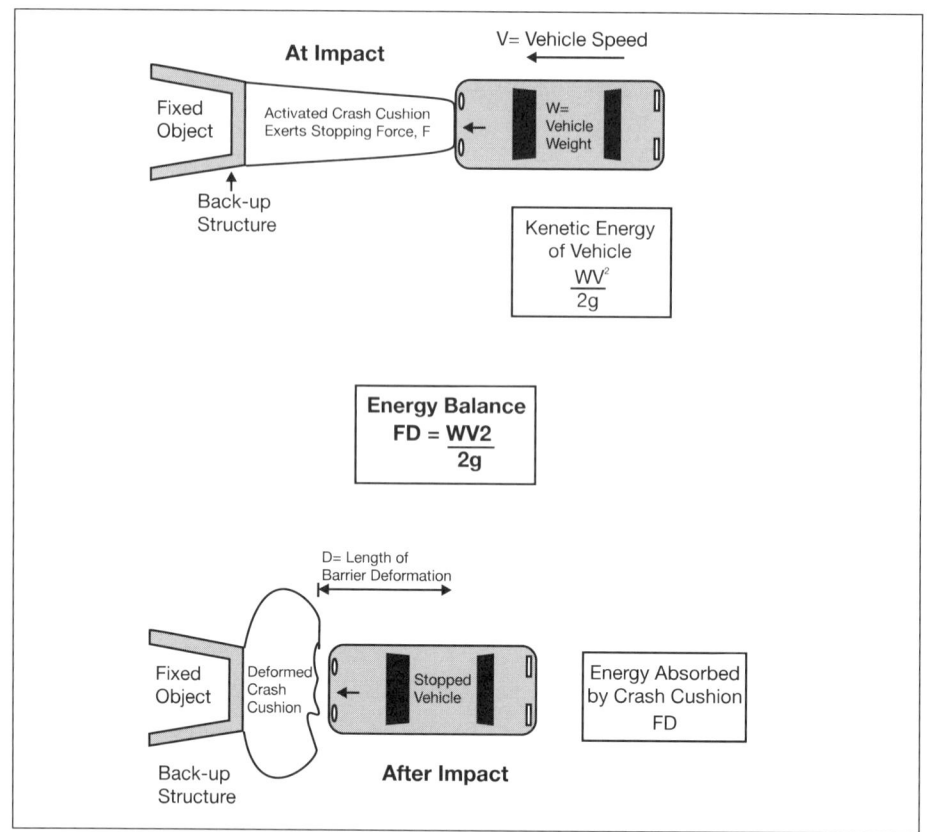

Figure 3-28 Compression Crash Cushion Physics.

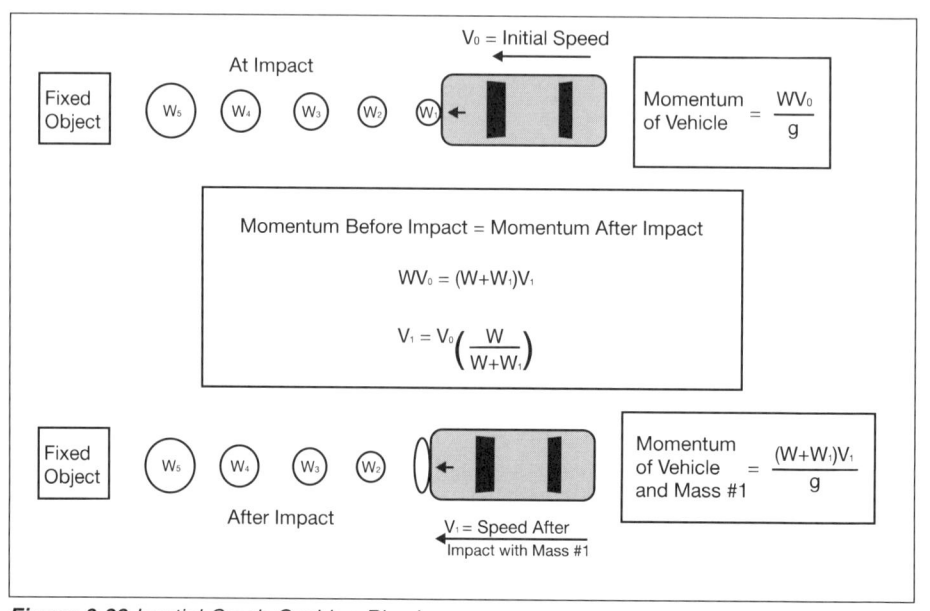

Figure 3-29 Inertial Crash Cushion Physics.

Traffic Barriers

momentum to the sand when it shatters the module. Because, after impact, the total momentum equals the momentum of the vehicle plus the sand, the vehicle speed is reduced sequentially as it impacts each barrel. Theoretically, the vehicle can never be completely stopped. Practically, after the vehicle speed is reduced to about 10 mph, the remaining momentum is imparted to the sand as the vehicle plows through the module.

Designs

Several different crash cushions are used today. The most common of these are:
- HI-Dro Sandwich System
- HI-Dro Cell Cluster
- Hex-Foam Sandwich System
- Guardrail Energy Absorbing Terminal (GREAT)
- Sand-filled Plastic Barrels
- Gravel-Bed Attenuator

The first four of these crash cushions are the proprietary products of Energy Absorption Systems, Inc. That company also make one of the two types of sand barrel products, the Energite System. The other sand barrel product, the Fitch Inertial System, is made by Roadway Safety Services, Inc.

The *HI-Dro Sandwich System* dissipates the kinetic energy of the impacting vehicle first by forcing water (or anti-freeze) up through orifices in the top of each in a series of plastic tubes and, second, by transferring the energy from the vehicle to the expelled water. Figure 3-30 shows a typical installation of the HI-Dro Sandwich System.

The *HI-Dro Cell Cluster* is used where speeds are below 45 mph and redirective capabilities are not needed. This unit, a smaller version of the HI-Dro Sandwich System, is used where space limitations prevent using a larger crash cushion. Figure 3-31 shows a typical installation of the HI-Dro Cell Cluster.

The *Hex Foam Sandwich System* is similar to the HI-Dro Sandwich both in outward appearance and in performance. This crash cushion absorbs the kinetic energy of an impacting vehicle primarily through the crushing of expendable cartridges containing a matrix of hexagonal cardboard shapes filled with polyurethane foam. Figure 3-32 shows a typical application of the Hex-Foam Sandwich System.

The *GREAT* attenuator is designed to shield the ends of median barriers, and other narrow fixed objects likely to be struck head-on. This crash cushions uses expendable Hex-Foam cartridges similar to those described above. These cartridges are packaged within a framework of Thrie-beam panels which telescope

98 Roadway Defects and Tort Liability

Figure 3-30 HI-Dro Sandwich Crash Cushion (Courtesy of Energy Absorption Systems, Inc.)

Figure 3-31 HI-Dro Cell Cluster (Courtesy of Energy Absorption Systems, Inc.)

Traffic Barriers 99

Figure 3-32 Hex Foam Crash Cushion (Courtesy of Energy Absorption Systems, Inc.,

Figure 3-33 GREAT™ Crash Cushion (Courtesy of Energy Absorption Systems, Inc.)

back under a head-on impact. For side impacts, the framework is restrained by leg pins and guidance cables, which allow the system to act as a longitudinal barrier to redirect the vehicle. Figure 3-33 shows a typical application of the GREAT attenuator.

As described earlier, *Sand-Filled Plastic Barrels* dissipate the kinetic energy of an impacting vehicle by transferring its momentum to the sand. Although both the Energite System and the Fitch Inertial System are patented systems, the sizes and weights of both systems are so similar that the modules are interchangeable. Because neither system can redirect a vehicle for a side impact, modules near the rear of the array should be placed to minimize the likelihood of a vehicle striking the corner of the shielded fixed object. Figure 3-34 shows a typical array of Energite barrels.

Both the Energite and Fitch Inertial systems have been tested extensively and have performed successful for years on U.S. roadways. Both manufacturers have developed standard arrays to be used with specific fixed objects. They also provide design charts for custom applications.

The *Gravel-Bed Attenuator* is designed specifically for large trucks. This feature is typically used on truck escape ramps along descending roadway grades where runaway trucks have been or are likely to be a problem. Design details for this attenuator are contained in the 1990 AASHTO publication, *A Policy on the Geometric Design of Highways and Streets.*[16]

Figure 3-34 Energite Sand Barrels (Courtesy of Energy Absorption Systems, Inc.)

Placement

Crash cushions should be placed on flat surfaces where the approach path of an impacting vehicle is free of any obstructions or irregularities. Roadside curbs or excessive slopes can cause a vehicle to become airborne with undesirable roll and pitch angles before impacting the crash cushion.

The surface on which the crash cushion is placed should be flat and smooth. This allows the crash cushion to compress uniformly during an impact.

> **TRAFFIC BARRIERS**
> **Rule of Thumb**
> *Hazards Protected by Traffic Barriers Should Have More Severe Impact Consequences than the Traffic Barrier*

3.5 Technical Aspects of Traffic Barrier Defect Cases

Many U.S. roadways abound with defective traffic barriers, particularly in local jurisdictions. In bringing a potential tort claim alleging a defective barrier, the direction is clear if the barrier violated the commonly recognized standards for design and placement at the time it was installed. If, on the other hand, the barrier was installed prior to the late 1960's, two particular questions arise depending on the jurisdiction:

1. *Design versus Maintenance Function* - Is the installation of roadside barriers within the design function or the maintenance function of the roadway agency? If installation is judged to be legally within the design function, the plaintiff has no cause of action for a defective barrier that was installed before the late 1960's either under any existing standard or in the absence of a standard. On the other hand, if installation is judged to be legally within the maintenance function, the plaintiff's argument can be directed to what is a reasonable number of years before a roadway agency exercises its duty to maintain a reasonably safe roadway by upgrading a defective safety device.

2. *Collision Repair* - Should a roadway agency be expected, as part of its legal duty, to provide a reasonably safe roadway by upgrading a defective barrier when it is being repaired after a major portion of the

barrier has been damaged in a previous collision? A common procedure followed by many roadway agencies when a barrier is damaged is to restore that barrier to its pre-impact condition regardless of whether it was defective or not.

Table 3.4 lists several types of defects for the four categories of traffic barrier. Each of these defects are discussed in the following paragraphs.

Table 3.4 Primary Traffic Barrier Defects

Guardrail Defects
- Beam Too Low
- Beam Too High
- Too Few Posts
- Length Too Short
- Inadequate Length of Need
- No End Treatment
- Not Anchored
- Backwards Splice
- Missing Bolts
- Faulty Terrain
- Angled Too Severe
- Wrong Kind
- Inadequate Clearance
- Maintenance Defects
- Exposed Back
- Unrepaired Damage

Crash Cushion Defects
- Improper Installation
- Leaking Sand Barrel
- Leaking HI-Dro Cell
- Unrepaired Damage

Bridge Rail Defects
- Too Weak
- Snag Elements
- Impaling Elements
- Untreated End
- Unrepaired Damage

Median Barrier Defects
- Flexible Barrier in Narrow Median
- Rigid Barrier in Wide Median
- Inadequate Transition Between Semi-Rigid and Rigid Barriers
- Untreated End
- Unrepaired Damage

Guardrail Defects

More than any other single element on U.S. roadways, guardrails can be found in a defective state. The following list (with associated figures) attempts to detail these defects. The focus is on the Blocked-out W-beam guardrail which, since the late 1960's, has had the commonly recognized standards of a 27-inch beam height, a 6.25-foot post spacing, and a crashworthy end treatment.

1. *Beam Height Too Low* - Guardrails that have a low beam height are common on U.S. roadways. Prior to the late 1960's, the most common height was 24 inches. Although this height might be adequate most of

Traffic Barriers

the time on low-speed roadways, it is not adequate to prevent vaulting and/or penetration on high-speed roadways.

Beam heights between 12 and 23 inches can often be found, particularly on local roadways. These heights are inadequate for almost any impact speed. These low heights usually occur either because of faulty installation or because the elevation of the traveled lanes was raised once or more by resurfacing without raising the guardrail. Figure 3-35 shows a guardrail mounted at less than 18 inches.

Figure 3-35 Guardrail Mounted Too Low.

2. *Beam Height Too High* - When guardrails are placed too high, the wheels of an impacting vehicle can snag on the posts, causing high impact forces on the occupants. Figure 3-36 shows a guardrail placed at about 36 inches.
3. *Too Few Posts* - Guardrails with too few posts are common on U.S. roadways. Prior to the late 1960's, the most common post spacing was 12.5 feet, usually in combination with a 24-inch beam height. Although this post spacing might be adequate most of the time on low-speed roadways, it is inadequate to prevent vaulting, penetration, and/or pocketing on high-speed roadways. [Pocketing describes an

Figure 3-36 Guardrail Mounted Too High.

impact where because of large beam deflections, the vehicle impacts directly on a post]. Figure 3-37 shows a guardrail with 12.5-foot post spacing.

4. *Length Too Short* - To function properly, guardrail must be long enough to develop beam strength. Generally, a current AASHTO design would need about 75 feet for this purpose. Guardrails are frequently found with only one 12.5-foot beam held up by two posts (see Figure 3-38) or two 12.5-foot beams held up by three posts. Even with 6.25-foot post spacing, these guardrails would not be adequate to prevent penetration, vaulting, and/or impalement.

5. *Inadequate Length of Need* - The principle of Length of Need is discussed in Section 3.1. Either because of ignorance or because of the latent effects of old standards, many guardrails along U.S. roadways either start too late or end too early to completely shield a roadside fixed object or steep slope. Figure 3-39 shows a hazardous body of water that is not shielded for opposing traffic.

Prior to 1967, all States had a side-slope height criterion for the placement of guardrail. This criterion ranged from 6 to 20 feet for side slopes steeper than 4:1.[17] If an old guardrail placed with a 20-foot side slope height criterion on a 2:1 slope has never been upgraded, chances are that 100 to 200 feet of hazardous roadside slope is unshielded, by current standards. An unfortunate aspect of this

Traffic Barriers 105

Figure 3-37 Guardrail with Too Few Posts.

Figure 3-38 Guardrail Too Short to Develop Beam Strength.

Figure 3-39 Insufficient Length of Need for Guardrail to Shield Dangerous Body of Water.

deficiency occurs when end treatments are upgraded without adding guardrail to cover the length of need, as shown in Figure 3-40. Figure 3-41 shows a common defect where the new guardrail is sufficient to shield the bridge rail end, but fails to shield the typical steep side slope on the bridge approach.

6. *No Crashworthy End Treatment* - Probably more than half of the guardrails in the U.S. have no crashworthy end treatment. When hit head-on at speeds above about 35 mph, these guardrails will often penetrate the passenger compartment. When impacted on the side near the end, these guardrails will often collapse, thereby exposing the vehicle and its occupants to the hazard being shielded. Figure 3-42 shows an untreated guardrail end in place. Figure 3-43 shows a parent's lament along an Interstate highway, and Figure 3-44 shows the unprotected end that killed their daughter.

7. *Not Tied to Bridge* - One of the more common guardrail applications in on the approach to a bridge rail. In the past, these guardrails were either not anchored to the bridge or didn't have a transition design. For both of these conditions, an impacting vehicle can be directed right into the rigid bridge rail end. Figure 3-45 shows an unanchored approach guardrail that will channel errant vehicles into the end of the bridge rail.

Traffic Barriers

Figure 3-40 Recent Upgrade of Guardrail End Treatment with Inadequate Length of Need.

Figure 3-41 New Guardrail Installation with Insufficient Length of Need to Shield Steep Side Slope.

Figure 3-42 Spearing Guardrail End Treatment.

Figure 3-43 A Parent Lament Along Interstate 70.

Traffic Barriers

Figure 3-44 Untreated Guardrail End that Speared the Colliding Vehicle.

Figure 3-45 Unanchored Approach Guardrail.

8. *Backward Splice Overlap* - One important detail of W-beam guardrail is how the 12.5-foot sections are overlapped. They should be spliced so that the upstream rail overlaps the downstream rail. If the rails overlap the opposite way, the edge of the downstream rail is exposed to impact and may impale a colliding vehicle. Figure 3-46 shows an improper overlap.
9. *Missing Splice Bolts* - A standard W-beam guardrail has nine bolts at each splice. One bolt holds the two sections to the post, and eight bolts hold the two sections together. When some of these bolts are missing, the guardrail is subject to failure on impact by the bolts shearing through the edge of the metal beam section. Figure 3-47 shows a guardrail splice that has missing bolts. This guardrail had failed on impact at another splice downstream.
10. *Faulty Approach Terrain* - Steep approach slopes, curbs, or otherwise uneven terrain can cause an impacting vehicle to either submarine under or vault over a guardrail. Figure 3-48 shows a guardrail where the ground in front of the rail has washed away leaving a hole.

Figure 3-46 Guardrail with Backward Splice Overlap.

Traffic Barriers

Figure 3-47 Guardrail with Missing Splice Bolts.

Figure 3-48 Hole in Front of Guardrail.

11. *Angled Too Severely* - Guardrail is a longitudinal barrier designed for impacts of 20° or less. When it is placed at an angle of 20° or more to the roadway, (see Figure 3-49) it will either fail on impact or cause intolerable decelerations in the passenger compartment.
12. *Wrong Kind of Guardrail* - This defect usually occurs when a 3-Strand Cable guardrail is placed at the hinge point of a steep slope on a high-speed roadway. Because the cables can deflect up to 11 feet, an impacting vehicle can be exposed to the slope and submarine under the cable.
13. *Inadequate Clearance to a Fixed Object* - When a guardrail is less than one foot from a large rigid fixed object and is not stiffened at that point, an impacting vehicle can snag on the corner of the fixed object.
14. *Exposed Back* - Occasionally guardrail is installed for one roadway or direction that has an exposed back toward another roadway or direction. Striking the back of a guardrail can cause severe vehicle deceleration. Figure 3-50 shows an example of this defect. A solution is to put another rail on the back of the exposed posts.
15. *Maintenance Defects* - Old guardrails are subject to wood rot, rusting, loose bolts, loose cable clamps, etc. without periodic maintenance. Also maintenance care is important to prevent snow build-up in front of a guardrail that can cause a vehicle to ramp over the guardrail.

Figure 3-49 Guardrail Angled Too Severely.

Traffic Barriers

Figure 3-50 Exposed Guardrail Back.

Bridge Rail Defects

Although many counties and other local jurisdictions continue to install weak (Figure 3-51) or otherwise uncrashworthy (Figure 3-52) bridgerails, the majority of defective bridgerails pre-date the concept of crashworthiness. Many old bridge rails are massive concrete parapet rails that have snagging elements (Figure 3-53). Other old concrete rails have been weakened with age (Figure 3-54). Most old bridges are very narrow.

Many older bridge rails were little more than a handrail made of pipe, decorative cast iron, concrete, or wood (Figure 3-55). Many of these rails will not contain an impacting vehicle even at 5 mph. Another tragedy is when bridge rail elements break off and go through the windshield or otherwise penetrate the passenger compartment. Although considerable effort has been spent on upgrading old bridge rails, many still remain (Figure 3-56).

The other major category of bridge rail defects relates to the end condition. When the bridge rail end is unprotected (Figure 3-57), when an approach guardrail is not sufficiently stiffened, or when an approach guardrail is not anchored to the bridge rail, violent impacts with the bridge rail end will occur. Figure 3-58 shows an unanchored approach guardrail that has probably channeled many errant vehicles into the bridge rail end.

Figure 3-51 Weak Bridge Rails.

Figure 3-52 Uncrashworthy Bridge Rails.

Traffic Barriers 115

Figure 3-53 Concrete Bridge Rail with Snagging Elements.

Figure 3-54 Old Concrete Bridge Rail Weakened with Age.

Figure 3-55 Old Wooden Bridge Hand Rail.

Figure 3-56 Old Uncrashworthy Bridge Rail.

Traffic Barriers 117

Figure 3-57 Unprotected Bridge Rail End (But Notice Guardrail Anchored to Downstream End).

Figure 3-58 Unanchored Guardrail Approach to Rigid Bridge Rail.

Median Barrier Defects

A major defect for median barriers is using the wrong design for the median width. A flexible barrier, used in a narrower (less than 22 feet) median can allow head-on impacts even though the barrier otherwise functions alright. A rigid barrier used in a wider median (greater than 22 feet) invites higher impact angles than the barrier is designed for, with higher resulting impact forces and a higher probability of penetration.

Other major median barrier defects include unrepaired damage, unprotected end conditions, and insufficient transitions between rigid and semi-rigid systems (see Figure 3-59).

Crash Cushion Defects

All crash cushion designs are sensitive to the height of impact. If because of terrain effects, improper installation, or improper maintenance, the center of gravity (c.g.) of the impacting vehicles is sufficiently higher that the elevation of the resultant resisting force of the barrier, the vehicle can trip end over end (Figure 3-60) or ramp up over (Figure 3-61) the crash cushion. This defect is most pronounced with sand barrels, either when they are improperly filled, or when they leak sand with time. The Fitch Inertial System will leak sand if the rivets pop on the vertical seam (see Figure 3-62). Both the Fitch and Energite barrels will leak

Figure 3-59 Defective Median Barrier Transition.

Traffic Barriers

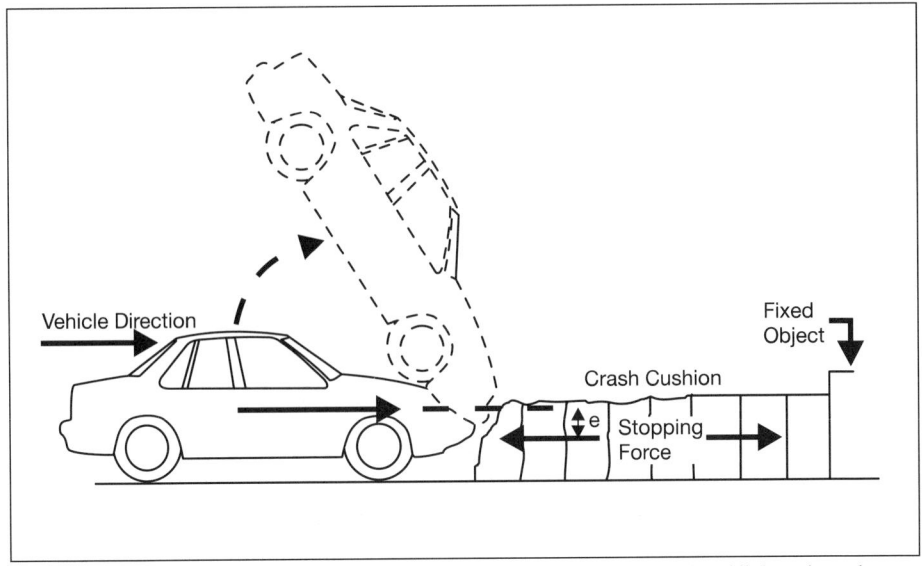

Figure 3-60 Vehicle Tripping that Results from the Impact Force being Higher than the Resisting Force of the Crash Cushion.

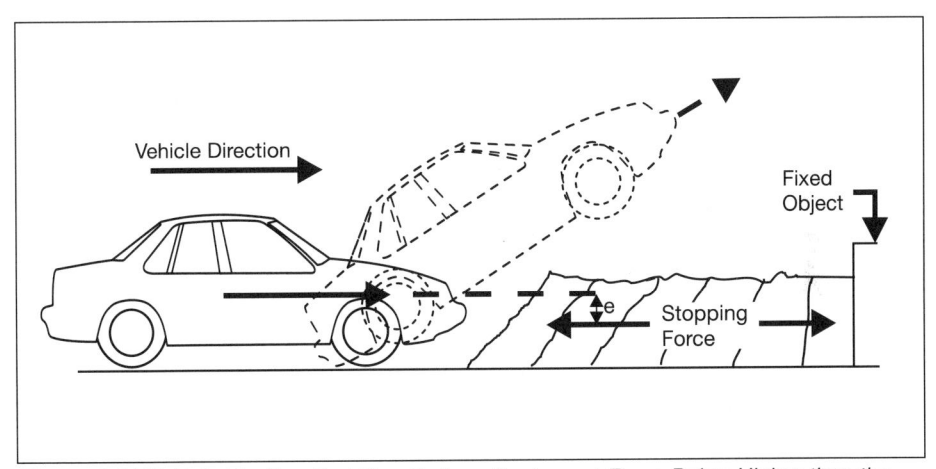

Figure 3-61 Vehicle Vaulting that Results from the Impact Force Being Higher than the Resisting Force of the Crash Cushion.

sand if incidental contact by maintenance vehicle punches a small hole (see Figure 3-63). The same problem can occur if water is allowed to leak or evaporate out of a HI-Dro system.

Another defect with the HI-Dro system occurs in the winter if anti-freeze is not used. The system, of course, will not function if the fluid is frozen.

120 Roadway Defects and Tort Liability

Figure 3-62 Fitch Inertial Sand Barrel Leaking Sand.

Figure 3-63 Hole in Energite Sand Barrel.

Traffic Barriers

All crash cushions, of course, will not function if left in the post-impact damaged stage. The energy absorbing element must be replaced, damaged elements repaired, and the barrier otherwise reconfigured to its crash-ready dimension.

3.6 Typical Defense Arguments

When defending tort claims that allege a defective traffic barrier, roadway agencies generally argue that no required standards exist for traffic barriers, and that the AASHTO *Roadside Design Guide*[4] and other similar documents are simply guidelines. In other words, if the roadway agency chose to implement a certain traffic barrier in one place but didn't use the same barrier at a similar location, the argument is that those decisions were entirely within their discretion.

Another major defense argument is the claim of design immunity. If, for example, the roadway was originally designed in 1954 before any significant traffic barrier guidelines or standards were recognized, the roadway agency could be immune from liability. Likewise, for those traffic barriers where technology has been constantly improving since about 1964, the roadway agency will argue that a 1970 design that is now known to be hazardous cannot be the subject of liability because of the design immunity statute.

At least one State has statutory protection against most barrier defects. In that State's tort claims act, the roadway agency is only liable for defects within the "improved" roadway, which only includes travel lanes and shoulders.

Beyond the available statutory arguments given above, roadway agencies often use other case-specific defenses that argue either that the driver was the sole negligent party or that the condition did not warrant improvement. Some of these arguments are:

1. The driver was negligent for running off the road.
2. The driver was negligent for speeding.
3. The driver was negligent for driving while intoxicated.
4. The driver was negligent for falling asleep at the wheel.
5. The subject location had no prior history of accidents.
6. The roadway agency lacked adequate funds to upgrade existing traffic barriers.
7. The roadway agency used a priority scheme for programming traffic barrier improvements, and this location was low on that priority list.
8. Traffic barrier improvements were being delayed because the entire section of roadway was planned for reconstruction.

Endnotes

1. Ross, H.E., Jr., Sicking, D.L., Zimmer, R.A., and Michie, J.D., *Recommended Procedures for the Safety Performance Evaluation of Highway Features*, Transportation Research Board, NCHRP Report 350, 1993.

2. Michie, J.D., *Recommended Procedures for the Safety Performance Evaluation of Highway Appurtenances*, Transportation Research Board, NCHRP Report 230, 1981.

3. Bronstad, M.E., and Michie, J.D., *Recommended Procedures for Vehicle Crash Testing of Highway Appurtenances, Transportation Research Board*, NCHRP Report 153, 1974.

4. American Association of State Highway and Transportation Officials, *Roadside Design Guide*, 1989.

5. Michie, J.D., and Calcote, L.R., *Location, Selection, and Maintenance of Highway Guardrails and Median Barriers*, Transportation Research Board, NCHRP Report 54, 1968.

6. Olson, Robert M., Post, Edward R., and McFarland, William F., *Tentative Service Requirements for Bridge Rail Systems*, Transportation Research Board, NCHRP Report 86, 1970.

7. American Association of State Highway and Transportation Officials, *Guide for Selecting, Locating, and Designing Traffic Barriers*, 1977.

8. Glennon, J.C. and Tamburri, T.N., *Objective Criteria for Guardrail Installation*, Highway Research Board, Record 174, 1967.

9. Ross, H.E., et. al., *Warrants for Guardrail on Embankments*, Highway Research Board, Record 460, 1973.

10. Sicking, Dean L., Gereshy, Asif B., and Ross, Hayes E., Jr., *Development of Guardrail Extruder Terminal*, Texas Transportation Institute, Paper No. 880484, 1989.

11. American Association of State Highway Officials, *Standard Specifications for Highway Bridges*, 1944.

12. American Association of State Highway Officials, *Standard Specifications for Highway Bridges*, 1961.

13. American Association of State Highway Officials, *Standard Specifications for Highway Bridges*, 1965.

14. American Association of State Highway Officials, *Standard Specifications for Highway Bridges*, 1973.

15. American Association of State Highway and Transportation Officials, *Guide Specifications for Bridge Railings*, 1989.

16. American Association of State Highway and Transportation Officials, *A Policy on Geometric Design of Highways and Streets*, 1990.

17. Highway Research Board, *Highway Guardrail - Determination of Need and Geometric Requirement,*, Special Report 81, 1964.

References

Ivey, Don L., Bronstad, M.E., and Griffin, Lindsay I., III, *Guardrail End Treatments in the 1990's*, Texas Transportation Institute, Paper No. 920910, 1992.

Federal Highway Administration, *Synthesis of Safety Research Related to Traffic Control and Roadway Elements*, Volume 1, Chapter 3, 1982.

Fitzpatrick, James F., et. al., *The Law and Roadside Hazards*, The Michie Company, 1974.

Federal Highway Administration, *Roadside Improvements for Local Roads and Streets*, 1986.

Federal Highway Administration, *Maintenance and Highway Safety Handbook*, 1977.

Federal Highway Administration, *A Handbook of Highway Safety Design and Operating Practices*, 1968, 1973, and 1978.

Transportation Research Board, *Designing Safer Roads*, Special Report 214, 1987.

Chapter 4

Stopping Sight Distance

Ability to see ahead is very important to the safe operation of a roadway. The path and speed of motor vehicles are subject to the control of drivers whose training is no more than elementary. If roadways are to be designed and maintained for safety, sufficient sight distance should be provided to allow drivers enough time and distance to control the path and speed of their vehicle to avoid unforeseen collisions with objects and other vehicles.

Since 1940, the geometric design policies of the American Association of State Highway and Transportation Officials (AASHTO)[1-7] have formally defined acceptable limits for stopping sight distance (SSD) based on a rational analysis of safety requirements. Adequate SSD is a function of roadway operating speeds and is achieved by designing non-restrictive horizontal and vertical roadway alignments and by avoiding or eliminating sight obstructions (vegetation, embankments, walls, buildings, etc.) on the inside of roadway curves.

4.1 Historical Perspective

For a tort action that alleges defectly-designed stopping sight distance, a design immunity statute will come into play in most States. The normal test under a design immunity statute is whether the roadway was designed according to the prevailing standard at the time it was built. For this reason, this section explores some of the early road building literature about roadway sight distance design.

Although the 1940 AASHTO standards for stopping sight distance were the first formally promulgated standards, this element of design was not ignored at the State and Federal levels prior to that time.[8] For example in a 1914 highway engineering textbook by Blanchard and Drowne,[9] the hazard of limited sight distance was recognized as follows:

> *Sharp curves are points at which collisions are very liable to occur, particularly if the view is obstructed. Sometimes, if it is impossible to increase the radius of the curve, a great improvement can be obtained by clearing away obstructions so that the curve can be seen throughout its entire length when approached in any direction.*

Two years later, a roadway construction textbook by Agg[10] gave some very limited recommendations for sight distance, as follows:

> *Steep grades, sharp curves and knolls that obstruct the view ahead should be avoided in the interest of safety. There should always be a clear view ahead for at least 250 ft. And if a curve exists on a hill, the grade should be flattened around the curve if possible so as to permit a quick stop in case of emergency.*

Missing in this recommendation is any reference to speed, eye height, object height, or the functional emergency model being employed.

By 1924, Agg[11] expanded his description of sight distance to include consideration of a head-on vehicular encounter on a roadway curve where a driver was viewing an opposing vehicle, as follows:

> *Horizontal and vertical curves, embankments, railroad crossings and intersections with other highways, constitute the dangerous portions of a public highway...To minimize danger at curves it is desirable to provide ample sight distance and to construct horizontal curves with long radii and ample superelevation. The sight distance should be such as to permit a view of an approaching vehicle 400 ft. away. That distance will permit both vehicles to be brought to a stop before colliding. Since the line of sight on a horizontal curve will depend upon whether the curve is in cut or not and upon the width of cleared right-of-way, no standard radius of curvature can be suggested that will provide the desired sight distance but it is easily computed in any case...The radius of curvature for vertical curves that will give a sight distance of 400 ft. is about 3,500 ft. and this applies regardless of the rate of grade if the algebraic difference in grades exceeds about 5 percent.*

The practice of roadway design in Michigan in 1926 was documented by Brightman,[12] indicating the need to see another car at 500 feet as follows:

> *The subject of visibility is one that cannot be overlooked and it is related to both horizontal and vertical alignment. In order that the motorist may always see far enough ahead for safety, the road should be so aligned that a clear vision of 500 feet ahead is available. This is worked out by long curves and the cutting away of banks which may hide the view in horizontal alignment. The vertical alignment is solved by the use of vertical curves in the grades which allow a car to be seen at a point 500 feet distant at all times.*

This same practice was adopted by AASHTO[13] in 1928.

Stopping Sight Distance

A 1935 highway engineering textbook[14] gave further recognition to the fact that sight distance needs were tied to speed as follows:

> On double-track paved roads, the sight distance should be such that a driver can observe an approaching vehicle without being startled when traveling at normal road speeds and with the corresponding degree of concentration of attention given the road. On account of increasing automobile speeds a minimum sight distance of 600 ft. is desirable.

Gutman[15] in 1936 reported on the German stopping sight distance standards for the 5000-mile Reichsautobahn system. These standards were based on a one second perception-reaction time, friction coefficients of 0.4 - 0.5 for stopping, a 3.9-foot driver eye height, and an 8-inch object height.

Conner,[16] in 1937, noted that AASHTO "recommends" a minimum sight distance of 800 feet on roadway curves. Both this recommendation and the earlier 1928 standard, clearly show that AASHTO was providing guidance on sight distance needs prior to 1940.

In 1938, the U. S. Bureau of Public Roads[17] in it's annual report included the following under a section titled, *Greatest Needs on Main Roads are Widening, Longer Sight Distances, and Reduction in Curvature*:

> Eliminating those curves that have become traffic hazards at the now normal driving speed and increasing sight distances by road straightening and by grading at the tops of hills are widespread needs on the existing main highways. These defects are found generally on roads in every part of the country and their danger to traffic is the consequence of an increase in vehicle speed far beyond what was envisioned 15 or 20 years ago and far in excess of the legal limitations that existed in most states.

In summary, sight distance was recognized as an important element of safe roadways prior to 1940. The emphasis on sight distance requirements was to allow the driver to see other vehicles in time to avoid them, but the aspects of eye height, object height, adequate perception-reaction time, and reasonable braking distances were not thoroughly understood.

4.2 Stopping Sight Distance Design Standards

Analysis of the operational and safety aspects of SSD requires an understanding of the concept of SSD as it relates to roadway operations. The geometric design policies published by AASHTO[2,3] discuss the need for SSD:

> If safety is to be built into highways, the designer must provide sight distance of sufficient length in which drivers can control

the speed of their vehicles so as to avoid striking an unexpected obstacle on the traveled way...

The minimum sight distance available on a highway should be sufficiently long to enable a vehicle traveling at or near the likely top speed to stop before reaching an object in its path. While greater length is desirable, sight distance at every point along the highway should be at least that required for a below average operator or vehicle to stop.

This short discussion alludes to many of the operational elements of stopping sight distance — namely, vehicle performance, driver ability, and the roadway alignment. This AASHTO operational model thus provides a reasonable starting point for considering the relationship between SSD and roadway operations.

SSD as defined by AASHTO is the sum of two distances: (a) the distance a vehicle travels between the time a driver sights an object and the time he applies the brakes; and (b) the distance a vehicle travels in braking to a stop. SSD is determined by the following equation:

$$SSD = 1.47Vt + \frac{V^2}{30(f \pm g)}$$

where:

V = initial speed, mph;
t = perception-reaction time, sec;
f = coefficient of friction; and
g = percent of grade divided by 100

AASHTO defines minimum SSD requirements in terms of a passenger car approaching a stationary object in its path. This basic functional model has remained unchanged since 1940. The following review of the evolution of AASHTO stopping sight distance policy illustrate the reasoning behind this model. It also demonstrates the need to go beyond this simple "abstraction" to gain insight into the safety relationships of SSD.

In 1940, AASHTO[1] formally recognized the need for a sight distance requirement to help drivers avoid collision circumstances. Although AASHTO recognized that a clear sight line to the pavement was desirable, analyses of how this requirement affected construction costs led to a compromise. A design object height of 4 inches was selected on the basis of optimizing the trade-off between the lowest possible object height and the amount of cut needed to achieve the required vertical curve length. Although the object height criterion is discussed in the AASHTO policy as it related to objects in the road, the selection of a 4-inch height clearly was not based on the frequency or severity of such

Stopping Sight Distance

objects. This conclusion is further borne out by later changes in AASHTO policy to a 6-inch object height when the same discussion was used in relating this height to roadway events.

Selection of other design parameters such as speed, perception/reaction time, eye height, and pavement friction was rational; individual design values were selected based on the currently known distributions of these physical values, which were periodically updated as indicated in Table 4.1. Yet, the underlying methodology was by design an abstraction — a simplified set of elemental factors used to derive a distance — with only an indirect link to the functional needs for sight distance.

The minimum standards for SSD promulgated by AASHTO have not changed since 1940. Figure 4-1 shows the values and how they are measured for roadway hillcrests. Figure 4-2 shows the values and how they are measured for roadway curves. Proper adjustments to the stopping sight distance standards shown in Figures 4-1 and 4-2 must be made where the roadway is on grade. Table 4.2 shows the AASHTO recommended adjustments for grade.

The minimum SSD values are for passenger-car operation and could be questioned for the safety of large truck operations. Although, large trucks require longer stopping distances from a given speed than do passenger cars, the greater eye height allows the truck driver to see farther over a vertical obstruction and, thereby, compensates for longer braking distance. This eye-height compensation, however, is not available to a truck driver on a roadway curve where the driver must see around obstructions such as vegetation on the inside of the curve.

4.3 Accident Experience with Restricted Stopping Sight Distance

Studies of the relationship between motor-vehicle accidents and SSD are very limited.[18] In most studies, SSD was only one of several roadway elements considered in combination. None of these studies offers any reliable method for determining the incremental accident effects of increasing SSD.

A study by Olson, et. al.[19] does provide a broad insight into the accident effects of SSD. A small, but well designed, study was conducted on 10 matched pairs of sites—one site was a hillcrest with very limited (118-308 feet) SSD and the other site in the pair was a nearby hillcrest with otherwise identical conditions except that it had very good (700 feet or greater) SSD. Another way of describing these comparisons is looking at hillcrests with 20-40 mph design speeds versus hillcrests with 75 mph or greater design speeds. As a group, study pairs had a 50% higher accident rate for the restricted SSD sites compared to the good SSD sites.

Table 4.1 Evolution of AASHTO Stopping Sight Distance Criteria

Year	Design Parameters — Eye Height (ft)	Object Height (in)	Perception/ Reaction Time (sec)	Assumed Tire/ Pavement Friction Demand (ft)	Assumed Speed for Design	Effective Change from Previous Policy
1940[1]	4.5	4	from 3.0 sec at 30 mph to 2.0 sec at 70 mph	from 0.50 at 30 mph to 0.40 at 70 mph	Design speed	NA
1954[2]	4.5	4	2.5	from 0.36 at 30 mph to 0.29 at 70 mph	Lower than design speed (28 mph at 30 mph design speed; 59 mph at 70 mph)	No net change in design distances
1965[3]	3.75	6	2.5	from 0.36 at 30 mph to 0.27 at 80 mph	Lower than design speed (28 mph at 30 mph design speed; 64 mph at 80 mph design speed)	No net change in design distances
1970[4]	3.75	6	2.5	from 0.35 at 30 mph to 0.27 at 80 mph	Minimum values same as 1965; desirable values design speed	Desirable values are up to 250 ft greater than minimum values
1984[5] 1990[6] 1994[7]	3.50	6	2.5	slightly lower than 1970 values for higher speeds	Minimum values same as 1965; desirable values design speed	Computed values always rounded up giving slightly higher values than 1970

Stopping Sight Distance 131

Design Speed (MPH)	AASHTO Stopping Sight Distance Requirements (ft.)		
	1940-1965	1965-1984	1984-present
30	200	200	200
40	275	275	275
50	350	350	400
60	475	475	525
70	600	600	625
Driver's Eye Height (ft.)	4.5	3.75	3.5
Object Height (in.)	4	6	6

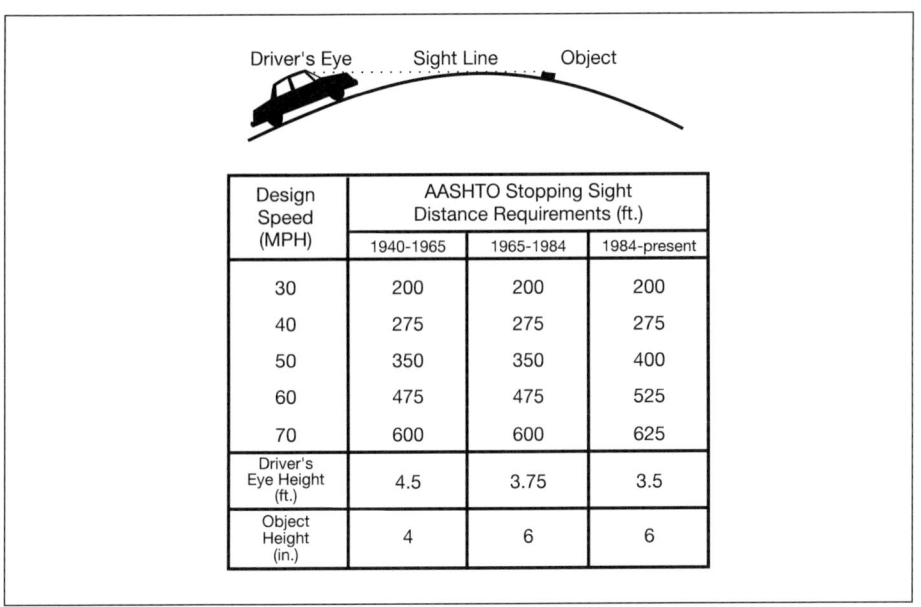

Figure 4-1 Sight Distance Standards Over Hillcrests (Courtesy of Criterion Press, Inc.)

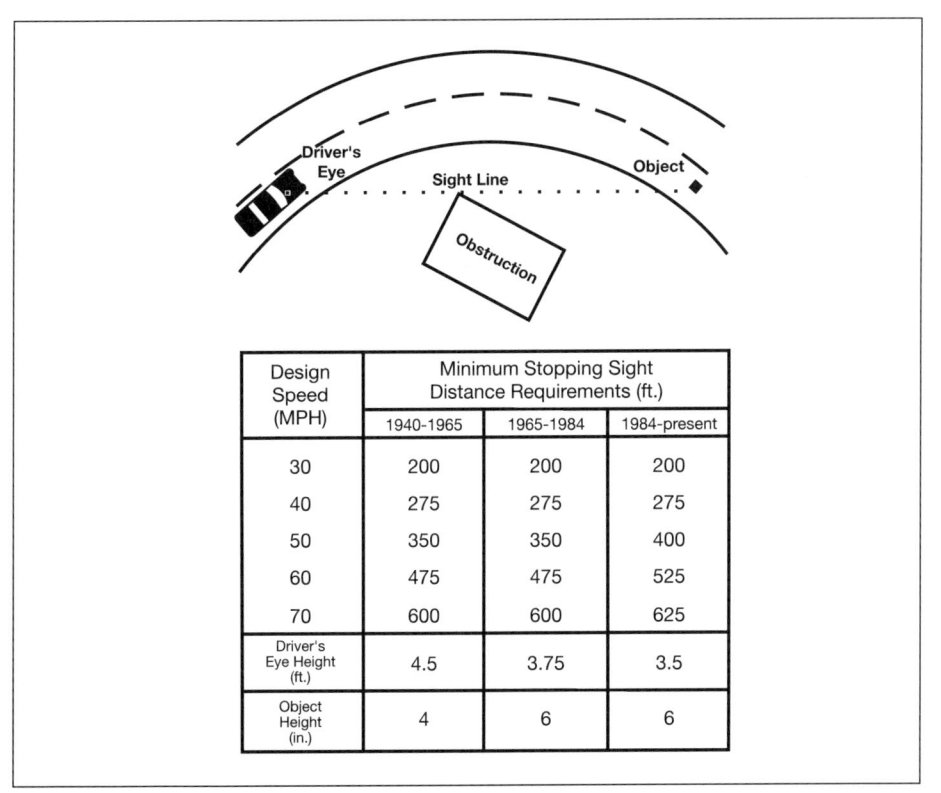

Design Speed (MPH)	Minimum Stopping Sight Distance Requirements (ft.)		
	1940-1965	1965-1984	1984-present
30	200	200	200
40	275	275	275
50	350	350	400
60	475	475	525
70	600	600	625
Driver's Eye Height (ft.)	4.5	3.75	3.5
Object Height (in.)	4	6	6

Figure 4-2 Sight Distance Standards for Roadway Curves (Courtesy of Criterion Press, Inc.)

Table 4.2 Effect of Grade on Stopping Sight Distance

Design Speed (mph)	Correction to SSD (feet)					
	Decrease for Upgrades of			Increase for Downgrades of		
	3%	6%	9%	3%	6%	9%
30	0	10	20	10	20	30
40	10	20	30	10	30	50
50	20	30	NA	20	50	NA
60	30	50	NA	30	80	NA
70	40	70	NA	50	100	NA
80	60	90	NA	70	150	NA

4.4 Functional Analysis of Stopping Sight Distance Requirements

Neuman, Glennon, and Leisch[20] present the results of a study that critically reviewed design practices for SSD. They developed a concept of SSD that focuses on various roadway operational requirements. From this operational concept of SSD, shortcomings and inconsistencies in the AASHTO design policy were revealed. Analysis of the functional requirements for SSD gives focus to the types of accidents and hazardous situations that result from limited SSD. The following points are useful in understanding the link between SSD and safety:

SSD accidents are event oriented. The mere presence of a segment of roadway with inadequate SSD does not guarantee that accidents will occur. SSD-related accidents occur only after an event or events create a critical situation. These events can be the arrivals of conflicting vehicles, the presence of objects in the road, poor visibility, poor road surface conditions, or all of these. Some of these events are a function of the roadway type (e.g., crossing conflicts at intersections do not occur on freeways); some are related to other geometric or environmental elements (e.g., requirement for severe cornering maneuver on wet pavement); and others may be totally random (e.g., presence of an object in the road).

The probability of critical events occurring within the influence of SSD restrictions defines the relative hazard of these restrictions. The relative hazard of various SSD-deficient locations can be estimated by examining the probabilities of critical events. Traffic volume, frequency of conflicts (rear-end, head-on, crossing, object in road), and time exposure of each vehicle to the restricted SSD are all useful in estimating these probabilities.

Severity as well as frequency is important. SSD situations that create severe although infrequent conflicts (e.g., head-on or angle collisions) may be just as important as situations with frequent, less severe, conflicts.

Object height is an important variable. AASHTO assumes that an object becomes visible when the driver first sees the very top of the object. But, the visibility of an object is a function of many factors such as the target area of the object, its contrast with the background, the ambient light, and any glare sources in the field of view.

Many uncontrollable or unquantifiable factors also contribute to accident causation. Driver performance characteristics such as perception/reaction time, vehicle characteristics such as braking ability, and certain imponderables such as the driver's state of mind, all contribute to increased accident potential. Although these factors are exclusive of the presence of a poor SSD location, their importance is undoubtedly heightened when the deficiency in SSD means the driver has less time to react to an event. This reduced time may make the difference between collision avoidance and an accident.

Figure 4-3 shows the complexity of SSD requirements when viewed as a function of all the elements discussed previously. Present AASHTO policy, which defines SSD requirements based on only one event and one set of conditions, promotes sufficient SSD for certain events or conditions but not for others.

Application of these functional relationships for SSD reveals situations for which greater SSD than the minimum AASHTO values might be advisable. These included not only approaches to intersections and sharp roadway curves but also on roadway curves. Truck operations on roadway curves with sight restrictions created by vertical obstructions (such as trees and walls) were found to be the situation where the AASHTO model least fit the SSD needs for the following two reasons:

1. When a vehicle brakes on a roadway curve, the frictional demand is greater than for the same braking level and speed on a straight roadway because the total deceleration is the resultant of the braking deceleration and the lateral cornering acceleration. Because of this compounding of frictional demand, AASHTO-level braking on curves can lead to loss of control, particularly for large trucks. Therefore, the need for hard braking should be reduced by the provision of longer sight distances.

2. Vertical obstructions on the inside of roadway curves create special problems for large trucks. For these situations, the greater eye height of the truck driver is of no value in compensating for the longer truck stopping distances. Therefore, trucks need greater SSD for stopping on curves because of both longer stopping distances and the need to keep resultant friction within a tolerable range.

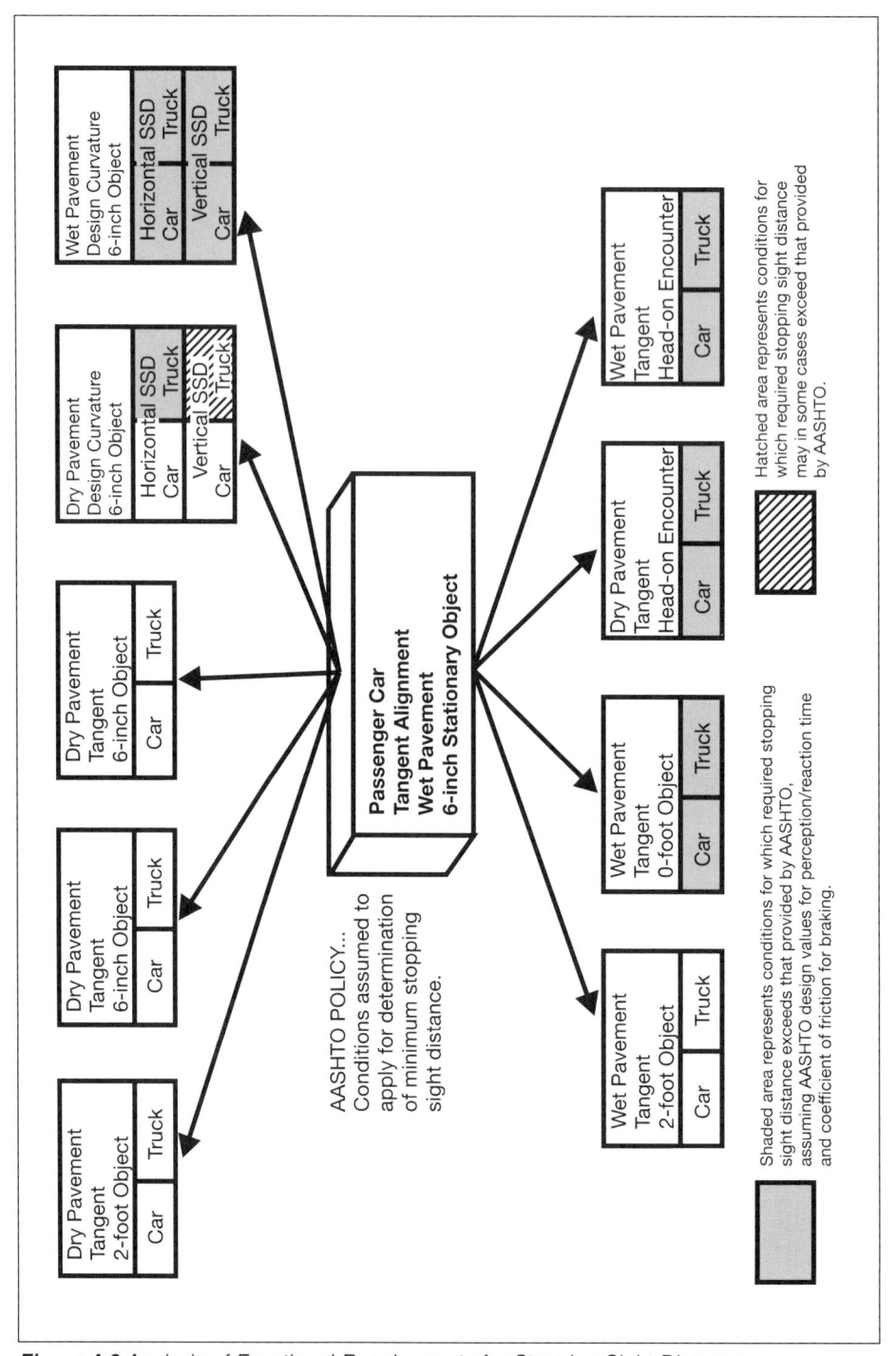

Figure 4-3 *Analysis of Functional Requirements for Stopping Sight Distance.*

> **STOPPING SIGHT DISTANCE DEFECTS**
>
> **Rules of Thumb**
>
> - *Most Tort Cases Involve an Existing Stopping Sight Distance that is Considerable Less than the AASHTO Standard*
> - *Deficient Sight Distances are Most Detrimental When They Hide a Nearby Intersection, Sharp Roadway Curve, Narrow Bridge, or Other Roadway Inconsistency*

4.5 Technical Aspects of Stopping Sight Distance Defects Cases

Thousands of roadway sections in the United States have severely restricted SSD for the legal speed of the roadway. These substandard sections are often an element in tort litigation against roadway agencies, most particularly in local rural jurisdictions. Common arguments are that the roadway agency was negligent for not designing the roadway properly or, given that the design was defective, that they were somehow negligent for failing to warn of the contingent hazards. Many times the sight restriction, in combination with other roadway defects, will be used to argue that the comparative negligence of one or both drivers in an accident was minimal because the roadway defects were the major causal factors.

Most often at issue in the restricted SSD tort case is the severity of the sight restriction in terms of how deviant the effective design speed is from the prevailing speed limit. For 45-65 mph speed limits, effective SSD design speeds that are 15-30 mph below the speed limit are normally the subject of tort claims. Typical roadway defects involved in restricted SSD cases are discussed below.

Hillcrests with Deficient SSD on Narrow Roadways

Many rural roads are 18-feet wide or narrower. Where a SSD-restricted hillcrest is present, opposing vehicles are at great risk of head-on collisions. At these location, ruts along the edge of roadway are common, being caused by wary local drivers. Unfamiliar drivers or unwary local drivers, however, occasional get involved in head-on accidents where the total time to view the opposing vehicle was only 2 seconds or less, as in Figure 4-4. One simple remedy for this condition is to widen the roadway at the hillcrest.

Figure 4-4 The Stopping Sight Distance for this Hillcrest is 20 mph on an 18-foot Roadway with a 45-mph Speed Limit.

Stopping Sight Distance

Sometimes this same condition is created by the roadway agency's unthoughtful maintenance practices. For example, under heavy snowfall when a snow-plow operator only cleans one lane of a two-lane roadway, head-on collisions can be caused at sight-restricted hillcrests. A similar situation can be created by a motor-grader operator who leaves windrow's of gravel at the hillcrest along the edge of an already narrow gravel roadway.

Hillcrests with Deficient SSD that Hide Nearby Intersections

When this condition exists, as shown in Figure 4-5, drivers making turning maneuvers at the intersection must sometimes do so without knowledge of whether a conflicting vehicle is approaching. When the available sight distance allows for less than four seconds of preview at the prevailing roadway speeds, critical conflicts can occur. These conflicts become more critical when the turning maneuver intersects a conflicting path for longer periods of time either because of a slower turning speed, because of a longer length of path overlap, or because of a more complex roadway situation where roadway curves or acute intersection angles may act to distract the turning driver.

Head-on left-turn collisions are particularly apparent at these locations. Typically, a left-turning driver will start his turn before reaching the intersection. If the intersection is very near the hillcrest, the available SSD will be less on the

Figure 4-5 This 40-mph Stopping Sight Distance Hillcrest Hides a Nearby Intersection on this 55-mph roadway.

approach to the intersection. This creates a double jeopardy whereby the drivers not only have minimal sight distance, but their separation distance is decreasing by the combined speed of both vehicles.

The most effective improvement for this condition, of course, is to flatten the hillcrest. Although this remedy should always be considered, it may not always be feasible. Sometimes a more reasonable improvement is to somehow move the intersection farther from the hillcrest. Traffic control measures that should be considered include reduced speed limits, advance intersection warning signs with advisory speed plates, or flashing intersection beacons. In some very tight situations such as making a left turn onto a humped bridge from an adjacent ramp, roadway agencies have used fisheye mirrors to allow turning vehicles to see over the hump.

Sharp Roadway Curves Hidden by SSD-restricted Hillcrests

This condition requires the driver to be able to see the pavement in the curve far enough in advance to properly adjust vehicle speed and path. When restricted SSD prevents the driver from seeing the roadway curve for more than about three seconds, the driver should be expected to have difficulty steering the roadway curve, particularly the unfamiliar driver at night. When this condition exists without adequate warning devices, it is clearly a roadway defect. The warning signs that are designed to overcome these deficiencies are the advance Curve or Turn signs, the Advisory Speed plate, and the Large Arrow or Chevron Alignment signs.

Hidden STOP Signs

Obviously a STOP sign needs to be visible for an adequate distance to allow a driver to see it and respond with a comfortable braking action. A STOP sign that cannot be seen for an adequate time should always be preceded by a STOP AHEAD sign. Every attention should be given also to erect the STOP sign so it is angled properly and is as large and high as reasonable. For more severe sight restrictions or where related accident seem to persist, additional devices including redundant STOP and STOP AHEAD signs on the left side of the roadway may be needed. Figure 4-6 shows a sharp roadway curve that hides a nearby stop sign, with no advance warning of the curve or the STOP sign.

A Project to Flatten a SSD-restricted Hillcrest Can Produce No Safety Benefit

The only way to eliminate deficient SSD on existing hillcrests is to rebuild the roadway by cutting the hillcrest down and /or flattening the approach grades. These are expensive improvements and probably not cost-effective unless the

Stopping Sight Distance 139

Figure 4-6 This Sharp Roadway Curve Hides a Stop Sign Only 200 Feet Away.

sight is severely restricted, the traffic volumes are high, and a severe hazard (busy intersection, sharp curve, etc.) is hidden by the hillcrest.

When roadways are being upgraded, caution is in order because cutting the hillcrest down is not always better. This is particularly true when an extremely deficient hillcrest is upgraded to provide a higher design speed that is still below the roadway operating speed. Figure 4-7, shows an actual "improvement" made on a 55-mph highway in an Indiana county. Profile 1 shows the original hillcrest with a minimum stopping sight distance of 150 feet. Profile 2 shows the "improved" hillcrest where a 9-foot cut was made to provide a minimum SSD of 275 feet. Profile 3 shows what the hillcrest would look like if the AASHTO stopping sight distance of 450 feet had been provided.

Notice that although the "improved" Profile 2 has a higher minimum SSD than the original Profile 1, it also increases the length of sight-deficient roadway from 600 to 1,000 feet. This example indicates the possible futility in cutting down some existing hillcrests. Not only would the construction be expensive for cutting about 9 feet into the hill to achieve Profile 2, but the safety benefit may be small or even negative. Although changing from Profile 1 to Profile 3 might produce some positive safety benefits, this improvement would require a 33-foot deeper cut at the subject location.

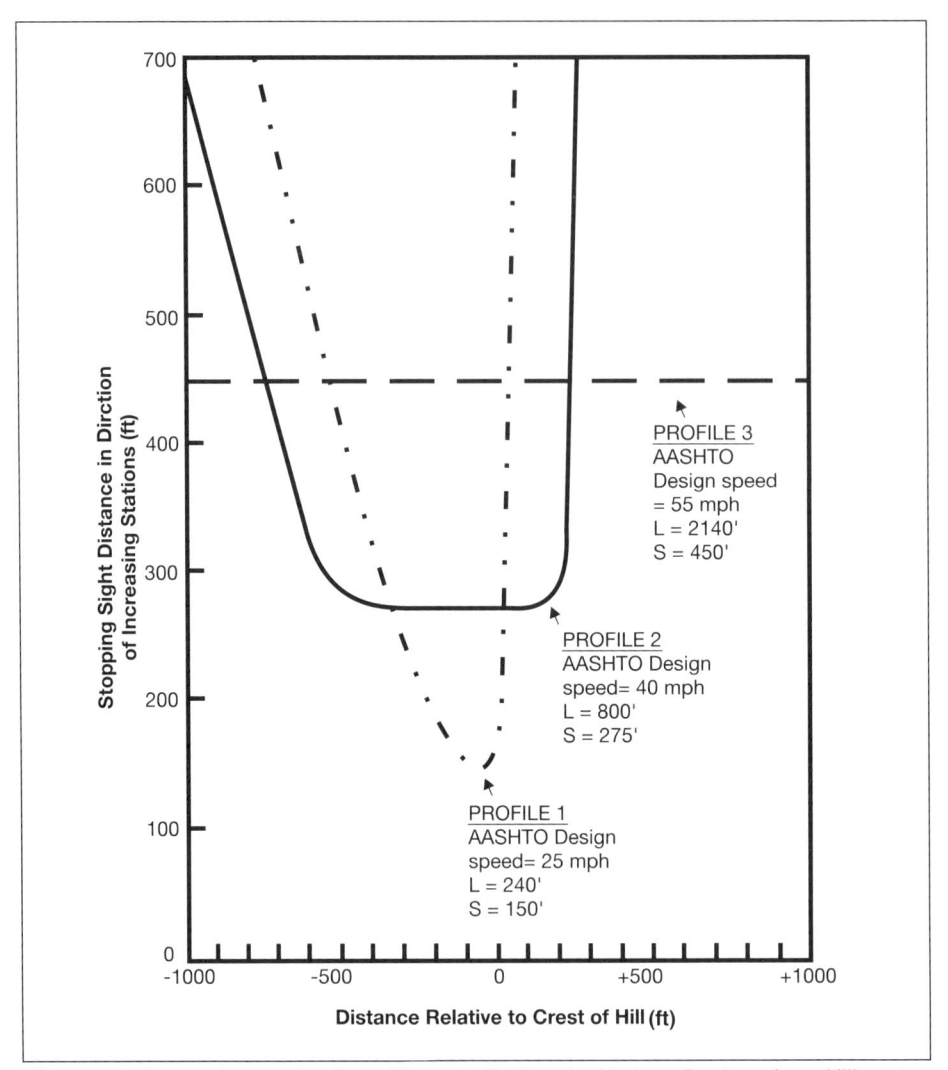

Figure 4-7 Comparison of the Sight Distance Profiles for Various Designs for a Hillcrest Joining Two 7% Grades.

Sharp Roadway Curves with Restricted SSD

Most of the attendant hazards associated with SSD-restricted hillcrests are also common with SSD-restricted roadway curves. Narrow roadways, hidden intersections, and hidden stop signs can combine to create a roadway defect. Unlike hillcrests, SSD restrictions or roadway curves are not produced by the roadway itself, but by roadside obstructions on the inside of the curve.

Stopping Sight Distance 141

Figure 4-8 Vegetation Severely Limits the Stopping Sight Distance on this Sharp Roadway Curve.

The most common sight restriction on roadway curves is where the lack of maintenance has allowed vegetation to grow close to or over the inside edge of the traveled way (see Figure 4-8). Because of the sharp curvature, sometimes coupled with a narrow roadway, a high proportion of drivers making the left-hand curve can be expected to travel over the centerline in conflict with opposing vehicles.[21] With the restricted SSD, opposing drivers are often on a foreseeable collision course that allows very little time for collision avoidance. Even if these kinds of roadway curves have otherwise adequate signing, this signing may not help avoid head-on accidents. The roadway agency must take care to trim the vegetation back as far as possible to allow adequate SSD.

Sharp Roadway Curves on Freeways with a Concrete Median Barrier

When a concrete median barrier is placed in a narrow median of a freeway, sight obstructions are imposed on sharp left-hand roadway curves. The worst case is for the left-lane driver trying to see traffic slowing ahead. Although the driver can usually see the vehicles, the brake lights are generally blocked from view by the concrete median barrier.

4.6 Accident Reconstruction Aspects

An accident reconstruction of the time-distance relationship between two opposing vehicles, to find when each driver first saw the other vehicle, requires a reconstruction of the vehicle speeds and a survey of the sight obstructions. For a hillcrest, either as-built plans or a separate survey of the road profile will allow graphical analysis of how far another vehicle can be seen over a hillcrest from any position on the approach grade and from any eye height. The same kind of analysis can be done for roadway curves with a scaled drawing both of the roadway curve and also the sight obstructions on the inside of the curve.

4.7 Typical Defense Arguments

When defending tort claims that allege defective SSD, roadway agencies generally may have one or two major legal defenses. The first of these legal defenses is the *design immunity* statute, which is available in most if not all States. Roadway design is generally held to be a discretionary function; carrying limited immunity. Under the design immunity statute, government bodies and their employees are immune from liability when the design was in general accord with the prevailing standards at the time the roadway was built. Because many roadway miles in the United States were originally built before roadway design standards were codified, roadway agencies often claim immunity from tort actions that only allege a design defect.

The second major legal defense available to roadway agencies in some States is the argument of *discretionary function*. Even though most roadway agencies have the express duty to warn of hazards, many States have tort claims acts that strictly interpret the wording of *the Manual on Uniform Traffic Control Devices (MUTCD)*.[22] The *MUTCD* has three specifically defined words as follows:

Shall: a mandatory condition
Should: an advisory condition
May: a permissive condition

Those States with strict interpretations say that, unless the *MUTCD* mandates a particular traffic control device, its use is discretionary, and therefore, immune from tort claims. Because the *MUTCD* only makes casual reference to sight restrictions, no devices appear to be mandated for this condition. For this reason, the *failure to warn* argument may be summarily dismissed unless some other provision can be used. For example, the *MUTCD* in talking about signs says that "Warning signs are essential where hazards are not self-evident."

Beyond the available statutory arguments given above, roadway agencies often use other case-specific defenses that argue either that the driver was the sole negligent party and/or that the restricted SSD condition had no causal connection to the accident. Some of these arguments are:

1. The driver was negligent for not slowing down for the sight-restricted condition.
2. The driver did not pull up to the exact center of the intersection before starting his turn.
3. The driver was negligent for crossing the centerline and colliding with another vehicle.
4. The driver was negligent for speeding.
5. The driver was negligent for driving while intoxicated.
6. The driver was negligent for not yielding the right of way.
7. The roadway agency had no notice of the defective roadway.
8. The subject location lacked any accident records indicating a safety problem.
9. The roadway agency lacked adequate funds to upgrade existing roadway locations.

Endnotes

1. American Association of State Highway Officials, *A Policy on Sight Distances for Highways*, 1940.
2. American Association of State Highway Officials, *A Policy on Geometric Design of Rural Highways*, 1954
3. American Association of State Highway Officials, *A Policy on Geometric Design of Rural Highways*, 1965.
4. American Association of State Highway Officials, *A Policy on Design Standards for Stopping Sight Distance*, 1970.
5. American Association of State Highway and Transportation Officials, *A Policy on Geometric Design of Highways and Streets*, 1984.
6. American Association of State Highway and Transportation Officials, *A Policy on Geometric Design of Highways and Streets*, 1990.
7. American Association of State Highway and Transportation Officials, *A Policy on Geometric Design of Highways and Streets*, 1994.
8. Hall, J. W., and Turner, D. S., *Stopping Sight Distance: Can We See Where We Now Stand*, Transportation Research Record 1208, 1989.
9. Blanchard, A. H., and Drowne, H. B., *Textbook on Highway Engineering*, First Edition 1914.
10. Agg, T. R., *Construction of Roads and Pavements*, First Edition, 1916.
11. Agg, T. R., *Construction of Roads and Pavements*, Third Edition, 1924.
12. Brightman, H. L., *Alignment and Grade as Affecting Location of the Modern Highway*, Twelfth Annual Conference on Highway Engineering, University of Michigan, 1926.

13. American Association of State Highway Officials, *Reports of Standing Committees before the American Association of State Highway Officials*, 1931.

14. Wiley, C. C., *Principles of Highway Engineering*, Second Edition, 1935.

15. Gutman, I., *Engineering Sets High Mark on German Highways*, Engineering News-Record, August 1936.

16. Conner, C. N., *Design Trends to Correlation with Traffic Needs*, Engineering News-Record, January 1937.

17. Bureau of Public Roads, U.S. Department of Agriculture, *Annual Report of Department of Agriculture*, 1938.

18. Glennon, John C., *Effect of Sight Distance on Highway Safety*, Transportation Research Board, State of the Art Report 6, Relationship Between Safety and Key Highway Features, 1987.

19. Olson, P. L., et. al., *Parameters Affecting Stopping Sight Distance*, Transportation Research Board, NCHRP Report 270, 1984.

20. Neuman, T. R., Glennon, J. C., and Leisch, J. E., *Functional Analysis of Stopping Sight Distance Requirements*, Transportation Research Record 923, 1984.

21. Glennon, J. C., Neuman, T. R., and Leisch, J. E., *Safety and Operational Considerations for Design of Highway Curves*, Federal Highway Administration, 1983.

22. Federal Highway Administration, *Manual on Uniform Traffic Control Devices*, 1988. (available through Lawyers and Judges Publishing Co.)

References

Transportation Research Board, *Highway Sight Distance Design Issues*, Transportation Research Record 1208, 1989.

Glennon, John C., *A Safety Evaluation of Current Design Criteria for Stopping Sight Distance*, Highway Research Record 312, 1979.

Glennon, John C., *The Traffic Safety Toolbox: Chapter 10: Geometric Design — Sight Distance*, Institute of Transportation Engineers, 1993.

Chapter 5

Intersection Sight Distance

Roadway intersections are places where the competition between two or more traffic movements can create traffic conflicts. The safety of an intersection depends on the way these conflicts are resolved. Any intersection should be designed and maintained so that the driver of an approaching vehicle has an unobstructed view of not only the whole intersection but also a length of the intersecting roadway sufficient to permit him to avoid collision with conflicting vehicles.

AASHTO[1-8] has promulgated intersection sight distance design standard since at least 1940. These standards have addressed the sight distance needs at open intersections, yield-controlled intersections, and stop- or signal-controlled intersections. The functional models for open and yield-controlled intersections address the needs of the through driver to slow, accelerate, or stop to avoid collision with intersecting vehicles. The functional models for stop- or signal-controlled intersections address the needs of a stopped driver to scan the approaches, accelerate, and clear the conflict area for left-turn, right-turn, and crossing maneuvers.

5.1 Sight Distance Requirements for Uncontrolled Intersections

Uncontrolled intersections should only be used where the probability of intersection conflict is low, and the intersection sight distance is good. For safety, AASHTO[1-8] has always recommended a sight triangle (Case II) that is defined by legs along each roadway equal in length to the AASHTO stopping sight distance (see Chapter 4) for the speed of each roadway (see Figure 5-1). In other words, for a minimum sight triangle, either driver upon passing the vertex of the defined triangle can stop before colliding with a conflicting vehicle that was also past its vertex. If obstructions, which cannot be removed, fix the vertexes of the sight triangle to points less than the stopping sight distance requirements, the roadway agency either needs to adjust speed limits downward or control the intersection with STOP or YIELD signs. To measure existing sight distances, AASHTO currently uses a 3.5-foot driver eye height and a 4.25-foot object height (roofline of a passenger car).

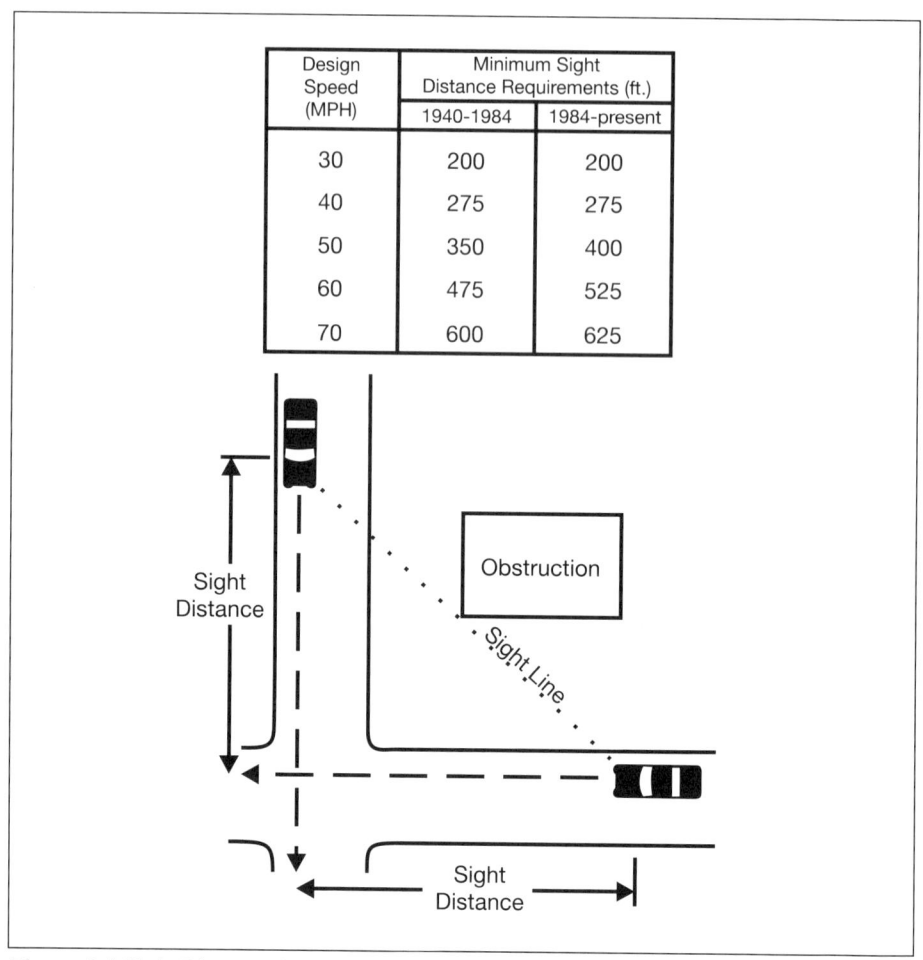

Design Speed (MPH)	Minimum Sight Distance Requirements (ft.) 1940-1984	Minimum Sight Distance Requirements (ft.) 1984-present
30	200	200
40	275	275
50	350	400
60	475	525
70	600	625

Figure 5-1 Sight Distance Standards for Uncontrolled Intersections. *(Courtesy of Criterion Press, Inc.)*

5.2 Sight Distance Requirements for Yield-Controlled Intersections

Table 5.1 is what AASHTO calls Case I intersection sight distance. These values are the distance traveled by a vehicle in three seconds and are thought to be minimal for a driver to see, perceive, and react by slowing or accelerating to avoid collision. Any intersections with sight distances lower than these values should have STOP signs on the minor roadway. Intersections with sight distances between Case I values and Case II (or stopping sight distance) values (Figure 5-1) should have yield signs on the minor roadway. To measure existing sight distances, AASHTO currently uses a 3.5-foot driver eye height and a 4.25-foot object height (roofline of a passenger car).

Intersection Sight Distance

5.3 Sight Distance Requirements for Stop - Controlled Intersections

When intersection sight distances are less than the Table 5.1 values, and cannot be improved, two-way STOP signs are the appropriate control. Once two-way STOP signs are in place, a different functional requirement comes into play for sight distance. Now, a driver at a stop sign needs to see far enough down each intersecting approach to allow enough time to scan both approaches, make a decision to proceed, accelerate into the intersecting roadway, and then either cross, turn left, or turn right without collision. Because AASHTO criteria for left and right turns use assumptions that generate sight distances that are much longer than the minimum gaps in traffic that most drivers are willing to accept, these criteria are very controversial, and are the subject of ongoing research. For this reason, the discussion here is limited to the crossing maneuver. These considerations, however, can be modified by the reader to account for the other maneuvers, if necessary.

Table 5.1 AASHTO Case I Sight Distances — Allowing Vehicles to Adjust Speed

Speed (mph)	Sight Distance (feet)
10	45
15	70
20	90
25	110
30	130
35	155
40	180
50	220
60	260
70	310

The AASHTO sight distances for crossing maneuvers from the stopped position at stop-controlled intersections is measured to the roofline (4.25 feet) of an oncoming passenger car from the standard eye height (3.5 feet) of a driver in a passenger car positioned 20 feet from the edge of the through roadway. Figure 5-2 lists the sight distances down the through roadway needed per 10-mph increment of speed on that roadway (e.g., sight distance for a passenger car crossing a 45-mph, 2-lane roadway is 45 x 100/10 = 450 feet).

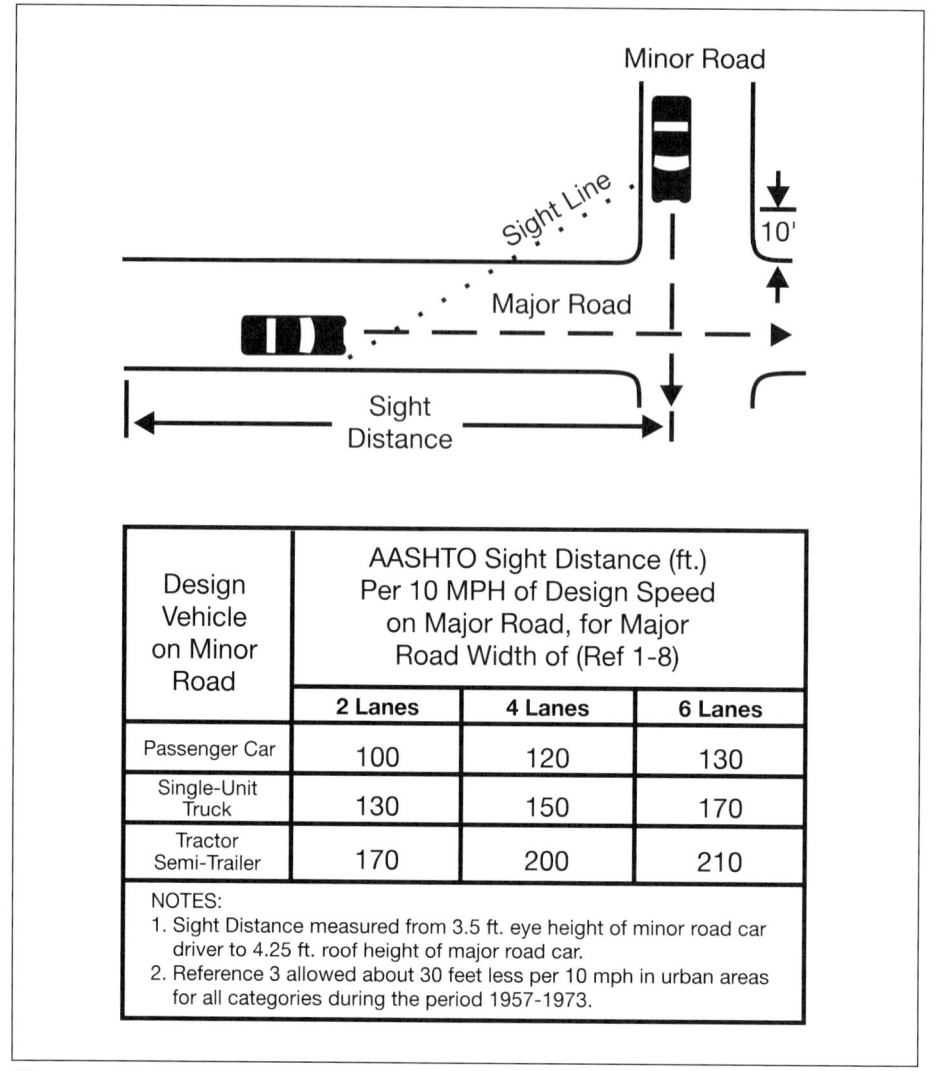

Design Vehicle on Minor Road	AASHTO Sight Distance (ft.) Per 10 MPH of Design Speed on Major Road, for Major Road Width of (Ref 1-8)		
	2 Lanes	4 Lanes	6 Lanes
Passenger Car	100	120	130
Single-Unit Truck	130	150	170
Tractor Semi-Trailer	170	200	210

NOTES:
1. Sight Distance measured from 3.5 ft. eye height of minor road car driver to 4.25 ft. roof height of major road car.
2. Reference 3 allowed about 30 feet less per 10 mph in urban areas for all categories during the period 1957-1973.

Figure 5-2 Sight Distance Standards for Stop-Controlled Intersection (Courtesy of Criterion Press, Inc.)

5.4 Rail-Highway Grade Crossing Sight Distance

Rail-highway grade crossings are similar to highway and street intersections except that one of the conflicting vehicles, the train, can require one-half mile or more to stop. At crossings with only a crossbuck sign, the motor-vehicle driver, therefore, needs enough sight distance in order to either stop or proceed ahead to clear the crossing when an approaching train becomes visible. The subject of rail-highway grade crossing sight triangles is discussed fully in Chapter 8.

> **INTERSECTION SIGHT DISTANCE OBSTRUCTIONS**
>
> **Rule of Thumb**
>
> *For High-Speed Roadways, the Sight Distance needed is Hundreds of Feet. Close-in Sight Distance is Usually of Little Consequence.*

5.5 Technical Aspects of Intersection Sight Distance Defect Cases

Thousands of intersections both urban and rural have severely restricted sight distance. These intersections are often a major element in tort litigation against roadway agencies. Quite frequently, the issue revolves around the placement of needed traffic controls. In other cases, the issue is the lack of maintenance (vegetation control). When the parties in one vehicle sue the parties in another vehicle, the defending driver often argues that the sight restriction was a significant factor that reduces his negligence.

The need for adequate sight distance at intersections cannot be stressed enough. Constant vigilance to prevent man-made objects from being placed and vegetation from growing in the necessary sight triangles will probably do more for traffic safety at the lowest cost than most other measures. Another principle to remember is that partial sight improvements that provide less than AASHTO requirements may be beneficial because not all traffic conflicts occur at the operating speed of the roadway. Also removing even smaller objects within an otherwise clear triangle can improve a driver's view at critical points along the approach. The guiding principle is to always clear as much as practical.

When the intersection sight obstruction is an embankment which initially only provides visibility of the very top of the intersecting vehicle, the driver may miss seeing the other vehicle in the daytime because of the lack of target value. At night, on a dark roadway, the top of that vehicle will not be visible at all. Because intersections must operate at night as well as day, serious consideration should be given to discarding the AASHTO roofline object-height criterion used to measure the adequacy of existing sight triangles and replacing it with a 2-foot headlight-height criterion.

Most often at issue in the restricted intersection sight distance case is the severity of the sight restriction in terms of how deviant the available sight distance is from the AASHTO requirement. Quite often, available sight distances are only 20 to 30% of the AASHTO values. Typical roadway defect cases involving restricted intersection sight distance are discussed below.

Uncontrolled Intersections with Deficient Sight Distance

Typically, uncontrolled intersections are at the crossing of low-volume local roads or streets. In rural areas, they are usually on unpaved roads that intersect other unpaved roads. In urban areas, they are usually on residential streets that intersect other residential streets. Uncontrolled intersections are not usually on main rural highways or major urban arterials.

At urban uncontrolled intersections, sight obstructions are usually either buildings, vegetation, embankments, or signs close to the corner. At rural uncontrolled intersections, sight obstructions are usually vegetation (including seasonal crops) or embankments.

In rural areas, particularly in the Midwestern, Southern and Western States, many uncontrolled intersections with 55-mph speed limits have less than 240-foot sight triangles. In urban areas, many uncontrolled intersections with 30-mph speed limits have less than 130-foot sight triangles. The reluctance of local jurisdictions to use stop control at these intersections is usually based on one or more of the following considerations:

1. The cost of placing additional signs
2. The difficulty in maintaining the signs
3. The desire to minimize vehicle delay
4. A lack of sensitivity to safety principles
5. The problem was not perceived because of low traffic volumes and mostly familiar drivers

Although the first three of these are seemingly rational reasons for not using traffic control, Table 5.1 defines a minimum sight triangle below which conflicting drivers have little or no opportunity to avoid collision. For this reason, local jurisdictions should consider the Table 5.1 values as absolute minimums, and either reduce speed limits or place STOP signs at severely sight-restricted intersections.

When an uncontrolled highway or street intersection has serious sight obstructions, an extreme burden is placed on drivers to: (1) expect an oncoming vehicle; (2) judge what direction it might be coming from; (3) anticipate its speed; (4) judge their own closing rate to the conflict area; and (5) adjust their own

Intersection Sight Distance

speed and position to avoid collision once the conflicting vehicle is seen. This is a burden that few drivers handle well, particularly when sight distance is highly restricted.

Surely for any intersection sight condition, there is a distribution of driver behavior such that some drivers would judge the available sight distance as adequate even when the available sight distance is only half of the AASHTO value based on oncoming vehicle speeds. The appropriate roadway operations strategy, therefore, calls for great care in maintaining intersection sight distance standards. Cutting trees, shrubs, and grass, grading embankments, removing walls, fences, and advertising signs, restricting parked vehicles, and placing STOP and YIELD signs should be everyday applications in the quest for safety.

In an attempt to provide some minimal sight distance at intersections, most urban jurisdictions have ordinances prohibiting property owners from placing obstructions within a certain corner sight triangle, usually defined with 60-75 foot legs along the near curbline. Most of these ordinances were apparently passed in the early 1900's because they generally only allow for maximum safe speeds of 15-20 mph. This form of control can be made more effective if: (1) statewide speed limits for urban, unmarked streets are lowered to the 20-25 mph range; (2) city sight obstruction ordinances specify 75-foot legs or greater along the near curbline; and (3) ordinance compliance is actively monitored and enforced.

As shown in Figure 5-3, control of sight obstructions in rural areas is usually very lax. County jurisdictions are reluctant to ask farmers to cut back crops or hedgerows for better corner visibility. Because uncontrolled intersections with serious sight obstructions are probably a major accident contributor on local rural roads, ordinances should be enacted and enforced to prevent plantings on the roadway right-of-way. In addition, agreements could be formulated with property owners to provide corner triangles free from planting.

After all reasonable measures to bring uncontrolled intersections into compliance with Table 5.1 values are exhausted, STOP signs remain as the only reasonable alternative for safe operation of these intersections. STOP signs are a low-cost measure and should eliminate sight-distance related accidents. In placing 2-way STOP signs, intersections should be checked for adequate sight distance for the stop-controlled condition. If that sight distance cannot be provided, 4-way STOP signs should be considered.

Given that intersection sight obstructions are an insidious problem of our local roadway system, Table 5.2 summarizes the countermeasures that could go a long way toward improving safety at thousands of currently uncontrolled urban and rural intersections.

152　　　　　　　　　　　　　　　　　Roadway Defects and Tort Liability

Figure 5-3 Rural Uncontrolled Intersection with 10-mph Sight Distance on a 55-mph Roadway.

Intersection Sight Distance

Table 5.2 Summary of Remedial Measures at Existing Uncontrolled Intersections

Urban	Rural
Enact maximum statewide (or citywide) speed limit of 20-25 mph for unsigned streets	Enact maximum statewide (or county-wide) speed limit of 45-50 mph for unsigned roads
Enact and enforce city ordinance prohibiting obstructions within 75-ft triangle (measured along near curb)	Formulate simple agreements with property owners to provide obstruction-free corner triangles as large as possible
Cut back vegetation and/or embankments to achieve Table 5.1 values or better	Cut back vegetation and/or embankments to achieve Table 5.1 values or better
Remove walls, fences, signs or other obstructions on right-of-way	Remove walls, fences signs or other obstructions on right-of-way
Use 2-ft object height where possible for nighttime view of headlights	Use 2-ft object height where possible for nighttime view of headlights
Restrict parking next to intersection for 75 feet or more	Use speed zoning on approach to intersection
Place 2-way stop signs where Table 5.1 values cannot be obtained in all 4 quadrants	Place 2-way stop signs where Table 5.1 values cannot be obtained in all 4 quadrants

Most States have adopted a requirement for drivers, which is usually called the *right-hand rule*. What this rule requires is when drivers approach an uncontrolled intersection, they are to yield to vehicles approaching from their right. This rule is often used in intersection accident cases as a argument for the negligence of the driver on the left. A driver on the left can also argue, however, that the rule does not apply to a severely sight-restricted intersection. The rule is really intended as a communication between drivers that can see each other's vehicle in time to avoid collision.

Stop-Controlled Intersections with Sight Restrictions

Sight restrictions at stop-controlled intersections are usually caused by vegetation that is too close to the intersecting roadway. In rural areas, these are usually crops, hedgerows, or uncut brush on the roadway right of way. In urban areas, the vegetation that blocks the driver's view is usually trees and bushes planted by the property owner.

Other obstructions in urban areas are embankments (a defect often formed when a roadway is widened), walls, fences, buildings, signs, or vehicles parked too close to the intersection. Other obstructions in rural areas include embankments and buildings.

When stop-controlled highway or street intersections have severe sight restrictions, an extreme burden is placed on the minor road driver to : (1) scan both directions; (2) judge the speed of oncoming vehicles: and (3) judge the adequacy of the available sight distance to allow proper acceleration to cross, turn right, or turn left without collision. This is a burden that few drivers handle well, particularly when sight distance is highly restricted.

Table 5.3 list some of the appropriate remedial measure for reducing or eliminating the risk associated with sight restrictions at stop-controlled intersections. Figure 5-4 is an example of where traffic signals were installed at an intersection that had deficient sight distance and was previously controlled by STOP signs. Figure 5-5 shows a unique solution to a sight restriction for vehicles entering a very busy 4-lane divided roadway from a stop-controlled side street.

Table 5.3 Summary of Remedial Measures at Existing Stop-Controlled Intersections

Cut back vegetation and/or embankments as far as possible	Remove walls, fences, advertising signs or other obstructions on right-of-way
Restrict Parking	Reduce through roadway approach speed limits
Paint stop line closer to the intersection when that position offers a clearer sight line. Install sign that says "Pull Up to See"	Install 4-way stop signs when adequate sight distance can't be achieved
Install traffic signals when warranted	Use 2-foot object height where possible for nighttime view of headlights

Sight Obstructions at Freeway Interchanges

The bridge rail for the bridge over a freeway sometimes obstructs the sight to the left of an off-ramp driver, particularly when the ramp is close to the bridge. If the ramp is stop-controlled, the AASHTO standard for stop-controlled intersections should apply.

Sight Obstructions at Driveways

A driveway driver has the same sight distance needs as an intersection driver to move into or across the intersecting roadway. However, the driveway driver often has to contend with signs, walls, and vegetation that block his view. The availability of AASHTO intersection sight distance at driveways is an aspect of roadway safety that should be addressed when a property owner applies for a driveway permit.

Intersection Sight Distance

Figure 5-4 Traffic Signal Installed to Compensate for a Severely Sight Restricted Intersection.

Figure 5-5 A Unique Traffic Control Solution for an Intersection Sight Restriction at a Stop-Controlled Side Road.

5.6 Accident Reconstruction Aspects

Reconstructing the time-distance relationship between two intersecting vehicles at an intersection can demonstrate the severity of the sight obstruction. This exercise first requires a physical reconstruction of the vehicle impact speeds and any pre-braking speeds and, second, requires a scaled drawing of the intersection and the sight obstructions. With this information, the positions of the two vehicles can be plotted on the drawing at any pre-impact time to determine whether the drivers could see each other around or over the obstruction. A similar but hypothetical analysis may be useful using the speed limits for each vehicle.

5.7 Typical Defense Arguments

Roadway agencies generally argue not only that they have no duty to clear sight obstructions other than on their own right-of-way, but also that the AASHTO intersection sight distance standards are only guidelines to be followed for the design of new intersections and do not apply to the maintenance of existing intersections.

Other case-specific arguments are that either the driver(s) was the sole negligent party and/or the roadway condition had no contribution to the causation of the collision. Some of these arguments are:

1. The sight obstructions should have been self-evident, and the drivers should have controlled their speeds accordingly.
2. The driver(s) were negligent for exceeding a reasonable and prudent speed, even though their speeds were significantly below the speed limit.
3. A driver was negligent for driving while intoxicated.
4. The drivers should have expected the sight-obstructed intersection because of the generally substandard nature of the roadway.
5. The *MUTCD* does not mandate STOP or YIELD signs at sight-obstructed intersections.
6. At a stop-controlled intersection, if the minor road driver had pulled up closer to the intersection (than the specified AASHTO eye position of 20 feet back), he would have had adequate sight distance.
7. The roadway agency had no notice of the defective condition.
8. The subject location lacks any accident records indicating a safety problem.
9. The roadway agency has limited funds that prevent it from putting signs everywhere.
10. If STOP or YIELD signs are erected, they will either be shot full of holes or stolen.

Endnotes

1. American Association of State Highway Officials, *A Policy on Intersections at Grade*, 1940.

2. American Association of State Highway Officials, *A Policy on Geometric Design of Rural Highways*, 1954.

3. American Association of State Highway Officials, *A Policy on Arterial Highways in Urban Areas, 1957*.

4. American Association of State Highway Officials, *A Policy on Geometric Design of Rural Highways*, 1965.

5. American Association of State Highway Officials, *A Policy on Design of Urban Highways and Arterial Streets, 1973*.

6. American Association of State Highway and Transportation Officials, *A Policy on Geometric Design of Highways and Streets*, 1984.

7. American Association of State Highway and Transportation Officials, *A Policy on Geometric Design of Highways and Streets*, 1990.

8. American Association of State Highway and Transportation Officials, *A Policy on Geometric Design of Highways and Streets*, 1994.

References

Glennon, John C., *The Traffic Safety Toolbox: Chapter 10: Sight Distance*, Institute of Transportation Engineers, 1993.

Kuciemba, Stephen R., and Cirilo, Julie Anna, *Safety Effectiveness of Highway Design Features: Volume 5: Intersections*, Federal Highway Administration, 1992.

Mason, John M. Fitzpatrick, Kay, and Harwood, Douglas W., *Intersection Sight Requirements for Large Trucks*, Transportation Research Record 1208, 1989.

Alexander, Gerson J., *Search and Perception-Reaction Time at Intersections and Railroad Grade Crossings*, ITE Journal, Institute of Transportation Engineers, 1989.

Chapter 6

Slippery Pavements and Hydroplaning Sections

The skidding of rubber-tired vehicles occurs when the forces at the tire-pavement interface exceed the ability of that tire and that surface, for the ambient condition, to develop frictional resistance. For dry conditions, the friction between most tires and pavement surfaces is sufficient to handle all but the more severe vehicle maneuvers without skidding. But, for wet conditions, the ability to develop tire-pavement friction can be significantly reduced for deficient pavements and/or tires.

In the middle 1970's, when government-mandated gasoline-efficient motor vehicles were introduced into the U.S. fleet, and when vehicle travel was curtailed by spiraling gasoline prices and reduces speed limits, all States experienced a real reduction in gas-tax revenues. Along with this trend, the U.S. also experienced double digit inflation. As a result, State and local communities were faced with the dilemma of matching spiraling material costs with lower revenues. The result was a rapid deterioration of roadway pavements.

As a result of these trends, the U.S. Congress in 1977 stepped in to bail out the States with what was called the 3R program (for Resurfacing, Restoration, and Rehabilitation). This program allowed up to 20% of the Federal Highway Trust Fund's annual expenditures to be allocated to what were basically maintenance projects. Up to that point, the Federal Highway Trust Fund had been dedicated to new construction. Although the majority of these 3R funds since 1977 have been dedicated to resurfacing roadways for structural integrity, rideability, and skid resistance, many roadways still exhibit low skid resistance.

Vehicle running speeds increased over the years until 1973. With this increase and the increase in the number and horsepower of vehicles, skidding accidents became a major concern of those involved with roadway safety. Although the reduced national speed limits and oil conservation measures of the 1970's may have curtailed the frictional demands related to speed, the relentless increases in traffic volume and both the percentage and size of large trucks has accelerated the degradation of the frictional supply provided by U.S. pavements. Now that Congress has given speed control back to the States, who have in turn adopted higher speed limits, many miles of deficient or marginal pavements could become critical to wet weather safety.

To say that the driver in a skidding accident was "driving too fast for conditions" or that the accident was caused by "driver error" may be inaccurate in many cases. Roadway agencies are charged with providing pavement surfaces that contribute enough skid resistance to handle most of the frictional needs of traffic under reasonably foreseeable conditions.

Almost all dry asphalt or concrete roadway pavements have reasonable skid resistance. For wet pavements, however, the degree of skid resistance is a function of the coarseness of the surface texture. A pavement composed of angular aggregate which "bites" into the tire can produce high wet skid resistance. But, for smooth pavements, even small amounts of water may significantly reduce the friction developed between a vehicle's tires and the pavement.

Low wet-pavement skid resistance can result from one or a combination of causes. Some of the more commonly recognized causes are:

1. **Low Microtexture.** The term microtexture describes the grittiness of a pavement that develops adhesion between the tire and the wet pavement. Good microtexture can penetrate the thin film of water that is not removed by the tire. Pavement mix designs that do not include course-grained sands or rough textured aggregates will have low microtexture. Microtexture is particularly important on all low-speed roadways when wet and on those high-speed roadways that have small amounts of water on the surface.

2. **Low Macrotexture.** Good macrotexture is defined by an open-graded pavement that has voids between the exposed aggregates to drain water from under the tire. Macrotexture is most important for high-speed roadways. Pavement mix designs that use already polished aggregates or those susceptible to polishing will either have low initial skid resistance or will quickly lose their skid resistance.

3. **Asphalt Flushing.** When a bituminous concrete (asphalt) pavement contains too much asphalt or the wrong grade of asphalt, the asphalt may work up to the surface, cover the aggregate, and create a smooth surface that is very slippery when wet.

4. **Dirty Pavements.** Accumulations of dust, oil drippings, and rubber can form a coating, causing the pavement to become slippery during the first few minutes of rain. Loose materials such as sand, stones, or mud on the pavement may cause a skidding hazard in any weather.

5. **Worn Tires**. Both tire tread and pavement macrotexture are important elements for draining water from under the tire. Worn or bald tires can cause loss of control on a pavement with otherwise adequate texture.

6. **Hydroplaning Conditions.** Hydroplaning occurs when the tire separates from the pavement and rides up on a film of water. In addition to macrotexture and tire tread depth, the other critical factors affecting the propensity to hydroplane are vehicle speed, tire pressure, and water depth.

Early research on the relation between skid resistance and roadway safety was reported by Moyer[1] in the 1930's and led to a focus of intensive research as roadway travel and speeds increased after World War II. This research dealt with several aspects of tire-pavement interaction including; developing methods of measuring skid resistance,[2,3] comparing and correlating different measuring methods[3,4] and relating road surface properties to skidding accidents.[3,5]

These research activities culminated in the First International Skid Prevention Conference in 1959.[3] Research progress from 1959 to 1977 was accumulated at the Second International Skid Prevention Conference in 1977.[5] The proceedings from both of these conference contain a wealth of material that remains useful and has served as the basis for further research.

6.1 Design and Construction of Pavements

On *portland cement concrete* pavements, the microtexture depends on the fine aggregates, and the macrotexture is formed during the finishing operation. Concrete surfaces are usually finished by brooming with stiff wide-spaced bristles.

A good concrete mix design should use a durable angular fine aggregate with a high silica content. The proportion of fine aggregate should be as high as practical to ensure proper placement, finishing, and texturing.

Traffic and weather may cause portland cement concrete pavements to lose their skid resistance long before the structural or riding qualities are lost. The skid resistance can be improved by sawing grooves, by mechanical abrading, or by resurfacing with asphalt.

The texture of a *bituminous concrete (asphalt)* pavement is provided by the grittiness of the individual aggregate particles and the gradation of the mix. For lasting skid resistance, the chosen aggregate should resist abrasion, polishing, and consolidation.

An asphalt pavement depends mainly on the course aggregate to produce skid resistance and to provide pavement drainage. For a good friction surface over a wide speed range, mixes that have a large proportion of non-polishing one-sized course aggregate and a large void content produce an open-textured surface that provides both good adhesion and good drainage under a tire.

6.2 Mechanics of the Tire-Pavement Interface

Any moving vehicle produces kinetic energy that increases with the square of its speed. Dissipation of this kinetic energy is required in order to bring the vehicle to rest. For a moving vehicle, energy is dissipated between the tires and pavement and within the braking system by the creation of friction forces opposing the motion of the vehicle. After wheel lock-up, the total friction force available to oppose the motion of the vehicle must be generated at the tire-pavement interface.

On flat grades, the locked-wheel coefficient of friction, f, is related to the stopping distance and speed of an automobile by the following expression:

$$D = \frac{S^2}{30f}$$

where

D = the locked-wheel stopping distance, feet
S = the vehicle speed, mph

This equation shows that the stopping distance for any speed can be decreased by increasing the skid resistance at the tire-pavement interface.

The rest of this section paraphrases a discussion by Kummer and Meyer,[6] who proposed a unified theory of tire-pavement friction based on earlier theories. This theory is widely accepted as the state of knowledge on tire-pavement friction.

The true interpretation of friction measurements made between tires and pavements is difficult, not only because the friction processes are molecular but also because many surface, tire, vehicle, and operational factors affect the measurement. These friction processes have not yet been described with any scientific rigor. Empirical study of sliding tires on pavement surfaces, on the other hand, tend to mask important processes which need to be measured if the controlling phenomena are to be understood. Researchers have generally taken a simplified theoretical approach combined with laboratory and field experiments to study tire-pavement friction.

Figure 6-1 shows a uniformly loaded tire sliding over a rough-textured surface. The measured friction force, F, is the sum of the adhesion force, F_a, and the hysteresis force, F_h. The *adhesion force*, a function of molecular activity, is considered to be the product of the interface shear strength and the contact area of the tire with a single mineral particle.

The *hysteresis force*, also a function a molecular activity, is caused by damping losses within the tire material from deformation of the tire as it rolls or slides across an irregular surface. The damping by the tire opposes the displacement of

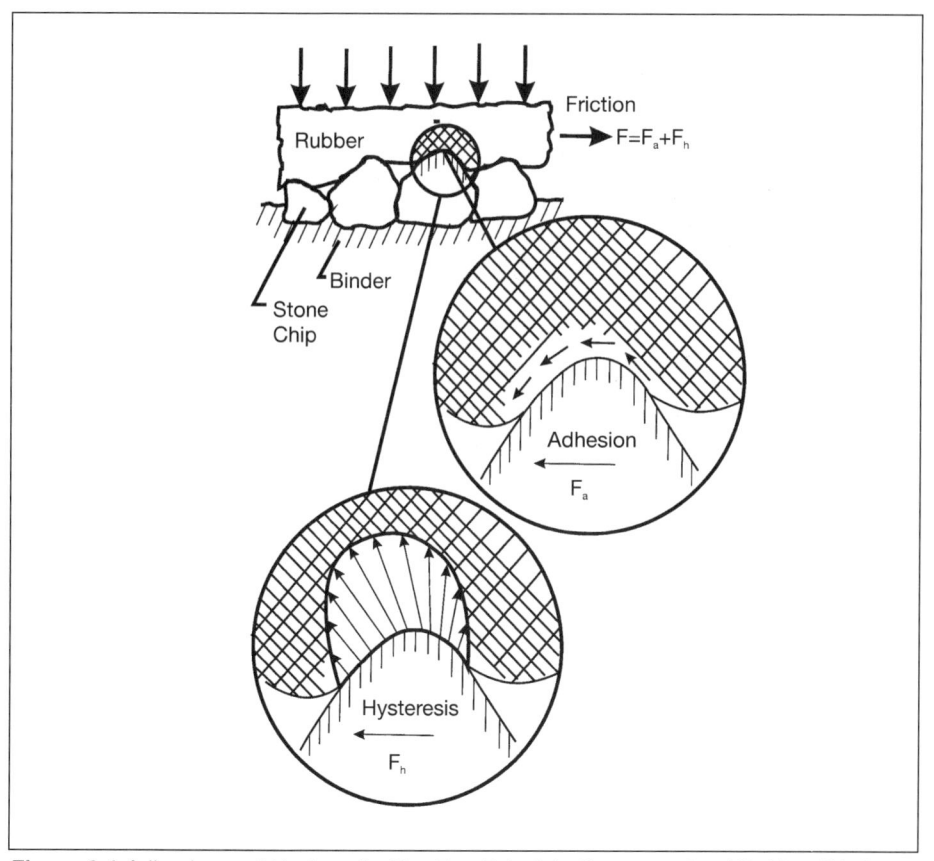

Slippery Pavements and Hydroplaning Section 163

Figure 6-1 Adhesion and Hysteresis: The Two Principle Components of Rubber Friction.

the tire upstream of the particle and the recovery downstream of the particle, producing the pressure distribution shown in Figure 6-1. The hysteresis force is represented by resolving the resultant force into a vertical component acting on the particle and a horizontal component, which opposes the sliding.

Both adhesion and hysteresis release energy in the form of heat loss. During the adhesion process, energy is released when the junctions between the tire and pavement surface molecules are formed and then ruptured. During the hysteresis process, energy is released due to the excitation of the chain-like tire molecules caused by deformation.

Under dry conditions, a high adhesion coefficient is obtained by making the ratios of contact area to nominal area and shear strength to pressure as high as possible. The area ratio can be increased by sliding a soft tire over a high-polished surface, where the tire can readily adapt to the microscopic surface roughness. The ratio of shear strength to pressure can be made large by cleaning all contaminants from the contact area or by reducing the pressure. When wet, smooth surfaces have very low adhesion because they trap water film, which weakens the shear strength. Higher adhesion is only obtained by increasing the surface roughness to displace the water.

A high hysteresis coefficient is obtained either by increasing the volume of the tire that is deformed or by increasing the energy dissipated by damping. A high volume of deformation requires a soft tire and angular particles that are tall relative to their width.

This discussion of the mechanics of adhesion and hysteresis, demonstrates that a measured coefficient of friction will primarily derive from hysteresis when a tire is slid over a knobby but well-polished and lubricated surface. When a tire slides on a clean and macroscopically smooth surface, the coefficient of friction will primarily derive from adhesion.

On dry surfaces, the adhesion component provides the greater contribution to total friction for speeds below 50 mph. Since the magnitude of the adhesion component depends on the contact area available to generate shearing forces, the intrusion of water into this contact area drastically reduces the adhesion component leaving only hysteresis forces to oppose sliding or skidding. Thus, removal of the water from the surface can significantly increase the skid resistance of pavements.

This discussion indicates the need to identify the test conditions under which friction measurements are made. When pavement friction is measured with a skidding tire, adhesion and hysteresis usually occur together, because few pavements are so smooth as to eliminate hysteresis or so well-lubricated as to eliminate adhesion. For these circumstances, the test conditions must be standardized to assure any consistency. When different pavement surfaces are measured for their relative friction levels, knowing exactly what the tire and operational factors are is not necessary, if the same tire, same equipment, and same test procedure are always used.

6.3 Frictional Requirements of Traffic

More rapid accelerations, higher travel speeds, and more demand for braking made possible by modern roadway and vehicle designs have continued to raise the frictional demands at the tire-pavement interface. Larger forces are required to keep the vehicle on its intended path. On the other hand, for wet pavements, the frictional capability of the tire-pavement interface decreases with increasing speed. In addition, higher traffic volumes and speeds and the increased percentage and size of large trucks promotes a faster rate of degradation in the frictional

Slippery Pavements and Hydroplaning Section

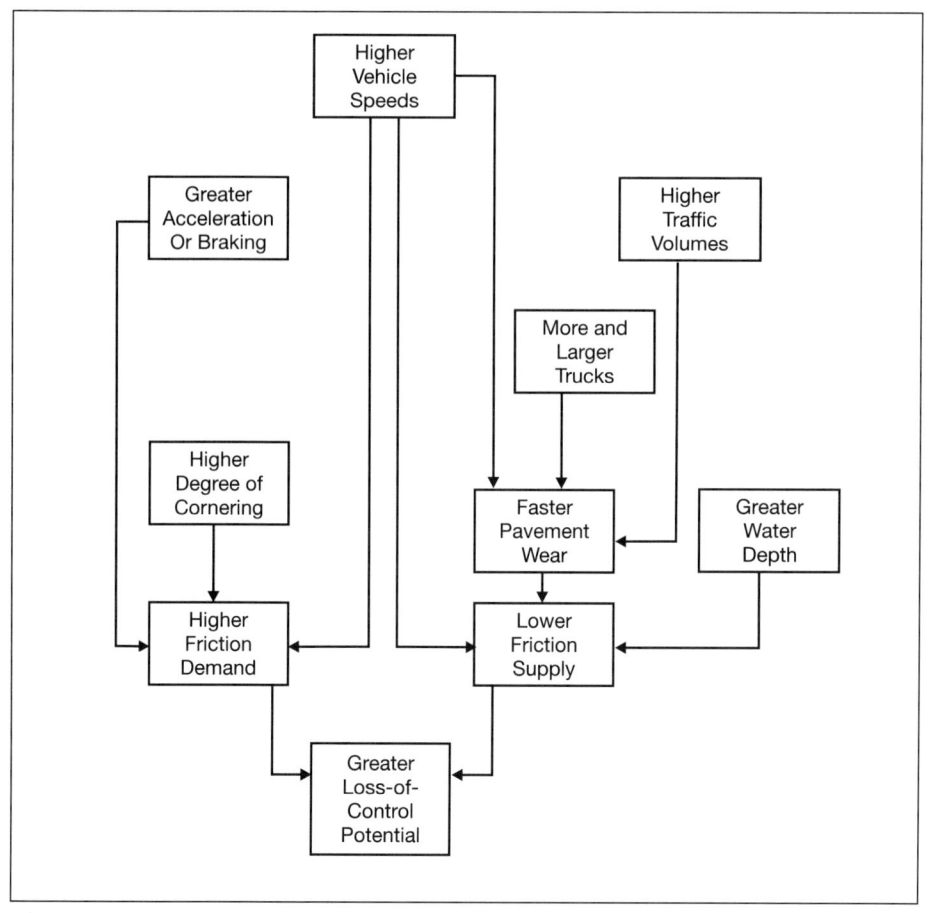

Figure 6-2 *Parameters that Produce Loss of Control on Wet Pavements.*

capability of the pavement. Figure 6-2 illustrates how these parameters interact to produce higher loss of control potential.

The tire-pavement friction level at which skidding is imminent depends mainly on the speed of the vehicle, the degree of cornering path, the magnitude of acceleration or braking, the condition of the tires, and the characteristics of the pavement surface. On wet pavements, speed is perhaps the most significant parameter not only because the frictional demand increases as the square of the speed, but also because the skid resistance decreases with increasing speed. Figure 6-3 depicts a generalization of these relationships showing, for a given degree of cornering, acceleration, or braking, how the factor of safety against skidding decreases rapidly with increasing speed, until the skid is imminent.

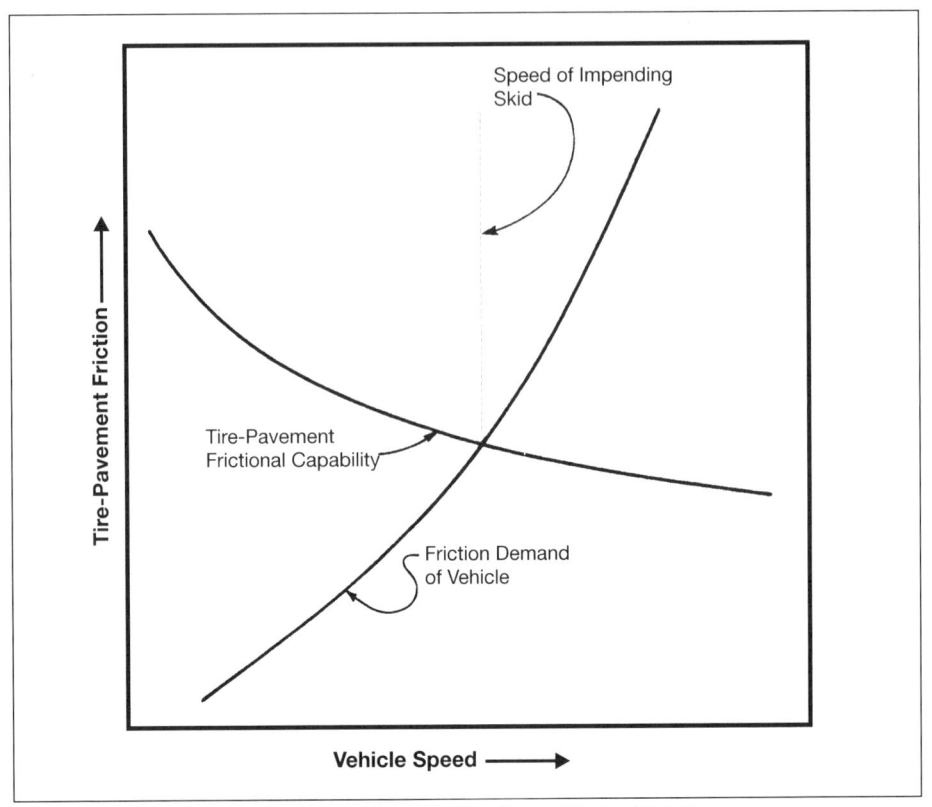

Figure 6-3 Friction Demand Related to Friction Capability for Wet Pavements.

The required level of skid resistance should be set to meet a majority of frictional demands foreseen for the operating speed of the roadway. When discussing the frictional needs of traffic, some classification is appropriate as follows:

1. **Normal Needs** - all driving, cornering, and braking maneuvers made by the majority of drivers under normal traffic operations.
2. **Intermediate Needs** - strong braking or steering corrections brought on by driver inattention, by misjudgment of a traffic event, or by marginal roadway design or traffic control elements.
3. **Emergency Needs** - Oversteering or wheel lockup brought on by reckless driving, by driver inattention, by sudden unforeseen traffic events, or by defective roadway design or traffic control elements.

Although the boundaries between these frictional levels cannot be clearly defined, all dry clean pavements probably satisfy the normal and intermediate frictional needs of traffic without leading to a skid, whereas emergency maneuvers will likely result in skidding. When wet, the average pavement surface

Slippery Pavements and Hydroplaning Section

should satisfy both the normal frictional needs of traffic and the intermediate friction needs at most speeds. For emergency maneuvers, a wet pavement cannot be expected to satisfy what dry surfaces cannot.

The distributions of driver behavioral measures can be used to determine frictional needs in passing, cornering, stopping for obstacles, stop signs, and signals, and steering to correct an errant path. Kummer and Meyer[6] suggest that a braking demand of 0.3 g's is the maximum level that driver's define as comfortable. Taragin[7] indicated that 89% of observed maneuvers on roadway curves required 0.3 g's or less in lateral acceleration. Giles[8] studied the combined frictional needs during cornering and braking and found that 4% of drivers required 0.4 g's or more.

Glennon[9] developed relationships between the frictional needs of traffic and speed for cornering, for passing, for stopping relative to sight distance restrictions, and for emergency corrections for errant paths. *Frictional requirements for roadway curves* were related to the degree of curve and superelevation rate. The frictional requirements for cornering were determined using a relationship between roadway curvature and the path radius exceeded by only 10% of drivers, developed from field studies by Glennon and Weaver.[10] A safety margin of 0.08 was included in the total friction requirement yielding the following equation:

$$f = \frac{V^2}{7.86R + 4030} + 0.08 - 0.7e$$

where

f = lateral acceleration, g's
V = vehicle speed, mph
R = roadway curve radius, feet
e = superelevation rate, ft/ft

Frictional requirements for passing maneuvers were also developed from field studies by Weaver and Glennon.[11] Using an impeding vehicle to stage passing situations, passing maneuvers were photographed from a trailing vehicle. Passing paths and speeds were determined using photogrammetric measurements. The frictional requirements for passing were determined as the vector sum of lateral and forward accelerations. Lateral acceleration requirements were for the path exceeded by only 10% of the passing vehicles. Forward acceleration for passing was estimated assuming a range of 40% of full-throttle at 40 mph to 60% of full-throttle at 80 mph. The relation between critical friction demand and speed is shown in Figure 6-4. A safety margin of 0.06 was included in the total friction requirement to allow for the uncertainty related to skid resistance

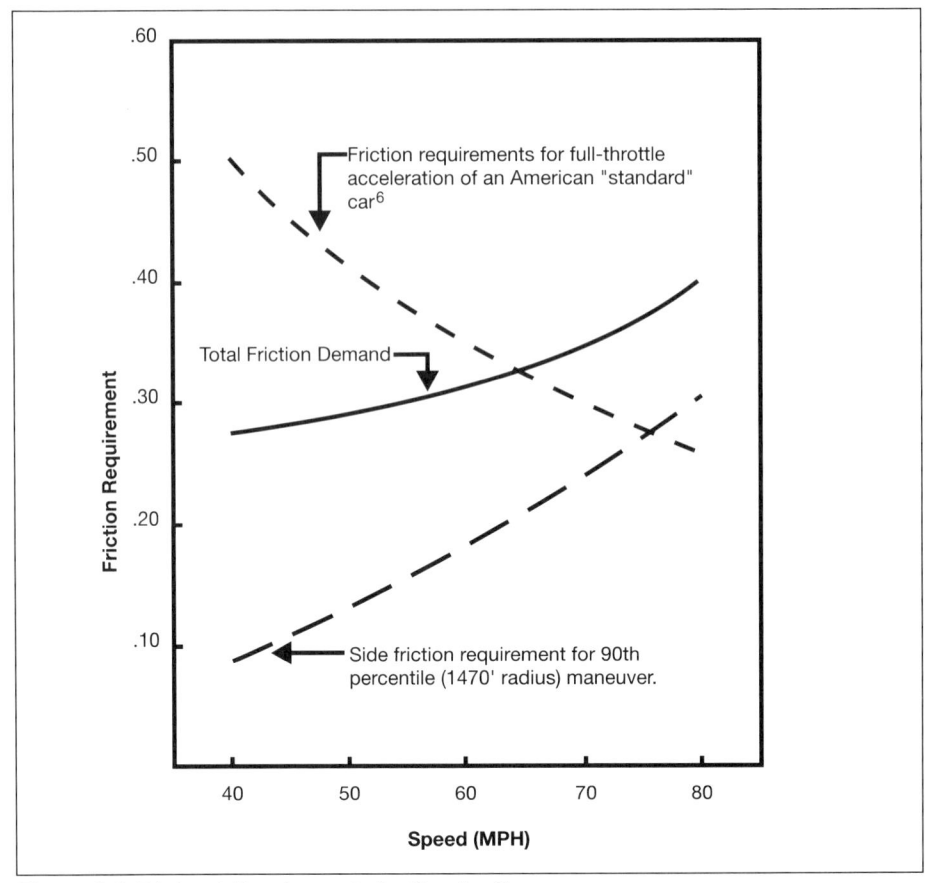

Figure 6-4 Frictional Requirements for Passing.[11]

measurements and to other unaccounted variables that may either increase the friction demand or decrease the skid resistance.

Frictional requirements for stopping were related to available stopping sight distance using the standard stopping distance method.[12] The frictional requirement presumes stopping to avoid hitting an object, animal, or other hazard on the roadway that becomes visible at the available stopping sight distance. The relation between friction demand and speed is shown in Figure 6-5 for various levels of sight distance. A safety margin of 0.08 was included to allow for the uncertainty related to skid resistance measurements and to unaccounted variables that may either increase the friction demand or decrease the skid resistance.

Frictional requirements for emergency path corrections were derived from the minimum vehicle path radius required to keep a vehicle on the roadway once the driver perceived a roadside encroachment path. These requirements are highly

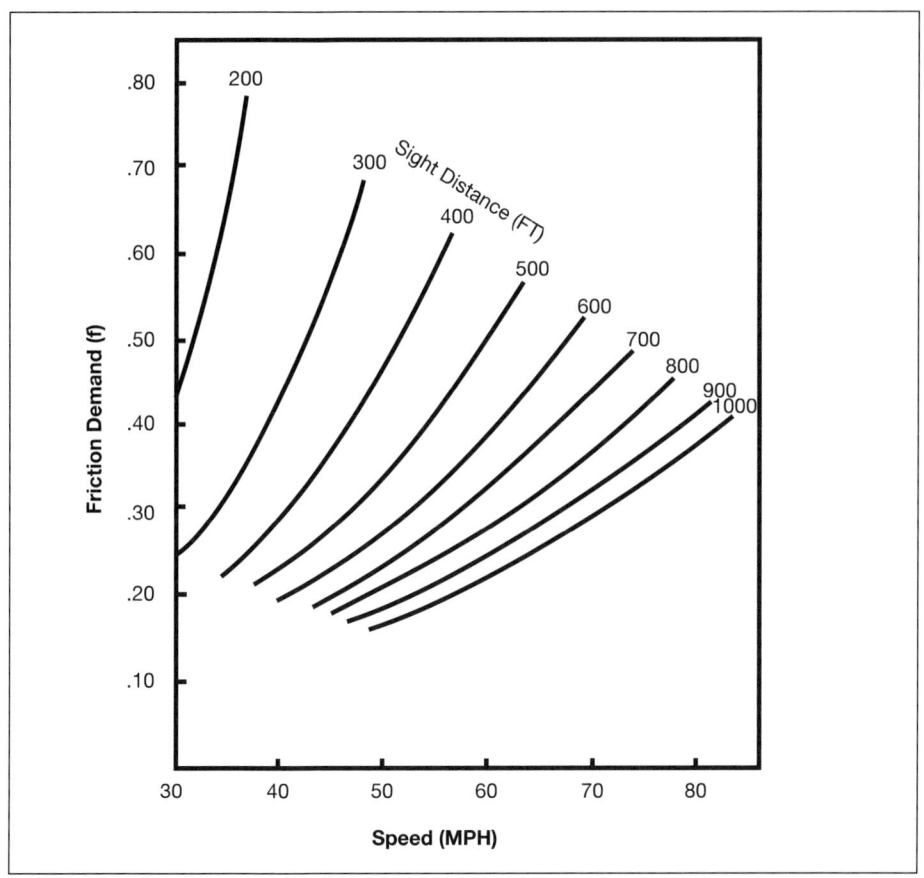

Slippery Pavements and Hydroplaning Section 169

Figure 6-5 Frictional Requirements for Emergency Stopping Related to Limited Sight Distance.[9]

sensitive to the encroachment angle and the initial distance from the edge of the paved roadway (including a paved shoulder if available). Emergency maneuvers to avoid these circumstances often exceed the skid resistance even on dry pavements. Therefore, the frictional requirements were developed for nominal values of the encroachment angle and the distance from the edge of the roadway. These frictional requirements are most critical for roadways without paved shoulders, requiring up to 0.33 g's at 55 mph and 0.80 g's at 60 mph.

6.4 Measurement of Pavement Skid Resistance

Several methods are available for measuring skid resistance. The locked-wheel trailer method (see Figure 6-6) standardized by the ASTM Method E-274[13] is the most widely accepted method in the U.S. This method ranks pavements in the same order as they would be found to be slippery by traffic, particularly if measurements are made at prevailing traffic speeds.

Figure 6-6 Locked-Wheel Skid Trailer.

By using a tire representative of those most commonly used in traffic, some engineers believe that the results of skid resistance measurements can be directly applied to the performance of vehicles in traffic. However, the variety of tires available makes this generalization tenuous. To overcome the variations associated with tires, a standardized tire becomes necessary if the relative contribution of skid resistance for different pavements is to be measured.

In the United States, ASTM Standard E-249[13] defines the standardized pavement test tire. ASTM Method E-274 describes how this tire is to be used when measuring the skid resistance of roadway surfaces by the locked-wheel trailer method. For a measurement, the trailer is towed at 40 mph over the dry pavement, and water is applied in a specified way in front of the test tire. The test wheel is then locked by braking, and after it has been sliding along the pavement long enough to permit the temperature in the contact patch to stabilize, the longitudinal force on the test tire is measured. This force multiplied by 100 and divided by the weight on that tire yields the Skid Number.

Results of a questionnaire survey[14] in the U.S. indicate that most State roadway agencies have developed a skid-resistance testing program to monitor the wet-surface friction conditions of their roadways. Some of these programs are routine and extensive, measuring most pavements. Other States perform tests only where skidding problems arise or where potential skidding problems are suspected. All agencies surveyed operate a locked-wheel trailer using the ASTM standard E-274. A majority of agencies send their trailers to national calibration centers every 1 to 2 years.

Because locked-wheel skid trailers can cost $200,000 or more and require frequent and expensive calibration, observational methods have often been recommended as an alternative way to judge skid resistance, particularly for under-funded local roadway agencies. Actually, roadway agencies have been using the observational method to judge the skid resistance of pavements at least since Joseph Barnett[15] conducted roadway curve studies in the 1930's.

Various observational methods have been recommended to evaluate skid resistance by looking at and feeling the pavement. Pavements are looked at to judge flushing (asphalt) and macrotexture. Pavements are felt to judge microtexture. A basic description of these factors and their observation is as follows:

1. **Flushing** - whether a bituminous pavement has serious flushing can easily be seen. If the asphalt extensively covers the aggregate, the pavement will have a low Skid Number and be in need of immediate rehabilitation. If the aggregate is exposed on at least 50% of the surface, the pavement may need resurfacing for other reasons but not because of flushing.
2. **Macrotexture** - the large-scale roughness or macrotexture can easily be seen from a standing position. It is usually classified by the average depth of the recesses between the peaks of the large aggregate.

One suggested[16] scale for macrotexture is given in Table 6.1. The *poor* macrotexture will tend to manifest low skid resistance at higher speeds and with greater water depths. The *fair* macrotexture will tend to have low skid resistance if combined with either moderate to high asphalt flushing and/or fair to poor microtexture as indicated in Table 6.2. The *excellent* macrotexture will tend to have good skid resistance.

Table 6.1 Suggested Observational Rating for Pavement Macrotexture

Macrotexture Measure	Judgement of Condition
less than 1/100"	Poor
2/100 – 4/100"	Fair – Good
greater than 4/100"	Excellent

3. **Microtexture** - the small-scale roughness, or microtexture, cannot be seen from the standing position. Microtexture is that feature of surfaces needed to penetrate the thin film of water left between the

tire and pavement after the bulk of water is drained by the either of the macrotexture or the tire tread. Microtexture is mostly associate with both the size of the small aggregate and the surface roughness of the larger aggregate. This feature of the texture is best evaluated by running a hand over the surface and comparing the feel with various sandpapers graded from course to fine.

Table 6.2 Suggested Observational Rating for Pavement Microtexture

Microtexture Description	Judgment of Condition	Author's Note
Smooth surface	Poor	Fine sandpaper or smoother
Fractured surface	Moderate	Medium sandpaper
Grainy surface	Excellent	Course sandpaper

Table 6.2 shows a suggested[16] scale for microtexture. The *poor* microtexture will tend to manifest low skid resistance at most speeds and water depths, when the macrotexture is from poor to fair. For *moderate* microtexture, the skid resistance will be mostly determined by the macrotexture or degree of flushing. Pavements with *excellent* microtexture tend to have good skid resistance both at lower speeds and also at higher speeds unless the macrotexture is almost non-existent or the asphalt flushing is high.

6.5 Accidents on Wet Pavements

In 1980, The National Transportation Safety Board (NTSB) used the National Highway Traffic Safety Administration's (NHTSA) Fatal Accident Reporting System to analyze fatal accidents on wet pavements.[17] The wet-pavement exposure was determined by using precipitation data from the National Weather Service. The study then developed a wet fatal accident index (WFAI), using the fatal accident data and the precipitation data to reflect the relative severity of the problem in each State. This index was represented by the following equation:

$$WFAI = \frac{\% \text{ of fatal accident on wet pavement}}{\% \text{ of wet time}}$$

From this analysis, the NTSB concluded that wet-pavement accidents occur 3.9 to 4.5 times more frequent than expected if wet pavements had no effect.

Blackburn, et. al.,[18] conducted an extensive before and after study to determine the relationship between wet-pavement accident rate and skid number for various roadway types. Figure 6-7 shows these relationships for urban streets and highways. Figures 6-8 shows these relationships for rural highways. A relationship between dry- and wet-pavement accidents rates was also developed as shown in Figure 6-9.

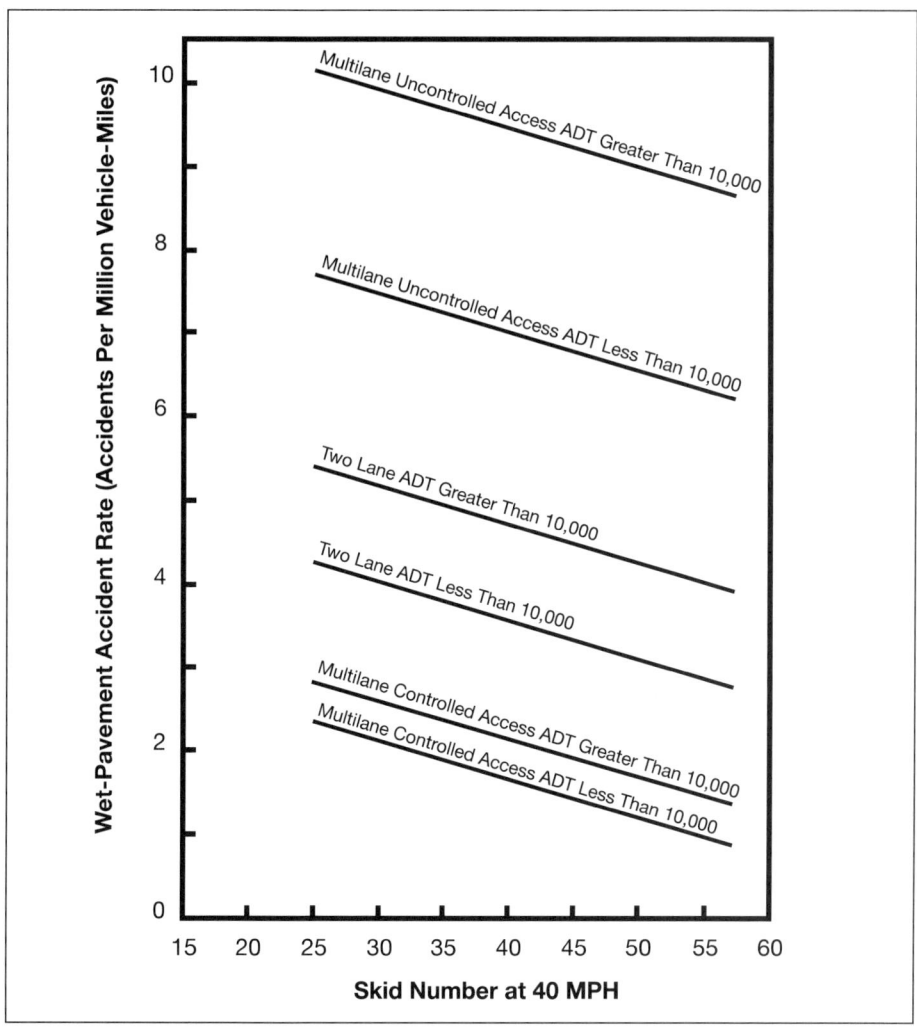

Slippery Pavements and Hydroplaning Section 173

Figure 6-7 Relationships Between Wet-Pavement Accident Rates and Skid Number 40 for Urban Roadways.

6.6 Standards for Skid Resistance

The most widely recognized and quoted recommendations for pavement skid resistance requirements come for a 1967 research report titled *Tentative Skid-Resistance Requirements for Main Rural Highways,* by Kummer and Meyer.[6] This report recommends the minimum Skid Numbers shown in Table 6.3 for main rural highways. These Skid Numbers (defined as the coefficient of friction times 100) are those measured by a skid trailer in accordance with ASTM Method E-274[13] at 40 mph in the most polished track of the roadway during summer or fall.

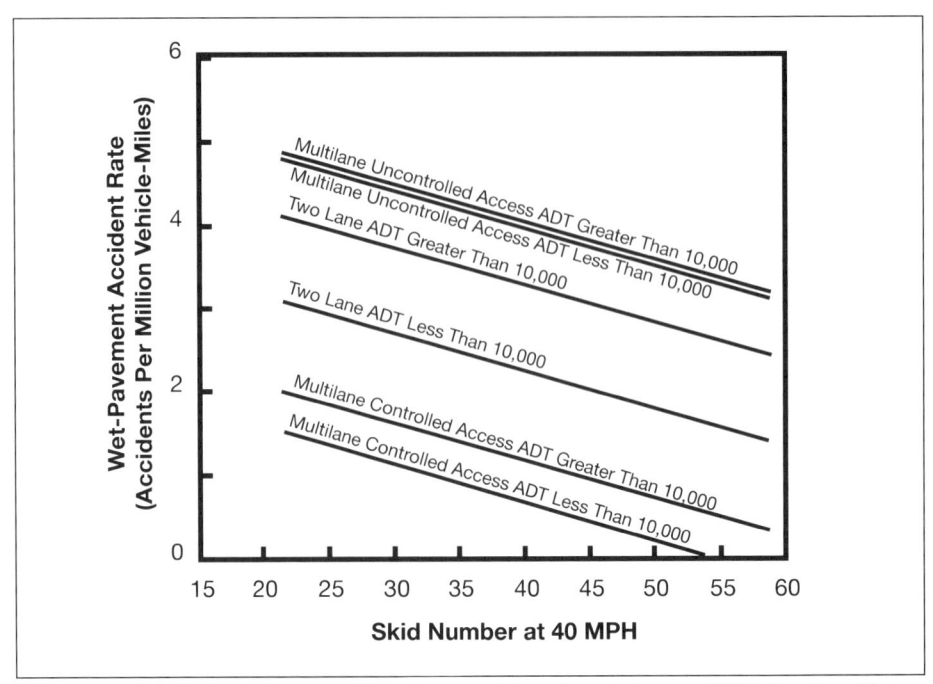

Figure 6-8 Relationship Between Wet Pavement Accident Rates and Skid Number 40 for Rural Roadways.[18]

The higher Skid Numbers for higher speed roadways are recommended to compensate for the degradation of available friction between tires and wet pavement as speeds increase. The recommended requirements are based on the normal friction demands of traffic derived from driver behavior studies.

Results of a questionnaire survey[14] in the U.S. revealed that a wide range of Skid Number criteria are used to evaluate the results of skid tests using the ASTM Standard. Some roadway agencies have no established criteria. Of the agencies that have a criterion, the most common value seems to be the Skid Number of 37 recommended by Kummer and Meyer.[6] Other agencies use Skid Numbers from 30 to 40. A general conclusion drawn from this survey is that most roadway agencies that use a Skid Number criterion either list roadways for future pavement surface rehabilitation or initiate pavement rehabilitation when the Skid Number drops to between 29 and 36. Clearly, a Skid Number of 35 to 37 is a desirable minimum. Also clear is that a Skid Number below 30 is considered deficient. Decisions made to either initiate pavement surface improvement or defer those improvements are usually based on either perceived or real economic considerations.

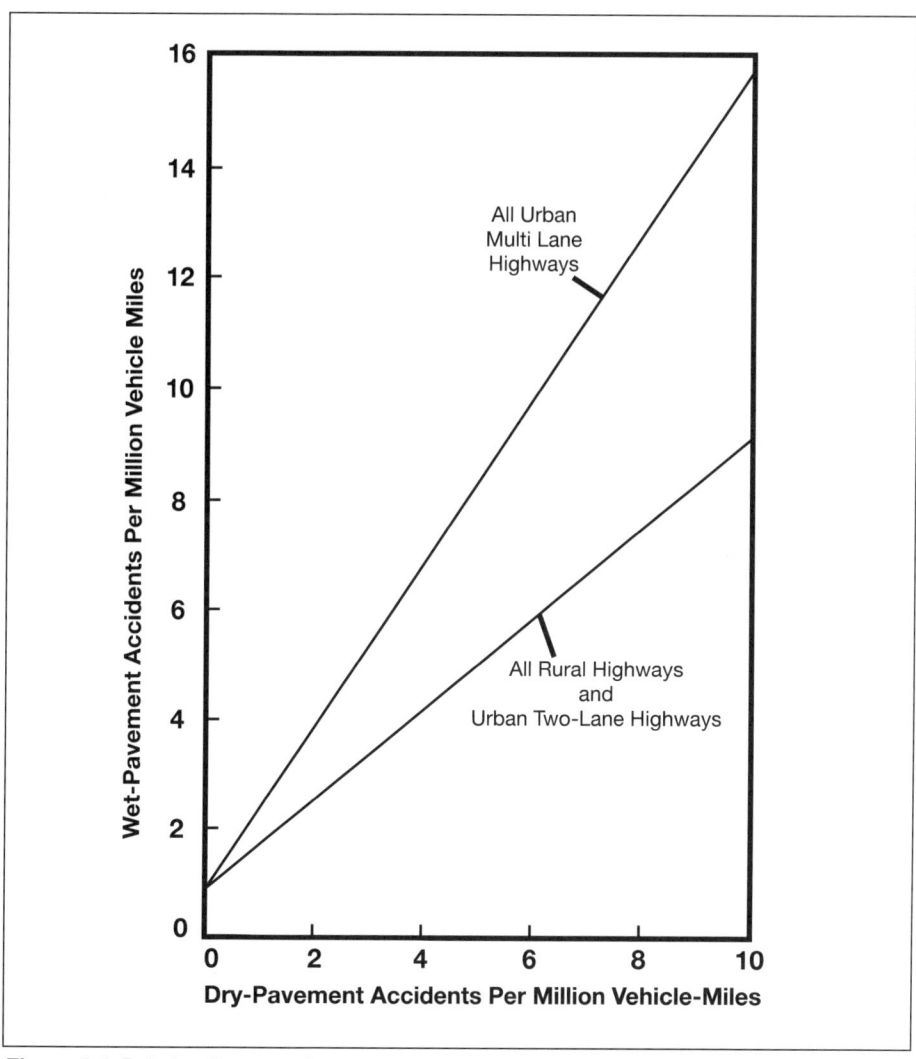

Figure 6-9 Relation Between Dry Pavement and Wet-Pavement Accident Rates.[18]

Table 6.3 Skid Resistance Requirements for Main Rural Highways[6]

Mean Traffic Speed (mph)	Skid Number (at 40 mph)
30	31
40	33
50	37
60	41
70	46
80	51

Figure 6-10 shows a percentile distribution of Skid Numbers at various speeds computed from a random sample of 500 pavements in one State.[9] These measurements were taken in 1964 employing a modified version of the ASTM trailer. Figure 6-11 shows a similar plot of 1962 data for Germany.[19] If these inventories could be regarded as typical, only about 60 - 65% of all pavement would satisfy a Skid Number criterion of 37, and about 10 - 20% of all pavements would not satisfy a Skid Number criterion of 30.

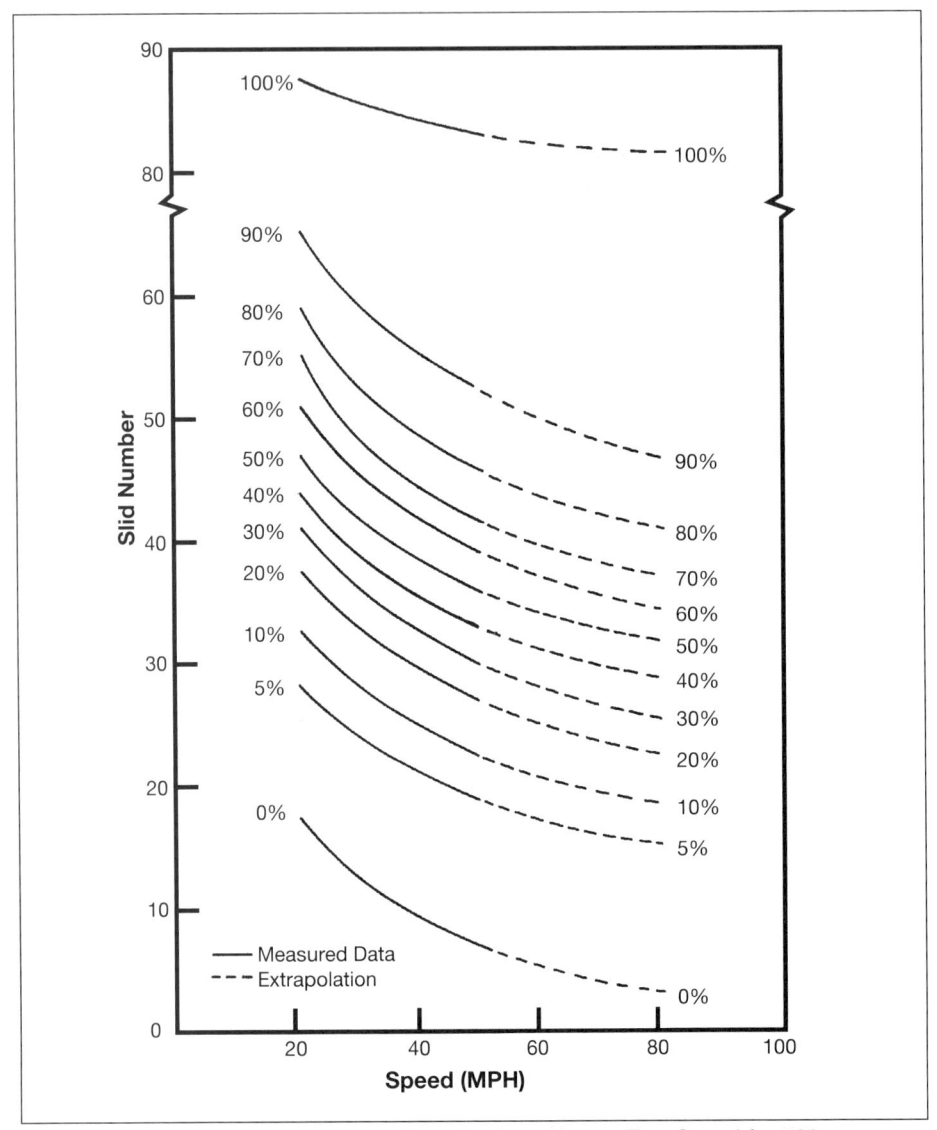

Figure 6-10 Percentile Distribution of Skid Numbers Versus Test Speed for 500 Pavements in One State.[9]

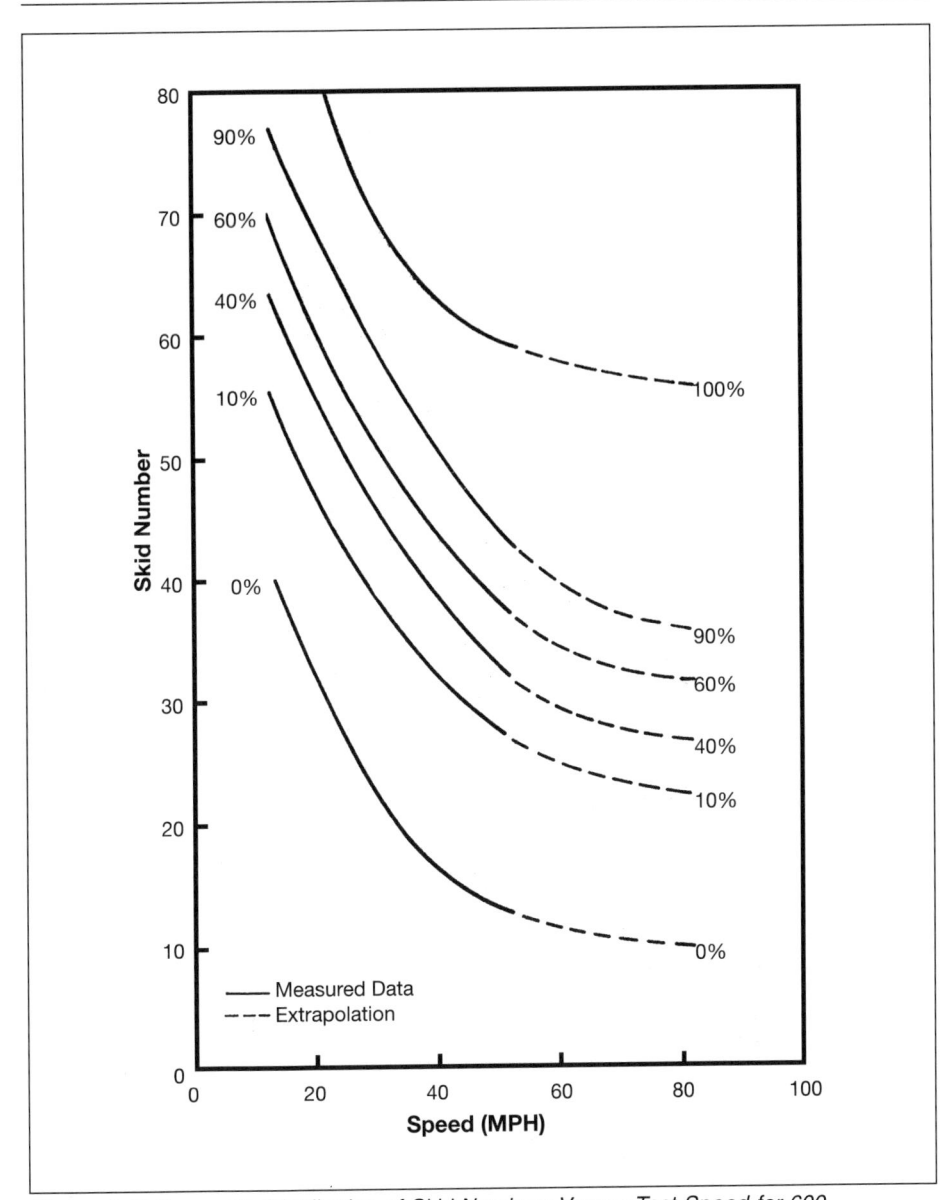

Slippery Pavements and Hydroplaning Section 177

Figure 6-11 Percentile Distribution of Skid Numbers Versus Test Speed for 600 Pavements in Germany.[19]

> **HYDROPLANING**
> **Rule of Thumb**
> *Hydroplaning Can be Expected at Speeds Above 45 MPH Where Water Ponds to Depths of 1/10 Inch or More Over a Roadway Length of 30 Feet or More.*

6.7 Hydroplaning

Hydroplaning is a phenomenon characterized by complete loss of directional control when a tire is moving fast enough that it rides up on a film of water and thereby loses contact with the pavement. Although several vehicle, roadway, and environmental factors affect the probability of hydroplaning, a general rule of thumb for rural highways is that hydroplaning can be expected for speeds above 45 mph where water ponds to a depth of one-tenth inch or greater over a distance of 30 feet or greater.

Hydroplaning was first recognized as a problem with aircraft landings, and became more apparent as landing speeds increased with the advent of jet engines. Likewise, roadway collisions involving hydroplaning became more apparent as roadway speeds increased, wider pavements were built, flexible pavements became more widespread, and greater pavement wear occurred because of greater traffic and heavier loads. Although hydroplaning is a very complex phenomenon, its likelihood is known to be associated with several vehicle, roadway, and environmental factors. The likelihood of hydroplaning on wet pavements increases with:

- Depth of Compacted Wheel Tracks
- Deterioration of Pavement Microtexture
- Deterioration of Pavement Macrotexture
- Decrease of Pavement Cross-slope
- Vehicle Speed
- Tire Tread Wear
- Ratio of Tire Load to Inflation Pressure
- Rainfall Intensity
- Rainfall Duration

In addition to these factors, the likelihood of vehicular loss of control from hydroplaning is increased by the degree of cornering, the amount of acceleration or deceleration, the type of vehicle (combination vehicles are more sensitive), and the magnitude of crosswinds.

A roadway agency concerned with minimizing the occurrence of hydroplaning can only exert minimal control over vehicle, driver, and environmental factors. It can, however, identify and correct roadway locations that demonstrate a susceptibility to hydroplaning. Although several kinds of roadway situations produce the opportunity for hydroplaning, the most prevalent places are where water is allowed to pond such as at the bottom of sag vertical curves and on pavements with noticeably compacted wheel tracks or ruts. These locations are usually corrected by asphalt mix overlays designed and constructed to increase the cross slope, eliminate rutting, improve the microtexture and macrotexture of the surface, and provide a surface resistant to wear and compaction. Where funding for resurfacing is not available over the short-term, roadway locations prone to hydroplaning can be treated with reduced speed limits and slippery pavement warning signs as low-cost interim countermeasures.

Mechanics of Hydroplaning

Horne[20] divided hydroplaning into three categories:

1. Viscous Hydroplaning
2. Dynamic Hydroplaning
3. Tire Tread Rubber Reversion Hydroplaning

Viscous and dynamic hydroplaning are important considerations for normal passenger car and truck operations on roadways. Tire tread rubber reversion hydroplaning only occurs when heavy vehicles such as large trucks or airplanes lock their wheels at high speeds on wet pavements with good macrotexture but little microtexture.

Viscous Hydroplaning only occurs on wet pavements with little or no microtexture.[21] With viscous hydroplaning, a thin film of water continues to separate the tire and pavement because of insufficient microtexture to break down the water film. Viscous hydroplaning may occur at any speed.

Dynamic Hydroplaning (see Figure 6-12) occurs when the tire is completely separated from contact with the pavement by a layer of water of 0.1-inch thickness or more. As a tire moves over a wet pavement surface, the bulk of the water is normally removed by the normal force of the tire squeezing the water from beneath the tire footprint through the grooves in the tire tread and the asperities (macrotexture) in the pavement. When significant water is allowed to accumulate on high-speed roadways with inadequate pavement texture, the tire (particular

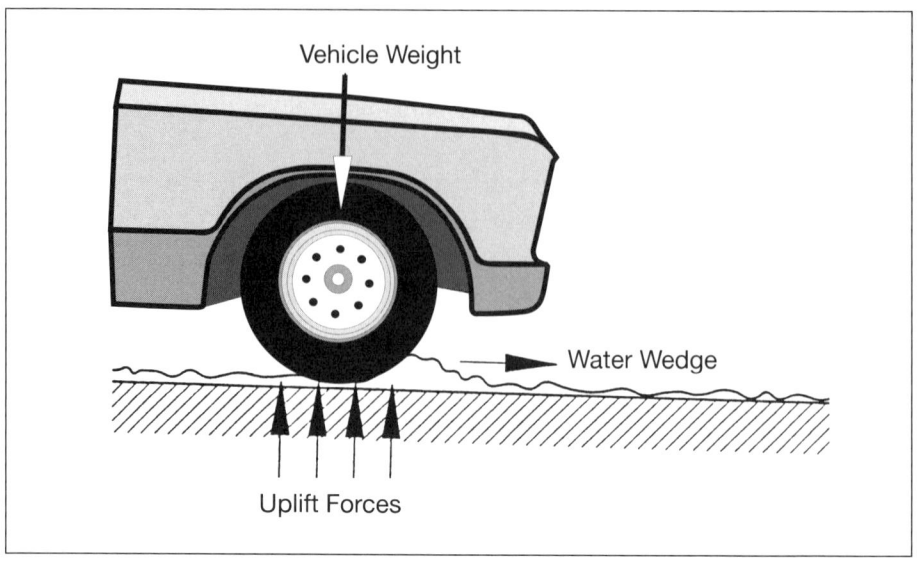

Figure 6-12 Dynamic Hydroplaning.

when worn) cannot evacuate enough water and a wedge forms in front of the tire, which rides up on the water much like water skiing. When dynamic hydroplaning occurs to the front tires, the driver will experience loss of steering and, unless the hydroplaning ceases quickly, a complete loss of control.

The depth of water required for full dynamic hydroplaning varies mainly with vehicle speed, tire tread design, tire tread depth, and pavement texture. Partial dynamic hydroplaning can occur on wet pavements at lower speeds with inadequate pavement texture and/or tire tread depth.

In the 1960's and 1970's, several empirical and analytical investigations were conducted on the subject of hydroplaning. While the empirical work[22-28] generally has been helpful in pointing out many of the key factors to hydroplaning, and has suggested some worthwhile solutions such as pavement grooving, the analytical efforts[29-35] have been less successful. Several equations have been developed that relate hydroplaning speed to factors such as tire inflation pressure and water depth. Although these equations describe the best-fit relationships from empirical data, they have not been derived based on a fundamental understanding of the physics involved.

Pavement Texture

The pavement texture characteristic most critical to hydroplaning is macrotexture, the extent of large-scale protrusions from the surface of the pavement. Macrotexture acts in combination with tire tread grooves to provide the escape channels for bulk water drainage from beneath the tire footprint.

Balmer and Galloway[36] conducted an extensive investigation on the ability of various pavement textures to reduce the risk of hydroplaning. They found that increasing the macrotexture depth from 0.03 inches to 0.15 inches raised the speed at which dynamic hydroplaning would occur by 10 mph for a tire inflation of 30 psi, a tread depth of 8/32 inches, and a water depth of 0.3 inches. They recommended a minimum texture depth of 0.06 inches for roadways with higher speed limits.

Pavement Drainage

The undisturbed depth of water in the tire path is important in determining the likelihood of hydroplaning for a given tire-pavement combination. The greater the depth of water, the greater will be the fluid inertial forces acting on the tire, and the greater will be the probability that the combined drainage capacities of the tire-tread grooves and the pavement macrotexture will be exceeded. A study[28] of the drainage design of roadway surfaces demonstrated that the frequency of dynamic hydroplaning can be radically reduced by designing surfaces with drainage characteristics that minimize water depth.

Rain water forms a layer of increasing thickness as it flows to the edge of a sloped pavement surface. This water can be a hazard to motorists because it reduces the friction between the tires and wet surface. The need for cross slope to drain rain water quickly from pavements is most important on multi-lane roadways where greater water depths can be expected on the lower lanes. Providing adequate macrotexture to allow the remaining water to escape under the tire is also important to achieve necessary tire-pavement adhesion, particularly on high-speed roadways.

The effects of various cross slope and roadway grade combinations on drainage lengths are given in Table 6.4 for a pavement width of 24 feet.[37] This table shows that steep roadway grades have substantially greater drainage lengths than flat grades, particularly for pavements with flatter cross slopes. Steeper roadway grades not only increase the flow-path length but also increase the water depth along the lower side of the road, relative to a zero roadway grade. The problem is further complicated by the presence of horizontal curves, vertical curves, intersections, and their combinations.

A major benefit of steep cross slopes is the reduction in the volume of water that can pond in pavement deformations, particularly where wheel rutting is prominent. Most pavements will drain surface water rapidly and reasonably complete if cross slopes are steep enough. But the cross slope should not be steeper than that accepted for driveability. On a straight roadway, driving on a steep cross slope presents some hazard because of the vehicle's tendency to pull toward the

Table 6.4 Effect of Cross Slope and Roadway Grade on Drainage Length[37]

Cross Slope (%)	Roadway Grade (%)	Drainage Length (ft.)	Increase in Drainage Length Compared to Zero Grade (%)
1	0	24	0
	1	33	38
	2.5	62	158
	5	118	392
	10	232	867
3	0	24	0
	1	25	4
	2.5	31	29
	5	45	88
	10	80	233

low edge of the pavement. To balance the need for drainage with the need to maximize driveability, AASHTO[38] recommends the cross slope rates given in Table 6.5.

Table 6.5 Recommended AASHTO Standards for Pavement Cross Slopes

Surface Type	Range in Cross Slope Rate (%)
High	1.5 – 2.0
Intermediate	1.5 – 3.0
Low	2.0 – 6.0

Pavement Wheel Ruts

Hydroplaning as a result of compacted wheel ruts is commonly recognized by roadway engineers. The 1990 publication, *A Policy on Geometric Design of Highways and Streets*,[38] by the American Association of State Highway and Transportation Officials (AASHTO) delineates the concern for pavement slipperiness by saying (p. 333):

> *The four main causes of poor skid resistance on wet pavements are rutting, polishing, bleeding, and dirty pavements. Rutting causes water accumulation in the wheel tracks. Polishing reduces, and bleeding covers, the microtexture. In both cases, the harsh surface features for penetrating the thin water film are diminished. Dirty pavements, pavements contaminated by oil drippings or layers of dust or organic matter, will lose their skid resistance.*

Although AASHTO does not give a maximum wheel rut depth criterion, States have adopted various values, usually about one-half inch.

Actually, the formation of compacted wheel tracks not only creates a depression for water to collect, but also tends to bring more asphalt binder to the surface. This bleeding (flushing) condition invariably decreases the macrotexture of the surface which is detrimental to good skid resistance at higher speeds. Compaction and wear also reduce the microtexture, and thus is detrimental to good skid resistance at all speeds.

In summary, the compaction of flexible pavements in the wheel tracks is an insidious roadway problem that obviously requires careful monitoring of existing pavements and the application of properly designed and constructed overlays or micro-surfacing to minimize rutting, flushing, and wearing of texture.

Vehicle Tires

The vehicle tire can be a critical factor influencing hydroplaning. Even on a reasonably maintained pavement surface, well-worn or under-inflated tires can increase the risk of hydroplaning considerably. Another factor affecting hydroplaning is tire tread pattern.

Four characteristics relate to the effectiveness of the tread grooves to minimize hydroplaning: tread depth; groove capacity; groove shape; and groove spacing. The amount of surface water evacuated is referred to as the tire's groove capacity. When the amount of pavement surface water exceeds the groove capacity, the excess water needs time to be displaced without building up in front of the tire and creating hydrodynamic uplift on the tire. Higher speeds reduce the time of displacement and increase the risk of hydroplaning.

Another tire factor related to hydroplaning is the *tire footprint*, or the area of contact between the tire and the pavement.[39-41] As the width of the footprint increases, the tire running on a wet pavement will interact with greater amounts of water, which in turn will increase the hydrodynamic forces. An increase in the length of the tire footprint has the opposite effect, because of the greater area of dry contact.

The important characteristic of the tire footprint, therefore, is the *aspect ratio*, which is calculated by dividing the width of the tire footprint by its length, as shown in Figure 6-13. Higher aspect ratios indicate a tendency toward hydroplaning. The aspect ratio is particularly important in looking a tractor-trailer truck hydroplaning. Aspect ratios for large trucks are affected by their load. For an empty truck, the aspect ratio is considerably higher than for a loaded truck. Accident statistics confirm that jackknifing of empty tractor-trailer trucks on wet pavements is often related to dynamic hydroplaning.[42]

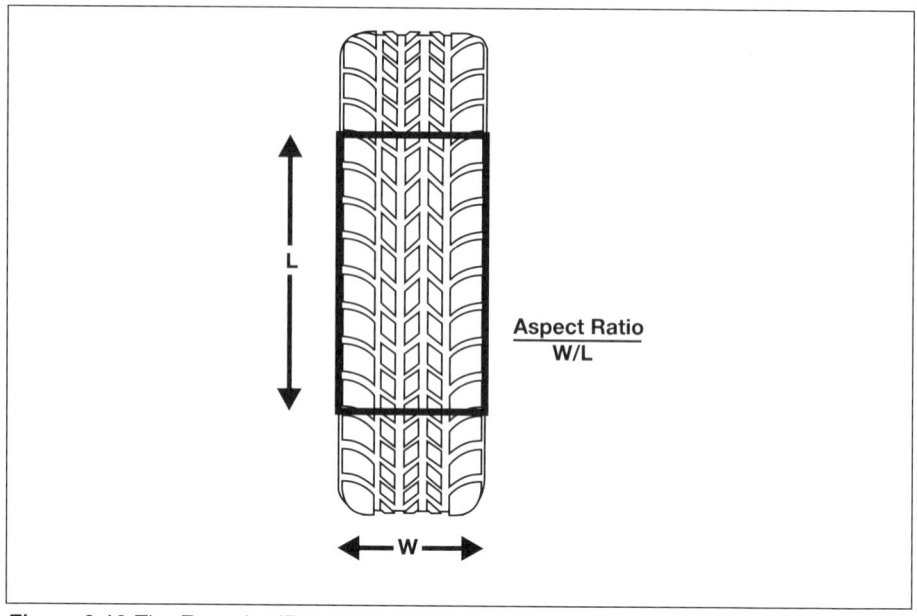

Figure 6-13 Tire Footprint (Pavement View).

For modern tires, inflation pressure and tread depth are the most significant factors related to hydroplaning. Inflation pressures that are 15 psi or more below recommended values for a particular tire cause the tire to deflect inward allowing any water wedge to penetrate deeper into the tire footprint. Tread depth is primarily a factor of how much wear a tire has experienced. When a tire is worn to an extent where the tread depth is less than the 2/32-inch minimum specified by most States, the risk of hydroplaning on wet high-speed roadways is relatively high.

6.8 Technical Aspects of Slippery Pavement or Hydroplaning Cases

Of 66 roadway agencies responding to a 1990 questionnaire,[14] thirty indicated that they had litigation as a result of wet-weather accidents. Only 16 agencies believed that litigation was a significant problem. Some of the comments were:

a. *100 claims have been related to friction number and/or wet weather*
b. *20 to 25 claims have been related to snow, ice, pooled water, frost*
c. *Three cases in 1986.*
d. *Two cases settled since 1981 at $750,000, with others pending.*
e. *Several favorable defense verdicts and 2 or 3 claims settled favorably.*
f. *Any litigation is a problem.*
g. *No settlements yet.*

Slippery Pavements and Hydroplaning Section

The slippery pavement is not a common aspect of tort claims. Proving both that a pavement had a low coefficient of friction and that condition caused a collision is usually difficult. Many investigators use a drag sled or tire to estimate friction. These hand-pulled devices largely measure static friction and are inaccurate measures of dynamic friction particularly at higher speeds. Stopping distance tests can be undertaken with exemplary vehicles and tires but, for higher speeds on wet pavements, these tests are extremely dangerous, particularly if traffic is not barricaded from the test area.

Most States conduct a inventory of pavement skid resistance. If available, these data will identify whether a pavement is slippery. Usually, Skid Numbers below 35 or coefficients of friction below 0.35 indicate a slippery pavement when speed limits are 50 mph or greater. Federal rules of evidence, (23 USC 409), however, which have been adopted in many States, preclude the discovery of collected data whose ultimate purpose is to make roadways safer. What is left in the slippery pavement case is the observational method of identifying slippery pavements. The observance of poor microtexture and/or poor macrotexture is a strong indication of a slippery pavement.

One of the more common indicators where slippery pavements are alleged to cause accidents is when the driver indicates that a comfortable to moderate steering or braking maneuver created his/her loss of control. Statements like "I was just slowing for traffic ahead when the car started sliding" or "I tapped the brakes as I was rounding the curve and my truck started swinging out" are common statements. Trucks are particularly vulnerable to jackknifing or trailer swing out when making minor corrective maneuvers on very slippery pavements.

Allegations about hydroplaning are more common than allegations about slippery pavements, per se. A plaintiff attempting to recover for injuries sustained from a motor-vehicle collision where pavement and/or other roadway conditions are alleged to have caused hydroplaning needs to establish an evidence base to support that claim. Most important is a complete survey of elevations, so that a small-interval contour map can be drawn of the pavement and/or immediate shoulder area. This contour map can then be used to compute both the watershed area draining to the accident area and the maximum depth of ponding that can occur. Other evidence that would be important is the testimony of neighbors and other frequent users of the roadway, who will say that ponding is a frequent occurrence at this location. Photographs of the ponding or videos tapes showing vehicles splashing water and/or tires lifting up are also useful.

The hydroplaning case usually comes about because a driver or passenger from a colliding vehicle says something like "I hit the water and had no steering" or "We hit some water and lost control." Dynamic hydroplaning results from hydrodynamic lift on the front tires causing the loss of steering authority. The roadway section that causes hydroplaning usually has a combination of one of the following factors:

1. *Inadequate Pavement Cross Slope* - cross slope is the tilt of the pavement to the side, usually down to the right side, but on freeway lanes sometimes down to the left. Also, for roadway curves to the left, the cross slope (or superelevation) tilts down to the left. Normal crown cross slopes are in the 1.5 - 2.5% range. This means that the pavement drops 0.15 - 0.25 feet vertically for every 10 feet laterally. Cross slopes less than 1.5% combined with steeper grades and other roadway features can contribute to hydroplaning.

2. *Sag Vertical Curve* - a sag vertical curve is a low spot designed into the vertical alignment of a roadway. It is the opposite of a crest vertical curve (or hillcrest). Sag vertical curves become potential hydroplaning sections when combined with steep downgrades and deficient cross slopes.

3. *Build-up of Turf Shoulders* - turf shoulders have the tendency over time to produce thatch and to collect dirt and debris, which all become part of a turf that builds up higher than the traveled pavement. When this happens, the pavement can pond water against the raised shoulder. This condition becomes more critical when combined with a deficient cross slope.

4. *Sheet-Flow Conditions* - Some sections of roadway can cause water to flow in sheets across the pavement during rainy weather. One way this happens is when culverts or ditches are clogged, and the drainage has nowhere to go except across the pavement. The most prominent sheet-flow condition combines some form of poor drainage with a relatively steep downgrade leading into the high side of a roadway curve. What happens is rain water is collected along the side of the traveled way, drains downgrade along that edge to a point where the cross slope changes from negative to positive at the beginning of the roadway curve. At this point, all of the collected water is dumped across the pavement to the other side.

5. *Rutted Pavements* - heavy volumes of traffic and/or frequent heavy trucks will both wear and compress the pavement in the normal wheel tracks. These wheel ruts will pond water during rainy weather particular where grades are flat, where cross slope is deficient, and/or at the bottom of sag vertical curves. When wheel ruts will hold 0.10 inch or more of water on high-speed roadways, hydroplaning can occur.

6.9 Accident Reconstruction Aspects

Two major questions are normally addressed with accident reconstruction analysis where hydroplaning or a slippery pavement is alleged. The first of these questions attempts to establish whether partial or full dynamic hydroplaning did in fact occur. The second question asks whether excessive speed was a major factor in causing hydroplaning.

For plaintiff arguments, eye-witness statements often are the sole basis of proving hydroplaning. Testimony of substantial water on the pavement and/or sudden loss of control usually underpin most tort claims. Another source of proof may be an accident reconstruction. Whatever collision occurred after the loss of control, whether it be with another vehicle, a fixed object off of the roadway, or a subsequent rollover, an adequate factual basis should allow computation of the impact or rollover speed of the skidding or hydroplaning vehicle. If, in addition, eye-witness testimony can establish an initial speed and position of the vehicle before the skidding or hydroplaning, then the effective coefficient of friction manifest between initial loss of control and impact or rollover can be calculated as follows:

$$f = \frac{V_0^2 - V_1^2}{30d} - g$$

where

f = coefficient of friction
V_o = initial speed of vehicle, mph
V_I = impact or rollover speed of vehicle, mph
d = distance between initial loss of control and impact, feet.
g = algebraic grade, rise or fall divided by path distance, feet/feet

For roadway speeds of 45 mph or greater, a calculated coefficient of friction less than 0.2 usually indicates full dynamic hydroplaning. If the calculated coefficient of friction is between 0.20 and 0.35, partial hydroplaning or a slippery pavement is probable.

For defense positions, proof of excessive speed can be a valuable argument. Again eye-witness testimony could be important to establish this argument. In addition, accident reconstruction could be useful. If, for example, the impact speed after the initial skid is found to be higher than the speed limit, an excessive speed argument should be useful, knowing also that the initial loss of control speed must be higher than the impact speed. If the initial skid distance is known, then a conservative estimate of the initial loss of control speed can be made by inserting a conservative estimate of the coefficient of friction into the following rearrangement of the previous equation:

$$V_o = \sqrt{V_1^2 + (30)(d)(f+g)}$$

6.10 Typical Defense Arguments

A basic question is what minimum level of skid resistance should a roadway agency maintain to be free of liability. State laws are clear that roadway agencies have the obligation to provide a roadway that is reasonably safe to the motoring public. For this purpose, the levels recommended by *NCHRP Report 37*[6] have received wide recognition and serve as a legal defense when the subject pavement can be shown to exceed these levels.

Another defense relates to the economic feasibility of treating all roadways in a skid resistance program when not only are funds insufficient to treat all roadway sections that need resurfacing but also many other competing needs are demanding the same available funds. This defense, however, is normally inadequate unless the roadway agency can show that the roadway defect in question has been programmed for improvement at a later date under a priority scheme that treats the greatest need first.

Other case-specific defenses argue that either the driver was the sole negligent party and/or the specific roadway section had no contribution to the causation of the collision. Some of these arguments are:

1. The driver should have slowed below the speed limit for wet pavement conditions.
2. Once the driver began to slide, he/she did not take the proper actions to regain control of the vehicle.
3. The driver was negligent for speeding.
4. The driver was negligent for driving while intoxicated.
5. The vehicle's bald tires were a major cause of the accident.
6. The vehicle's low tire pressure was a major cause of the accident.
7. The roadway agency had no notice of the slippery roadway section.
8. There was not enough water present to create full dynamic hydroplaning.
9. The rut depth did not exceed the maintenance criterion for repaving.
10. The roadway agency has many miles of roadway in need of repaving and funds are not adequate.
11. The subject section of roadway has been selected to be resurfaced on a priority programming schedule.

12. The roadway agency had no notice of the defective roadway surface.
13. Wet weather accident records did not show a skidding problem.
14. The roadway agency had no standard for skid resistance.

Endnotes

1. Moyer, R.A., *Skidding Characteristics of Automobile Tires on Roadway Surfaces and Their Relation to Highway Safety*, Bulletin 120, Iowa Engineering Experiment Station, Ames, Iowa, 1934.

2. Shelburne, T.E., and Sheppe, R.L., *Skid Resistance Measurements of Virginia Pavements*, Highway Research Board, Report 5B, 1948.

3. Virginia Council of Highway Investigation and Research, *Proceedings of the First International Skid Prevention Conference*, 1959.

4. Smith, L.L., and Fuller, S.L., *Florida Skid Correlation Study of 1967: Skid Testing with Trailers*, American Society for Testing and Materials, Special Technical Publication 456, 1969.

5. Transportation Research Board, *Proceedings of the Second International Skid Prevention Conference*, published in Transportation Research Records 621, 622, 623, and 624, 1976.

6. Kummer, H.W., and Meyer, W.E., *Tentative Skid Resistance Requirements for Main Rural Highways*, Highway Research Board, NCHRP Report 37, 1967.

7. Taragin, A., *Driver Performance on Horizontal Curves*, Highway Research Board Proceedings, Volume 33, 1954.

8. Giles, C.G., *Some Recent Developments in Work on Skidding Problems at the Road Research Laboratory*, Highway Research Record 46, 1964.

9. Glennon, John C., *A Determination Framework for Wet Weather Speed Limits*, Texas Transportation Institute, Research Report 134-8F, 1971.

10. Glennon, John C., and Weaver, Graeme D., *The Relationship of Vehicle Paths to Highway Curve Design*, Texas Transportation Institute, Research Report 134-5, 1971.

11. Weaver, Graeme D., and Glennon, John C., *Frictional Requirements for High-Speed Passing Maneuvers*, Texas Transportation Institute, Research Report 134-7, 1971.

12. Glennon, John C., *Evaluation of Stopping Sight Distance Design Criteria*, Texas Transportation Institute, Research Report 134-3, 1969.

13. American Society for Testing and Materials, *Skid Resistance of Paved Surface Using a Full Scale Tire*, ASTM Method E-274, Book of ASTM Standards, 1986.

14. Dahir, Sabir H.M., and Gramling, Wade L., *Wet-Pavement Safety Programs*, Transportation Research Board, NCHRP Synthesis 158, 1990.

15. Barnett, J., *Safe Side Friction Factors and Superelevation Design*, Highway Research Board Proceedings, 1936.

16. Ivey, D.L., et. al., *Texas Skid Initiated Accident Reduction Program*, Texas Transportation Institute, Research Report 910-1F, 1992.

17. National Transportation Safety Board, *Safety Effectiveness Evaluation*, Report NTSB-SEE-80-6, 1980.

18. Blackburn, R.R., et. al., *Effectiveness of Alternative Skid Reduction Measures*, Federal Highway Administration, Report FHWA-RD-79-22, 1978.

19. Schulze, K.H., and Blackman, L., *Friction Properties of Pavements at Different Speeds*, ASTM Special Technical Publication 326, 1962.

20. Horne, Walter B., *Tire Hydroplaning and its Effects on Tire Traction*, Highway Research Record 214, 1968.

21. Browne, A.L., *Mathematical Analysis for Pneumatic Tire Hydroplaning*, American Society for Testing and Materials, ASTM STP 583, 1975.

22. Horne, W.B., and Joyner, U.T., *Pneumatic Tire Hydroplaning and Some Effects on Vehicle Performance*, SAE International Automotive Engineering Congress, January 1965.

23. Albert, B.J. and Walker, J.C., *Tire to Wet Road Friction at High Speeds*, Proceedings, Institute of Mechanical Engineers, Volume 180, Part 2A, No. 4, 1966.

24. Yeager, R.W. and Tuttle, J.L., *Testing and Analysis of Tire Hydroplaning*, SAE Paper 720471, 1972.

25. Mosher, L., *Results of Studies of Highway Grooving and Texturing by Several State Highway Departments*, Conference on Pavement Grooving and Traction Studies, Paper 27, Langley Research Center, 1968.

26. Farnsworth, E.E. and Johnson, M.H., *Reduction of Wet Pavement Accidents on Los Angeles Metropolition Freeways*, California Division of Highways, 1971.

27. Stocker, A.J., Dotson, J.T., and Ivey, D.L., *Automobile Tire Hydroplaning—A Study of Wheel Spin-down and Other Variables*, Texas Transportation Institute, Research Report 147-3F, 1974.

28. Gallaway, B.M. et. al., *Pavement and Geometric Design Criteria for Minimizing Hydroplaning*, Federal Highway Administration, Report No. FHWA - RD-79-31, 1979.

29. Daughaday, H., and Tung, C., *A Mathematical Analysis of the Hydroplaning Phenomena*, Cornell Aeronautical Laboratory, Report No. AG-2495-1, 1969.

30. Moore, D.F., *International Journal of Mechanical Sciences*, December 1967.

31. Martin, C.S., *Hydrodynamics of Tire Hydroplaning*, Georgia Institute of Technology, Final Report, Project B-608, 1966.

32. Eshel, A., *A Study of Tires on a Wet Runway*, Ampex Corp., Research Report 67-24, 1967.

33. Tsakonas, S., Henry, C.J., and Jacobs, W.R., *Hydrodynamics of Aircraft Tire Hydroplaning*, NASA Report CR-1125.

34. Boness, R.J., *Automobile Engineer*, Volume 58, No. 7, June 1968.

35. Bathelt, H., *Calculations of the Aquaplaning Behavior of Smooth and Profiled Tires*, ATZ 75, Volume 10, 1973.

36. Balmer, G.G., and Gallaway, B.M., *Pavement Design and Controls for Minimizing Automotive Hydroplaning and Increasing Traction*, American Society for Testing and Materials, ASTM STP 793, 1983.

37. Gallaway, Bob M., and Rose, Jerry G., *The Effects of Rainfall Intensity, Pavement Cross Slope, Surface Texture, and Drainage Length on Pavement Water Depth*, Texas Transportation Institute, Research Report No. 138-5, 1971.

38. American Association of State Highway and Transportation Officials, *A Policy on Geometric Design of Streets and Highways*, 1990.

39. Horne, W.B., *Predicting the Minimum Dynamic Hydroplaning Speed for Aircraft, Bus, Truck, and Automobiles Tires Rolling on Flooded Pavements*, ASTM Committee E-17 presentation, Texas Transportation Institute, 1984.

40. Ivey, D.L., *Truck Tire Hydroplaning - Empirical Confirmation of Horne's Thesis*, Texas Transportation Institute, 1984.

41. Horne, W.B., Yager, T.J., and Ivey, D.L., *Recent Studies To Investigate Effects of Tire Footprint Aspect Ratio of Dynamic Hydroplaning Speed*, American Society for Testing and Materials, ASTM STP 929, 1986.

42. Ivey, Don L., et. al., *Tractor Semi-Trailer Accidents in Wet Weather*, Texas Transportation Institute, 1985.

References

Highway Research Board, *Skid Resistance*, NCHRP Synthesis 14, 1972.

American Association of State Highway and Transportation Officials, *Guidelines for Skid Resistant Pavement Design*, 1976.

Chapter 7

Pavement Edge Drops

Pavement edge drops (see Figure 7-1) have been recognized as a potential roadway safety problem for many years. In the 1954 publication, *A Policy on Geometric Design of Rural* Highways,[1] by the American Association of State Highway Officials (AASHTO), the subject of pavement edge drops was covered as follows (p. 205):

> *Unstabilized shoulders frequently are hazardous because the elevation of the shoulder at the pavement edge may be several inches lower than the pavement.*

Figure 7-1 Pavement Edge Drop.

This same passage was expanded to add further caution in the 1965 edition[2] of the same AASHTO Policy as follows (p. 239):

Unstabilized shoulders frequently are hazardous because the elevation of the shoulder at the pavement edge tends to become one-half to several inches lower than the pavement.

These statements in the AASHTO Policy probably came from the general perceptions and experiences of the policy writers, because no quantitative research on the effects of pavement edge drops was available before 1977. As will be discussed in following paragraphs, research since 1977 has demonstrated that the probability of severe consequences resulting from pavement edge-drop traversals is a function of the speed and path angle of the vehicle and the height and shape of the pavement edge drop.

7.1 Characteristics of a Pavement Edge Drop Traversal

When a vehicle strays from the travel lane onto the shoulder, the normal reaction of a driver is to return to the safe haven of the travel lane. The driver's urgency to make this return is usually heightened by the sudden drop when the shoulder elevation is significantly lower than the travel lane. Depending on the severity of the vehicle's exit angle, the driver's level of surprise, and the driver's steering and/or braking response, seven general outcomes are possible, as described below and shown in Figure 7-2:

1. **High Exit Angle** - during a pavement edge-drop encounter, if the exit angle is high, recovery is very unlikely and some kind of collision on the roadside is probable (*Outcome 7*).
2. **Low to Moderate Exit Angle** - if the exit angle is low to moderate, the driver may be able to attempt a recovery (*Outcomes 1-6*).
3. **Steer Parallel** - for a low exit angle, if the driver knows about the hazardous dynamic effects of a pavement edge drop and is able to respond quickly and appropriately, he may be able to steer parallel to the pavement edge and slow to a speed where the pavement can be remounted safely (*Outcome 1*). This may be a difficult task for larger edge-drop heights, for rough and uneven shoulders, and/or where the driver is completely surprised by the edge-drop excursion.
4. **Shallow Return Angle** - if the driver steers a low return angle, the vehicle tires will tend to scrub along the pavement edge and either remount the pavement (*Outcomes 3-5*) or possibly rebound out of control (*Outcome 2*) depending usually on the driver's subsequent steering response. As will be seen later, the driver's ability to achieve

Pavement Edge Drops

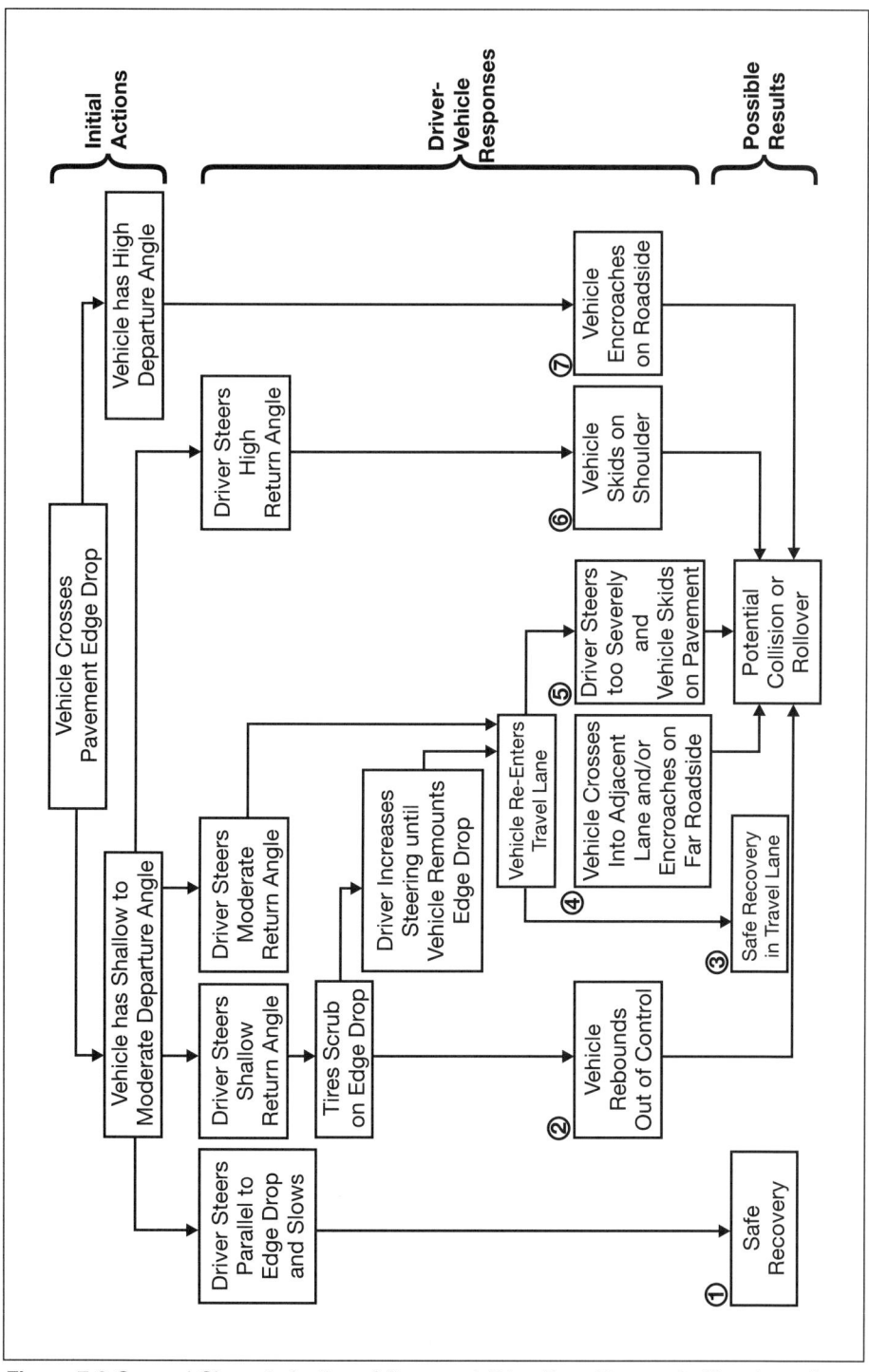

Figure 7-2 *General Characterization of Pavement Edge Drop Traversals (Courtesy of Criterion Press, Inc.)*

a safe recovery is largely a function of the vehicle speed and the edge-drop geometry.

5. **Moderate Return Angle** - for lower speeds and edge-drop heights, if the driver steers a return angle that is just high enough to avoid tire scrubbing, the vehicle will usually remount the pavement and recover within the proper travel lane (*Outcome 3*). For return angles that are greater than the minimum necessary to remount, particularly at higher speeds, the vehicle may either encroach on the adjacent travel lanes and/or the far roadside or skid out of control, depending on the driver's steering response (*Outcomes 4 and 5*).

The dynamic response of a vehicle produced by scrubbing re-entry is the primary focus in evaluating the hazard associated with pavement edge drops. For higher speeds, higher edge-drop heights, and more severe edge-drop slopes, the undesirable dynamic responses are either excessive lateral encroachments onto adjacent (often opposing) lanes and/or opposite roadsides (*Outcome 4*) or loss of control and/or rollover because of excessive steering corrections to the right (*Outcome 5*).

A scrubbing re-entry occurs when a tire that contacts the pavement edge has insufficient normal velocity to overcome the retarding force produced at the tire-edge contact. The term *scrubbing* describes the parallel movement of a tire sidewall rubbing along the pavement edge. During this process, the tire develops a large resistance to remounting the pavement and the driver, who is attempting to recover to the travel lane, will continue to increase the steering in further attempts to remount the pavement. The contact between the tire and the pavement edge will continue until the front-wheel steer angle is sufficient to overcome the retarding force and to create enough side force at the unobstructed front tire to lift the obstructed (scrubbing) front tire over the edge drop. Once the obstructed front tire has remounted the pavement, the large steer angle produces a large yaw acceleration with the vehicle pivoting about the obstructed rear tire. As the car yaws to the left, the tire sideslip angle increases and the side force on the front tires build even higher resulting in unstable dynamics until the rear tire remounts the pavement. These eccentric forces combine to produce rapid lateral movement. This lateral movement will continue unless the driver reverses the steer angle and the vehicle has time to respond to a steering reversal. Figure 7-3 shows the tire marks left by a vehicle with an unsuccessful scrubbing re-entry.

Pavement Edge Drops 197

Figure 7-3 Tire Marks Left by a Vehicle with a Scrubbing Re-entry from a Pavement Edge Drop.

7.2 Writings and Research on Pavement Edge Drops

Uncertainty and misinformation seem to dominate the common knowledge base on pavement edge drops. Some studies point to edge drops as serious hazards, while others seem to indicate they are not much to worry about. Because of these seeming contradictions, standards in some States try to limit edge drop heights to very low values, while standards in other states are much more lenient.

In no other roadway defect's arena are the liability issues so heavily tied to the results of research studies and other writings. Because only ten studies seem to be prevalent, a review of these documents is crucial to a clear understanding of the subject. The following paragraphs discuss each of these studies in chronological order giving a brief summary of methodology, pertinent conclusions, and a critique of the report or paper.

Ivey and Griffin (1976)

This study[3,15] examined 15,968 single-vehicle accidents that occurred in North Carolina during 1974. Computerized police narratives were interrogated for 19 key words noting some kind of surface discontinuity as a contributing factor. Of these accidents, 566 were associated with a surface discontinuity. Some 154 accidents appear to have been related to a pavement edge drop.

Although this study indicates a relatively small percentage of accidents that involve pavement edge drops, it did not attempt to determine the relative exposure to edge-drop conditions. If only roadway sections with pavement edge drops were considered, the contribution of the edge drops would be considerably more substantial.

Nordlin, Parks, Stoughton, and Stoker (1976)

This study[4] was undertaken by the California Department of Transportation in response to concerns about some accident reports where pavement edge drops had been cited as a contributing factor. The purpose of the study was to test the adequacy of the Department's existing maintenance standards, which called for a maximum edge drop of 1.5 inches for unpaved shoulders and 0.75 inches for paved shoulders. Fifty full-scale tests were run on a rounded asphalt edge-drop using a former race driver. Edge-drop heights of 1.5, 3.5, and 4.5 inches were tested. Various vehicles with various tires were driven at 60 mph through an arced trajectory with either two or four wheels off the pavement.

Results indicated that all tests were performed without erratic vehicle responses. In all tests, the edge drop was at the edge of a 5-foot shoulder and the vehicle was steered to a recovery within the 12-foot travel lane. In other words, a 17-foot recovery width was indirectly defined as safe.

These tests demonstrate that, even with speeds of 60 mph and edge-drop heights of 4.5 inches, a pre-warned professional driver can recover within 17 feet from a shallow-angle departure. Whether this kind of maneuver is truly safe and whether it can be performed by a non-professional driver, who is surprised when his right-front tire suddenly drops along a roadway on a dark night, was not addressed by this study. The study also did not address the probability of the vehicle being in a scrubbing mode or the consequences of re-entry from a scrubbing mode.

Klein, Johnson, and Szostak (1977)

This study[5] conducted 73 full-scale tests using *naive* drivers to determine the statistical distribution of consequences when these drivers encountered a 4.5-inch vertical face (0.5-inch corner rounding) pavement edge drop and tried to recover to the traveled way. Fifty-three percent of these tests produced tire scrubbing on the edge drop. Of the tests that produced scrubbing, 56% of the vehicles exceeded the 12-foot lane boundary after remounting the pavement. The probability of exceeding the lane boundary for scrubbing re-entries increased as vehicle speed increased. None of the non-scrubbing re-entry vehicles exceeded the lane boundary.

Pavement Edge Drops

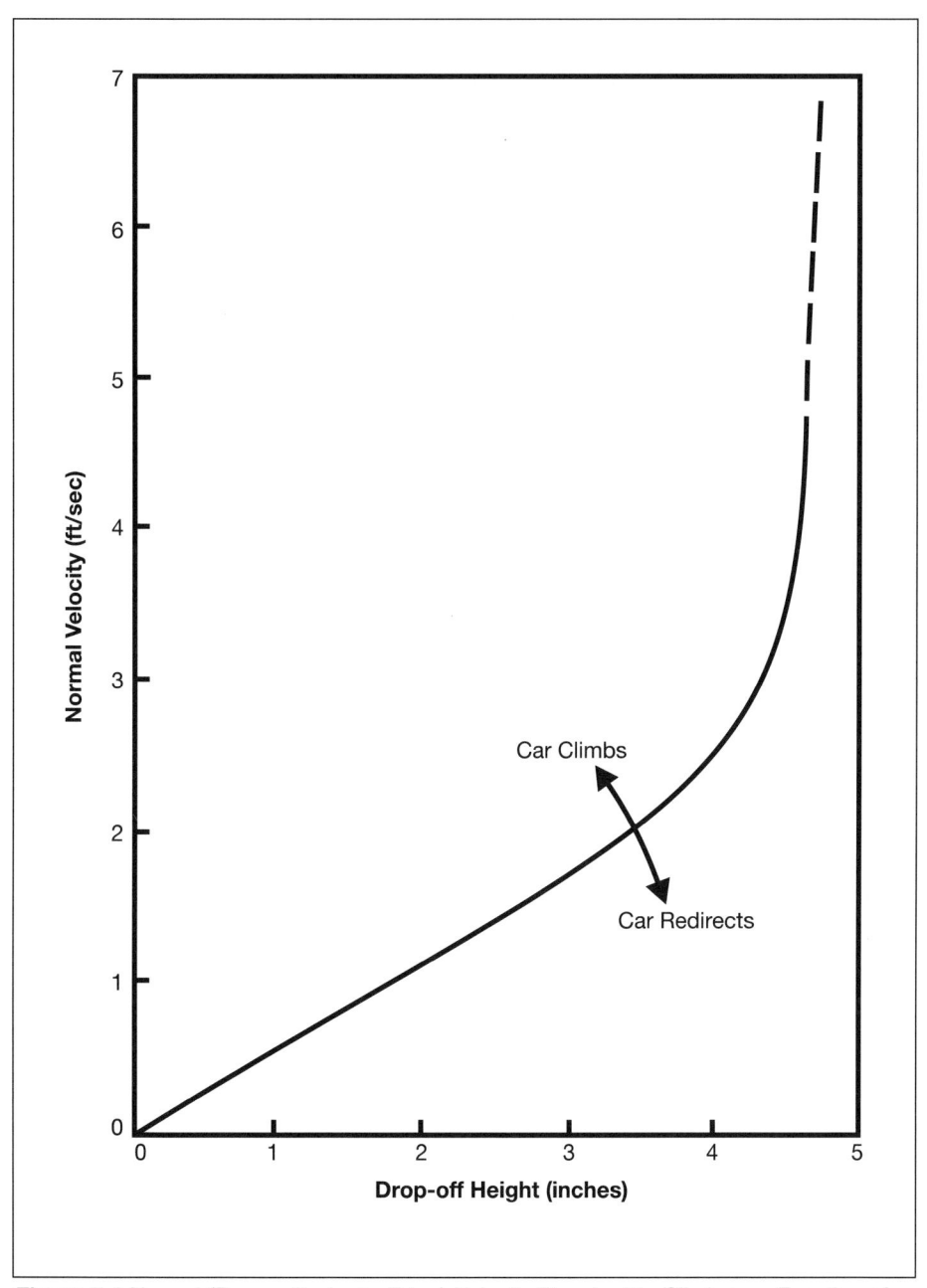

Figure 7-4 Normal (Perpendicular to Edge) Velocity Required to Climb as a Function of Pavement Edge Height.[5]

These results may be somewhat misleading because the study report did not document the complete distribution of speeds employed in the tests. Limited examples from three tests, indicated a potential for loss of control with a 3.5-inch edge-drop for vehicle speeds as low as 40 mph. Another shortcoming is that the study did not account for any element of driver surprise. Drivers were fully pre-warned about the study circumstances.

This study also conducted several full-scale shallow-angle return tests to evaluate the potential for remounting a vertical edge-drop when the vehicle was not initially scrubbing. Various edge-drop heights were tested with various combinations of vehicle return angle and speed to produce a relationship between edge-drop height and the minimum normal speed necessary to remount the pavement. [Normal speed is that component or vector of the vehicle speed that is perpendicular to the pavement edge.]

This relationship between vertical edge-drop height and the minimum normal speed necessary to remount the edge drop is shown in Figure 7-4. This figure indicates that if the return angle is too small for the given vehicle speed and edge-drop height, the vehicle will not remount the pavement but will be redirected into a scrubbing mode. The relationship also indicates that the probability of scrubbing increases dramatically for edge-drop heights above 4 inches.

The results of Figure 7-4 have been widely accepted and made part of most research conducted since this 1977 study. These results were also partly validated in 1984 by Graham and Glennon[6] using computer simulation techniques to study vertical pavement edge drops. Unfortunately, no such relationship has been established for any other edge-drop shape.

This study found the scrubbing re-entry maneuver to be the most hazardous type of recovery. The reason for this hazard might is best explained by another relationship developed in the study as shown in Figure 7-5. This figure, for a vertical edge drop with a 0.5-inch corner rounding, shows that as the edge-drop height exceeds about 3 inches, very large steering wheel inputs are required to remount the pavement. Under these conditions, when the tire remounts the pavement, the resulting large steer angle produces a rapid lateral movement.

Zimmer and Ivey (1982)

This study,[7,16] conducted by the Texas Transportation Institute for the Texas State Department of Highways and Public Transportation developed maintenance guidelines for maximum pavement edge drops.

The study consisted of 327 full-scale tests runs using a pre-warned professional driver. The pavement edge drop consisted of an 1.5-inch asphalt lift, roughly tapered at 35 degrees from vertical, laid on top of a concrete pad with a

Pavement Edge Drops

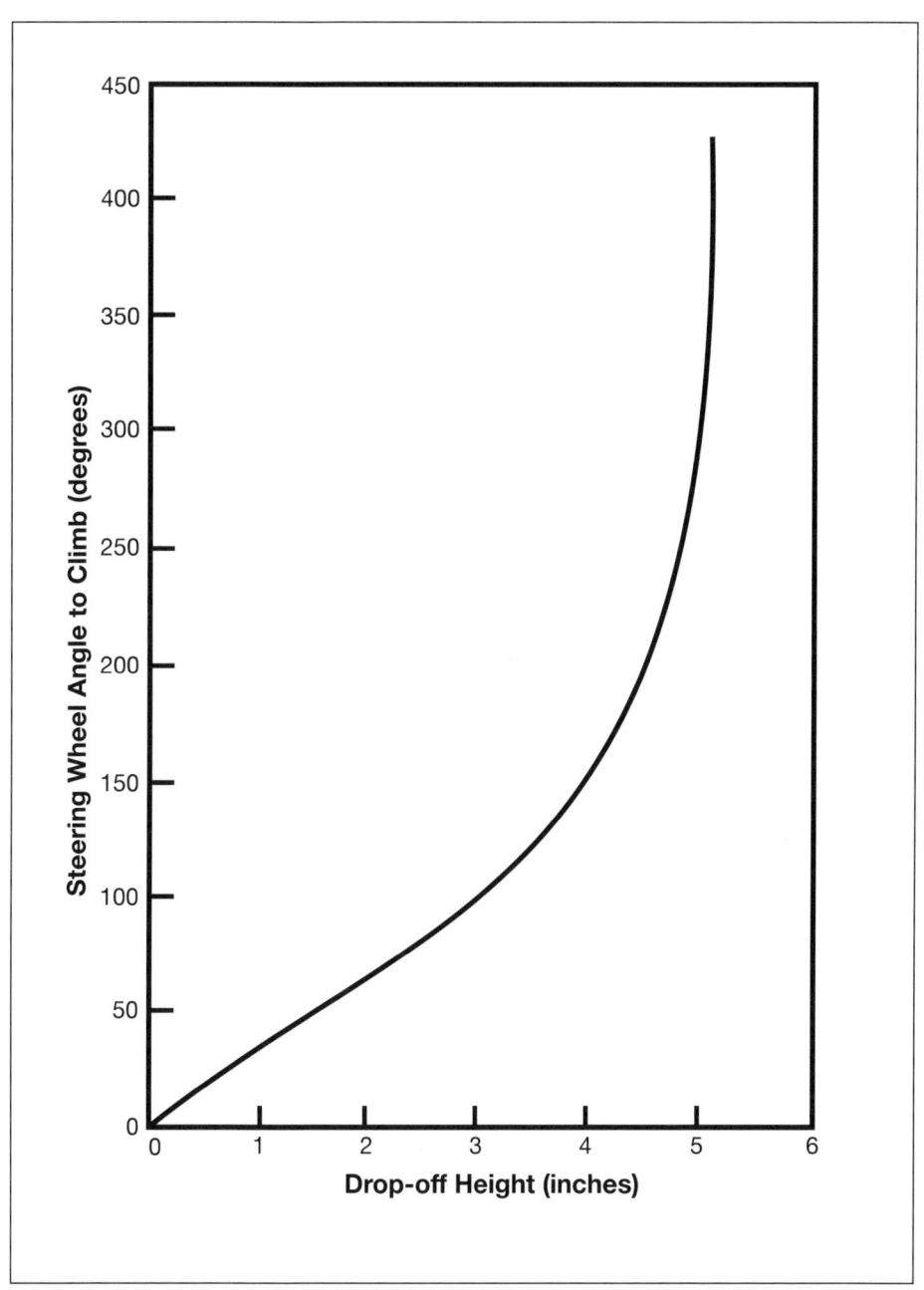

Figure 7-5 *Steering Wheel Angle Required to Climb Various Pavement Edge Heights from a Scrubbing Mode.*[5]

vertical face. Total edge-drop heights of 1.5, 3.0, and 4.5 inches were tested using various vehicles and tires with vehicle speeds of 35, 45, and 55 mph.

To evaluate the dynamic responses generated by the three edge-drop heights, for both scrubbing and non-scrubbing modes, a subjective severity scale was used. The figure of merit was the average severity rating, from one to ten, judged by the professional driver, the only driver to complete the entire matrix.

For the non-scrubbing re-entry tests, the severity ratings were averaged over all speeds and tire types. These results indicated only minor sensitivity to vehicle type and edge-drop height. The sensitivity to vehicle speed was not documented.

For the scrubbing re-entry tests, the results reported a definite sensitivity to vehicle speed and edge-drop height when severity ratings were averaged across vehicle and tire types. Severe vehicle dynamics were indicated for the 4.5-inch edge-drop at speeds of 45 mph and greater. Other informal (undocumented) tests conducted on a 45-degree beveled, 5-inch edge drop were reported as reasonably safe.

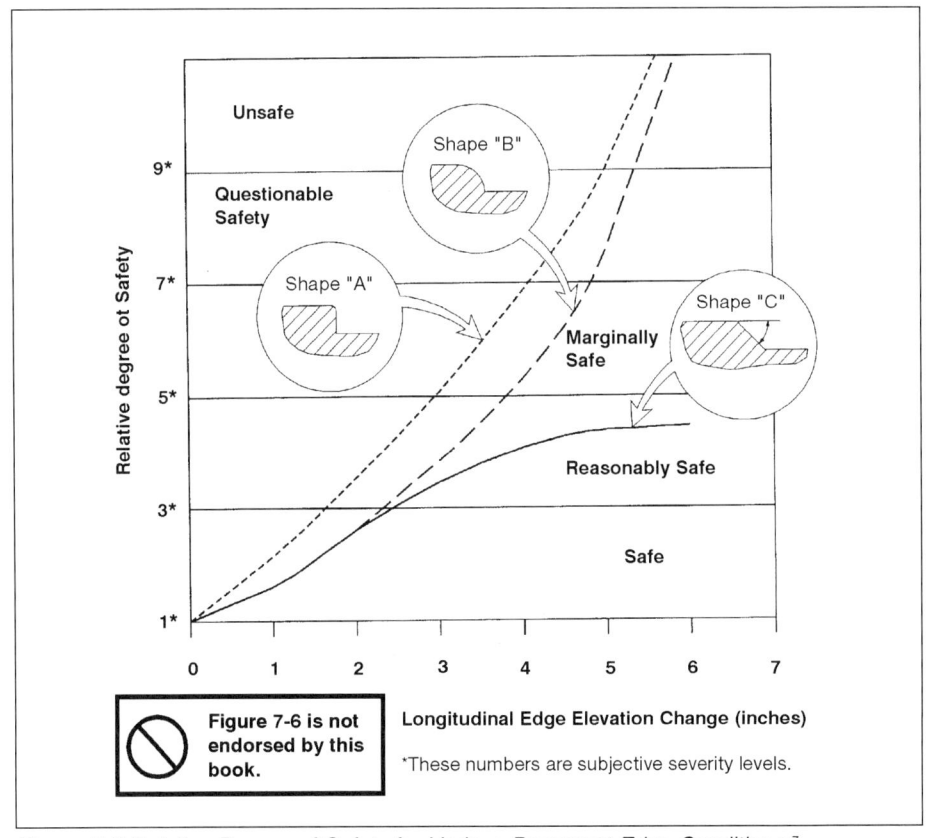

Figure 7-6 *Relative Degree of Safety for Various Pavement Edge Conditions.*[7]

This study produced the well-circulated graph shown in Figure 7-6. From this research, the authors suggested that vertical edge drops of 2 to 3 inches in height might indicate that need for maintenance. For a rounded shape, the maintenance criterion was suggested at 2.5 to 3.5 inches.

The non-scrubbing re-entry results of this study appear to support the California Department of Transportation study cited earlier. Both studies indicate that reasonably safe non-scrubbing re-entry maneuvers can be made by pre-warned professional drivers on rounded edge-drops with heights up to 4.5 inches at vehicle speeds of 55-60 mph. Although this study also showed that pre-warned non-professional drivers can perform almost as well as a pre-warned professional driver at edge-drop heights up to 3 inches, this comparison was not verified for higher heights. Therefore, this research did not answer whether a non-professional driver, particularly one surprised when his right-front tire suddenly drops, can safely perform a non-scrubbing re-entry at rounded edge-drop heights of 3 inches or higher.

The results of the scrubbing re-entry tests are also somewhat questionable for explaining the critical responses of drivers who suddenly encounter a pavement edge drop. How the average subjective ratings of a single pre-warned professional driver translate to a measure of criticality for ordinary drivers experiencing actual roadway conditions was unanswered.

Although the results of this work are limited, when compared to other research, they do indicate that rounding or tapering of a vertical pavement edge drop can reduce its hazard . The optimal shape for a given height, however, was not determined.

Graham and Glennon (1984)

This study[6] for the Federal Highway Administration was focused on pavement edge drops in construction zones. At issue was the determination of edge-drop heights above which delineation, warning, and/or barrier protection may be warranted.

The research employed both the HVOSM[8] computer simulation model and some simplified analytical modeling to determine critical operational boundaries for vehicular encounters with vertical pavement edge drops. Shallow return-angle approaches and scrubbing re-entries for two different vehicles on 2- and 3-inch edge drops were studied. The initial simulation runs showed the general validity of the HVOSM model to duplicate the relationship between edge-drop height and the minimum normal velocity required to remount the pavement, as shown by Klein, Johnson, and Szostak.[5]

A limited number of scrubbing re-entry runs were performed with the HVOSM to study the severity of responses as a function of vehicle speed, vehicle type, and edge-drop height. These results indicated that even at relatively low speeds and edge-drop heights, the scrubbing re-entry maneuver is likely to produce either an excursion beyond the normal 12-foot travel lane, or a lateral acceleration that would exceed the frictional capability of many pavements, or both. Dynamic responses to the edge drops were nearly identical for mid-size and compact cars.

The study used analytical methods to define the return-angle boundaries for successful non-scrubbing recovery from a vertical edge drop. From the relationship developed by Klein, Johnson, and Szostak (shown in Figure 7-4), the minimum return angle to avoid scrubbing was solved to form the lower boundary of successful recovery for any combination of vehicle speed and edge-drop height, as shown in Figure 7-7. For the upper boundary, the maximum return

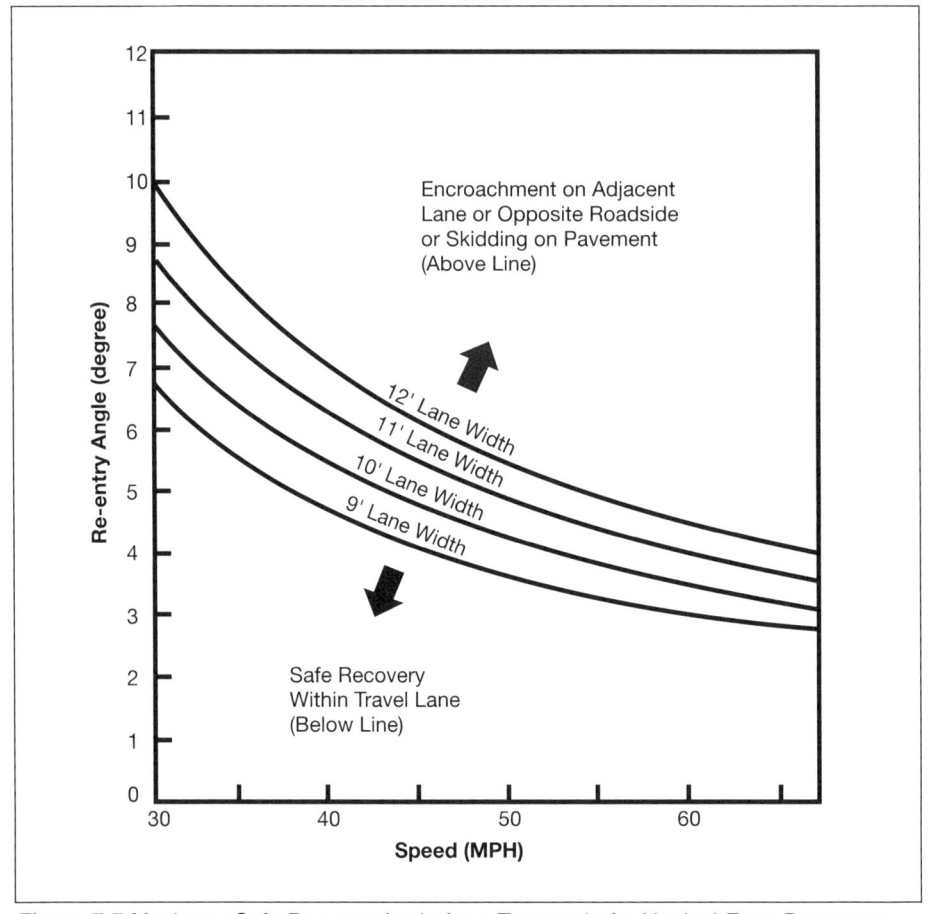

Figure 7-7 Maximum Safe Re-entry Angle for a Traversal of a Vertical Face Pavement Edge Drop as a Function of Speed and Lane Width.[6]

Pavement Edge Drops

angle was solved as a function of speed and lane width, as shown in Figure 7-8. This analysis was constrained to recovery within the lane using a maximum 0.3 g's tolerable lateral acceleration and a 0.7-second driver perception-reaction time for steering reversal.

Figure 7-9 shows the plotted boundaries for two combinations of edge-drop height and lane width. Although the exact form of the analytical model and its constraining assumptions can be debated, Figure 7-9 does show that the window of safe re-entry angles decreases with vehicle speed, edge-drop height, and lane width.

The results of this study are limited to vertical edge drops with no rounding or taper. They may be further limited by the veracity of the assumption used to model driver responses. However, these analytical results appear to generally agree with the full-scale test results of the Klein, Johnson, and Szostak study.[5]

Glennon (1985)

This study[9,17] prepared for the Transportation Research Board under a research program mandated by the U.S. Congress, was a synthesis of the state-of-the-art concerning the safety of pavement edge drops. The purpose of

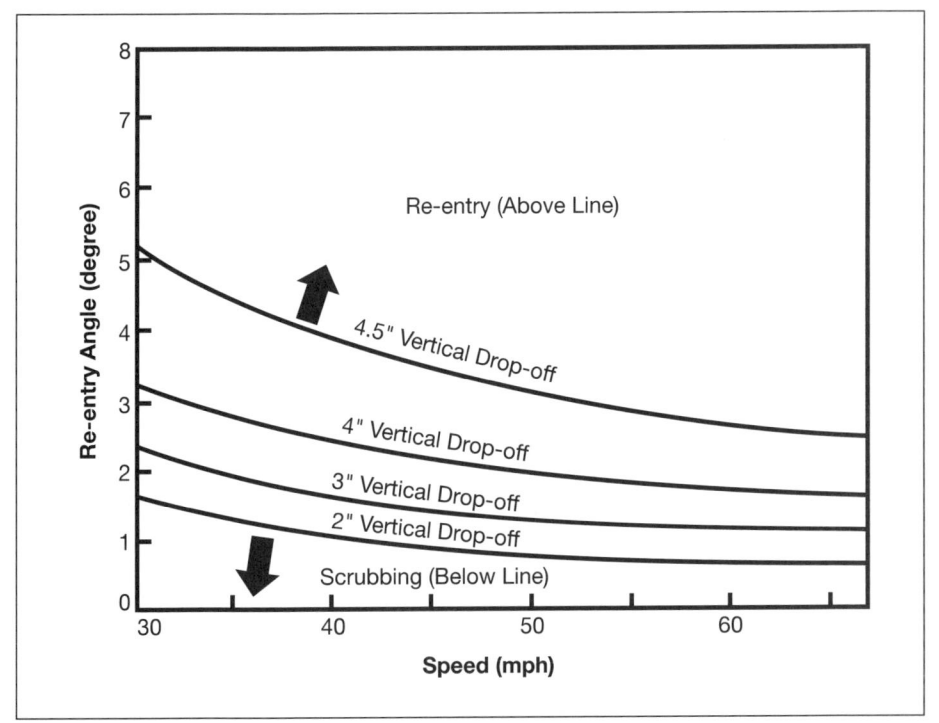

Figure 7-8 Minimum Safe Re-entry Angle for a Traversal of a Vertical Face Edge Drop as a Function of Speed and Edge Height.[6]

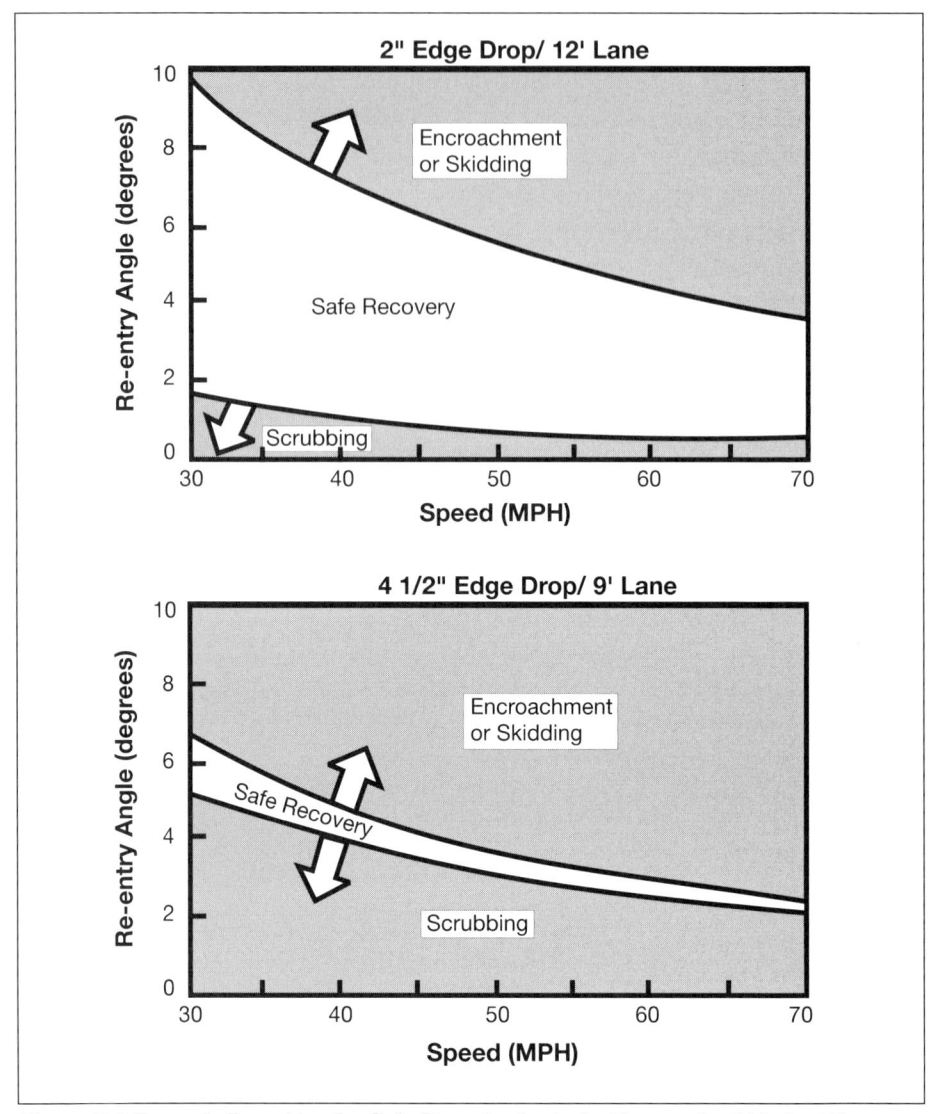

Figure 7-9 *Example Boundries for Safe Re-entry Angle for Traversals of Vertical Face Edge Drops as a Function of Speed, Edge Height, and Lane Width.*[6]

the study was to develop practical guidelines for the treatment of pavement edge drops in the process of repaving existing roadways under the 1977 Federal 3R (Resurfacing, Rehabilitation, and Restoration) funding authorization.

Based on a critical analysis of previous studies, this 1985 study concluded the following:

1. The probability of a scrubbing re-entry from a shoulder encroachment at a pavement edge drop increases with edge-drop height and as the edge shape approaches vertical.

Pavement Edge Drops

2. The probability of an unsuccessful recovery from a shoulder encroachment at a pavement edge drop increases as the edge-drop height and vehicle speed increases and as the lane width decreases.
3. The severity (yaw acceleration, lateral displacement, etc.) of a scrubbing re-entry maneuver increases as the edge shape approaches vertical and as the edge-drop height and vehicle speed increase.
4. A five-inch edge-drop height is a practical maximum to prevent the severe consequences (usually rollover) associated with vehicle undercarriage drag on the pavement.

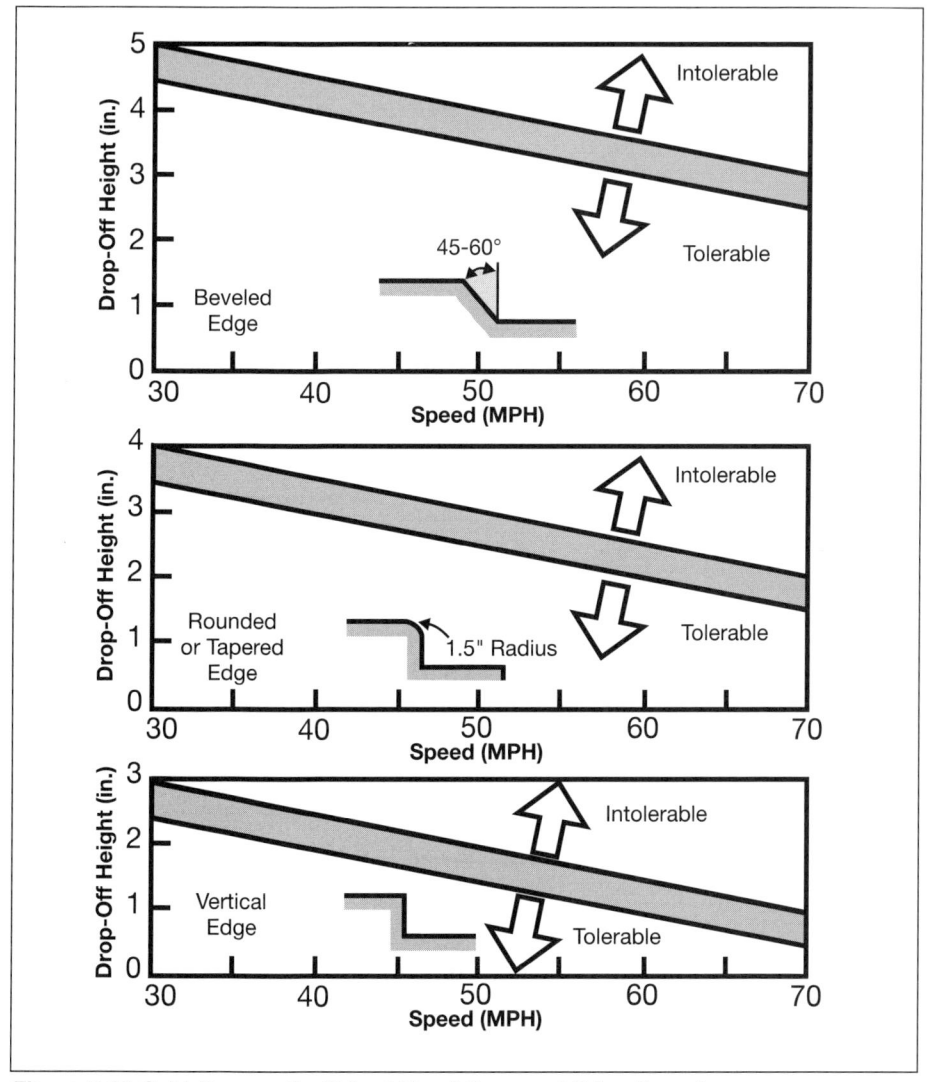

Figure 7-10 *Guidelines on the Tolerability of Pavement Edge Drops.*[9]

Using these general conclusions, and the fairly sparse measures of lane exceedance and lateral acceleration found in the previous studies, this 1985 study developed the suggested guidelines shown in Figure 7-10. As seen in this figure, the suggested tolerable edge-drop height decreases with increasing speed and as the edge shape approaches vertical. The shaded band on each graph was intended to allow for (a) some degree of uncertainty associated with the available data, and (b) the variance in criticality associated with lane widths whereby narrow lanes should have lower tolerable edge-drop heights than wide lanes.

The suggested guidelines shown in Figure 7-10 were based on the studied sensitivity of various sized passenger cars and pickup trucks. Lower tolerable edge-drop heights might be appropriate for roadways that carry significant portions of tractor-trailer trucks or motorcycles.

Olson, Zimmer, and Pezoldt (1986)

Like the 1985 Glennon study, this study[10] was also prepared for the Transportation Research Board under the same research program funded by the U.S. Congress. The study undertook 185 full-scale tests to evaluate the performance of 50 ordinary (naive) drivers on their encounter with pavement edge drops. Various vehicles sizes and speeds were used to test both vertical face and 45° beveled edge drops of various heights with both hard and soft surfaced shoulders. The primary measure of failure was the intrusion of a vehicle beyond its own 12-foot lane during recovery.

The conclusions of this study were as follows:

1. The data taken in earlier studies using professional drivers significantly over-estimated the performance to be expected from ordinary drivers.
2. For 4.5-inch edge drops with vertical faces and minor corner rounding, neither professional nor naive drivers could recover within the adjacent lane at 30 mph with any of the vehicles used.
3. Based on the results of this study, edge drops with a vertical face should be controlled to no more then 3 inches, and then should be posted for speeds no greater then 25 mph.
4. For high-speed roadways, edge drops should either be controlled to heights less then 3 inches or the pavement edge should be shaped at a 45-degree or flatter bevel to facilitate vehicle recovery from a shoulder excursion.
5. Assuming that recovery from a scrubbing encounter is the determining factor for tolerable pavement edge drop heights, the recommendations of Zimmer and Ivey[7] are not adequate.

This study appears to be the most definitive to date. It's results appear to generally agree with the suggested guidelines of the 1985 Glennon[9] study. However, the sensitivity of vehicle recovery from various edge shapes was not completely resolved.

Ivey and Sicking (1986)

This analytical paper[11] was undertaken to develop a relationship between effective heights for vertical edge drops and a variety of other shapes including both rounded and tapered edges. The idea was to try to translate the results of full-scale tests on vertical edges (such as Zimmer and Ivey[7,16]) to those other shapes. The authors presented the relationships shown in Figure 7-11, and then presented a fairly detailed mathematical exercise in an attempt to validate these relationships.

The initial development of the relationships shown in Figure 7-11, came by simply solving for the contact point between a scrubbing tire and the pavement edge using the following assumptions:

1. *Round Tire* - the tire has a 3.6-inch edge radius (as seen looking at the rear view of the tire).
2. *Rigid Tire* - the tire is rigid and cannot be deformed by the forces acting on the tire at the pavement edge.
3. *Parallel Tire* - the tire contact has a zero-degree orientation to the pavement edge.

The remainder of the Ivey and Sicking paper is a fairly extensive mathematical development, along with a discussion of computer simulation studies, for the express purpose of validating Figure 7-11.

Figure 7-11 or some similar figure, seems now to have gained fairly wide acceptance by roadway agencies. As intuitively appealing as these relationships may be, however, their validity is tenuous at best. The assumptions that the tire cross-section has a circular edge, that the tire is rigid, and that the tire sidewall is exactly parallel to the edge are not realistic. Tire edge shapes are basically square. Then too, the tire that is running parallel to and in contact with a pavement edge drop will have contact forces tending to deform the tire to the shape of the pavement edge. The extent of deformation is a function of the vehicle speed, and steering angle, and also of the tire size, geometry, composition, wear, and inflation pressure.

Although the authors also claim to validate Figure 7-11 through mathematical derivation and computer simulation, these exercises only prove that validity using circular argument. The entire development has a continuum of many unsupported interdependent assumptions, seeming bent on making the mathematical

Condition	Pavement Edge Profile	Effective Edge Height Δe, Inches
1	4" drop	4.0
2	2" / 4"	2.5
3	4" / 4"	1.5
4	45° / 6"	0.75
5	45° / 4"	0.75
6	45° / 2"	0.75
7	45° / 4"	0.75
8	30° / 4"	0.50
9	15° / 4"	0.20

Figure 7-11 is not endorsed by this book.

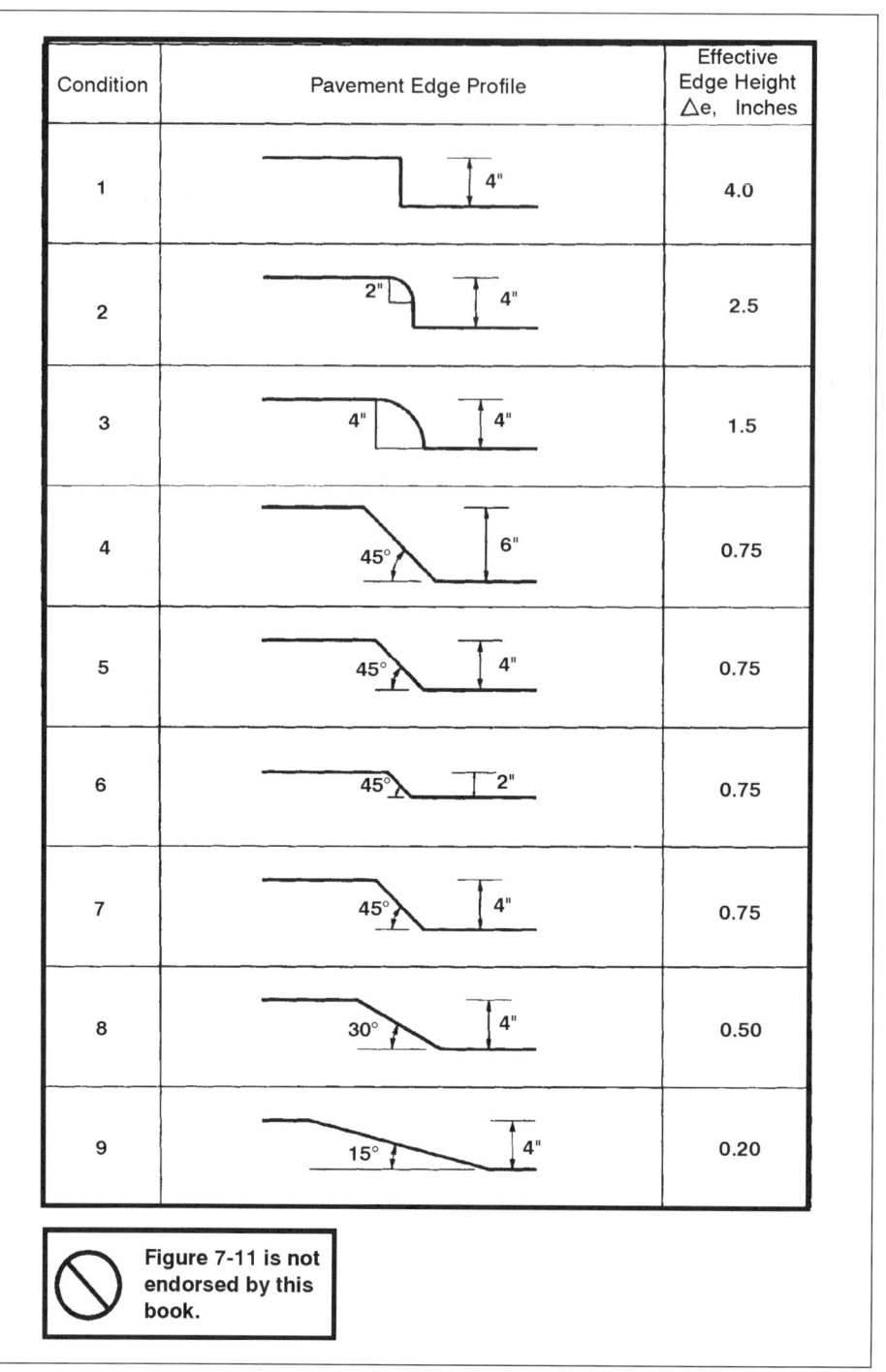

Figure 7-11 Purported Effective Edge Heights for Various Edge Slopes.[11]

and simulation results agree with Figure 7-11, which was developed with the assumptions of an undeformed tire with a circular edge.

Any further discussion of these mathematical and simulation exercises would take too much space in this book, so the interested reader is encouraged to interrogate this paper further.

Ivey, Mak, Cooner, and Marek (1988)

This paper[12] addresses the development of guidelines for warning and protective devices in construction zones. These guidelines were developed for and are now used by the Texas State Department of Highways and Public Transportation.[13] Based on an apparent synthesis of Zimmer and Ivey,[7] Ivey and Sicking,[11] and additional analyses about vehicle drag (undercarriage contact on the pavement edge) and rollover, they recommended the following guidelines:

Edge Condition I. For an edge drop with an 18° slope or flatter, use a shoulder drop-off sign (Figure 7-12b) and vertical panels (Figure 7-12d) when the edge drop height is 2 inches or more.

Edge Condition II. For an edge drop with a 19 to 45° slope, use a shoulder drop-off sign (Figure 7-12b) and vertical panels (Figure 7-12d) when (a) the edge drop height is between 2 and 5 inches, and (b) the edge drop height is 5 inches or greater when the lateral clearance is 20 feet or more. When the lateral clearance is less than 20 feet and the edge drop height is between 5 and 24 inches, use drums (Figure 7-12c) with steady-burn lights. [As an alternative, consider using an edge fillet to change the slope to Edge Condition I]. When the lateral clearance is less than 20 feet and the edge drop is greater than 24 inches, check the need for a positive barrier.

Edge Condition III For an edge drop with a 50 to 90° slope, use a shoulder drop-off sign (Figure 7-12b) and vertical panels (Figure 7-12d) when (a) the edge drop is less than 2 inches, and (b) the edge drop height is 2 inches or greater and the lateral clearance is 20 feet or more. When the lateral clearance is less than 20 feet and the edge drop height is between 2 and 24 inches, use drums (Figure 7-12c) with steady-burn lights. When the lateral clearance is less than 20 feet and the edge drop height is greater than 24 inches, check the need for a positive barrier.

Although this paper presents the results of Zimmer and Ivey[7] and Ivey and Sicking[11] as background to the recommended practices, the authors never make a direct connection between that previous research and the recommendations. Actually the recommended practices seem more in line with the results of Graham and Glennon,[6] Glennon,[9] and Olson, Zimmer, and Pezoldt,[10] which the paper never mentions. Part of the discussion on the development of these recommended

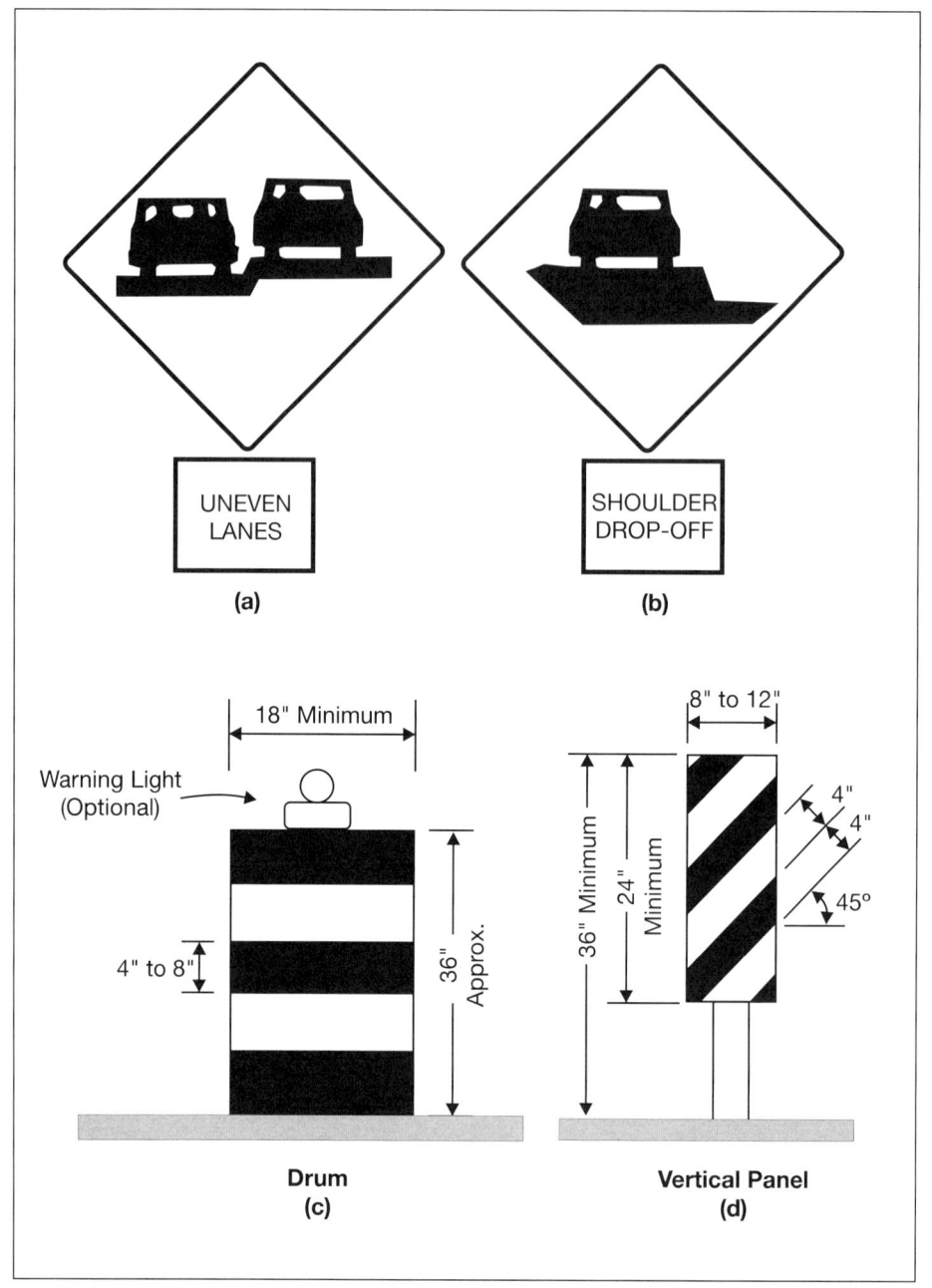

Figure 7-12 Traffic Controls Recommended for Pavement Edge Drops in Construction Zones.[12]

Pavement Edge Drops

practices indicates that numerous revisions were made to the draft recommendations based on inputs from State and Federal reviewers.

Part of the input to the recommended practices for portable concrete barriers was a mathematical analysis of the effects of undercarriage drag (see Figure 7-13) on edge drops. Figure 7-14, from the paper, shows a percentile distribution of undercarriage clearance for 267 manufactured vehicles. Clearly, about one-third of these vehicles would have undercarriage drag on edge drops of 5 inches. Edge drops of 6 inches would catch about two thirds of these vehicles.

The reported result of the mathematical analysis was that the only dynamic effect of undercarriage drag was that the deceleration caused by the contact increases the probability of a rear-end collision with a following vehicle. This result seems contrary to observations of actual accidents on 55-mph roadways, where undercarriage drag is almost always associated with rollover.

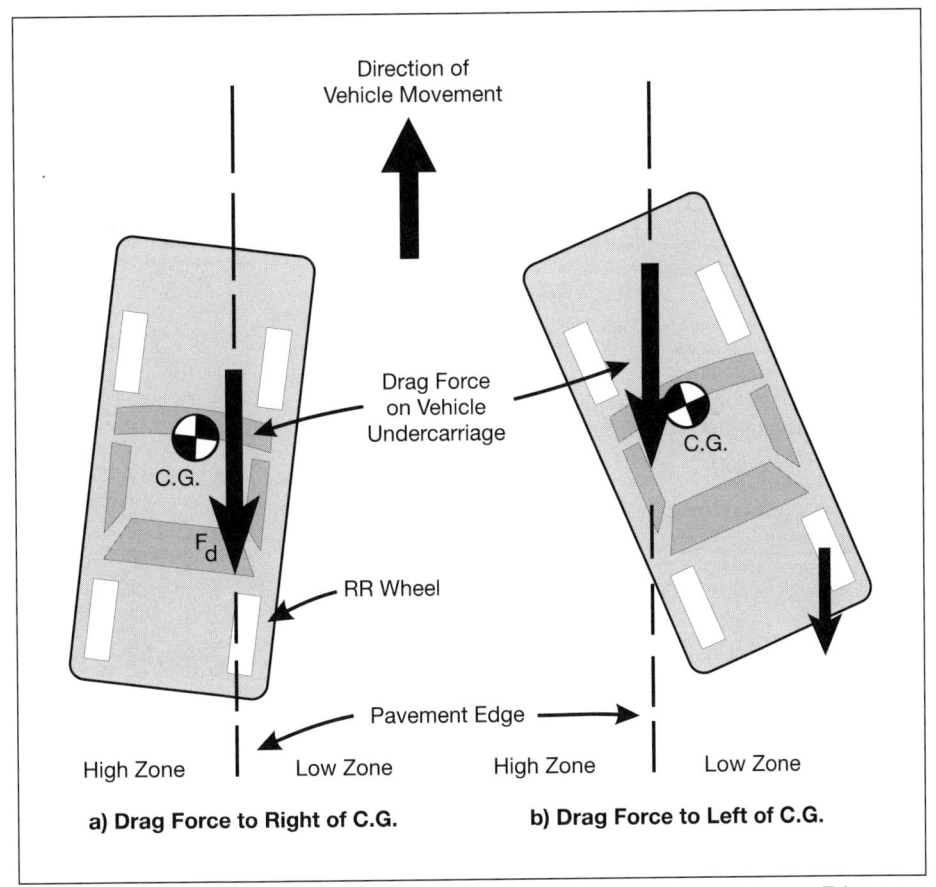

Figure 7-13 Forces on an Automobile with Undercarriage Drag on a Pavement Edge Drop.

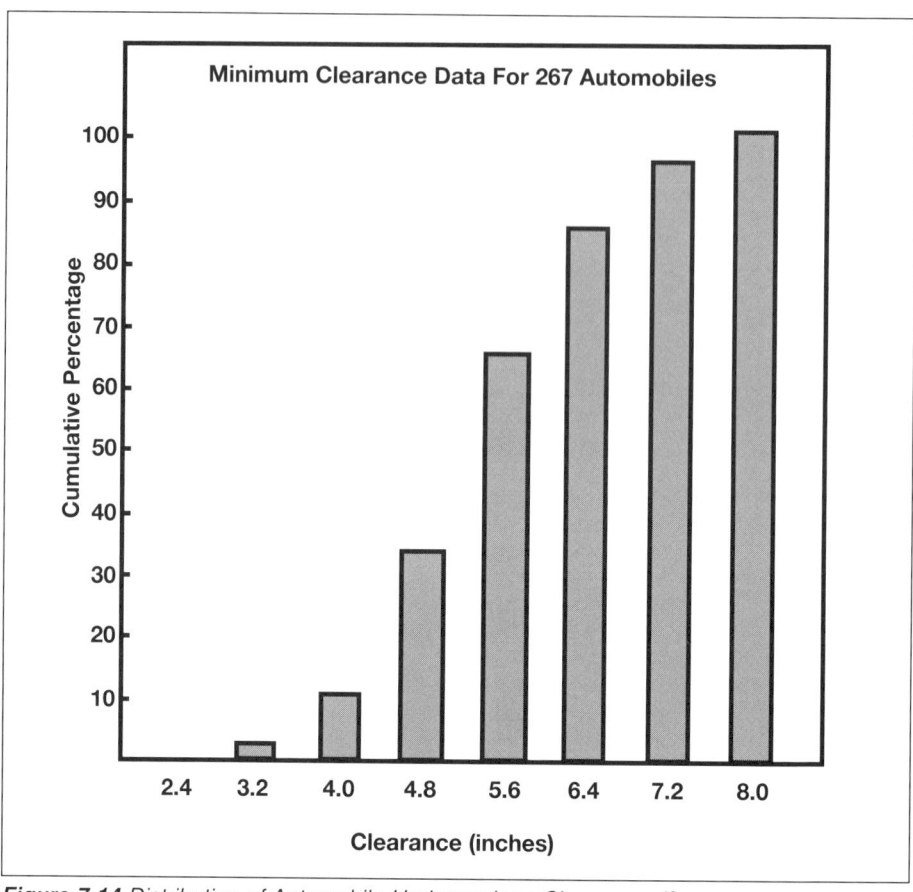

Figure 7-14 Distribution of Automobile Undercarriage Clearances.[12]

Humphreys and Parham (1994)

This pamphlet[14] produced for the AAA Foundation for Traffic Safety recommends changes in roadway resurfacing contracts to eliminate or mitigate the creation of hazardous edge drops. Making reference to Zimmer and Ivey,[7] Ivey and Sicking,[11] and a survey of State practices, this work recommends the following:

1. The most effective way of solving the problems associated with pavement edge drops is to simply eliminate the issuance of pavement resurfacing contracts where shoulder work is not included.
2. If, for whatever reason, the roadway agency believes in *no shoulder work* contracts, one of the following recommendations should be implemented in resurfacing contracts:
   a) Require that shoulder materials be pulled up to the new surface as a non-pay item;

b) Require that appropriate signing remain installed along the roadway to warn drivers of a low shoulder condition;
c) Require that a 45° asphalt fillet be placed along the edge of the roadway as a part of the resurfacing;
d) Both Recommendations b and c; or better still,
e) Both Recommendations a and c.

These recommendations appear sound and in general agreement with previous studies. The background presentation that was limited only to the results of Zimmer and Ivey[7] and Ivey and Sicking,[11] however, doesn't completely support the recommendations. Although both of these references talk about the desirability of 45° slopes, they both otherwise have questionable results as discussed earlier. The studies of Graham and Glennon,[6] Glennon,[9] and Olson, Zimmer, and Pezoldt,[10] could have added support to the recommendations.

> **PAVEMENT EDGE DROPS**
> **Rules of Thumb**
> - *Vertical or Near-Vertical Edge Drops of 2 inches or More can Cause Sling-Shot Accidents Even at Low Speeds*
> - *Edge Drops of 6 inches or More Will Cause Undercarriage Contact Often Resulting in Rollover*

7.3 Recommended Standards for Pavement Edge Drops

In 1987, the Transportation Research Board[18] determined that *pavement edge drop hazards are greater than previously believed* and *pavement edge drops are a common source of tort claims against roadway agencies.* Considerable debate has occurred in the courtrooms across the nation not only about what constitutes a hazardous edge drop but also what responsibility roadway agencies have for minimizing hazardous edge drops and/or warning of their existence. In an attempt to clear the air on this debate, the following discussion will attempt to summarize the knowledge base for the purpose of suggesting a nationwide standard for the treatment of edge drops.

Recommended Practices for Normal Maintenance

Pavement edge drops at unpaved shoulders are a recurring hazard, particularly along narrower two-lane roadways with heavy truck traffic. Trucks not only blow away shoulder material during dry weather, but they also commonly disturb shoulder material by running with one wheel of a tandem overhanging the edge, particularly on the inside of roadway curves. Then too, unstabilized shoulder material is susceptible to rutting by all vehicles during wet weather.

Many State roadway agencies have adopted a 3-inch edge height as a maximum level before maintenance should be done, based on Zimmer and Ivey[7] and other publications[11,12,14,16,] that have adopted the results of that study. Later research, however, has both disputed the Zimmer and Ivey results and provided general guidance that suggests that a 1-1/2 to 2 inch criterion would be a more appropriate for maintenance on 55- to 75-mph roadways.

When edge drops of more than 1-1/2 inches are a frequently recurring problem along a particular stretch of roadway, one or more of the following treatments should be employed to either replace or supplement a normal maintenance practice of simply adding more shoulder material when the edge drop exceeds 1-1/2 inches:

1. Place low-shoulder warning signs
2. Add stabilizing agents to shoulder material that is otherwise susceptible to rutting.
3. Either pave the entire shoulder or pave at least a 2- to 3 foot strip of shoulder adjacent to the travel lane (particularly along the inside of roadway curves with pavements narrower than 22 feet).

Recommended Contract Provisions for Resurfacing Roadways

The most effective way to mitigate present and future hazards associated with pavement/shoulder edge drops is to only issue pavement resurfacing contracts where placing a stabilized shoulder flush with the pavement surface is an integral part of the contract. In addition, all resurfacing contracts where the shoulder is unpaved should require a 45° or flatter bevel along the pavement edge to mitigate any future edge-drop effects.

Recommended Practices for Construction Zones

Table 7.1 shows a recommended practice for treating edge drops in traffic-maintained construction zones. These recommendations are generally patterned after those of Ivey, Mak, Cooner, and Marek,[12] which indicate an increasing need

Pavement Edge Drops

Table 7.1 Traffic Control Needs in Construction Zones with Edge Drop Condition

Edge Drop Height	Lateral Position of Edge Drop					
	IN WHEEL TRACK	IN LANE	ON LANE LINE	AT EDGE OF PAVEMENT	AT EDGE OF SHOULDER	OUTSIDE OF SHOULDER UP TO 30 FT.
1 to 1-1/4"	UNEVEN PAVEMENT SIGNS	UNEVEN PAVEMENT SIGNS	UNEVEN PAVEMENT SIGNS	LOW SHOULDER SIGNS	DO NOTHING	DO NOTHING
1-3/8" to 2"	DISALLOWED	DISALLOWED	CHANNELIZING DEVICES WITH STEADY-BURN LIGHTS	CHANNELIZING DEVICES WITH STEADY-BURN LIGHTS	CHANNELIZING DEVICES WITH STEADY-BURN LIGHTS	DO NOTHING
2-1/8" to 5-7/8"	DISALLOWED	DISALLOWED	CHANNELIZING DEVICES WITH STEADY-BURN LIGHTS	CHANNELIZING DEVICES WITH STEADY-BURN LIGHTS	CHANNELIZING DEVICES WITH STEADY-BURN LIGHTS	CHANNELIZING DEVICES
6" or more	DISALLOWED	DISALLOWED	DISALLOWED	POSITIVE BARRIER	POSITIVE BARRIER	CHANNELIZING DEVICES OR POSITIVE BARRIERS

for traffic control as the edge drop height increases and as the lateral placement of edge drop has relatively higher exposure to vehicle contact. The signing for one-inch edge drops is generally consistent with the Texas[13] practice.

Edge drops placed within a travel lane during resurfacing should be discouraged both because of the high exposure to contact and also because of the nearness of conflicting traffic. Positive barriers should be placed where 6-inch or higher edge drops are close to traffic to eliminate the high probability of vehicle rollover on contact.

As recommended in the *Traffic Control Devices Handbook*,[19] edge drops in construction zones should be limited in duration and length. This handbook also recommends that a fillet or wedge (see Figure 7-15) be placed along an edge drop, if it is left overnight. Additional consideration should be given to tapering the edge of a resurfacing overlay, particularly if it is within or immediately adjacent to the travel lanes.

FILLET OF MATERIAL

A fillet of material is a "wedge" of gravel, or other material placed in a manner that will provide stability for errant vehicle and is used to reduce the drop-off as a result of an excavation, as illustrated in Figure 6-8. It can be used when work in the excavation is discontinued for a short period of time, as at night, and removed when work will start again. Frequently, this wedge is composed of the same material which is either being excavated or back filled (such as crushed rock base course).

Fillet of Material

Roadway

Fillet of Material

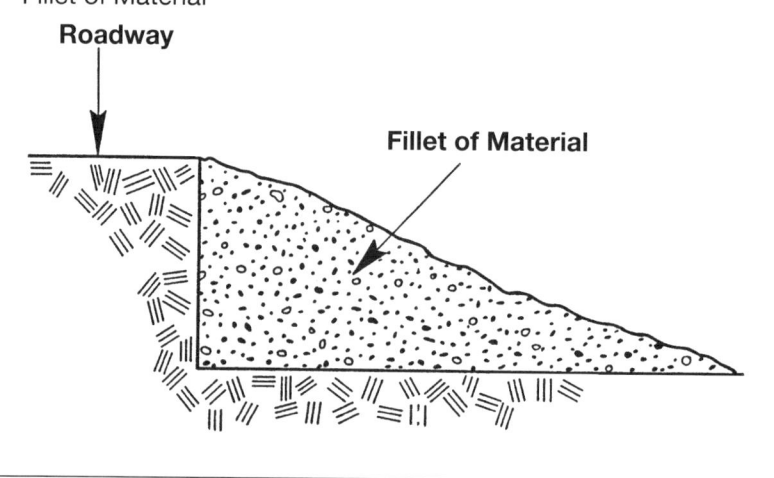

Figure 7-15 Traffic Control Devices Handbook Recommendations.[19] *(Courtesy of Criterion Press, Inc.)*

7.4 Technical Aspects of Pavement Edge Drop Cases

Pavement edge drops have gained increased attention over the past few years as a primary factor in tort claims against roadway agencies. Edge drops most commonly occur along two-lane roadways at the edge between a paved surface and an unpaved shoulder. This condition results either because of (a) a lack of timely maintenance, (b) a soft shoulder that allows a edge drop to form, or (c) a pavement was overlaid under contract without any contract specification for treating the shoulder.

Edge drops also occur during construction projects either where pavement overlays are being placed, and all lanes have not been completed, or where excavation is underway to construct new lanes or shoulders next to the existing travel lanes. Edge drops can occur in construction zones at several places across the roadway as follows (listed in order of increasing hazard):

1. At 12 feet or more from the edge of the nearest travel lane.
2. At the outside edge of a paved shoulder, often without pavement markings or other traffic control devices separating the travel lane and the shoulder.
3. At the edge of the traveled way.
4. Between adjacent lanes in the same direction or between opposing lanes.
5. Within a travel lane (a practice to be discouraged).

Both shoulder and lane edge drops can cause drivers to have unexpected collisions. Most commonly, the vehicle will be affected in one of three ways: (a) move abruptly across the travel lanes and either collide with other vehicles or collide with roadside hazards off the opposite edge of the roadway; (b) overturn on the roadway or roadside; or (c) collide with roadside hazards beyond the edge drop.

The research cited earlier has generally shown that the driver's ability to recover from a pavement edge-drop excursion is a function of edge-drop height and shape, vehicle speed and path angle, and the width of lane available for recovery. Then too, certain vehicles such as motorcycles, sub-compact cars, vehicles pulling trailers, and tractor-trailer trucks have a much greater sensitivity to edge drops.

Knowing not only that 30-40% of all accidents involve single vehicles running off the road,[20] but also that only one accident occurs for every 6 roadside encroachments, there comes a realization that many different drivers encroach on the roadside for many different reasons. For that reason, the roadway community over the last 30 years has generally subscribed to the *forgiving roadside concept*, which says that the roadside should be designed and maintained to minimize the severity of occupant injuries associated with roadside accidents.

When investigating the tort liability of a roadway agency for an edge drop condition, that condition can be found clearly contrary not only to the *forgiving roadside concept* but also to the function of shoulders as a place of pavement support, safe emergency refuge for vehicles, and safe recovery from casual encroachments. For any edge drops above about one inch, the edge drop could have been a significant contributor to the accident. Although a driver has some general responsibility to keep his vehicle under control, this control becomes tenuous when the driver is suddenly surprised (particularly at night) by the inadvertent dropping of one or more wheels off of the pavement.

A plaintiff attempting to recover for injuries resulting from a collision, where an negligent edge drop is alleged, needs to establish an evidence base to support the claim. The important evidence is as follows:

a) The height and shape of edge drop at increments along the vehicle's shoulder excursion from exit to re-entry (the most important area is about 50 to 150 feet before re-entry)

b) The lateral placement of wheel tracks on shoulder (Did the tires scrub on the edge?)

c) Any evidence of scrubbing found on the pavement edge or the inside sidewalls of the tires.

d) The driver's recollection of surprise, fear, or feeling of futility.

e) The witness statements of the vehicle speed.

f) An accident reconstruction of the vehicle speed.

g) Any evidence of unstable shoulder material.

h) Any pavement irregularities that could have caused the roadside encroachment.

i) The roadway agency maintenance or construction zone standards regarding the maximum allowable edge drops.

j) The local maintenance practice regarding the treatment of edge drops.

The most important evidence in a edge-drop case is the measurement of the edge-drop height and shape at several locations along the edge-drop excursion (indicated by the accident report). Some investigators will measure the edge-drop geometry at the re-entry point. Other investigators will measure the edge-drop geometry at the exit point. Others will measure the maximum or minimum edge-drop conditions. The best way is to measure the edge-drop geometry is at 10-20 foot intervals along the entire off-road excursion. That way, the most critical edge-drop condition can be determined.

Pavement Edge Drops

Although most of the research has concentrated on the vehicle dynamics associated with scrubbing re-entry, other accident situations can be causally related to edge drops. Most clearly, the vehicle rollover associated with vehicle undercarriage contact on larger edge drops can be cited. Then too, when drivers attempt to recover from a shoulder or roadside excursion, they will steer a circular path that requires lateral acceleration. If the radius of this path is relatively high for the vehicle speed, a contact with a vertical discontinuity such as an edge drop could trigger loss of control (much like hitting a large bump while steering a roadway curve).

7.5 Typical Defense Arguments

When a roadway agency defends an edge-drop case they will usually allege either that the driver was the sole negligent party and/or the specific edge drop had no contribution to the causation of the collision. More specifically these arguments are:

1. The driver was negligent for running off the travel lanes.
2. The driver was negligent for not having slowed the vehicle, run parallel to the edge drop, and then eased the vehicle back to the traveled way at a slow speed (a practice suggested in the driver's handbook in most States).
3. The driver was negligent by oversteering the vehicle.
4. The driver was negligent for driving while intoxicated.
5. The driver was negligent for speeding.
6. Direct evidence is lacking to indicate that the vehicle was scrubbing on the edge drop. Therefore, the only conclusion about loss of control is that the driver committed an error.
7. The roadway agency had no duty to repair or warn of this pavement edge drop because they had no notice of the condition.
8. The roadway agency had no duty to repair or warn of this pavement edge drop because it was not high enough to constitute a hazard, as per Zimmer and Ivey,[7] or Ivey and Sicking,[11] or Ivey, et.al.[16]
9. The roadway agency had no duty to repair the edge drop because it did not violate their own maintenance or construction zone standard.
10. The roadway agency had no duty to repair or warn of this pavement edge drop because it was tapered or rounded enough to mitigate any hazard, as per Ivey and Sicking.[11]
11. There are no nationally recognized standards that require roadway agencies to treat edge drops.

Endnotes

1. American Association of State Highway Officials, *A Policy on Geometric Design of Rural Highways*, 1954.

2. American Association of State Highway Officials, *A Policy on Geometric Design of Rural Highways*, 1965.

3. Ivey, D.L., and Griffin, L.I., *Driver/Vehicle Reaction to Road Surface Discontinuities and Failures*, 16th Congress of the International Federation of the Societies of Automotive Engineers, Tokyo, Japan, 1976.

4. Nordlin, E.F., Parks, D.M., Stoughton, R.L., and Stoker, J.R., *The Effect of Longitudinal Edge of Paved Surface Drop-off on Vehicle Stability*, California Department of Transportation, 1976.

5. Klein, R.H., Johnson, W.A., and Szostak, H.T., *Influence of Road Disturbances on Vehicle Handling*, National Highway Traffic Safety Administration, 1977.

6. Graham, J.L., and Glennon J.C. *Work Zone Design Considerations for Truck Operations and Pavement/Shoulder Drop-offs*, Federal Highway Administration, 1984.

7. Zimmer R.A., and Ivey D.L., *Pavement Edges and Vehicle Stability - A Basis for Maintenance Guidelines*, Texas Transportation Institute, 1982.

8. McHenry, R.R. and Deleys, N.J., *Automobile Dynamics - A Computer Simulation of Three Dimensional Motions for Use in Studies of Braking Systems and of the Driving Task*, Cornell Aeronautical Laboratory, 1970.

9. Glennon, J.C., *Effect of Pavement/Shoulder Drop-offs on Highways Safety: A Synthesis of Prior Research*, Transportation Research Board, 1985.

10. Olson, P.L., Zimmer, R.A., and Pezoldt, V., *Pavement Edge Drop*, Transportation Research Board, 1986.

11. Ivey, D.L., and Sicking, D.L., *The Influence of Pavement Edge and Shoulder Characteristics on Vehicle Handling and Stability*, Transportation Research Board, Record 1084, 1986.

12. Ivey, D.L., Mak, K.K., Cooner, H.D., and Marek, M.A., *Safety in Construction Zones where Pavement Edges and Drop Offs Exist*, Transportation Research Board, Record 1163, 1988.

13. Texas State Department of Highways and Public Transportation, *Texas Manual on Uniform Traffic Control Devices for Streets and Highways*, 1988.

14. Humphreys, J.B., and Parham, J.A., *The Elimination or Mitigation of Hazards Associated with Pavement Edge Drop-offs During Roadway Resurfacing*, AAA Foundation for Traffic Safety, 1994.

15. Griffin, L.I., *The Influence of Roadway Surface Discontinuities on Safety (Chapter 2 - Accident Data Relationships)*, Transportation Research Board, State-of-the-Art Report, 1984.

16. Ivey, D.L., Johnson, W.A., Nordlin, E.F., and Zimmer, R.A., *Influence of Roadway Surface Discontinuities on Safety (Chapter 4 - Pavement Edges)*, Transportation Research Board, State-of-the-Art Report, 1988.

17. Glennon, J.C., *Relationship Between Safety and Key Highway Features (Chapter 3 - Effect of Pavement/Shoulder Drop-offs on Highway Safety)*, Transportation Research Board, State-of-the-Art Report 6, 1987.

18. Transportation Research Board, *Designing Safer Roads*, Special Report 214, 1987.
19. Federal Highway Administration, *Traffic Control Devices Handbook*, 1983.
20. American Association of State Highway and Transportation Officials, *Roadside Design Guide*, 1988.

Chapter 8

Rail-Highway Grade Crossings

Reducing accidents at rail-highway grade crossings has long been a subject of public concern. No other kind of motor-vehicle accident has such a high severity, making this a safety issue of primary significance. The ratio of persons killed and injured to the number of accidents at grade crossings is 40 times the same ratio for all motor-vehicle accidents.

Of the 182,000 public crossings in the United States, about 64,000 have active warning devices that give the driver a real-time warning of an approaching train.[1] Of the remaining 118,000 crossings, about 9,000 have no warning devices and about 109,000 have some form of passive protection (static signs and markings). These devices inform the driver of the location of the crossing. But in the absence of active devices, the driver is left without positive guidance about approaching trains.

In 1973, the U.S. Congress passed the Federal Highway Safety Act that earmarked categorical federal funding for safety improvements at rail-highway grade crossings. Current annual federal spending is in the $250 million range, with most of these funds spent on major improvements such as grade separations, roadway realignment, flashing signals, and automatic gates.[2] In contrast, very little money is spent to upgrade the tens of thousands of low traffic crossings in the U.S., which need improvement but can only justify spending for passive traffic control devices.

This chapter focuses not only on the key features of grade crossings that relate to safety, but also on the most widely accepted and/or most cost-effective means to minimize crossing accidents. The reader, wishing a more comprehensive understanding, should consult References 1-6 listed at the end of this chapter.

8.1 Accident Statistics

The U.S. Department of Transportation estimates the annual number of vehicle-train accidents at public crossings at 5,800.[1] In addition to the 600 fatalities, these accidents produce about 2,300 injuries. This injury figure, however, does not tell the whole story. Many of the injuries in other kinds of accidents are minor in consequence, whereas the collision between a 3,000-lb. vehicle and a several-hundred-ton train is much more likely to produce severe trauma.

About five times as many accidents at rail-highway grade crossings do not involve trains.[3] These accidents include rear-end collisions with vehicles stopped at the crossing, fixed-object collisions with crossing devices, and ran-off-road collisions by drivers losing control on rough crossings or on severely humped crossings. Therefore, remedial measures for reducing vehicle-train collisions must also not compound non-train-involved accidents.

Other available statistics on vehicle-train collisions pinpoint causative factors as follows[1]:

1. Crossings with either passive traffic control or no devices have about 50% of all vehicle-train injury and fatal accidents;
2. Trucks are over-represented in vehicle-train accidents;
3. Drivers over 65 years of age are over-represented in vehicle-train accidents;
4. The number of vehicle-train accidents is about one-third higher in the winter months as in the summer months; and
5. During daylight, 75% of vehicle-train accidents at crossings with passive traffic control devices involve a motor vehicle being struck by a train, while at night, almost 50% involve a motor vehicle running into the side of a train.

Because 65% of public grade crossings have either passive or no control, and because 45,000 of these crossings have two or less trains per day and less than 500 vehicles per day,[1] economic justification for other than passive devices is probably not possible. Knowing the number of grade crossings and the number of vehicle-train accidents, the average time-rate for vehicle-train accidents at a crossing is easily computed as one accident every 31 years. Stated differently, each year one accident occurs for every 31 crossings. Likewise, a crossing will average one fatality about every 300 years. These rates simply point to the rarity of this type of accident, and stress the need for inexpensive countermeasures.

8.2 Legal Responsibilities of Railroads and Roadway Agencies

Courts have upheld that railroads have a general duty for the safety of the motoring public that uses rail-highway grade crossings. Their duties to provide a safe environment at a crossing include the placement and maintenance of standard crossbucks, the clearance of sight obstructions on their right-of-way, the maintenance of relatively flat and smooth crossings, and the placement of additional traffic control devices where needed at hazardous crossings.

Roadway agencies also have a general duty for the safety of the motoring public that uses rail-highway grade crossings. Their duties to provide a safe environment at a crossing include the placement and maintenance of advanced warning signs and pavement markings, clearance of sight obstructions on their right-of-way, the maintenance of safe approach roadways, and the placement of additional traffic control devices where needed at hazardous crossings.

8.3 Legal Requirements of Drivers at Crossings

The roadway driver is a key component of rail-highway grade crossing operation and is charged with obeying traffic control devices, traffic laws, and the rules of the road. Roadway and railroad engineers, who design crossings, should be aware of the several characteristics, capabilities, needs, and obligations of the driver. This information will help them, through the proper engineering of crossings, to assist drivers to drive safely.

The Uniform Vehicle Code (UVC),[4] a model set of motor vehicle laws, describes the actions that a driver is required to take at crossings as described below:

- *Approach Speed (Sec. 11-801)*

  No person shall drive a vehicle at a speed greater than is reasonable and prudent under the conditions and having regard to the actual and potential hazards then existing. Consistent with the foregoing, every person shall drive at a safe and appropriate speed when approaching and crossing an intersection or railroad grade crossing...

- *Passing (Sec. 11-306)*

  No vehicle shall be driven on the left side of the roadway under the following conditions:
  when approaching within 100 feet of or traversing any...railroad grade crossing unless otherwise indicated by official traffic control devices...

- *Stopping (Sec. 11-701)*

  Obedience to signal indicating approach of train. Whenever any person driving a vehicle approaches a railroad grade crossing under any of the circumstances stated in this section, the driver of such vehicle shall stop within 50 feet, but not less than 15 feet from the nearest rail of such railroad, and shall not proceed until it is safe to do so. The foregoing requirements shall apply when:

  - a clearly visible electric or mechanical signal device gives warning of the immediate approach of a railroad train;

  - a crossing gate is lowered or when a human flagger gives or continues to give a signal of the approach or passage of a railroad train;

- a railroad train approaching within approximately 1,500 feet of the highway crossing emits a signal audible from such distance and such railroad train, by reason of its speed or nearness to such crossing, is an immediate hazard;

- an approaching railroad train is plainly visible and is in hazardous proximity to such crossing.

The UVC also prohibits any vehicle from driving around or under any gate while it is being opened or closed (Sec. 11-701b).

Each State has its own regulations that may vary somewhat from the UVC. Information on State regulations for crossings are contained in *Compilation of State Laws and Regulations on Matters Affecting Rail-Highway Crossings.*[5]

8.4 Stages of the Rail-Highway Grade Crossing Maneuver

The track-crossing maneuver faced by a roadway driver can be pictured (see Figure 8-1) as three stages or zones as described in Figure 8-2 and the following text:

Stage I - Approach Zone

At some distance before the crossing, this roadway zone is where the driver must formulate and act on decisions needed to avoid colliding with an approaching train. The driver is usually informed of the presence of a crossing by advance warning signs, by pavement markings, or sometimes by seeing the crossing devices or a train.

The advance warning devices compete for the attention of a driver who may be listening to the radio, talking on a cellular phone, or talking with a passenger. Drivers who are familiar with the crossing may be cautious when they know train traffic is heavy. On the other hand, a driver may not give much attention to a crossing that has shown little train activity in the past.

This Approach Zone precedes the Critical Stopping Zone. For crossings with passive traffic control, the driver who cannot see a nearby train while in the approach zone, will have difficulty avoiding a collision.

Stage II - Critical Stopping Zone

Drivers approaching a crossing reach the Critical Stopping Zone where a decision must be made to stop if a train is approaching or on the crossing. If the decision is delayed much beyond the beginning of this zone, the available stopping distance will be too short to avoid hitting the train.

Rail-Highway Grade Crossings 229

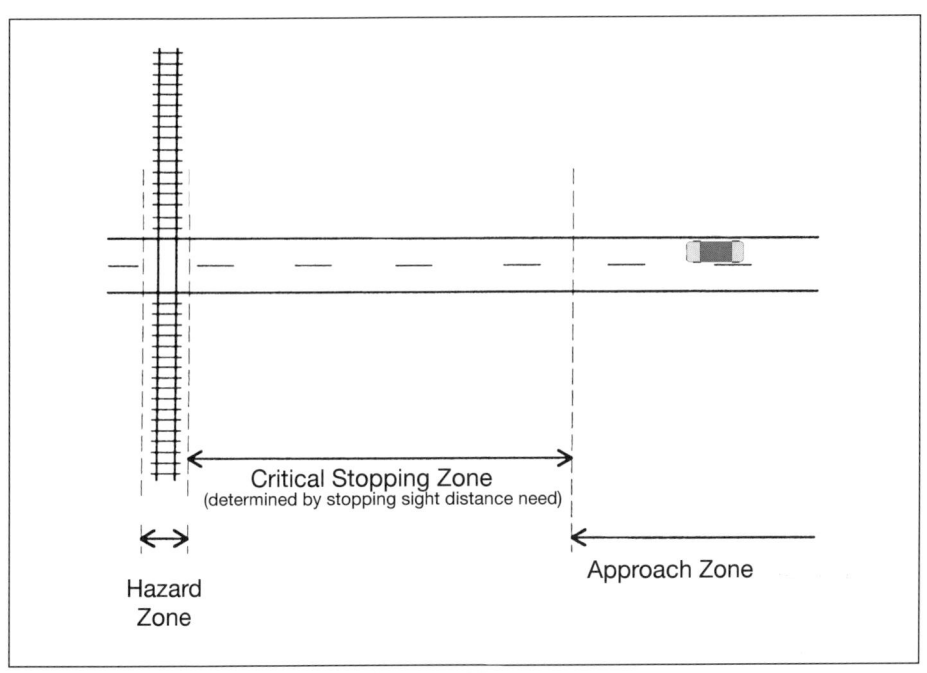

Figure 8-1 Stages of Rail-Highway Crossing Maneuvers.

HIGHWAY ZONE	CROSSING CONTROL	DRIVER INFORMATION NEEDS	SAFE DRIVER RESPONSE
Approach Zone	(Railroad Crossing sign)	• Adequate sight triangle for train detection • Advance warning devices • Sound of train	• Search for trains • Listen for trains • Slow if indicated
	(Crossing signal) OR STOP	• View of control devices • Advance warning devices	• Watch for flashers or stop sign • Slow as indicated
Critical Stopping Zone	(Railroad Crossing sign)	• Immediate or prior detection of train	• Stop for train
	STOP	• Immediate or prior view of advance warning of stop sign	• Stop for stop sign
	(Crossing signal)	• Immediate or prior view of flashers and train	• Stop for train
Collision Zone	STOP	• Sight distance to accelerate and clear crossing from stop	• Stop for trains • Listen for trains • Proceed when safe
	(Railroad Crossing sign) OR (Crossing signal)	• View of train • Knowledge of hidden oposite direction trains or presence of gates.	• Wait for nearby trains to pass • Proceed when safe

Figure 8-2 Driver Information Needs at Rail-Highway Crossings (Courtesy of Criterion Press, Inc.)

The beginning of the Critical Stopping Zone is defined by the stopping sight distance needed for the roadway speed. Stopping sight distances and the dimensions of sight triangles, discussed extensively later in this chapter, are major factors in the safety of crossings.

Adequate sight triangles and/or the proper traffic control give the majority of drivers the information needed to make the decision to stop. At crossings with passive traffic control, driver's should have an adequate view of the crossbuck that, by its design, informs them of the crossing and requires, as a regulatory device, that they approach the crossing so they can stop their vehicle if needed.

Active traffic controls devices such as flashing signals and gates are placed at the more hazardous crossings. They are activated by an approaching train and are designed to assist the driver in making the decision to stop for an approaching train before the Critical Stopping Zone is reached.

Stage III - Collision Zone

In this final zone, the driver must safely cross the tracks. At crossings with adequate sight triangles and passive traffic control, the driver who has seen the approaching train in time can stop short of the collision zone. At crossings with active traffic control, the driver who has heeded the activated devices in time can likewise stop short of the collision zone. Of course, once the train has cleared and no other trains are approaching, the driver can cross safely.

At crossings with passive traffic control and inadequate sight triangles, drivers can pass the beginning of the Critical Stopping Zone unaware of an approaching train and enter the collision zone unable to avoid hitting the train. Other driver dilemmas at passively controlled crossings can include, indecision whether to stop or go, stopping too close to the track, or stalling on the track.

At crossings with active traffic control, care should be taken to avoid false activations when no trains are approaching. Otherwise, drivers will lose respect for the signals and/or gates and may enter the collision zone when a train is approaching. The same is true for overly long activation lead times.

8.5 Vehicle-Train Conflict Circumstances

The vehicle-train conflict is similar to the vehicle-vehicle conflict at roadway intersections. Quite often, sight distance is restricted. Sometimes distracting elements are present. Then too, driver perception of an ensuing conflict is difficult at night. But the most critical factors are the vehicle speeds and separation distances at the time the driver is able to perceive a potential collision. If, at this time, the speeds are too great and the distances too small, a collision is inevitable.

Reducing the criticality of vehicle-train conflicts requires driver's awareness of a potential conflict so they have enough perceptual lead time to avoid collision. This requirement is sometimes difficult to satisfy because of two distinct confounding effects. First, driver awareness of a potential collision may be difficult because humans are poor judges of closing rate (relative speed). When driver's see an approaching train, they must decide whether to stop for the train or to proceed through the crossing ahead of the train. This decision may require split-second judgment of several complex factors including the train speed and distance from the crossing. The result may be indecision or a wrong decision.

The second confounding effect, and perhaps more important, is the driver's expectancy. The driver enjoys the freedom of mobility, and the ability to maneuver quickly and independently. With this freedom, the driver must avoid conflict with trains that are restricted to rails and are incapable of anything more than planned and deliberate maneuvers. Although drivers generally *respect* trains, the problem is that drivers rarely *expect* trains. Because most drivers seldom encounter a train, they tend to expect the absence rather than the presence of trains. For this reason, traffic control devices must counteract driver's false expectancy and alert them to actions that are necessary when a train approaches.

8.6 Driver Needs

Any effective program for improving safety at rail-highway grade crossings should address the driver's needs in approaching and traversing the crossing. To this end, all feasible steps should be taken to assist the driver's task by conveying the proper messages and encouraging the proper attitude.

The design of a grade crossing should examine the decisions that drivers must make, their proficiency in making them, their attitudes about various situations encountered, their informational needs, their ability to digest that information, and their response to particular bits of information. The objective is to elicit proper behavioral responses by drivers. Behavioral responses are a result of perceptions, real or imagined. Each driver's ability to respond appropriately depends largely on how accurately their perceptions reflect the real world. The accuracy of a driver's awareness to the driving environment is a function not only of sensory capabilities, but also of the degree to which they are presented with perceptual aids, in the form of devices such as signs, signals, and markings.

When drivers approach a crossing, they first need to know that the crossing is there and, second, if a train is on the crossing, approaching the crossing, or not in the vicinity of the crossing. They also may need guidance on how to safely approach and traverse the crossing.

At passively-controlled crossings, the driver's task can be made more hazardous by any of the following:

1. **Sight-Restricted Crossing**. A driver approaching a grade crossing will reach the vertex of a critical sight triangle where a decision must be made to stop if an approaching train is also within the triangle. The required sight triangle (see Section 8.8) is a function of both the roadway speed and track speed. When existing sight restrictions at a crossing allow the driver to pass that critical point without seeing a nearby train, the driver has difficulty either braking or proceeding through the crossing without hitting with the train.

   When a deficient sight triangle (see Figure 8-3) is allowed to exist at a grade crossing, all of the burden is put on the driver to know that the restriction is hazardous and to determine what speed is appropriate for that crossing. This is a task that many drivers do not handle well.

Figure 8-3 Typical Passively-Controlled Rail-Highway Grade Crossing with Severe Sight Restrictions.

Rail-Highway Grade Crossings

2. **Roadway Intersections Close to the Crossing.** Over 35% of public rail-highway crossings have a roadway intersection within 75 feet of the tracks, as shown in Figure 8-4. Usually this occurs when a major roadway is close to and runs parallels to the track, and a minor

Figure 8-4 Roadway Intersection Close to a Rail-Highway Grade Crossing.

roadway intersects both the major roadway and the track. These crossings generally have a higher occurrence of accidents because of several inherent deficiencies.

The major difficulty for these crossings lies with the driver who is turning off of the major roadway toward the crossing. This driver will not have a preview of a train that is approaching from behind until the vehicle turns onto the minor roadway and is squared up toward the crossing. In effect, the available sight distance triangle (see Section 8.8) may often only be safe for roadway speeds of 5 mph or less.

This kind of crossing also presents particular problems for long vehicles on the minor roadway that have to stop on the tracks before entering the major roadway. These vehicle have additional difficulty turning from the major roadway when a train is coming because they may end up blocking major road traffic when they stop for the train.

3. **Sharp Crossing Intersection Angle.** Sharp crossing angles present a similar problem as sight-restricted crossings and crossings with nearby intersections. Drivers, particular in large trucks, approaching a crossing either cannot or will not look back over their shoulder to search for a train. Figure 8-5 is a panoramic view by a truck driver on a crossing with a 25° crossing angle, showing that the driver can only really see to the front and immediately alongside his truck.

4. **Poor Visibility of the Train at Night**. Train cars are not reflectorized and are often dirty and painted dark colors, giving them little contrast with the background. As a result, approaching drivers cannot see a train that is already on a dark crossing as they approach at night.

5. **Lack of Preview of the Crossing**. Drivers may not see the crossing as they approach, either from a lack of conspicuous advance warning signs and pavement markings, or an obstructed view of the crossing itself due to unusual geometrics or sight obstructions.

6. **Rough Crossings**. When a crossing surface is rough or uneven, a driver's attention may be devoted primarily to choosing the smoothest path over the crossing, rather than scanning both track directions for an approaching train. This undesirable behavior will be particularly pre-conditioned if the driver is consistently exposed to rough or uneven crossings.

7. **Steep Crossing Approaches**. Steep crossing approaches have several attendant hazards. They distract drivers from searching for trains because of the extra attention needed to both negotiate the grade and to look for conflicting oncoming vehicles hidden by the crest. They

Rail-Highway Grade Crossings

Figure 8-5 *Panoramic View by Truck Driver at a Rail-Highway Grade Crossing with a 25° Crossing Angle.*

cause particular dangers for drivers under icy winter conditions. And, they can cause longer trucks to high-center on the crossing (see Figure 8-6). For these reasons, many States prohibit crossing grades steeper than about six percent. However, crossings sometimes violate State laws regarding maximum grade due to periodic track maintenance, which raises the level of the track.

8. **Appearance of Abandonment.** Crossings where advance warnings, pavement markings, crossbucks, and crossing surfaces are in disrepair may look as if they have been abandoned. Under these conditions, drivers may ignore searching for oncoming trains because they do not expect any.

8.7 Traffic Control Devices for Rail-Highway Grade Crossings

Traffic control devices for rail-highway grade crossings include all signs, signals, and markings along roadways approaching and at the crossings. The function of these devices is to encourage safe and efficient operation of roadway traffic. With due regard for safety and for the integrity of operations by roadway and railroad users, the roadway agency and the railroad company are entitled to jointly occupy the right-of-way to conduct their functions. This requires joint

Figure 8-6 *Steep Approach Grade to a Rail-Highway Grade Crossing.*

responsibility for traffic control between the roadway agency and the railroad company. The design, installation and operation of traffic control devices shall be in accordance with Part 8 of the *Manual on Uniform Traffic Control Devices (MUTCD)*,[6] which is part of Title 23 of the U.S. Code and is the required standard for all States. The remainder of this section mainly paraphrases the *MUTCD*.

Crossbuck Signs

Passive traffic controls consisting of signs, pavement markings, and grade-crossing illumination, identify and direct attention to the location of a grade crossing. The crossbuck, is a white reflectorized sign (see Figure 8-7), with the words RAILROAD CROSSING in black lettering. As a minimum, one crossbuck shall be used on each roadway approach to every grade crossing, alone or in combination with other traffic control devices. With two or more tracks between the signs, the number of tracks shall be indicated on an auxiliary sign of inverted T shape mounted below the crossbuck as indicated in Figure 8-7.

Where physically possible and also visible to approaching driver's, the crossbuck shall be installed on the right-hand side of the roadway on each approach to the crossing. Where an engineering study finds restricted sight distance or unfavorable road geometry, crossbucks shall be placed back to back or otherwise located so that two faces are displayed to that approach. The reflectorized backed-up crossbuck also serves as a flash panel to allow drivers to see moving trains at night on dark crossings.

Figure 8-7 *Crossbuck at a Rail-Highway Grade Crossing.*

Advance Warning Signs

A railroad advance warning sign (see Figure 8-8) is required by the *MUTCD* on each roadway in advance of every grade crossing except:

1. On low-volume, low-speed roadways crossing minor spurs or other tracks that are infrequently used and which are flagged by train crews.
2. In the business districts of urban areas where flashing signals are used.
3. Where physical conditions preclude an even partially effective display of the sign.

The W10-2, 3, and 4 signs shown in Figure 8-8 may be installed on roadways that are parallel to railroads to warn turning drivers that a railroad crossing is ahead.

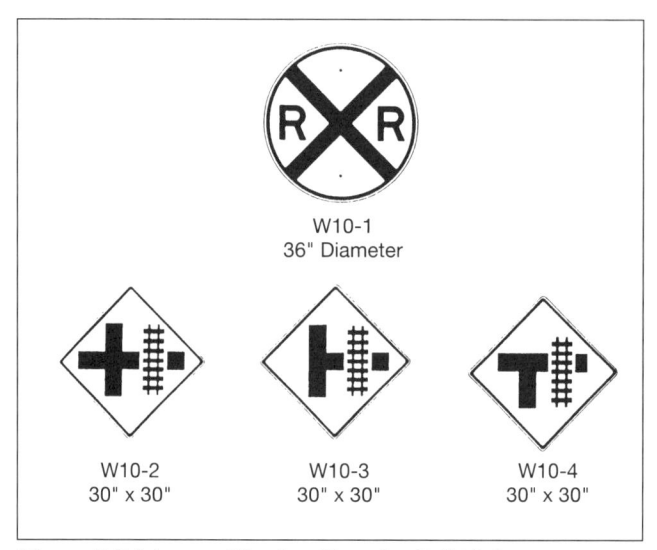

Figure 8-8 Advance Warning Signs for Rail-Highway Crossings.[6]

Pavement Markings

The *MUTCD* requires pavement markings in advance of a grade crossing, consisting of an X, the letters RR, a no-passing zone marking (2-lane roads), and certain transverse lines (see Figure 8-9). Identical markings shall be placed in each lane on all paved approaches to grade crossings where signals or automatic gates are located, and at all other grade crossings where the prevailing speed of highway traffic is 40 mph or greater. A portion of the pavement marking should be directly opposite the advance warning sign. If needed, supplemental pavement markings may be placed between the advance warning sign and the crossing.

Rail-Highway Grade Crossings

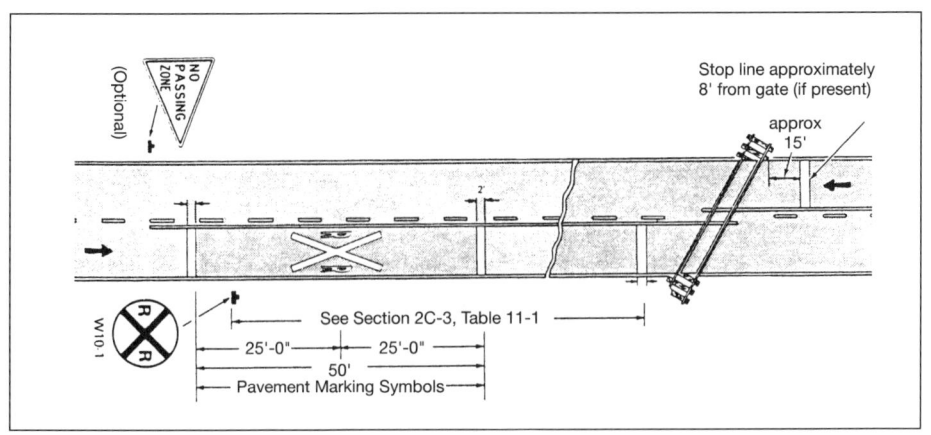

Figure 8-9 Advanced Pavement Markings Required at Rail-Highway Crossings.[6]

Pavement markings shall also be placed at crossings where engineering studies indicate a significant potential conflict between vehicles and trains. At minor crossings or in urban areas, these markings may be omitted if engineering studies indicate that other devices provide suitable control.

Turn Prohibition Signs

At signalized roadway intersections within 200 feet of a grade crossing, where the intersection traffic signals are preempted by the approach of a train, all existing turning movements toward the grade crossing should be prohibited by proper placement of a NO RIGHT TURN sign or a NO LEFT TURN sign or both.

Stop or Yield Signs

STOP or YIELD signs can be used for controlling grade crossings (see Figure 8-10). These signs are most appropriate at grade crossings with lower roadway traffic volumes that have the following characteristics:

1. Two or more trains per day.
2. Restricted sight triangles for approaching roadway drivers.
3. Adequate sight triangles for stopped roadway drivers.
4. Other compelling reasons found by an engineering study.

STOP AHEAD or YIELD AHEAD signs shall be installed in advance of a STOP or YIELD sign.

Flashing Light Signals

When showing the approach or presence of a train, the flashing signal (see Figure 8-11), displays two red lights flashing alternately in a horizontal line. A

240 Roadway Defects and Tort Liability

Figure 8-10 Stop Sign Used at a Rail-Highway Grade Crossing.

Figure 8-11 Post-Mounted Flashing Light Signal.

Rail-Highway Grade Crossings

bell, suitable for warning pedestrians and bicyclists, may be included in the assembly and operated in conjunction with the flashing signals.

Flashing signals should be placed to the driver's right on all approaches to a crossing. Additional pairs of lights may be mounted on the same supporting post and directed toward drivers from roadways closely adjacent to and parallel to the railroad. At crossings with roadway traffic in both directions, back-to-back pairs of lights shall be placed on each side of the tracks. Two sizes of lenses, 8-inch and 12-inch diameter, are available for flashing signals. The larger lens, with better visibility, is the preferred alternative, particularly for higher-speed roadways.

Where required for better visibility to approaching traffic, particularly on multi-lane approaches, cantilevered flashing signals are used (see Figure 8-12). Cantilever signals are also suitable for high-speed rural roadways, high-volume two-lane roadways, or specific locations with distractions.

Figure 8-12 Cantilevered Flashing Light Signals.

Automatic Gates

Automatic gates are used as an adjunct to flashing signals. They consists of a drive mechanism and a reflectorized red and white striped gate arm with lights, which when down extends across the approach roadway about 4 feet above the pavement. The flashing signal may be either supported on the same post with the gate or separately. A gate in the down position is shown in Figure 8-13.

Figure 8-13 Flashing Light Signal with Automatic Gates.

For a normal operation, both the flashing signals and the lights on the gate in the upright position are activated with detection of an approaching train. The gate shall start down not less than 3 seconds after the lights start flashing, shall reach horizontal before the arrival of the train, and shall remain down until the train clears the crossing. When the train clears, and no other train is coming, the gate shall rise to its upright position normally in not more than 12 seconds, following which the flashing signals and the lights on the gate shall cease flashing. For specific locations, the timing of the gate should accommodate any slow moving trucks.

Train Detection

To advise drivers and pedestrians of the approach or presence of trains, active traffic control devices are actuated by electronic train detection. Generally the method is automatic, using the track circuit, which is usually designed on the fail-safe principle, using closed loops.

Where trains operate at speeds of 20 mph or higher, circuits controlling automatic flashing light signals shall operate a minimum of 20 seconds before arrival of any train. Where the speeds of trains vary considerably, special devices or circuits should be used to give uniform notice in advance of all train movements. Special control features should also be used to eliminate the effects of station stops and switching operations within approach control circuits. A reasonably complete description of the types and elements of detection circuits can be found in the *Railroad-Highway Grade Crossing Handbook*.[2]

Traffic Signals Near Grade Crossings

When roadway intersection traffic signals are within 200 feet of a grade crossing, control of the traffic flow should provide drivers a measure of safety at least equal to what existed prior to the installation of the signals. The normal sequence of traffic signals should be preempted by the approach of trains to avoid trapping vehicles on the crossing.

Where a signalized roadway intersection is adjacent to a grade crossing without active traffic control, the possibility of vehicles being trapped on the crossing remains, and preemption of the traffic signal is usually required. Where preemption is used, flashing signals at the grade crossing are desirable, and usually require only a minor addition to the cost of the railroad circuits required for the preemption function.

> **RAIL-HIGHWAY CROSSING DEFECTS**
> **Rules of Thumb**
> - *Adequate Sight Triangles are Key to the Safety of Passively-Controlled Crossings*
> - *Large Trucks Will Have Trouble with Marginally Acceptable Sight Triangles*
> - *Sharp Roadway Approach Angles and Intersections Close to Crossings Severely Hamper the Effective Sight Distance*

8.8 Sight Distance Requirements

A roadway driver approaching a rail-highway grade crossing needs adequate sight distance to see an approaching train and to make the right decision on whether to proceed or stop. Also, when a driver stops at the crossing, particularly in a large truck, a clear line of sight is needed to give enough time to accelerate and clear the crossing. The sight distance can be constrained by horizontal and vertical alignment, terrain, vegetation, trains parked on a siding, and man-made structures.

Moving Roadway Vehicle

The speeds of the train and the roadway vehicle define the length of the legs of a clear sight triangle, which the moving driver needs to be able to see a train and either stop or clear the crossing before collision. Distance along the roadway must, as a minimum, be the safe stopping sight distance for a given approach speed. Distances along the track, as a minimum, are determined by the time needed for the roadway vehicle to clear the track traveling from the critical stopping sight distance point. In other words, the contingency is provided. If both vehicles arrive at the apexes of the triangle at the same time, and the roadway driver decides to proceed instead of stop, the roadway vehicle will just clear the crossing without impact.

Sight triangle requirements are given in the *Railroad-Highway Grade Crossing Handbook,*[2] the *Traffic Control Devices Handbook,*[7] and *A Policy on Geometric Design of Highways and Streets.*[8] Reference 8 has slightly different dimensions for the sight-triangle legs than the other two sources. The dimensions from the other two sources are more widely quoted and will be presented here.

Referring to Figure 8-14, the minimum sight-triangle leg, H along the roadway is measured from the nearest rail to the driver of the roadway vehicle. It is the sum of the minimum stopping sight distance (as defined in Reference 8 for all sections of roadway) and the minimum clearance distance between the nearest track and the driver after the roadway vehicle stops. This allows the vehicle to avoid collision by stopping before encroaching on the crossing area. The equation for this distance is:

$$H = 1.47 V_v t_{pr} + \frac{V_v^2}{30f} + D + d_e$$

where

H = minimum sight-triangle leg along the roadway, ft
V_v = velocity of vehicle, mph
t_{pr} = perception/reaction time of roadway driver, sec, (assumed: t_{pr} = 2.5 sec)
f = coefficient of friction used in braking (see Table 8-1 for assumed values)
D = clearance distance from front of the roadway vehicle to the nearest rail, ft, (assumed: D = 15 ft), and
d_e = distance from driver to the front of the roadway vehicle, ft, (assumed: d_e = 10 ft).

Rail-Highway Grade Crossings

Train Speed (MPH)	Departure from Stop	Vehicle Speed (MPH)						
		10	20	30	40	50	60	70
		\multicolumn{7}{c}{Distance Along Track from Crossing, T (ft.)*}
10	240	145	105	100	105	115	125	135
20	480	290	210	200	210	225	245	270
30	720	435	310	300	310	340	370	405
40	960	580	415	395	415	450	490	540
50	1200	725	520	495	520	565	615	675
60	1440	870	620	595	620	675	735	810
70	1680	1015	725	690	725	790	860	940
		\multicolumn{7}{c}{Distance Along Highway from Crossing, H (ft.)*}
All	25	70	135	225	340	490	660	865

*Highway distances for moving vehicle based on AASHTO Stopping Sight Distances plus clearance.

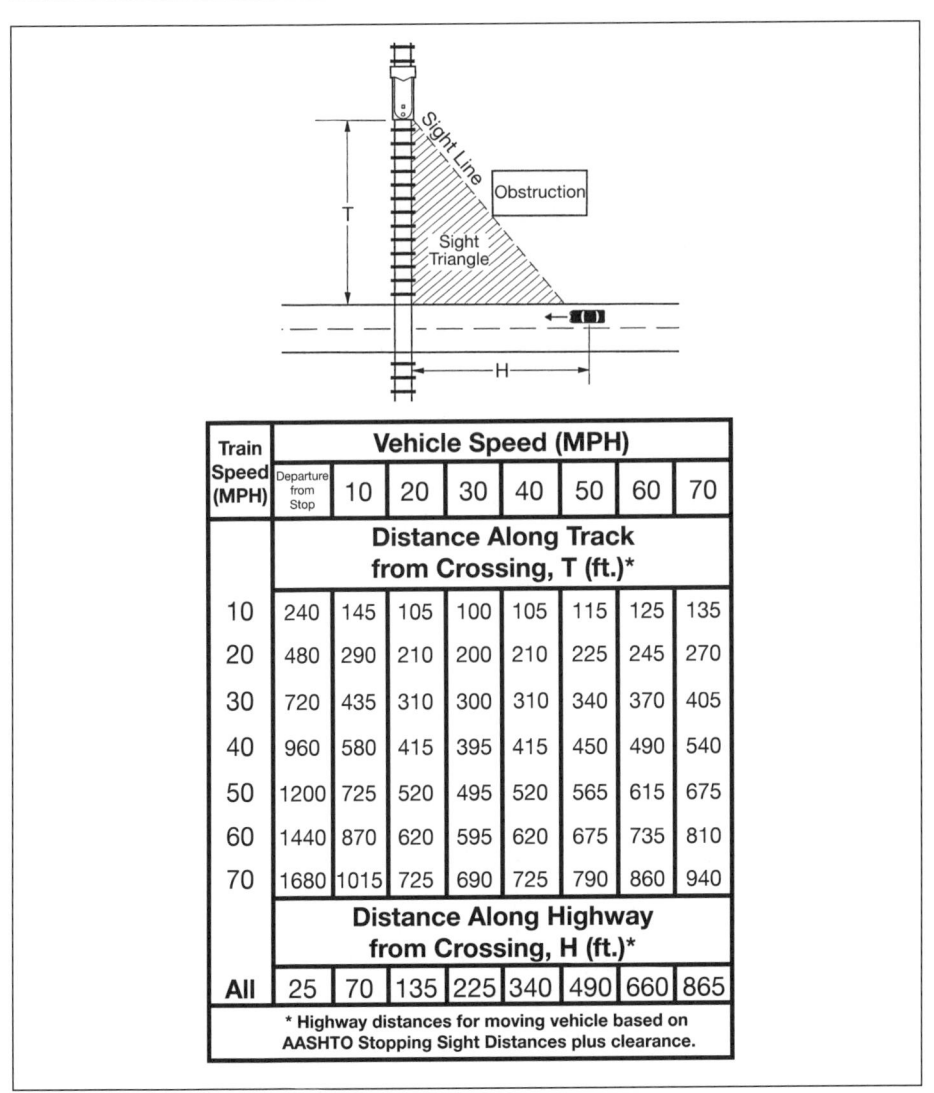

Figure 8-14 Rail-Highway Grade Crossing Sight Triangles *(Courtesy of Criterion Press, Inc.)*

Table 8-1 Coefficients of Friction, f [8]

Speed	f
10	0.40
20	0.40
30	0.35
40	0.32
50	0.30
60	0.29
70	0.28

The coefficient of friction values, f, are from the Reference 6 criteria for stopping sight distance (see Table 8-1). These values, compiled from several cited studies, represent the deceleration rates for a passenger car in locked-wheel braking on a wet pavement.

The minimum sight-triangle leg, T, along the track is measured from the nearest edge of the roadway lane being considered to the front of the train. It is the product of the train speed and the time required by a roadway vehicle both to traverse the highway leg, H, and to clear the crossing. The equation for this distance is:

$$T = \frac{V_t}{V_v} \left\{ 1.47 V_v t_{pr} + \frac{V_v^2}{30f} + 2D + L + W \right\}$$

Where

T = minimum sight-triangle leg along the railroad tracks needed by a moving roadway driver, ft,
V_t = velocity of train, mph,
V_v = velocity of roadway vehicle, mph,
t_{pr} = perception-reaction time of roadway driver, sec, (assumed: $t = 2.5$ sec),
f = coefficient of friction used in braking (see Table 8-1 for assumed values),
D = clearance distance from the roadway vehicle to the nearest rail, ft, (assumed: $D = 15$ ft),
L = length of roadway vehicle, ft, (assumed: $L = 65$ ft), and
W = distance between outer rails, ft, (assumed for a single track: $W = 5$ ft).

Values for moving-vehicle sight triangle legs calculated by these equations are shown in the table on Figure 8-14. Adjustments should be made for longer vehicles, slower acceleration capabilities, multiple tracks, steeper grades, and sharp crossing angles.

An additional hazard exists at crossings that have either roadway or track curvature at or near the crossing or have less than a 90° crossing angle. The maximum purview that a driver can be expected to scan is 90° to each side of the vehicle heading. Therefore, as the alignment becomes more acute, the effective sight triangle becomes more restricted simply because of the vehicle orientation to the track. So, even if no physical sight obstructions (such as trees or buildings) are present, the effective sight triangle can be severely restrictive for an acute alignment.

This same 180° driver purview should be applied to analyzing the effective sight triangle for a driver who is turning from a roadway that is parallel and close to the track. This driver will not have a view of the train that is approaching from

Rail-Highway Grade Crossings

behind until the vehicle turns onto the intersecting roadway and is squared up toward the crossing. This situation becomes more hazardous both as the distance between the parallel roadway and the track decreases and as the approach angle decreases from 90°

Stopped Roadway Vehicle

The sight triangle needed by a roadway driver stopped at the crossing is a function of both the train speed and the time needed for the roadway vehicle to begin moving, accelerate, and clear the tracks. It includes consideration of the train speed, the perception-reaction time of the roadway driver, the maximum speed of the roadway vehicle in starting gear, the acceleration rate of the roadway vehicle, and the length of the roadway vehicle. The equation for the sight-triangle leg, T, along the track is:

$$T = 1.47 V_t \left\{ \frac{V_g}{a_1} + \frac{L + 2D + W - d_a}{V_g} + J \right\}$$

where

T = minimum sight-triangle leg along the railroad tracks needed by a stopped roadway driver, ft,
V_t = velocity of train, mph,
V_g = maximum speed of roadway vehicle in first gear, fps, (assumed: V_g = 8.8 fps),
a_1 = acceleration of roadway vehicle in first gear, fpsps, (assumed: a_1 = 1.47 fpsps for a large loaded truck),
L = length of roadway vehicle, ft, (assumed: L = 65 ft for a large truck),
W = distance between outer rails, ft, (assumed for a single track: W = 5 ft),
D = clearance distance from the front of the roadway vehicle to the nearest rail, ft, (assumed: D = 15 ft),
J = sum of perception-reaction time of the roadway driver and time required to activate the clutch or an automatic shift, sec, (assumed: J = 2.0 sec), and
d_a = distance that a roadway vehicle travels while accelerating to maximum speed in first gear, ft, ($d_a = V_g^2/2a_1$).

Values for stopped-vehicle sight triangle legs calculated by this equation are shown in the table on Figure 8-14. Adjustments should be made for longer vehicle lengths, slower acceleration capabilities, multiple tracks, skewed crossings, and other than flat highway grades.

8.9 Rail-Highway Grade Crossing Improvements

Figure 8-15 is a general list of elements and actions that can be implemented to improve the safety of grade crossings. This section gives some additional discussion of specific alternatives at the most common problem locations and also other special application.

ELIMINATE CROSSING
- Closure
- Grade Separation

PHYSICAL IMPROVEMENTS
- Clear Sight Obstructions
- Realign Crossing to 90°
- Improve Nearby Curves, Grades, or Intersections
- Reconstruct or Repair Crossing Surface

CROSSING CONSPICUITY
- Overhead Lighting

TRAIN CONSPICUITY
- Train Color
- Train Headlighting
- Train Reflectivity
- Far-Side Crossing Flash Panels

ADVANCE TRAFFIC CONTROL
- Train Activated Warnings
- Rumble Strips
- Reduce Speed Limit
- Advisory Speed Plates
- Redundant Signs
- Warning Signs
- Pavement Messages

CROSSING CONTROL
- Stop Signs
- Flagman
- Flashers
- Gates
- Redundant Devices
- High Level Flashers

TRACK CONTROL
- Reduce Train Speed

Figure 8-15 Rail-Highway Crossing Safety Improvements (Courtesy of Criterion Press, Inc.)

Eliminate Crossings

Whatever the specific hazards are at a grade crossing, the first alternative that should be considered is elimination. Elimination can be done either by grade separating the crossing, closing the crossing to roadway traffic, or closing the crossing to railroad traffic through the abandonment or relocation of the rail line. Elimination of a crossing usually provides the highest level of crossing safety because the point of intersection between roadway and railroad is removed. However, the impact at adjacent crossings should be examined to make sure the net impact is not adverse.

Grade separations are recommended as an alternative for heavily traveled crossings. However, costs and benefits should be carefully weighted because

Rail-Highway Grade Crossings

grade separations are very expensive to construct and maintain. In some cases, a feasible alternative may be to separate grades at one crossing in a community and close most of the remaining crossings.

Other alternatives to reduce rail-highway grade crossing hazards include either relocation of the railroad track or roadway or consolidation of the railroad. These alternatives also provide a solution to other railroad impacts on communities such as delay and noise. However, the costs associated with relocation or consolidation can be quite high.

Railroad relocation to the outer limits of the community may be a viable alternative for alleviating these concerns while retaining the advantages of having railroad service. Relocation generally involves the complete rebuilding of railroad facilities.

Consolidation of railroad lines into common corridors or joint operations over the same tracks may allow for the removal of some tracks through a community. Railroad consolidation may provide benefits similar to those of railroad relocation, but possibly at lower costs.

Roadway relocations sometimes provide improved roadway traffic flow around communities and other developed areas. Planning for roadway relocations should consider routes that would eliminate grade crossings by avoiding the need for access over railroad tracks, or by providing grade separations.

The Federal Railroad Administration (FRA) has a current initiative to reduce the number of grade crossings in the U.S. by 25%. This FRA initiative is detailed in the 1994 publication, *A Guide to Crossing Consolidation and Closure*.[9] Closure of a rail-highway grade crossing to roadway traffic should always be considered as an alternative. Roadway traffic moving over several closely-spaced crossings in a community (see Figure 8-16) can be consolidated at a nearby crossing with flashing signals and automatic gates or over a nearby grade separation. Alternative routes should be within a reasonable travel time and distance from a closed crossing. The alternate routes should also have sufficient capacity to accommodate the diverted traffic safely and efficiently.

When a crossing is permanently closed to roadway traffic, the existing crossing should be obliterated by removing the crossing surface, pavement markings, and all traffic control devices both at the crossing and approaching the crossing. Generally, the railroad company is responsible for removing the crossing surface and traffic control devices located at the crossing (crossbuck signs, flashing signals, and gates). The roadway authority is responsible for removing traffic control devices in advance of the crossing. Nearby roadway traffic signals that were connected with crossing signals at the closed crossing should have their phasing and timing readjusted. The roadway authority is also responsible to alert drivers

Figure 8-16 Several Closely Spaced Rail-Highway Grade Crossings.

that the crossing is now closed using barricades, T-intersection signs, Road Closed signs, or other *MUTCD* approved devices.

Remove Devices at Abandoned Crossings

Rail-highway grade crossings on abandoned railroad lines, such as shown in Figure 8-17, present a different kind of safety and operational problem. Drivers, who consistently drive over abandoned crossings that have traffic control devices, may develop a careless attitude because they never see a train. Driver's may then carry over this behavior to crossings that have not been abandoned; perhaps resulting in a collision with a train. Thus, the credibility of grade-crossing traffic control may suffer.

Operational and safety problems may also exist at abandoned crossings. Many drivers will slow in advance of every crossing, especially those with passive traffic control. If the track has been abandoned, unnecessary delays will

Rail-Highway Grade Crossings

Figure 8-17 Maintained Crossbuck at a Roadway Crossing an Abandoned Railroad Right-of-Way.

result, particularly for buses and trucks carrying hazardous cargo that are required by Federal and State laws to stop at every crossing. In addition, these vehicles may be involved in rear-end collisions because their stops are not always expected by following drivers. The desirable action for abandoned crossings is to remove all traffic control devices related to the crossing and to remove or pave over the tracks.

Another type of inactive rail line is one whose service is seasonal. For example, rail lines that serve grain elevators may only have trains during harvest season. The lack of use during the rest of the year may cause the same safety and operational problems described above.

Improve Sight-Restricted Crossings

Without question, sight restrictions, particular at passively-controlled crossings, are the most pervasive hazard at rail-highway grade crossings. Crossings with one or more quadrants that do not meet the sight-triangle requirements of the *Railroad-Highway Grade Crossing Handbook*[2] are probably more frequent than those that do meet these requirements. Not only are sight restrictions prevalent, but they are also quite often severe. Not at all unusual is to find a sight-restricted quadrant on a 55-mph roadway where the safe speed is less than 10 mph.

Clearly, the more severely sight-restricted crossings can present an unreasonable burden on the approaching driver when an oncoming train is nearby. The provision of adequate sight triangles cannot be stressed enough. Constant vigilance to prevent man-made objects from being placed and vegetation from growing in the necessary sight triangle can probably do more for grade-crossing safety at the lowest cost than most other measures. Another principle to remember is that partial sight improvements that provide less than the *Railroad-Highway Grade Crossing Handbook*[2] requirements may be beneficial because not all traffic conflicts occur at the maximum operating speeds of the roadway or track. Even smaller objects within an otherwise clear sight triangle, should be removed because they can sometimes block a driver's view of a train at the point in time where he/she decides to look in that direction. The guiding principle is to always clear as much as possible.

If, after clearing a quadrant of all obstructions on either the railroad or roadway right-of-ways, no other clearance is possible and the crossing is still severely sight-restricted for moving vehicles, then additional traffic controls should be considered. First consideration should always be flashing signals. But, because of the high cost (about $50,000), flashing signals may only be justified at crossings where roadway and train traffic is sufficiently high.

The most common traffic control solution for severe moving-vehicle sight restrictions at low-traffic crossings is the installation of a stop or yield sign. The conditions for stop or yield signs installation are spelled out in the *MUTCD* as discussed earlier. Although the *MUTCD* lists conditions to use the stop sign, only one of which is a severe sight restriction, the severe sight restriction is the most compelling reason for stop signs at crossings where flashing cannot be installed immediately. Stop or yield signs are also used at these kinds of crossings as an interim measure where a flashing signal is planned.

When stop or yield signs are to be installed at a crossing, the stopped-vehicle sight triangle criterion should be checked (see Figure 8-14). Sometimes a curved-track approach will limit the available sight distance from the stopped position. If the crossing is severely restricted according to the *Railroad-Highway Grade Crossing Handbook*[2] requirements for stopped control, a flashing signal should be reconsidered. However, the stop sign will usually provide a safer environment than the crossbuck alone where the moving-vehicle sight triangle is severely restricted. Because, the stopped-vehicle sight triangle is based on a poorly accelerating long heavy truck clearing the track(s) from the stopped position, the stop sign may provide safety for passenger cars and smaller trucks even when the stopped-vehicle sight triangle is less than the required dimension.

Rail-Highway Grade Crossings

Where grade crossings have marginally inadequate sight triangles and cannot justify either stop signs or flashing lights, other traffic control measures can be applied to heighten the awareness of the driver. These include advisory speed plates on the advance warning signs and the use of redundant crossbucks and advance warning signs.

Bring Crossings Up to Standard

Many crossing cannot operate as safely as possible because of missing traffic control devices, poor installation, or lack of maintenance. A crossing with traffic control devices that are incomplete or in disrepair may be hazardous because the driver gets the impression that the crossing has been abandoned. The following is a list of improvements in this category:

1. Cut vegetation in sight triangle and in front of traffic control devices.
2. Install missing traffic control devices.
3. Replace non-reflecting signs.
4. Repair or replace damaged sign and signal elements.
5. Repaint worn pavement markings.
6. Replace signs that are mounted too low.
7. Replace burned-out signal lamps.
8. Re-aim signals for optimum driver visibility. (The signal unit on the right is usually aimed to cover a distance down the approach. The signal unit on the left are usually aimed near the crossing.)

Install Flashing Light Signals

Although flashing light signals are desirable at all grade crossings to reduce crossing accidents, their cost may be prohibitive for some low-traffic crossings. National warrants for the installation of flashing light signals have not been developed. Some States have established criteria based on exposure factors or priority indices. Other considerations include the following:

1. Volume of vehicular traffic
2. Volume of railroad traffic
3. Speed of vehicular traffic
4. Speed of railroad traffic
5. Volume of pedestrian traffic
6. Accident record
7. Sight distance restrictions

Install Automatic Gates

Automatic gates generally reduce accident potential when added to flashing signals. One or more of the following factors may indicate the need for gates:

1. Multiple tracks where a train on or near the crossing can obscure the movement of another train approaching the crossing.
2. High-speed train operation combined with limited sight distance.
3. A combination of high speed and moderately high volume roadway and railroad traffic.
4. Presence of school buses, transit buses, or farm vehicles in the traffic flow.
5. Presence of trucks carrying hazardous materials, particularly when the view down the track for a stopped vehicle is obstructed (curve in track, etc.).
6. Continuance of accidents after installation of flashing signals.
7. Presence of passenger trains.

Add Cantilever Signals

Where improved visibility to approaching drivers is required, cantilevered flashing signals can be used. Cantilevered signals may be appropriate when any of the following conditions exists:

1. Multi-lane roadways.
2. Roadways with paved shoulders or parking lanes that would require a post-mounted flashing signal to be more than 10 feet from the edge of the travel lane.
3. Roadside foliage obstructing the view of post-mounted flashing signals.
4. A line of roadside obstacles such as utility poles obstructs the view and a minor lateral adjustment of the poles would not solve the problem.
5. Distracting backgrounds such as an excessive number of neon signs.
6. Horizontal or vertical roadway curves at locations where cantilevered flashing signals over the traffic lane are needed to provide sufficient visibility for the required stopping sight distance.

A typical installation consists of one pair of cantilevered lights on each roadway approach, supplemented with a pair of lights mounted on the supporting mast. However, two or more pairs of cantilevered flashing signals may be desirable for multi-lane approaches, as determined by an engineering study. The cantilevered lights can be placed over each lane so that the lights are mutually visible from adjacent driving lanes.

Install Active Advance Warning Signs

An active advance warning sign (AAWS), has one or two 8-inch yellow hazard identification beacons mounted above the advance warning sign used in conjunction with flashing signals at the crossing. The AAWS gives roadway drivers advance warning that a train is approaching the crossing. The beacons are connected to the railroad track circuitry and are activated by an approaching train. The beacon should operate continuously with the crossing signals.

An AAWS should be considered at crossings where the flashing signals cannot be seen before the point from which a safe stop can be made. The AAWS can be supplemented with a message, either active or passive, that indicates the meaning of the device, such as "Train When Flashing."

Install DO NOT STOP ON TRACKS Sign

When an engineering study shows a high potential for vehicles stopping on the tracks, a DO NOT STOP ON TRACKS sign should be used. The sign may be located on the near or far side of the crossing, whichever provides better visibility to the roadway driver. On multi-lane and one-way roadways, a second sign should be placed on the near or far left side of the crossing.

Install Turn Prohibition Signs

At signalized roadway intersections within 200 feet of a crossing, where traffic signals are preempted by an approaching train, all turning movements toward the crossing should be prohibited. Turn prohibition signs should be visible only when the turn prohibition is in effect; thus a blank-out or internally illuminated sign, or other type of changeable message sign may be used. These signs are activated by the same train detection circuitry as the flashing signals.

Install Warning Bells

A crossing bell is used to supplement other active traffic control devices and is an effective warning to pedestrians and bicyclists. The bell, mounted on top of one of the flashing signal support posts, is activated along with the flashing signals. Silencing the bell when the train reaches the crossing or when the gates are down may be desired to accommodate residents of suburban areas.

Install Crossing Illumination

At crossings where substantial railroad operations are conducted at night, particularly where train speeds are low, where crossings are blocked for long periods, or where accident history indicates that drivers have difficulty seeing trains or control devices at night, illumination at and adjacent to the crossing can be installed. Regardless of the presence of other control devices, illumination will aid drivers in observing the presence of railroad cars on a crossing where the grade of the roadway approach allows the headlights of an oncoming vehicle to shine under or over the cars.

Recommendations for grade crossing illumination are contained in the *American National Standard Practice for Roadway Lighting*[10] and the *Roadway Lighting Handbook*.[11] Luminaries should be so located and their light so directed so they will not interfere with either the railroad signal system or the field of view for the locomotive crew.

Improve Roadway Horizontal Alignment

Roadways should intersect railroad tracks at a right angle with no nearby intersections or driveways. This layout enhances the driver's view of the crossing and tracks and reduces conflicting vehicular movements from crossroads and driveways. One State limits the minimum crossing angle to 70 degrees. At sharp-angled crossings, drivers must look over their shoulder to view down the tracks. Because of this awkward movement, some drivers may only glance quickly or not at all, and thereby miss seeing an approaching train.

To the extent practical, crossings should neither be located on roadway or railroad curves. Roadway curvature limits the driver's view of a crossing ahead and the driver's attention may be directed toward negotiating the curve rather than looking for a train. Railroad curvature inhibits a driver's view down the tracks from both a stopped position at the crossing and on the approach to the crossing. Crossings located on both roadway and railroad curves have both maintenance problems and poor rideability for roadway traffic due to conflicting superelevations. Similar difficulties arise when the superelevation of the track is opposite to the grade of the roadway.

Generally, improvements to the roadway horizontal alignment are expensive. Special consideration should be given to crossings that have complex horizontal geometrics. These crossings usually warrant the installation of active traffic control systems or, if possible, may be closed to roadway traffic.

Improve Roadway Vertical Alignment

Rail-highway grade crossings should be as level as possible to insure adequate sight distance, rideability, and braking and acceleration performance for roadway vehicles.

Track maintenance can result in raising the track as new ballast is added to the track structure. Unless the roadway profile is properly adjusted, this practice results in a *humped* crossing that may adversely affect safety and operation of roadway traffic over the railroad. Humped crossings can cause particular problems for vehicles with low underclearances, because these vehicles can high-center and become trapped on the tracks.

Rail-Highway Grade Crossings

Alternatives to this problem include a design standard that deals with maximum grades at the crossing, prohibiting truck trailers with certain combinations of underclearance and wheelbase to use the crossing, setting trailer design standards, or minimizing the rise in track due to maintenance operations.

The American Railway Engineering Association, *Manual for Railway Engineering*[12] recommends that the crossing surface be in the same plane as the top of rails for a distance of two feet outside of rails and that the surface of the roadway be not more than three inches higher nor six inches lower than the top of nearest rail at a point 30 feet from the rail.

Improve Roadway Cross-Section

The width of the roadway at the crossing should be sufficient to include all travel lanes and adjacent shoulders plus two feet.

Intersections and driveways should be avoided near crossings. Because a driver's attention may be distracted by other vehicles entering or exiting the roadway, the driver might not use appropriate caution at the crossing. Parking should be avoided near crossings for the same reason and also because parked vehicles may restrict a driver's view of an oncoming train or a crossing warning device.

School buses, transit buses, and vehicles transporting hazardous materials are required to stop at all crossings before proceeding across the tracks. Pullout lanes can be provided to remove these vehicles from the through lane so they can stop without creating hazard to following vehicles. Consideration should be given to low-cost improvements such as the removal of parking near the crossing and the closure of low volume intersecting roadways and driveways.

Improve Crossing Surfaces

Drivers, preoccupied by a rough surface at grade crossings, may have their attention diverted from an approaching train. The driver's attention may be devoted primarily to choosing the smoothest path over the crossing, rather than searching for an approaching train. This kind of behavior may also be conditioned by the driver being frequently exposed to rough crossing surfaces. In other words, a driver could slow the vehicle and divert an inordinate amount of attention to the surface, even at smooth crossings because his/her expectancy is that all crossings are rough. Rough crossing surfaces are also dangerous because some drivers may encounter the surface unexpectedly and lose control of their vehicle.

The *Federal-Aid Highway Program Manual*[13] provides for improvements of crossing surfaces as one means of eliminating hazards at a crossing. Projects for surface improvements are eligible for federal-aid funding. The reader is referred to the *Railroad-Highway Grade Crossing Handbook*[2] and the *Railroad-Highway Grade Crossing Surfaces*.[14]

Improve Train Conspicuity

Locomotives come equipped with headlights and horns as warning devices to help roadway drivers be aware of oncoming trains, particularly at passively-controlled crossings. While these are important warning methods, several factors may limit a driver's ability to detect an approaching train:
1. The train's headlight may be lost in a myriad of fixed lighting near the crossing or may be mistaken as a roadway vehicle.
2. Under adverse weather, the driver's visibility may be severely limited.
3. Sight distance may be limited by roadway alignment or obstacles in the sight triangle.
4. Closed windows, engine noise, exhaust noise, or a car radio, air conditioner, or heater may mask the sound of a train horn.

Conspicuity is the property of attracting attention by visual means. It is a vital element at grade crossings where so much of the burden for safe performance is placed on the driver's ability to detect the train by visual means. Train conspicuity is of particular concern at crossings with passive traffic control devices. Assuming adequate sight distance, the driver must be able to see the train and estimate its speed throughout a wide range of backgrounds and visual distractions. To increase train conspicuity, highly visible reflective panels on the side of trains have been suggested. Another treatment is to provide movement in the locomotive headlamp. Some lamps currently have a movement that sweeps the tracks. Other railroad companies are utilizing strobe lights or rotating beacons mounted on lead locomotives. However, none of these concepts have been adopted for general use.

Two promising grade crossing improvements aimed at improving train conspicuity are mirrors and flash panels. Mirrors can be so located that they reflect the train headlight down the roadway approach and alert the approaching driver of the oncoming train. Flash panels, are far-side reflective panels that show the presence of a train on a dark crossing at night by reflecting vehicle headlights back to the driver through the gaps between the train cars. This alternating on-off effect approximates an active warning device. One effective way to accomplish flash-panels is to back up existing crossbucks and/or put reflective sheeting on the crossbuck post. This low-cost improvement would also serve to provide redundant signing at the crossing.

Reduce Roadway Speed Limit

Reducing the roadway speed limit allows a driver to make the decision about whether to stop or proceed at a point closer to the crossing than at higher speeds. This alternative has two advantages. First, it reduces the sight triangle needed by

Rail-Highway Grade Crossings

the driver to safely traverse the crossing. Second, it has the effect of providing greater sight distance down the track from the now closer critical decision point.

Reduce Track Speed Limit

Because the sight triangle leg along the track is a function of train speed, reducing the train speed has the effect of reducing the sight triangle needed by a driver to safely traverse a crossing.

Install Combinations of Improvements

Sometimes a single improvement at a grade crossing may not substantially improve the safety of the crossing. For example, clearing sight-restricting brush at a crossing may not be enough to provide drivers with adequate sight distance at a crossing because of other, more permanent, sight restrictions at the crossing. However, a combination of removing the brush and installing stop signs may create an adequate sight triangle at the crossing. Other possible combinations include reducing both track and roadway speeds, and backing up the crossbucks in combination with any of the other measures previously mentioned.

8.10 Technical Aspects of Rail-Highway Grade Crossing Defect Cases

The plaintiff, who is injured in a collision with a train at a defective rail-highway grade crossing, may not be clear about who is the appropriate defendant. Who should make safety improvements at rail-highway crossings often depends on legal responsibilities. Many major improvements (mainly flashing signals and automatic gates) are normally financed by the U.S. Department of Transportation, with funds administered by State departments of transportation and given to the railroad companies, who then install the improvements. Railroad companies are often reluctant to spend their own money for improvements, but have argued unsuccessfully (Easterwood v. CSX) that the Federal government has preempted their responsibility for grade crossing safety.

When considering needed safety improvements at crossings with passive traffic control, some responsibility clearly falls on the railroad company, some responsibility clearly falls on the roadway jurisdiction, and some responsibility is not clear (See Table 8-2). For some elements, the railroad company has clear responsibility, such as installing and maintaining crossbucks[6] or installing and maintaining a smooth crossing with relatively flat approach grades. For other elements, the roadway jurisdiction has clear responsibility, such as installing and maintaining advance warning devices.[6] However, when it comes to placing additional passive devices within the common right-of-way shared by the railroad and the roadway, the responsibility and/or authority is not always clear.

Table 8.2 Responsibility for Safety Improvements at Passive Rail-Highway Grade Crossings

Railroad Responsibility
  Install and Maintain Crossbucks
  Install and Maintain Smooth Crossing Surfaces
  Install and Maintain Adequate Crossing Width
  Provide Relatively Flat Crossing Grades
  Clear Sight Obstructions on Railroad Right-of-Way
  Set Track Speed Limits
  Provide Train Conspicuity

Roadway Jurisdiction Responsibility
  Install and Maintain Advance Warning Devices
  Clear Sight Obstructions on Highway Right-of-Way
  Set Highway Speed Limits
  Close or Consolidate Crossings
  Install and Maintain Adequate Road Width
  Provide Adequate Sight Distance to Signs

Improvements with Arguable Responsibility
  Clear Sight Obstructions on Private Property
  Install Flashing Signals
  Install Crossing Gates
  Install Stop or Yield Signs
  Install Flash Panels (e.g., back up crossbucks)
  Install Mirrors to Reflect Train Headlights
  Remove Signs on Abandoned Crossings
  Install Special Warning Signs
  Install Lighting at the Crossing

This uncertainty is manifest most particularly with the need for stop signs, for backed-up crossbucks (or flash panels), and for other redundant signs and markings. For stop or yield signs, the railroad companies argue that they have no *authority* to stop roadway traffic. On the other hand, local roadway jurisdictions most often feel that they neither have *responsibility* or *authority* to place devices on the railroad right-of-way. For backed-up crossbucks, the need may not be evident to most local jurisdictions and to some railroad companies (although at least one railroad company uses this technique at all their passive crossings). Neither usually wants to spend the money.

Perhaps this area of uncertain responsibility dictates that some involvement by the State or Federal government is necessary to address these sorely neglected grade crossings. For example, by administrative decision, the U.S. DOT could probably designate a relatively small portion of one-year's funding for rail-highway grade crossing improvements to be spent on both (1) installing crossbucks

Rail-Highway Grade Crossings

at every crossing that now has no protection, and (2) backing up every existing crossbuck to provide redundant signs and flash panels at every passive crossing. Likewise, the need for stop signs at passively-controlled crossings with serious sight restrictions (already identified by most or all State inventories) could be addressed with either current funding or a special short-term program. Of course, clear program warrants need to be established for the installation of stop signs.

Most rail-highway grade crossing cases have a combination of defective elements that lead to stronger arguments about, design defects, failure to maintain, failure to warn, or a combination of these. The elements that are common to defective rail-highway grade crossing cases are as follows:

Sight Triangle Obstructions

The most common defective crossing case involves deficient sight triangle at a passively-controlled crossing (see Figure 8-3). Many times, one or more quadrants of the crossing will have only 10-25% of the sight distance recommended by the *Railroad Highway Grade Crossing Handbook*[2] shown in Figure 8-14. Given this hazardous condition, the normal plea is that the railroad company and/or the roadway agency failed to exercise their common-law duty to provide reasonable safety for the motoring public. Even though the driver has a legal duty both to exercise proper lookout and to yield to oncoming trains, however, he/she will have great difficulty exercising this proper care when an approaching high-speed train can neither be seen nor heard.

The first question in this kind of case is whether the sight obstruction could have been removed. Many times, however, the sight obstruction is neither on the roadway or the railroad right-of-way. Therefore, some form of additional traffic control and/or warning should have been installed to compensate for the restricted sight.

Nearby Intersections

As stated earlier, more than 35% of all rail-highway grade crossings are within 75 feet of an intersection. Most normally, the roadway crossing the tracks tees into the parallel roadway (see Figure 8-4). Without adequate traffic controls, these crossings are hazardous, particularly for large trucks and buses.

When a passively-controlled crossing is near a roadway intersection, the plaintiff often argues that flashing signals should have been installed. This argument becomes stronger as the separation distance decreases and as additional defects are present, such as a sharp crossing angle, a steep approach grade, sight triangle obstructions, a rough crossing surface, a steep approach grade, or missing or improper advance warning signs on the parallel roadway. Another argument is that these crossings should have DO NOT STOP ON TRACKS signs.

When an actively-controlled crossing is near a stop-controlled intersection, additional hazards are attendant, particularly for large trucks and buses. A common dangerous occurrence is that, at the same time an approaching train activates the flashing signal and/or automatic gates, a vehicle approaching the intersection can be stopped on the track either because it is a long vehicle or because it is the second or third vehicle stopped in a cue. For short distances between the crossing and intersection, the plaintiff will argue that a gate should have been in place to keep vehicles from moving onto the track as a train approaches. When the roadway traffic is fairly high at the crossing, and that traffic is controlled by a stop sign at the intersection with the parallel roadway, the plaintiff may argue that pre-empted traffic signals should have been placed at the intersection. These crossings should also have turn prohibition signs, in addition to DO NOT STOP ON TRACKS signs.

When an actively-controlled crossing is near an intersection with traffic signals, collisions sometimes occur between moving trains and vehicles that have been stopped by the traffic signal. Again large trucks and buses are particularly vulnerable. These collisions are sometimes a result of improperly timed crossing and/or traffic signals. Traffic signals should always be automatically pre-empted by the crossing signals so that vehicles on the crossing can be cleared before the arrival of the train. These crossings should also have turn prohibition signs, in addition to DO NOT STOP ON TRACKS signs.

Sharp Angle Crossing Approach

Much like the crossing with a nearby intersection, roadway approaches that form a sharp angle with the track (see Figure 8-5) restrict the driver's effective sight triangle. Figure 8-18 shows what can happen when a driver can't see an approaching train over his shoulder. Most of the same arguments that apply to sight-obstructed crossing and crossings with nearby intersections are similar for these crossings. Again large trucks and buses are particularly vulnerable.

For a sharp angle crossing, the plaintiff will often plea that a realignment of the crossing should have been considered. However, in many cases, this kind of improvement may not be the most cost-effective. Another major plea would say that the effective sight distance is so restricted that either the crossing should have been closed or that flashing signals should have been installed.

Poor Maintenance of Signs

Because both advance warning signs and crossbucks are mandatory traffic control devices that drivers expect to see at every crossing, their effectiveness depends on adequate maintenance to ensure their visibility and appearance. If they are hidden by vegetation or have lost their reflectance, the driver may not

Rail-Highway Grade Crossings 263

Figure 8-18 Collision at a Sharp Angled Crossing.

be adequately warned of the crossing. If they are hidden by vegetation, have lost their legibility, or are hanging loose on their posts, they can lose the respect of the driver, who may think the crossing is abandoned.

The most common tort claims relate to the nighttime visibility of crossbucks. Not only is the crossbuck itself mandatory but its reflectivity is also mandatory. Crossbucks that have lost a major percentage of their reflectivity cannot effectively delineate a crossing.

Lack of Advance Warnings

If advance warning signs and pavement markings are not present on the approach to a crossing, the plaintiff involved in motor-vehicle to train collision may file a tort claim alleging a failure to warn using mandatory traffic control devices.

Poor Flashing Signal Visibility

Several factors can affect the signal visibility. For example, newer signal roundels have 12-inch lenses whereas older roundels often have 8-inch lenses (see Figure 8-19). The 12-inch lenses are the only size installed today because they have 144/64 or 2.25 times the target area of the smaller lenses. However, many crossings still exist with the older 8-inch roundels.

(a) 8" Roundels *(b) 12" Roundels*

Figure 8-19 *Comparison of Flasher Lens Sizes.*

Rail-Highway Grade Crossings

The flashing light roundels have a very limited viewing cone in order to concentrate the light toward an approaching driver. For this reason, not only do the roundels need to be aimed properly to cover an adequate length of approach roadway, but also extra roundels, as shown in Figure 8-20, need to be installed and aimed properly when an approach roadway is curved or intersects the crossing roadway.

A relatively common flasher visibility plea in a motor-vehicle to train collision case is that the signal was not aimed so it could be seen on a substantial portion of the approach roadway. A typical improvement at crossings, which have high traffic volume, multiple lanes, high truck traffic, and/or curved approaches, is the installation of cantilevered signals.

Figure 8-20 Extra Roundels Mounted to Give Visibility Along a Close Intersecting Roadway.

Steep Crossing Grade

A plaintiff may have a legitimate tort claim if he/she was involved in a crossing collision where the crossing approach grade either was steeper than allowed by State statute or exceeded the AREA[12] standard (see Section 8.9).

The most common crossing collision tort case where a steep grade is the key roadway defect is when the roadway driver failed to see an oncoming train because he/she was trying to negotiate a steep grade with an icy surface. Another type of collision where a steep grade is the key alleged defect is when a large truck got stuck on the crossing because it high-centered (see Figure 8-6).

A steep crossing is also alleged as a contributory defect in many cases. The plaintiff will generally argue that the steep grade is one of several factors that increased the driver's work load to the point where he/she was distracted from searching for the train. The extra work load related to a steep grade came from, (1) simply paying extra attention to speed and position in negotiating the grade, and (2) scanning the horizon for head-on conflicts with oncoming vehicles hidden over the crossing.

Rough Crossings

The most common plea related to rough crossings is when the rough crossing caused the driver to lose control and have a collision either off the road or with another vehicle. Another type of collision that is seen, is where a rail or other part of a rough crossing surface contacts the vehicle undercarriage and causes a sudden deceleration of the vehicle. As discussed above, for steep crossing grades, rough crossing surfaces are also sometimes alleged as one of several factors that distracted the driver from his/her primary task of searching for oncoming trains.

8.11 Accident Reconstruction Aspects

Both train and motor-vehicle speed, stopping distance, and stopping time are often at issue in litigation involving rail-highway crossing accidents. By calculating these factors, issues of negligence, collision avoidance, and the viability of eye-witness testimony can be addressed.

Train Braking Performance

Train speed can be reasonably analyzed and explained using basic principles of physics and a working knowledge of braking system operation. For a complete discussion of this subject, the reader is referred to *Train Accident Reconstruction and FELA and Railroad Litigation*[15] by Loumiet and Jungbauer. Figure 8-21 from that book gives a general idea about the relationships between train speed and stopping distance for three different trains during emergency braking. All of the data for these curves were generated using the *Train Braking Simulator*[16] computer program. The long-loaded freight train had four locomotives and 100 fully-loaded cars. The short-empty freight train had one locomotive and five empty cars. The passenger train had two locomotives and ten cars.

Rail-Highway Grade Crossings 267

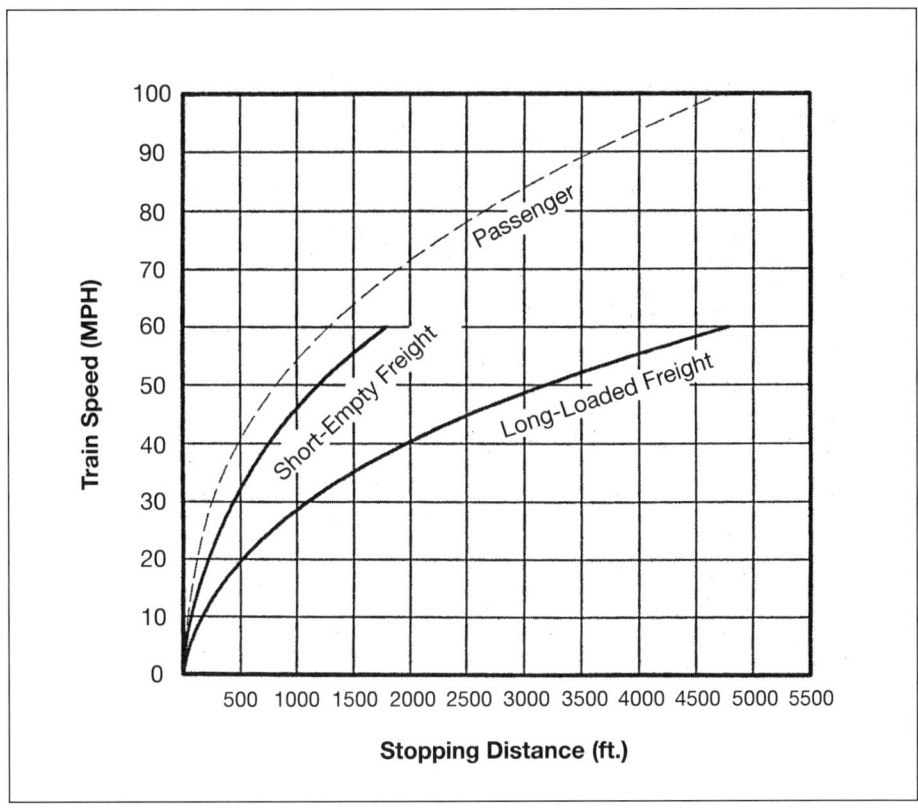

Figure 8-21 Train Speed vs. Emergency Stopping Distance.[15]

The reader should exercise caution using Figure 8-21 for a specific application. Several factors affect the braking performance of a train and, therefore, a given curve may not accurately represent the braking performance of a particular train.

Crush Damage Method

Analyzing the crush to motor-vehicles, particularly passenger cars, can sometimes yield information either about the motor-vehicle speed or the train speed. Since the 1960's, many crush tests have been conducted on a variety of automobiles to produce empirical relationships between the vehicle impact speed and the crush dimensions produced by an impact with a fixed barrier. Therefore, accident reconstructionists are familiar with the procedure of measuring crush damage to vehicles involved in a collision to assist in speed calculations. Most of these empirical relationships are derived under the assumption that all of the vehicle's kinetic energy is dissipated in the work performed to crush the vehicle.

The energy dissipated in crushing a vehicle can be calculated using a surrogate term called the equivalent barrier speed (EBS). The EBS is used to calculate an equivalent kinetic energy in the Conservation of Energy equation, which reduces to the following equation for a collision between a motor-vehicle and a train:

$$W_T S_{TI}^2 + W_M S_{MI}^2 = W_T S_{TP}^2 + W_M S_{MP}^2 + W_T (EBS_T)^2 + W_M (EBS_M)^2$$

where

W_T = weight of train, lbs.
W_M = weight of motor vehicle, lbs.
S_{TI} = impact speed of train, mph
S_{MI} = impact speed of motor vehicle, mph
S_{TP} = post-impact speed of train, mph
S_{MP} = post-impact speed of motor vehicle, mph
EBS_T = Equivalent Barrier Speed of train crush, mph
EBS_M = Equivalent Barrier Speed of motor-vehicle crush, mph

Many reconstructionists say they can estimate the *motor-vehicle impact speed* by assuming that it is equal to the motor vehicle EBS, but this assumption is ill-founded unless the train is stopped. To suggest that all of the damage to the motor-vehicle in collision with, say, a 40-mph train comes only from the pre-impact energy of the motor-vehicle is ludicrous. To derive such a relationship requires the assumptions (1) either that the train EBS is zero, the motor-vehicle post-collision speed is zero, and the train pre-impact speed equals the train post-impact speed, or (2) that the train pre-impact energy equals the sum of the train post-impact energy, the train crush energy, and the motor-vehicle post-impact energy. Neither of these assumptions can be supported. If, for example, the change in the speed of a 5,000-ton train at 50 mph is only 0.1 mph, the energy lost by the train would be about 20 times that of the EBS of the motor vehicle, hardly a factor to ignore. Also, in many cases, the post-impact kinetic energy of the motor vehicle will be greater than the crush energy of the motor-vehicle, making the exclusion of this factor a gross error.

Using crush damage to estimate the *motor-vehicle impact speed* when a motor vehicle impacts the side of a train stopped on the crossing is another matter. Loumiet and Jungbauer[15] show that the crush energy of the motor vehicle, in this case, is a reasonable estimate of the pre-impact kinetic energy, such that the impact speed of the motor vehicle equals the equivalent barrier speed of the motor vehicle as follows:

$$S_{MI} = EBS_M$$

Rail-Highway Grade Crossings

Loumiet and Jungbauer[15] also show that the *train impact speed* can be reasonably estimated by the motor-vehicle equivalent barrier speed when the train impacts a motor vehicle stopped on the crossing as follows:

$$S_{TI} = EBS_M$$

Another accident reconstruction aspect of collisions between trains and motor-vehicle is the time and distance relationships between the vehicles. A common plaintiff question is whether the train could have been stopped or at least slowed enough to avoid collision. Another common plaintiff question is whether the motor-vehicle driver had adequate time, given a restrictive sight triangle, to see, perceive, react, and stop before reaching the crossing. A common defense question is why the motor-vehicle driver didn't perform emergency braking to stop short of the crossing given the available sight distance.

All of these questions require standard motion equations with inputs about train speed and position and motor-vehicle speed and position at various points along the approaches to the crossing.

8.12 Typical Defense Arguments

In most rail-highway grade crossing defect cases, a railroad company is a defendant. Occasionally, the roadway agency is brought in as a defendant when claims are made for missing or faulty advance warnings or other defects on the approach roadway.

At passively-controlled crossings where sight restrictions, sharp crossing angles, nearby roadway intersections, and nearby roadway curves restrict driver's ability to see and/or respond to an oncoming train, the railroad defendant will usually argue that the driver has the sole responsibility to yield the right of way to an oncoming train. Regardless of the difficulties that the driver may have, they will say that he/she should have slowed enough that the defects of the crossing would be totally compensated for by the driver. This argument, of course, puts the total burden for safety on the driver.

Train engineers are required by law to sound the train whistle usually continuously from a point one-quarter mile before the crossing until the train occupies the crossing. Although, the plaintiff sometimes has evidence to the contrary, train crews will almost always testify that the train whistle was blown exactly as required. Then the railroad defendant will argue that the subject driver was required and should have responded to the train whistle by stopping.

When a plaintiff argues that the crossing was so defective that it should have had automatic signals and/or gates, the railroad company will argue that the need was not necessarily apparent and even if it was, their responsibility is pre-empted by the State and Federal governments activities directed toward rail-highway grade crossing improvements.

Other case-specific arguments include either that the driver was the sole negligent party and/or that the crossing condition had no contribution to the causation of the collision. Some of these arguments are:

1. The driver was negligent for speeding.
2. The driver was negligent for driving while intoxicated.
3. The driver was driving 5-10 mph and had plenty of time to stop, but he/she never looked.
4. The obstruction to sight was neither on the railroad or roadway rights-of-way.
5. The condition of the crossing and its traffic control is not the responsibility of the roadway agency.
6. The condition the crossing approach is not the responsibility of the railroad company.
7. Because the crossing had a low volume of roadway traffic, it did not economically justify the placement of any traffic control devices other than crossbucks and advanced warning signs.
8. The traffic control at the crossing complied with the MUTCD.
9. The crossing did not have a high calculated hazard index and was, therefore, not high on the State priority list for improvement.

Endnotes

1. Federal Railroad Administration, U.S. Department of Transportation, *Rail-Highway Crossing Accident/Incident and Inventory Bulletin*, Calendar Year 1988.

2. Tustin, B.H., et. al., *Railroad-Highway Grade Crossing Handbook*, Federal Highway Administration, U.S. Department of Transportation, September, 1986.

3. Glennon, John C., and James R. Loumiet, *Low-Cost Safety Improvements At Rail-Highway Crossings,* Criterion Press, Box 6852, Leawood, KS, 66206.

4. National Committee on Uniform Traffic Laws and Ordinances, *Uniform Vehicle Code and Model Traffic Ordinances*, 1992..

5. Russell, Eugene R., et. al., *Compilation of State Laws and Regulations on Matters Affecting Rail-Highway Crossings*, Federal Railroad Administration, Report FHWA-TS-83-203, April, 1983.

6. Federal Highway Administration, *Manual on Uniform Traffic Control Devices*, 1988. (available through Lawyers and Judges Publishing Co.)

7. Federal Highway Administration, *Traffic Control Devices Handbook*, 1983.

8. American Association of State Highway and Transportation Officials, *A Policy on Geometric Design of Highways and Streets*, 1990.

9. Federal Railroad Administration, *Highway-Railroad Grade Crossings - A Guide to Crossing Consolidation and Closure*, 1994.

10. Illuminating Society of America, *American National Standard Practice for Roadway Lighting*, July 1977.

11. Federal Highway Administration, *Roadway Lighting Handbook*, Implementation Package 78-15, December 1978.

12. American Railway Engineering Association, *Manual for Railway Engineering*, 1981.

13. Federal Highway Administration, *Federal-Aid Highway Program Manual*, updated periodically.

14. Hedley, William J., *Railroad-Highway Grade Crossing Surfaces*, Federal Highway Administration, Implementation Package 79-8, August 1979.

15. Loumiet, James R., and Jungbauer, William G., *Train Accident Reconstruction and FELA and Railroad Litigation*, Second Edition, Lawyers & Judges Publishing Co. 1995.

References

Loumiet, James R., *Train Braking Simulator*™, Computer Program, Version 2.1, Criterion Press, Inc., 1992. (available through Lawyers and Judges Publishing Co.)

Fitzpatrick, Kay, Mason, John M. Jr., and Glennon, John C., *Sight Distance Requirements for Trucks at Railroad-Highway Grade Crossings*, Transportation Research Record 1208, 1989.

Knoblauch, Karl, Hucke, Wayne, and Berg, William, *Rail Highway Crossing Accident Causation Study,* Federal Highway Administration, 1982.

National Transportation Safety Board, *Passenger/Commuter Train and Motor Vehicle Collisions at Grade Crossings*, 1985.

Glennon, John C., and Loumiet, James R., *Estimating Train Speed from Braking Distance*, Journal of the National Academy of Forensic Engineers, December 1992.

Chapter 9

Roadway Curves

Roadway curves are a necessary and important element of nearly all highways and streets. Their form has evolved from what appeared reasonable to the builder's eye to the more modern geometrically designed form of a circular curve with superelevation (banking), cross-slope transitions, and sometimes spiral transitions.

Despite a reasonably well conceived design procedure, which considers a tolerable level of lateral acceleration on the driver, roadway curves continually show a tendency to be high-accident locations. Several studies over the years have indicated both that roadway curves exhibit higher accident rates than straight roadway sections, and that accident rate increases as curve radius decreases. But, curve radius may be just one element that is interdependent with other elements that together contribute to accident rate. The roadway curve is one of the most complex features of streets and highways. The several elements listed in Table 9.1 are all aspects of roadway curves that may relate to safety.

9.1 Vehicle Cornering

When a vehicle moves in a circular path, it requires side friction on the tires to maintain that path. Physically, there is a minimum radius of curvature that a vehicle can negotiate at any given speed. A sharper turn cannot be held because the tires will not develop enough centripetal force to provide the necessary radial acceleration. Application of the pertinent laws of mechanics yields the centripetal force equation of the cornering vehicle as:

$$f = \frac{V^2}{15R} - e \qquad (9\text{-}1)$$

where

f = side friction factor, dimensionless
V = vehicle speed, mph
R = radius of vehicle maneuver, feet
e = superelevation rate, feet per foot

Table 9.1 Elements of Roadway Curves

A. Horizontal Alignment Elements
1. Radius of Curvature
2. Length of Curve
3. Superelevation Runoff Length
4. Distribution of Superelevation Runoff Between Tangent and Curve
5. Presence and Length of Spiral
6. Stopping Sight Distance Around Curve

B. Cross-sectional Elements
1. Superelevation Rate
2. Road Width
3. Shoulder Width
4. Shoulder Slope
5. Roadside Slope
6. Clear Zone Width

C. Vertical Alignment Elements
1. Coordination of Edge Profiles
2. Stopping Sight Distance on Approach
3. Presence and Length of Contiguous Grades
4. Presence and Length of Contiguous Vertical Curves

D. Other Elements
1. Distance to Adjacent Curves
2. Distance to Nearest Intersection
3. Presence and Width of Contiguous Bridges
4. Level of Pavement Friction
5. Presence and Type of Traffic Control Devices
6. Type of Shoulder Material

Equation 9-1 was originally employed for the design of railroad curves. Its first widespread recognition as a basis for roadway curve design was in "A Policy on Intersections at Grade,"[1] published in 1940 by AASHTO. As indicated by "A Policy on Geometric Design for Highways and Streets,"[2] published in 1990 by AASHTO, Equation 9-1 has continued to be the basis for roadway curve design. The use of this standard cornering equation depends on the assumption that all points in the vehicle have the same radial acceleration. This is equivalent to assuming that the vehicle has a "point mass."

9.2 Tire Pavement Skid Resistance

The side friction factor at which a cornering skid is imminent depends mainly on the speed of the vehicle, the degree of the cornering path, the condition of the tires, and the characteristics of the pavement surface. On wet pavements, vehicle speed is perhaps the most significant parameter not only because the frictional

Roadway Curves

demand increases with the square of the speed but also because the frictional capability of the tire-pavement interface decreases with increasing speed. Figure 9-1 depicts a generalization of this relationship that illustrates for a given degree of cornering how the factor of safety against skidding decreases rapidly with increasing speed, until the skid is imminent.

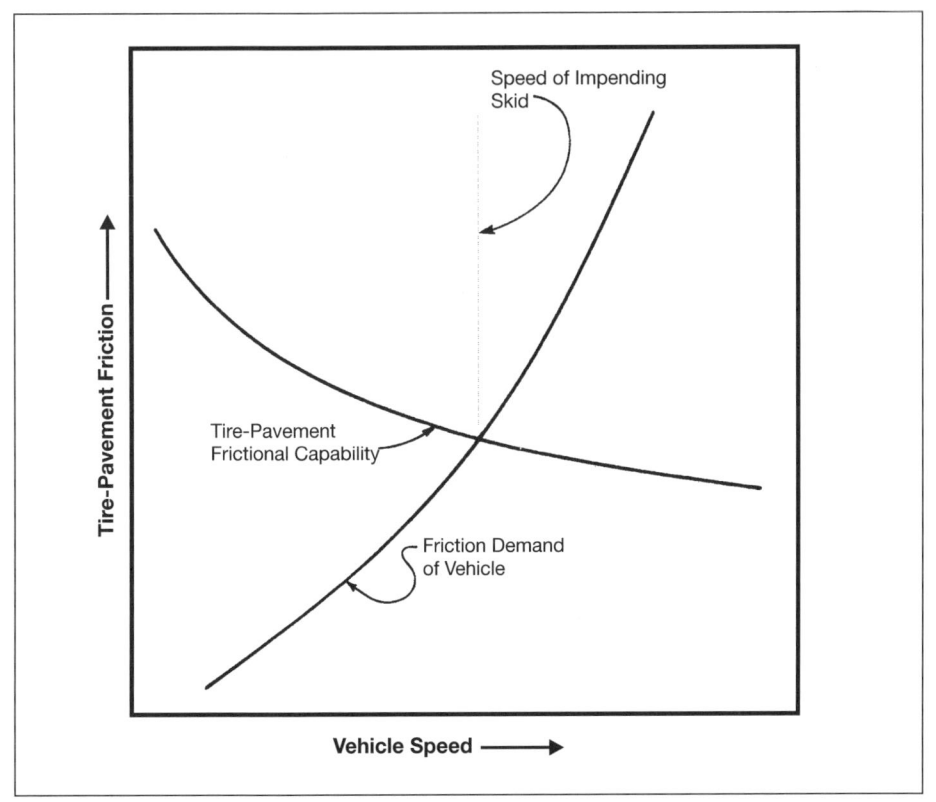

Figure 9-1 Friction Demand Related to Friction Capability for Wet Pavements.

Figure 9-2 shows a percentile distribution of Skid Numbers (defined as the coefficient of friction times 100) at various speeds computed for a random sampling of 500 pavements in one State.[3] These measurements were taken in 1964 employing a modified version of the ASTM standard trailer with standard ASTM test tires. The typical cornering capability assumed by the AASHTO Policy is also plotted on Figure 9-2. Note that about 45 percent of the pavements in this one State do not supply the typical capability level assumed by the AASHTO Policy. Figure 9-3 shows a similar percentile plot of skid numbers for 600 pavements in Germany.[4]

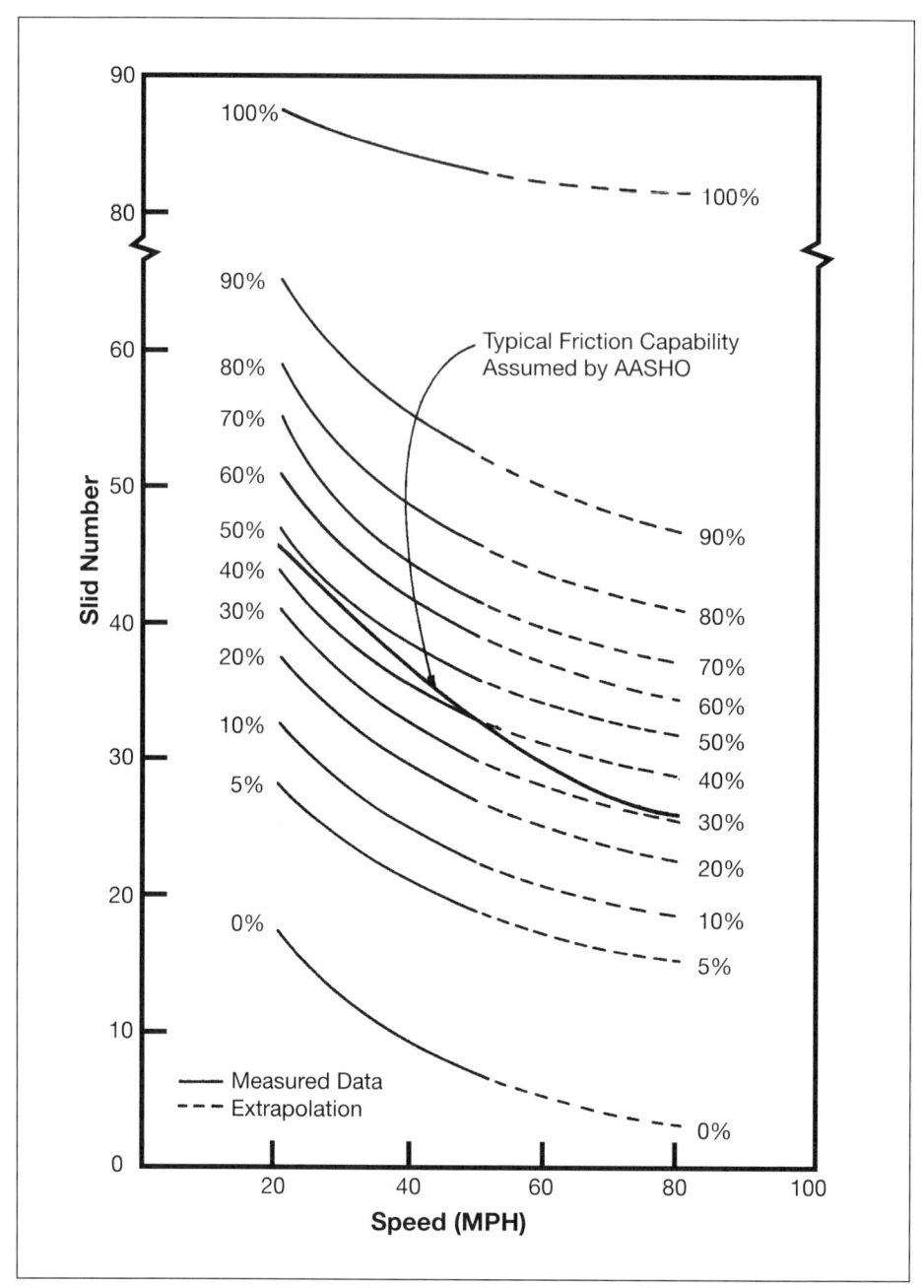

Figure 9-2 Percentile Distribution of Pavement Friction versus Speed for 500 Pavements in One State.[3]

Roadway Curves

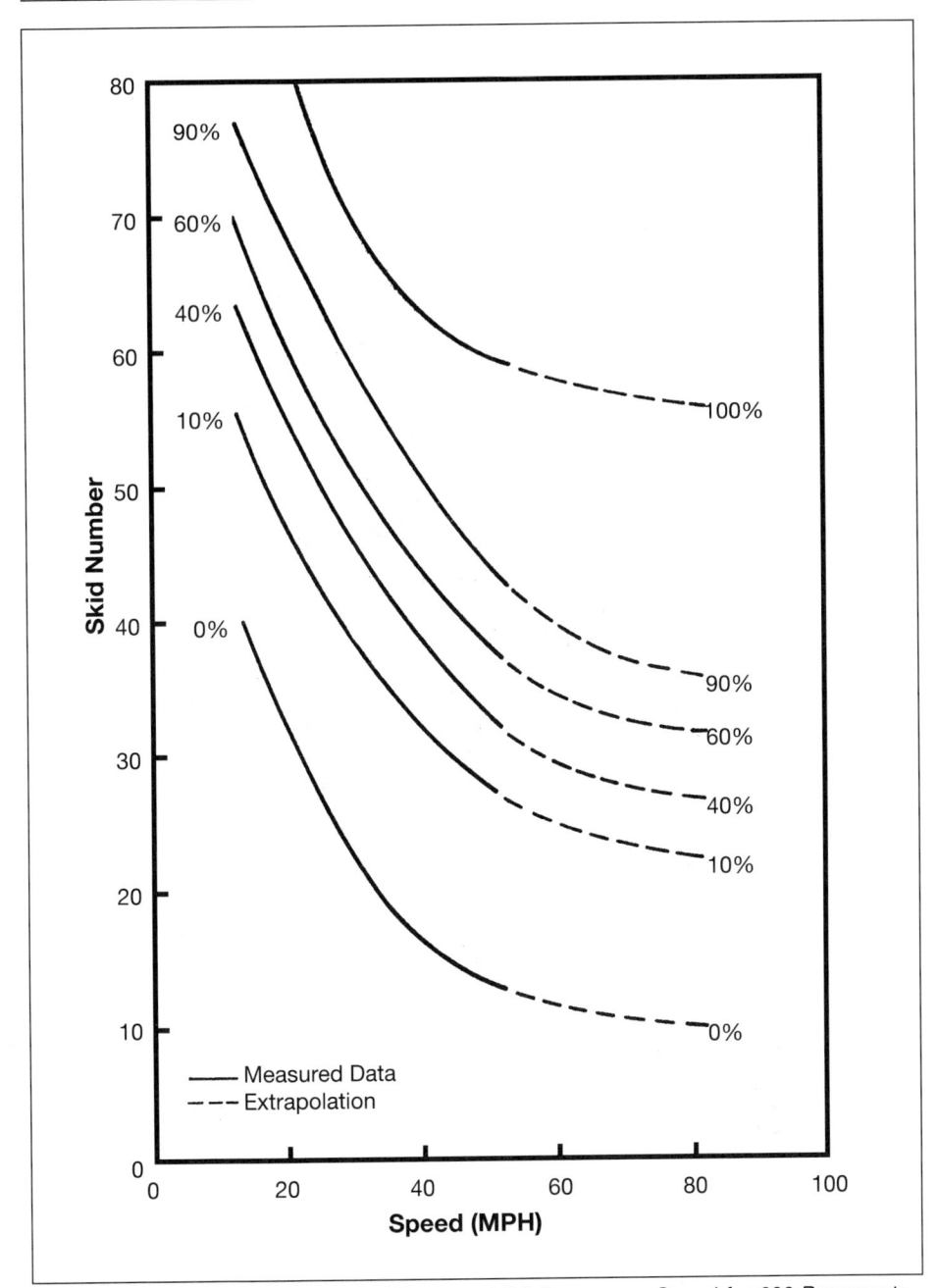

Figure 9-3 Percentile Distribution of Pavement Friction versus Speed for 600 Pavements in Germany.[4]

9.3 Roadway Curve Design Standards

The commonly accepted roadway curve standards are those presented in the AASHTO design policies.[2] The procedure solves for roadway curve radius and superelevation using the standard cornering equation (Equation 9-1). For a given design (or safe) speed, a safe design lateral acceleration (side friction) demand is input into the equation to solve for acceptable combinations of superelevation and minimum curve radius. A discussion of these factors follows.

Safe Side Friction Factors

According to the 1990 AASHTO Policy,[2] for a given curve radius, superelevation, and vehicle speed, the standard cornering equation yields the side friction demand by a cornering vehicle. As speed increases for a particular curve design, the side friction demand increases until the frictional capability of the tire-pavement interface is exceeded and loss of control results.

Roadway curves cannot, of course, be designed directly on the basis of the maximum friction capability of the tire-pavement interface. As in all engineering work, safety factors must be introduced. That portion of the frictional capability that can be used with safety by the vast majority of drivers then becomes the value for design. The AASHTO Policy expands on this thought by stating:

> *Values which are properly related to pavements that are largely deteriorated or poorly maintained—glazed, bleeding or oil slicked—should not control design because these conditions are avoidable and design should be based on acceptable structures attainable with reasonable cost.*

In selecting safe side friction factors for design, AASHTO has taken that point at which side force causes the driver to feel discomfort and act instinctively to avoid higher speed. These values, based on several driver behavior studies over the past 60 years, are shown in Table 9.2.

Table 9.2 Safe Side Friction Factors for Roadway Curve Design[2]

Design Speed	Safe Side Friction Factor
20	.17
30	.16
40	.15
50	.14
55	.13
60	.12
65	.11
70	.10

Maximum Superelevation Rates

For a particular design speed, the maximum superelevation rate and banking the assumption for the safe side friction factor in combination determine the minimum radius. The maximum rates of superelevation on roadway curves are controlled by several factors: (a) climatic conditions, such as frequent snow and ice; (b) type of area, whether rural or urban; (c) frequency of slow-moving vehicles; and (d) terrain conditions, such as flat, rolling, or mountainous. Consideration of these factors jointly has led to different conclusions by the various State roadway agencies. The maximum superelevation rate for open highways in common use is 0.12. Where ice and snow are factors, experience indicates that a superelevation rate of 0.08 is a logical maximum to minimize slipping across the pavement when stopped or when attempting to gain momentum from a stopped position. Some agencies have adopted a maximum rate of 0.10, based on avoiding the excessive outward friction forces required to drive slowly around the curve, a condition resulting in erratic operation. In urban areas, where it is difficult to warp crossing pavements for drainage without introducing negative superelevation for some turning movements, a maximum rate of 0.06 is common.

Minimum Curve Radius

The minimum radius of curvature, a limiting value for a given design speed, is determined from the maximum rate of superelevation and the safe side friction factor. Sharper curvature for that design speed would call for superelevation beyond the practical limit, or for a side friction demand more than that assumed safe, or both. Table 9.3 shows the minimum curve radii for various combinations of design speed and superelevation.

Table 9.3 Minimum Radius for Roadway Curves as a Function of Design Speed[2]

Design Speed (mph)	Minimum Radius (feet) for given Superelevation (ft/ft)			
	.04	.06	.08	.10
20	127	116	107	99
30	302	273	252	231
40	573	509	468	432
50	955	849	764	694
55	1,186	1,061	960	877
60	1,528	1,348	1,206	1,091

9.4 Roadway Curve Accident Characteristics

Most research studies dealing with roadway curves have come to the same basic conclusion, namely that curves are hazardous. Such conclusions, however, are meaningless by themselves. The ultimate question is, under what conditions are roadway curves particularly hazardous? Findings from some studies indicate answers to these questions. Kihlberg and Tharp[5] discovered that curves in combination with intersections resulted in greater accident rates. Billion and Stohner[6] found that overall poor alignment resulted in higher accident rates. Babkov and Coburn[7] reported accident rates for various curve radii. They found those roadway curves with radii less than 2800 feet are 20 to 50 percent more hazardous than straight roadway sections. Jorgensen[8] reported an approximate 15 percent higher accident rate for radii less than 1900 feet. Taylor and Foody,[9] studying curve delineation, found that the length of curve as well as its radius has an influence on accident rates.

A 1983, four-state study, by Glennon, Neuman, and Leisch,[10] is the most comprehensive analysis of roadway curve safety ever undertaken. This study compared the accident experience on 3304 rural two-lane curve segments to 253 rural two-lane straight segments. Each segment was 0.6 mile (one kilometer) long and was selected to minimize variance associated with intersections, bridges, nearby urban development, and nearby curvature. Figure 9-4 summarizes the significant characteristics of accidents on roadway curve segments in this data base. These data were also used to compute an effective rate over a portion of the roadway curve segments that included only the length of curve plus a 0.03-mile tangent transition at each end of the curve. This computation yielded the following conclusions:

1. The average accident rate for roadway curves is about three times the average accident rate for straight roadway segments.
2. The average single-vehicle ran-off-road accident rate for roadway curves is about four times the average single-vehicle ran-off-road accident rate for straight roadway segments.
3. Roadway curves experience a higher proportion of wet pavement accidents than do straight segments.
4. Roadway curves have a higher proportion of fatal and injury accidents than do straight segments.

Although these conclusions may vary by radius and length of curve, they do show that roadway curves are considerably more hazardous than straight segments and that single-vehicle run-off-road accidents are a prevalent aspect of curves.

Roadway Curves

The Glennon, Neuman, and Leisch study[10] also offered the additional following insights based on additional accident and operational observations and analyses:

1. **Relationship of Accident Types to Traffic Volume.** The proportion of accidents, which are single vehicle ran-off-road, increases substantially as average daily traffic decreases.
2. **Roadside Design is the Major Accident Factor.** Roadside character (roadside slope, clear-zone width, coverage of fixed objects) appears to be the most dominant contributor to the probability that a roadway curve has a high reported accident rate.
3. **Other Major Accident Factors.** Although roadside character is the dominant accident factor on roadway curves, most curves with high-accident rates usually have one or more other factors that contribute

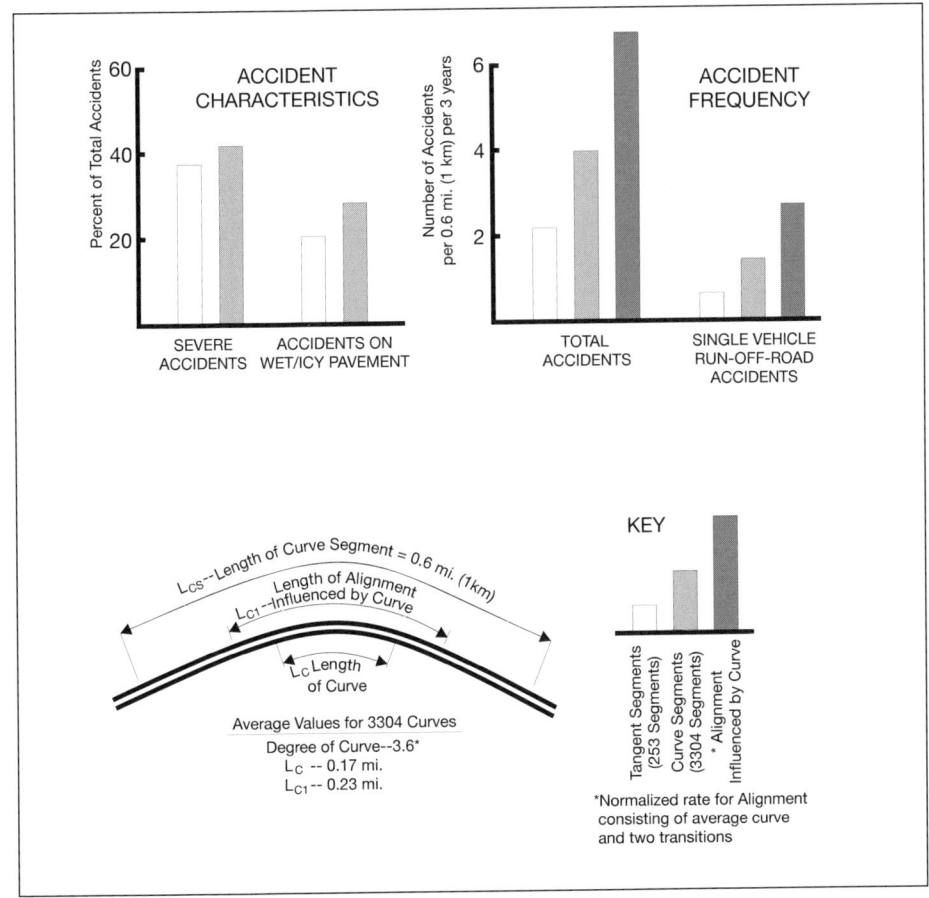

Figure 9-4 Accident Characteristics for Roadway Curves.[10]

to the total hazard (e.g., sharper curvature, longer curve lengths, narrower shoulders, and lower pavement skid resistance).

4. **Maximum Lateral Acceleration.** In traversing a roadway curve, a significant proportion of drivers produces path radii less than the roadway curve radius, regardless of their speed [This behavior is termed *path overshoot*]. Therefore, many drivers traveling at design speed or greater will exceed the AASHTO design friction factor.

5. **Driver/Vehicle Curve Transition Behavior.** All vehicles effect a spiral path transition in proceeding from a straight roadway segment to a roadway curve. This path behavior generally occurs over the full lane width, centered about the point of curvature. Although the severity or length of spiraling path behavior varies among drivers, it is independent of vehicle speed. Drivers with more severe spiraling rates tend to produce greater path overshoot and, therefore, higher levels of lateral acceleration.

6. **Driver/Vehicle Speed Behavior.** Higher-speed drivers approaching sharper roadway curves do not adjust their open roadway speeds to match a safe or comfortable speed for the roadway curve until the curve is imminent. Speed reduction begins about 200 to 300 feet in advance of the curve, and continues in the initial portion of the curve. [Note that these are daytime observations only. Nighttime conditions are probably more severe].

7. **Substandard Roadway Curves.** Existing roadway curves, which are significantly substandard for the prevailing roadway speeds, may pose considerable safety problems. Because drivers do not totally decrease their open roadway speeds to match the safe speed of a substandard roadway curve, high-speed drivers with extreme path behavior will generate very high lateral accelerations.

8. **Roadside Slopes.** Roadside slope traversals on roadway curves appear more severe than on straight segments. Severity is defined by the vehicle path angle to the slope, which is a function of roadway curvature. More severe traversals lead both to generally higher vertical decelerations and higher potential for rollover. These results suggest that, for comparable safety levels, roadside slopes on roadway curves need to be flatter than those on straight segments.

9. **Pavement Irregularities.** Vehicular control stability on roadway curves is very sensitive to pavement washboard and short pavement bumps or dips.

10. **Stopping Sight Distance.** AASHTO[2] stopping sight distance requirements may be inconsistent when applied to roadway curves because of higher resultant pavement friction demands created when a vehicle is both cornering and braking. Also, when the sight restriction is a vertical rock cut, wall, or line of trees, truck drivers lose their eye-height advantage, which AASHTO policy assumes to always compensate for the longer braking distances of trucks.

9.5 Signing for Roadway Curves

The *Manual on Uniform Traffic Control Devices*[11] *(MUTCD)* governs the application of warning signs for roadway curves. The Turn Sign, Curve Sign, Reverse Turn Sign, Reverse Curve sign, and Winding Road Sign are the advance warning signs known as the alignment series (see Figure 9-5). Shaped arrows indicate the direction, severity, and number of curves ahead. These signs are placed in advance of the roadway curve to forewarn drivers of the curvature. Table 9.4 is used to determine how far in advance of a roadway curve to post the sign. Except where two successive curves qualify as a reverse curve combination, advance posting of one curve should never precede the previous curve.

Table 9.4 Suggested Minimum Advance Placement for Curve Warning Signs[11]

Posted Speed (mph)	Advance Distance (feet) for Advisory Speed (mph)				
	10	20	30	40	50
30	150	100	—	—	—
40	275	250	175	—	—
50	425	400	325	225	—
60	575	550	500	400	300

Curve and Turn Signs and Advisory Speed Plates

Whether a particular roadway curve needs signing should be based on the safe speed (design speed) of the curve. This speed is determined by using Figure 9-6,[12] or by using a device called the ball-bank indicator (also called a slope meter). The curves given in Figure 9-6 are used where the radius, r, and superelevation, e, of the roadway curve are known. The radius and superelevation can either be measured directly or obtained from construction plans.

Figure 9-5 Alignment Series of Warning Signs.[11]

The ball-bank indicator (see Figure 9-7) is a curved level that determines the safe speed around a roadway curve by conducting trial speed runs. The device is mounted in a car and calibrated. Trial runs are made at successively higher speed until a 14° reading is recorded at speeds below 20 mph, or a 12° reading is recorded at speeds between 20 and 35 mph, or a 10° reading is recorded at 35 mph or greater.

Either of these methods can be used to determine if a Curve sign or Turn sign is needed. Curve signs are for safe speeds of 30 mph or greater. Turn signs are reserved for safe speeds less than 30 mph. These methods are also

Roadway Curves 285

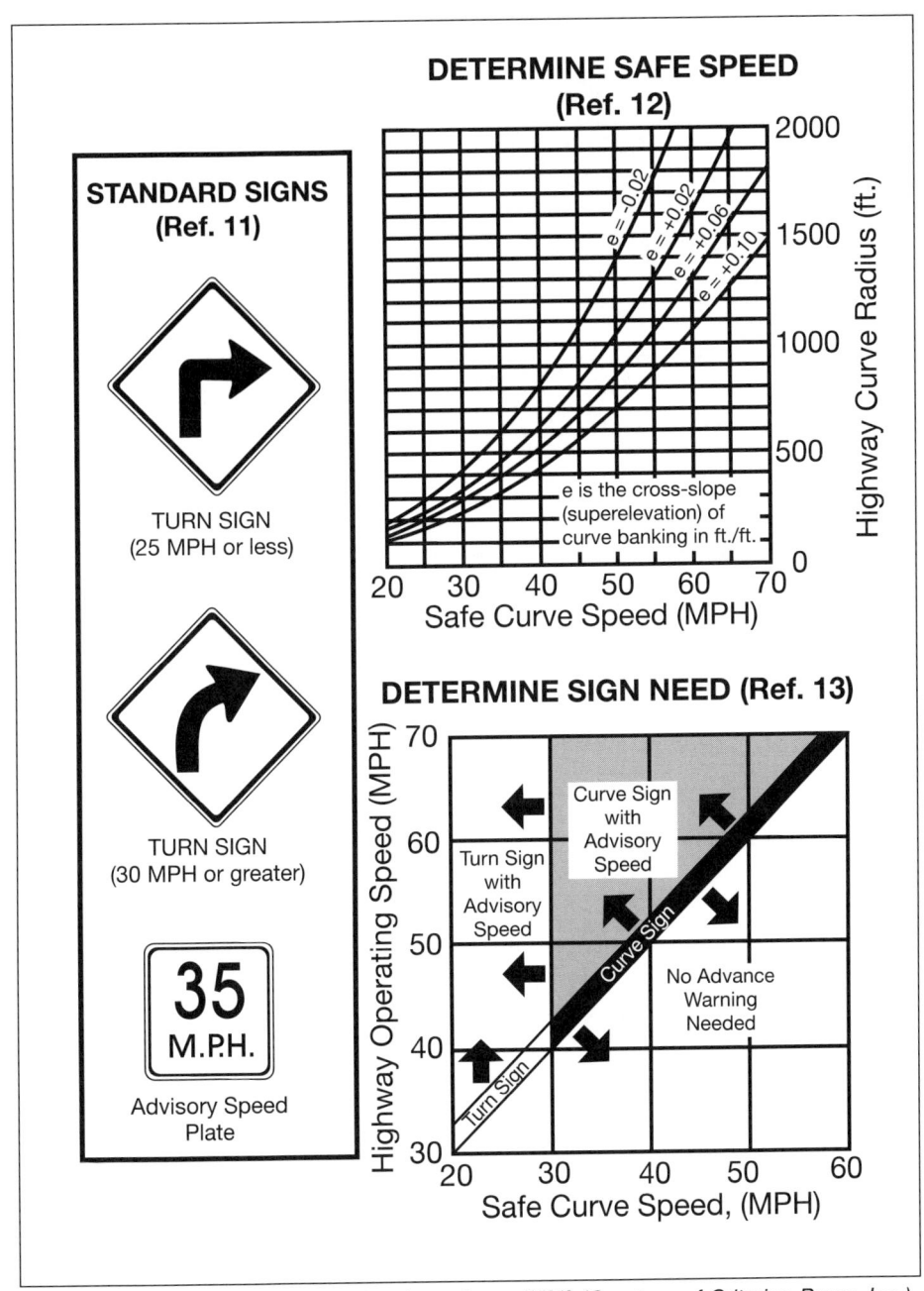

Figure 9-6 Advanced Signing for Roadway Curve.[11,12,13] (Courtesy of Criterion Press, Inc.)

Making a Safe Speed Test

1. Mount the indicator on the dashboard.
2. Place the car on a level surface to calibrate the indicator.
3. Take a reading on the curve by starting at a relatively slow speed and gradually increasing the test speed until the maximum safe speed is found. All readings should by taken with the car at a constant speed and centered in the lane throughout the test.

Safe Speed	Ball-Blank Reading that indicates maximum Safe Speed
Below 20 MPH	14°
20-35 MPH	12°
Above 35 MPH	10°

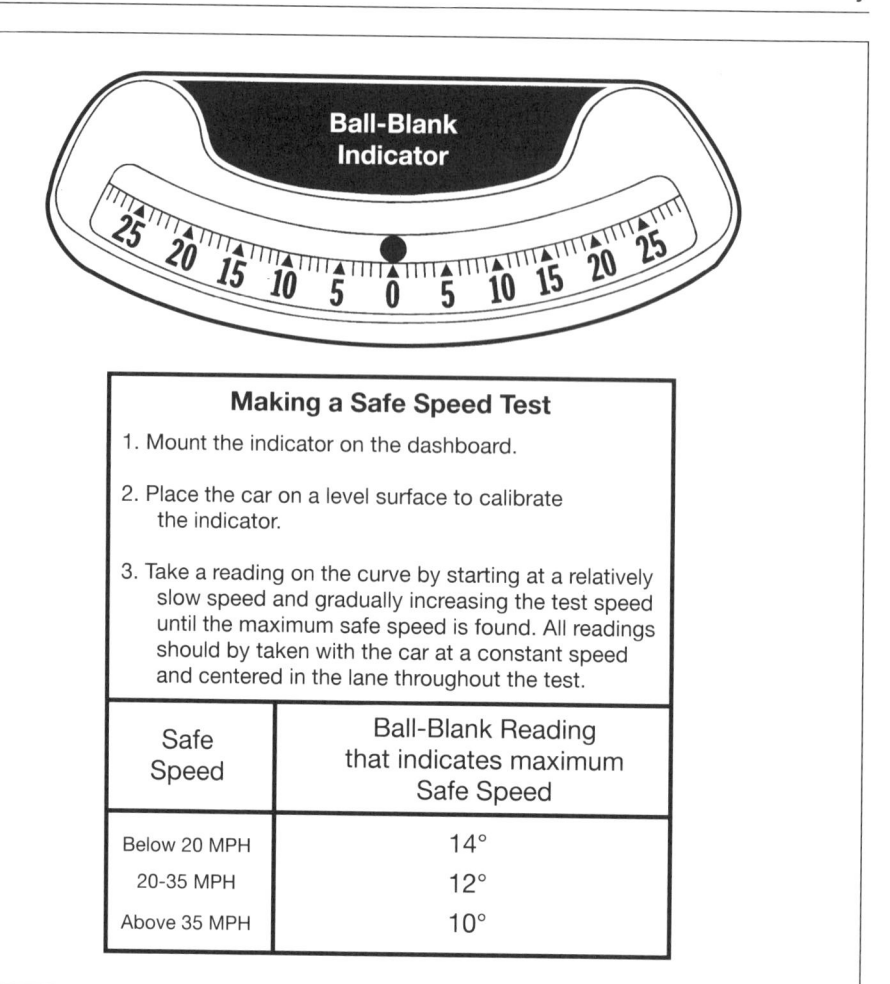

Figure 9-7 Ball-Bank Indicator for Measuring Roadway Curves.[12] (Courtesy of Criterion Press, Inc.)

appropriate for determining the speed on an Advisory Speed plate, which can be placed under the warning sign when the safe speed is less than the legal speed limit.

Figure 9-6 (in the lower-right corner) also shows a recommended practice for the application of Turn signs, Curve signs, and Advisory Speed Plates in Kansas.[13] This practice or one similar is used by most State roadway agencies.

Combination Curves

A combination curve consists of two successive curves. They may be connected with or without a short tangent section, and they may be either in the same or opposite direction. The recommended speed for a combination of curves

Roadway Curves

in the same direction is the lower of the safe speed for the two curves. If either of the curves requires a Turn sign, the Turn sign should be used for the combination.

A reverse roadway curve is a combination consisting of two curves in opposite directions separated by a tangent of less than 600 feet. Either the Reverse Curve sign or Reverse Turn sign should be used depending on the lower safe speed for the two curves. When reversing roadway curves are more than 600 feet apart, they should be signed separately.

Another combination roadway curve sign, the Winding Road sign, may be used for a succession of three or more curves.

Supplementary Curves Signs

Two other kinds of warning signs are used on sharp roadway curves to supplement advance warning signs. These are the Large Arrow and the Chevron Alignment signs shown in Figure 9-8. The Large Arrow sign is erected on the outside of the roadway curve at right angles to approaching traffic. It is usually used to supplement the Turn sign, but is also used for sharper curves with Curve signs on high-speed roadways.

The Chevron Alignment sign is used for special emphasis as an alternate or supplement to standard delineators and/or the Large Arrow sign. When used, it is erected in multiples on the outside of the roadway curve. Spacing of the signs should allow the driver to always see at least two signs while approaching and traversing the roadway curve.

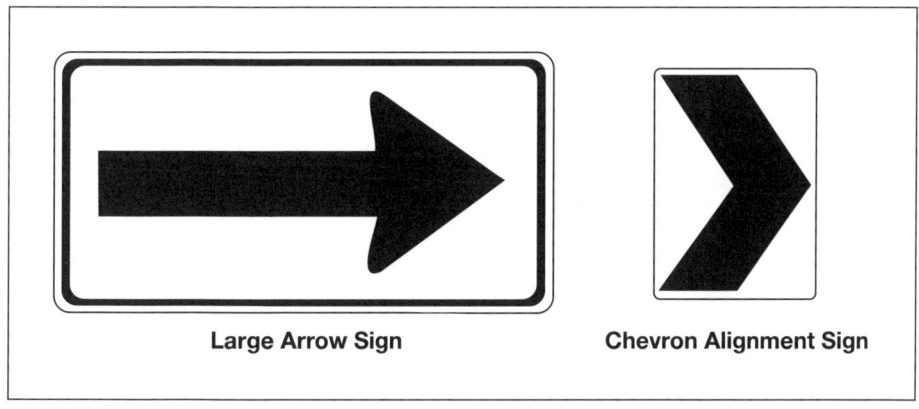

Figure 9-8 *Supplemental Alignment Signs.*[11]

> **ROADWAY CURVE DEFECTS**
>
> **Rule of Thumb**
>
> *The Most Common Roadway Curve Defect is a Sharp Curvature With Deficient Warning and/or Delineation and Sometimes Hidden by a Hillcrest*

9.6 Delineation for Roadway Curves

Part 3D of the *MUTCD*[11] covers the application of delineators on roadway curves. Delineators are reflective buttons mounted on posts on the outside of the roadway curve to indicate the alignment. Delineators are effective aids for night driving and are considered guidance devices rather than warning devices. An important advantage of delineators is that they remain visible when the roadway is wet or snow-covered.

Where used, delineators shall be mounted on supports at 4 feet above the outside shoulder. Spacing should be adjusted on approaches and throughout roadway curves so that several delineators are always visible to the driver. Table 9.5 shows the suggested spacing.

Table 9.5 Suggested Spacing for Delineators on Roadway Curves[11]

Radius of Curve (feet)	Delineator Spacing (feet)
150	30
250	40
400	55
600	70
800	80
1000	90
2000	130
3000	160

9.7 Technical Aspects of Roadway Curve Defect Cases

Thousands of roadway curves in the United States have design (safe) speeds that are considerably lower than the legal speed limit of the roadway. These substandard roadway curves are often an element in tort litigation against roadway agencies. Common arguments are that the roadway agency was negligent for not

Roadway Curves

designing the roadway properly, or given that the design was defective, that they were negligent for failing to warn of the hazard. Because of *design immunity* statutes in most States, the typical roadway curve case is one of, first, showing that the roadway curve was substandard but, second and more important, arguing that the roadway agency failed to properly warn of the hazard.

Most often at issue in the roadway curve tort case is the severity of the roadway curve in terms of how deviant the design (safe) speed is from the prevailing speed limit. For 45- to 70-mph speed limits, roadway curves with design (safe) speeds that are 15 mph or more below the speed limit are generally the subject of tort claims. For 30- to 44-mph speed limits, a deviation of 10 mph or more below the speed limit is typical. Normally, the basic argument is the lack of adequate warning. Either the subject curve had no warnings or it only had an advance Curve or Turn sign, without any supplementary devices. The warnings that are most appropriate for these substandard roadway curves are the advance Curve or Turn sign, Advisory Speed plates, and either the Large Arrow or Chevron Alignment signs.

The second most common substandard design element in roadway curve cases is when the curve is hidden, usually beyond a sharp hillcrest (refer to Chapter 4 for a discussion of AASHTO stopping sight distance standards). Contrary to the general design practice, when considering the approach to a roadway curve, the object height of importance is zero. In other words, with no other aids present, drivers need to see the surface of the roadway curve well before arriving at the curve in order to properly adjust their speeds. A sharp curve hidden over a sharp hillcrest without adequate warning devices is a roadway defect. The warning signs that are designed to overcome these deficiencies are the advance Curve or Turn signs, the Advisory Speed plate, and the Large Arrow or Chevron Alignment signs.

Another common difficulty with substandard roadway curves, particularly on higher-speed roadways, is when that curve is considerably sharper than the proceeding alignment. Because driver's most recent driving experiences largely conditions their expectancy, a sharp roadway curve at the end of a long straight roadway may greatly violate their expectancy. Another violation of driver expectancy is a sharp roadway curve located on a generally winding roadway where all the other curves have higher design speeds. These roadway deficiencies call for the positive guidance of extra warning and delineation to overcome these violations of driver expectancy.

Most roadway curve cases have a combination of defective elements that lead to stronger arguments about failure to maintain, or failure to warn, or both. In addition to the elements already discussed, several other elements are fairly common to roadway curve cases. These elements are discussed below.

Narrow Roadway

A common feature on substandard roadway curves on two-lane roads is a substandard width of 16 to 20 feet. AASHTO[2] generally considers 24-foot widths desirable, and 22-foot widths as acceptable. The narrower the roadway, of course, the more difficulty a driver will have in tracking the curve without running off the inside edge or over the centerline, depending on the direction of curvature. For sharper roadway curves, AASHTO recommends that the pavement be wider on the curve than on its approach. This extra width is particularly important for articulated trucks that need more width because of off-tracking.

For these narrow roadway curves, not only are all the signs important, but also the shoulder maintenance becomes important. Pavement edge drops, windrows of gravel, or other defective conditions seriously degrade the ability of drivers to negotiate these curves safely.

Sight Obstructions on Roadway Curves

Often on low-volume rural highways or urban streets a sharp curve on a narrow roadway will have vegetation growing up to or over the inside edge of the roadway, as shown in Figure 9-9. Because of the sharp curvature and narrow width, a very high proportion of drivers making the left-hand curve can be expected to travel over the centerline while on the curve.[10] With the vegetation, the stopping sight distance is often a small fraction of the AASHTO[2] standard (see

Figure 9-9 Sight-Restricted Roadway Curve.

Roadway Curves

Chapter 4) for the prevailing speeds. Under these circumstances, opposing drivers are often on a foreseeable collision course that allows very little time for collision avoidance. Even if these kinds of roadway curves have otherwise adequate signing, the signing may not be enough. Care is required to trim the vegetation back as far as possible to allow adequate stopping sight distance.

Tangent Roadway With False Visual Cues

Many mainline roadway curves intersect with another roadway that continues beyond the curve on the same line with the tangent approach roadway. This condition (see Figure 9-10) is known to have safety problems. First, drivers intending to proceed off the main roadway onto the intersecting roadway often proceed straight ahead both without signaling and without knowing they are violating the right-of-way of opposing vehicles. This situation is exacerbated by the very large intersection area that exposes this vehicle to head-on collision for a relative long period. Second, with absent or poor delineation, unfamiliar drivers at night are often fooled into thinking the mainline roadway proceeds straight ahead. This is particularly a problem when utility lines and tree lines continue along the tangent alignment rather than the curve. Off-road or head-on collisions occur when drivers realize they have two alternative paths and cannot decide which path is right. The third problem with this condition is also associated with

Figure 9-10 False Cues That the Main Roadway Proceeds Straight Ahead Instead of Curving Right.

the large intersection area. Because of the unique nature of what is essentially a flat-angle intersection, the convergence of edgeline and centerline striping makes delineation of the curved alignment through the intersection almost impossible.

Although this kind of intersection design has been discouraged by AASHTO since at least 1940, many have been built and continue to be accident prone. Every effort should be made to realign these intersections and roadway curves. For existing roadway curves, extra delineation and pavement marking treatments along with standard warning signs are important to overcome false visual cues, to delineate appropriate paths, and to clarify right-of-way assignment.

Low or Negative Superelevation

Because of faulty construction or resurfacing, the superelevation (banking) of an existing roadway curve may be low or even negative. Except for very mild curves on crowned roadways, roadway curves are usually banked such that the outside edge of the curve is higher than the inside edge. Generally, superelevation rates range from 0.04 to 0.08 ft/ft., which means that the outside edge of a two-lane roadway is usually one to two feet higher than the inside edge. Of course, within a limited range, the higher the superelevation rate the higher the safe speed.

When the superelevation rate or a sharp roadway curve is less than the normal expected by drivers, those drivers will tend to overdrive that particular curve more than usual. If driver's normally expect a 0.06 ft/ft superelevation rate but are surprised by a sharp curve with a negative superelevation rate, they may enter that curve at a speed too high to maintain steering stability. Although extra signing may be appropriate as an interim measure, this kind of condition demands resurfacing to correct the superelevation deficiency.

Rough or Uneven Pavements

Washboard, potholes, warped pavements, and short dip or bumps on roadway curves can cause loss of control, particularly on sharper curves on higher speed roadways. Motorcycles are particularly susceptible to loss of control for these conditions. Every effort should be made to minimize rough or uneven pavement conditions.

Slippery Pavements

Sharp roadway curves on higher-speed, higher-traffic roadways are prone to skidding accidents on wet pavements, not only because of the greater demand for friction but also because the traffic demand can rapidly degrade the friction capability of the pavement. In addition to proper signing and warning, these locations should be checked periodically for adequate skid resistance and then resurfaced if necessary.

Other Elements

Other roadway features or conditions that are less frequently associated with roadway curve cases or are otherwise discussed in other chapters of this book include: hydroplaning conditions; reverse curves that are too close together; improper transitioning from the normal crown to a superelevated curve; loose material on the pavement; roadside hazards; sharp cross-slope breaks between the travel lane and the shoulder, etc.

9.8 Accident Reconstruction Aspects

When considering vehicle speed as a factor in a roadway curve collision, a common calculation is the *critical speed*. The critical speed is usually determined by inputting the following factors into the standard cornering equation (Equation 9-1): the roadway curve radius; the maximum superelevation; and an estimate of the available coefficient of friction. For example, if an automobile ran off a 500-foot radius curve with an 0.08 superelevation and a 0.70 coefficient of friction, the critical speed is calculated as 76 mph. Although many accident reconstructionists would conclude that the vehicle in this accident had to be traveling at 76 mph or greater, this conclusion has a potential fallacy.

Most drivers will not accept steady-state lateral accelerations greater than about 0.3g (a 0.3 friction demand). When drivers are surprised by a roadway curve with a design (safe) speed considerably below their operating speed, they will either steer sharply, brake sharply, or both trying to avoid loss of control. Loss of control can occur either (1) because the instantaneous vehicle path radius is too sharp and causes a spike in the lateral acceleration demand that exceeds the available friction supply, or (2) because drivers panic and lock their brakes when their lateral acceleration exceeds 0.3g's. Because of these other probable events, the calculated theoretical critical speed may not even closely approximate the loss-of-control speed.

9.8 Typical Defense Arguments

When defending tort claims that allege defective roadway curves, roadway agencies generally may have one or two major legal defenses. The first of these legal defenses is the *design immunity* statute, which is available in most if not all States. Roadway design is generally held to be a discretionary function; carrying limited immunity. Under the design immunity statute, roadway agencies and their employees are immune from liability when the design was in general accord with the prevailing standards at the time the roadway was built (including major reconstruction). Because a vast majority of roadway miles in the United States were built before roadway design standards were codified, roadway agencies often claim immunity from tort claims that only allege a design defect.

The second major legal defense available to roadway agencies in some States is the argument of *discretionary function*. Even though most roadway agencies have the express duty to warn of hazards, some States have tort claims acts that strictly interpret the wording of the *Manual on Uniform Traffic Control Devices (MUTCD)*.[11] The *MUTCD* has three specifically defined words as follows:

Shall: a mandatory condition
Should: an advisory condition
May: a permissive condition

Those States with strict interpretations say that unless the *MUTCD* mandates a particular traffic control device, that its use is discretionary and, therefore, immune from tort claims. Because most of the *MUTCD* discussion about roadway curve warning signs uses the *may* term, claims for failure to warn, even on very sharp roadway curves, are usually futile in these States with strict interpretations of the *MUTCD*.

Beyond the available statutory arguments given above, roadway agencies use other case-specific defenses that argue that either the driver was the sole negligent party and/or the specific roadway curve condition had no contribution to the causation of the collision. Some of these arguments are:

1. The driver was negligent for running off the road.
2. The driver was negligent for crossing the centerline and colliding with another vehicle.
3. The driver was negligent for speeding.
4. The driver was negligent for driving too fast for conditions.
5. The driver was negligent for driving while intoxicated.
6. The driver should have expected the alleged condition because of the generally substandard nature of the roadway.
7. The roadway agency had no notice of the defective condition.
8. The subject roadway curve lacks any accident records giving notice to a safety problem.
9. The roadway agency has limited funds that prevent it from putting signs everywhere.
10. If signs are erected, they will either be shot full of holes or stolen.
11. The use of an advance warning sign without advisory speed plates or any other supplementary signs, delineators, or pavement markings is sufficient warning, even of very sharp roadway curves.

Endnotes

1. American Association of State Highway Officials, *A Policy on Intersections at Grade*, 1940.

2. American Association of State Highway and Transportation Officials, *A Policy on Geometric Design of Highways and Streets*, 1990.

3. Glennon, John C., *State of the Art Related to Safety Criteria for Highway Curve Design*, Texas Transportation Institute, Research Report 134-4, 1969.

4. Schulze, K.H., and Blackman, L., *Friction Properties of Pavements at Different Speeds*, ASTM Special Technical Publication No. 326, 1962.

5. Kihlberg, J.K., and Tharp, K.J., *Accident Rates as Related to Design Elements of Rural Highways*, Transportation Research Board, NCHRP Report 47, 1968.

6. Billion, C.E., and Stohner, W.R., *A Detailed Study of Accidents as Related to Highway Shoulders*, Highway Research Board Proceedings, 1957.

7. Babkov, V.F., and Coburn W., *Road Design and Traffic Safety*, Traffic Engineering and Control, September 1968.

8. Jorgensen, Roy and Associates, *Cost-Safety Evaluation of Highway Rehabilitation Policies*, 1972

9. Taylor, W.C., and Foody, T.J., *Curve Delineation and Accidents,* Ohio Department of Highways, 1968.

10. Glennon, J.C., Neuman, T.R. and Leisch, J.E., *Safety and Operational Considerations for Design of Rural Highway Curves,* Federal Highway Administration, 1983.

11. Federal Highway Administration, *Manual on Uniform Traffic Control Devices*, 1988. (available through Lawyers and Judges Publishing Co.)

12. Federal Highway Administration, *Traffic Control Devices Handbook*, 1983.

13. Kansas Department of Transportation, *Handbook of Traffic Control Practices for Low Volume Rural Roads*, 1981.

References

Zeeger, Charles V., et. al., *Safety Effectiveness of Highway Design Features: Volume 2: Alignment*, Federal Highway Administration, 1992.

Glennon, John C., *Effect of Alinement on Highway Safety*, Transportation Research Board, State of the Art Report 6, 1987.

Neuman, Timothy R., Glennon, John C., and Saag, James B., *Accident Analysis for Highway Curves*, Transportation Research Record 923, 1983.

Glennon, John C., Neuman Timothy R., and McHenry, Brian G., *Prediction of the Sensitivity of Vehicle Dynamics to Highway Curve Geometrics Using Computer Simulation*, Transportation Research Record 923, 1983.

Glennon, John C., *Effect of Alinement on Highway Safety: A Synthesis of Prior Research*, Transportation Research Board, 1985.

Glennon, John C., and Weaver, Graeme D., *Highway Curve Design for Safe Vehicle Operations*, Highway Research Record No. 371, 1972.

Glennon, John C., and Weaver, Graeme D., *The Relationship of Vehicle Paths to Highway Curves*, Texas Transportation Institute, Research Report No. 134-5, 1971.

Perchonak, K., et. al., *Methodology for Reducing the Hazardous Effects of Highway Features and Roadside Objects*, Federal Highway Administration, 1978.

Messer, C.J., et. al., *Highway Geometric Design Consistency Related to Driver Expectancy*, Federal Highway Administration, 1981.

Chapter 10

Construction and Maintenance Zones

Standard traffic control devices are particularly important during roadway construction and maintenance operations. Because of changing and unexpected traffic conditions, drivers are more dependent on traffic control devices to guide them safely through what would otherwise be a hazardous area.

The effectiveness of traffic control devices in construction and maintenance zones depends on their ability to satisfy the driver's need for information. Both the message content and the placement of traffic control devices is important. The positive guidance principles of providing clear and simple standard messages that command driver's attention at a point where they have adequate time to properly respond are the basis of any good traffic control plan.

Past experience indicates that the most serious failures to meet driver needs result from:

1. Contradictory information.
2. Misleading information
3. Messages with incorrect distances.
4. Non-standard traffic control devices.
5. Incorrect signs.
6. Transitions that are too short or are curved too sharply.

Construction and maintenance operations require temporary traffic control zones that cover the entire length of roadway between the first advance warning sign up to and including the last device where traffic returns to its normal path and operation. Most temporary traffic control zones can be divided into four distinct areas (see Figure 10-1): the advance warning area; the transition area; the activity area; and the termination area.

Principles, practices, and legal requirements for traffic control for construction and maintenance zones are well-defined by *Part 6* of the 1988 *Manual on Uniform Traffic Control Devices*[1] *(MUTCD)*. The latest revision of *Part 6* was printed in 1993 as a separate document titled, *Standards and Guides for Traffic Controls for Street and Highway Construction, Maintenance, Utility, and Incident Management Operations.*[2] Tort liability cases involving construction and maintenance most frequently claim that the roadway agency, the construction contractor, or both violated one or more provisions of the *MUTCD*.

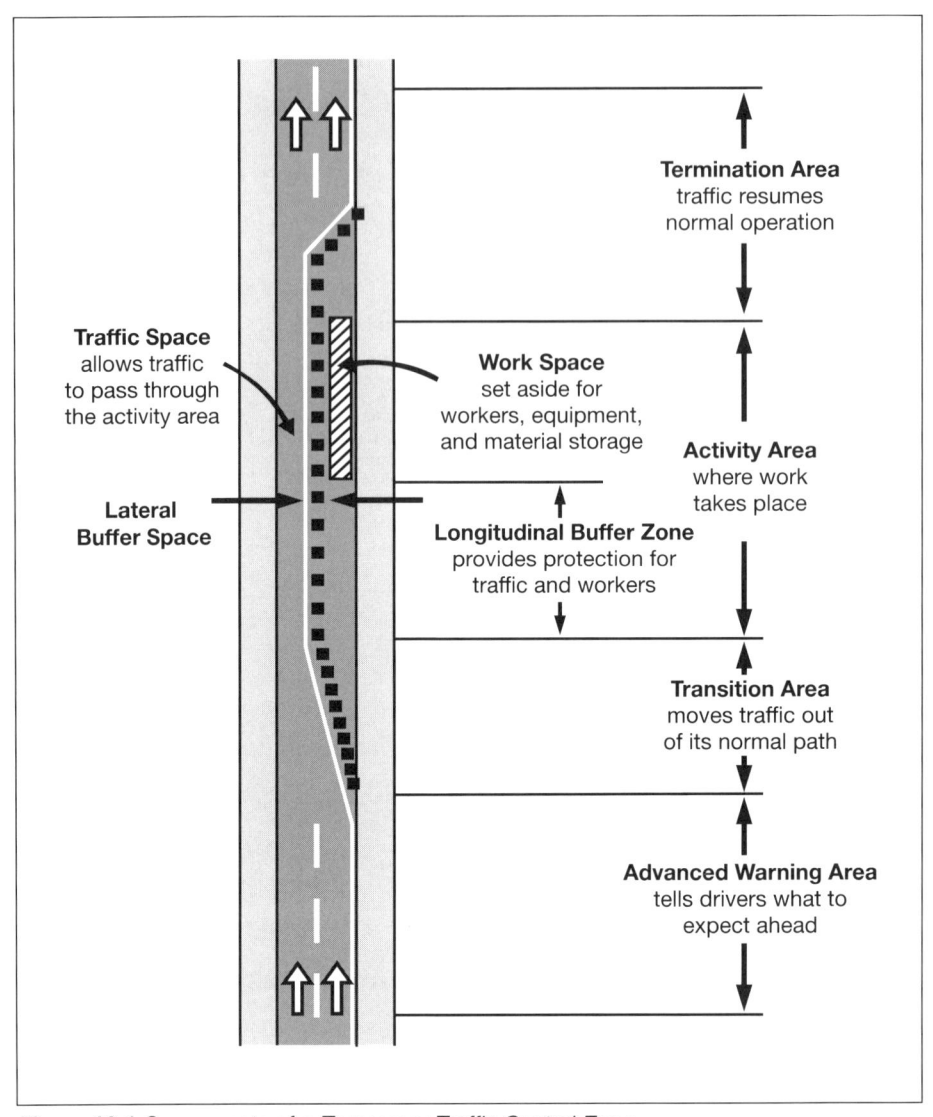

Figure 10-1 Components of a Temporary Traffic Control Zone.

10.1 Fundamental Principles

The most commonly accepted principles for construction and maintenance zone safety are stated in *Section 6A-5* of the 1988 *MUTCD*[1] and *Section 6B* of the 1993 revision.[2] Paraphrased below these principles are:
1. Assign a high priority to traffic safety.
2. Apply the same basic safety principles governing the design of permanent roadways.

Construction and Maintenance Zones

3. Prepare a traffic control plan that is appropriate to the complexity of the worksite. All responsible parties should be intimate with this plan.
4. Avoid reduced speed zoning as much as possible.
5. Avoid frequent and abrupt changes in geometrics that might surprise drivers.
6. Provide drivers with positive guidance both approaching and throughout the worksite.
7. Remove traffic control devices that are contradictory or inconsistent with intended paths or operations.
8. Employ flagging only when all other traffic control methods are inadequate.
9. Assign responsibility for safety at worksites to individuals who are trained in the principles of safe traffic control.
10. Monitor traffic operations in construction zones, under heavy and light traffic, during day and night, and under various weather conditions, to insure all traffic control measures are operating effectively and that all devices are clearly visible.
11. Analyze traffic accidents and conflicts as indicators of needed changes in traffic control. For example, skid marks or damaged traffic control devices may indicate a particular deficiency.
12. Maintain a roadside recovery area free of unnecessary construction equipment, materials, and debris that might cause serious injuries to occupants in ran-off road vehicle impacts.

10.2 Traffic Operational Elements

A formal Traffic Control Plan (TCP) is a required document for all Federal-aid roadway construction projects.[3] Other construction projects should also have a TCP, as indicated by *Section 6A-3* of the 1988 *MUTCD*. The TCP is a plan for guiding traffic through the construction or maintenance area. It should provide for the safety of pedestrians, motorists, and workers.

The TCP describes the construction zone layout, the traffic operational elements, and the traffic control devices needed to safely move traffic through the construction zone during each phase of construction. The details of the TCP should be consistent with the complexity of the project. For example, TCP's can range from a very comprehensive multi-phased plan for a high-traffic urban freeway to a single-phased plan for a two-lane rural road that simply refers to either a typical drawing in the *MUTCD*, a roadway agency standard plan sheet, or a simple drawing in the contract.

The following discussion includes the more important operational elements of construction and maintenance zones including tapers, detours, diversions, one-way traffic control, flagging operations, mobile operations, pedestrian safety, and worker safety.

Tapers

One of the most important elements in a construction or maintenance zone is the taper that uses cones, drums, pavement markings, etc. to transition traffic either out of or back into its normal path. An inadequate length of taper will almost always produce hazardous traffic operations. The real test of taper length is watching traffic move through the transition area. A brief period of observation will usually give a clear indication of whether drivers are having trouble. Frequent brake lights and erratic steering will show the need for increased taper length.

Table 10-1, from the *MUTCD*, shows the types of tapers and their recommended lengths. The maximum spacing of channelizing devices in a taper should be a distance, in feet, equal to the speed limit, in mph. The various types of tapers are described more fully below.

The *merging taper* is used to close a lane on a multi-lane roadway and to direct traffic in the closed lane to merge into the adjacent lane. Adequate taper length is needed by drivers to both find a gap in the adjacent traffic stream and to move into it safely. Figure 10-2 shows the dimensions and channelizing device spacing for this taper.

Table 10.1 Taper Length Criteria for Work Zones[1]

Type of Taper	Taper Length
Upstream Tapers	
Merging Taper	L Minimum
Shifting Taper	1/2 L Minimum
Shoulder Taper	1/3 L Minimum
Two-way Traffic Taper	100 feet Maximum
Downstream Tapers	100 feet per lane
(use is optional)	

Formulas for L	
Speed Limit	Formula
40 MPH or Less	$L = \dfrac{WS^2}{60}$
45 MPH or Greater	$L = W \times S$

L = Taper Length in feet
W = Width of offset in feet
S = Posted speed or off-peak 85 percentile speed in MPH

Construction and Maintenance Zones

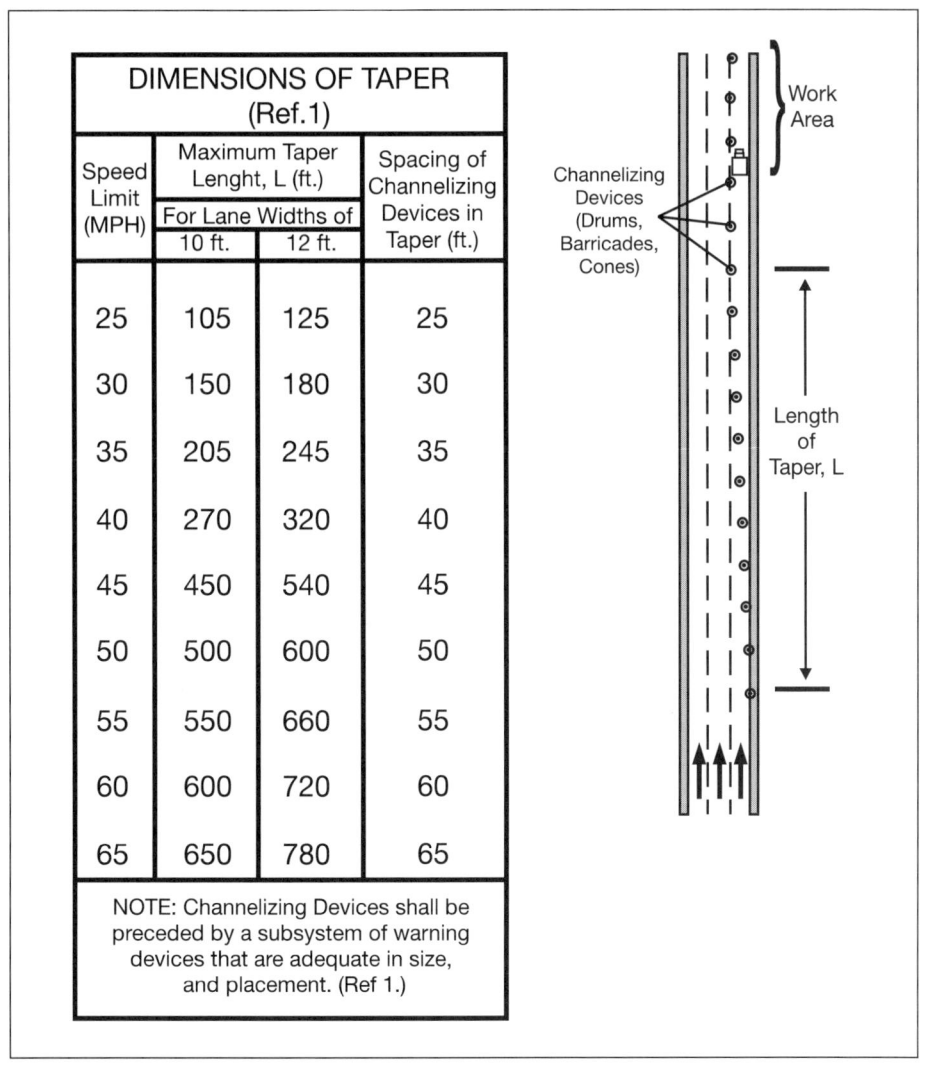

DIMENSIONS OF TAPER (Ref.1)			
Speed Limit (MPH)	Maximum Taper Lenght, L (ft.) For Lane Widths of		Spacing of Channelizing Devices in Taper (ft.)
	10 ft.	12 ft.	
25	105	125	25
30	150	180	30
35	205	245	35
40	270	320	40
45	450	540	45
50	500	600	50
55	550	660	55
60	600	720	60
65	650	780	65
NOTE: Channelizing Devices shall be preceded by a subsystem of warning devices that are adequate in size, and placement. (Ref 1.)			

Figure 10-2 Length of Taper Recommended for Channelization.[1] *(Courtesy of Criterion Press, Inc.)*

The *shifting taper* is used where a lateral shift is needed, but merging is not required. Although shifting tapers normally only need about one-half the length of a merging taper, longer distances may be beneficial where space is available.

The *shoulder taper* may be beneficial on high-speed roadways with work being done on improved shoulders of 8 feet or greater width. A shoulder closure is treated the same as any other closed portion of the roadway, and the work area on the shoulder should be preceded by a shoulder taper. A shoulder taper is normally one-third the length of a merging taper. If the shoulder, through practice or because of construction, is used as a travel lane, a merging or shifting taper should be used, as appropriate.

The *downstream taper* is used in termination areas to show the driver that access is now available to the original lane or path that was closed. This taper has a minimum length of 100 feet per lane, with channelizing devices spaced about 20 feet apart.

The *one-lane, two-way taper* is used in advance of work areas that require alternating one-way traffic. Typically, traffic is controlled by a flagger or a temporary traffic signal. A 100-foot taper, with channelizing devices spaced at 20 feet apart, should be used to guide drivers into the one-way section. However, if a portable concrete barrier is used as part of the taper, barrier taper standards (see Chapter 3) should be followed.

Detours and Diversions

At detours, traffic is directed onto another roadway to bypass the worksite. Detours should be signed clearly over their entire length so that drivers can easily determine how to return to the original roadway.

At diversions, traffic is directed onto a temporary roadway placed within or next to the traveled way. Typical diversions include a median crossover on a divided roadway or a shift to a new temporary alignment when reconstructing a bridge on a two-lane roadway.

One important aspect of detours and diversions is the transition area from the closed roadway to the temporary or alternative alignment. This transition, usually accomplished with a reverse roadway curve, should have a design speed as close to the roadway operating speed as possible. The introduction of very sharp curvature at a transition usually creates unsafe operations.

One-Way Traffic Control

When one lane is closed on two-lane, two-way roadways, the remaining lane is used for traffic in both directions. Alternating one-way traffic control can be operated with flaggers, a flag-carrying car, a pilot car, temporary traffic signals, or by using STOP or YIELD signs. The proper method is a function of traffic volume, length of one-way operation, project duration, and available sight distance (see 1993 *MUTCD* Revision[2]).

Flagging

Flagger stations shall be located far enough ahead of the work activity so that approaching drivers can stop safely before entering the work space. Flaggers need to be clearly visible to approaching traffic at all times. For day work, the flagger's vest, shirt, or jacket shall be orange, yellow, strong yellow green or fluorescent versions of these colors. For night work, similar outer garments shall be retroreflective orange, yellow, white, silver, or strong yellow green. To be clearly visible, the flagger should always stand alone either on the shoulder or in the barricaded lane. The flagger should only stand directly in front of traffic after it has stopped.

Flagging devices include STOP/SLOW paddles, lights, and red flags. The STOP/SLOW paddle, which gives drivers more positive guidance than red flags, is the primary flagging device. Red flags should be limited to emergency situations and to low-speed, low-volume traffic locations, which can be controlled by a single flagger.

Flaggers need to be adequately trained in safe traffic control practices and instilled with a sense of responsibility for the safety of motorists, workers, and pedestrians. Flagging procedures are illustrated in Figure 10-3.

Mobile Operations

These operations may be fast-moving, slow-moving, or intermittent-stop activities. Fast-moving operations are where the speed of the operation is 15 mph or less under the posted speed limit. A typical fast-moving operation involves a truck performing pavement striping. If traffic volume is low and sight distance is good, a well-marked vehicle is sufficient warning. For higher traffic volumes and speeds, a back-up vehicle equipped with signs, beacons, and/or an Arrow Display should follow the operation to warn traffic and protect workers.

Typical slow-moving and related intermittent-stop operations include herbicide spraying, pavement sweeping, and snowplowing. For slower-speed operations, stationary warning signs that are periodically retrieved and moved up with the operation may be feasible. For higher-speed roadways, a back-up vehicle equipped with signs, beacons, and/or an Arrow Display should follow the operations. For low-speed streets, a single vehicle equipped with signs and beacons may suffice.

Pedestrian Safety

Every effort should be made to separate pedestrian movement from both worksite activities and adjacent traffic. When possible, signing should direct pedestrians to safe street crossings in advance of the construction zone (see Figure 10-4).

The movement of work vehicles and equipment across pedestrian paths should be controlled by flaggers or temporary traffic control. Cuts in work areas that cross pedestrian walkways should be minimized because they often create rough and muddy paths.

Where pedestrians are vulnerable to impact by errant vehicles, they should be separated and protected by a crashworthy barrier system. Vertical curbs,

Figure 10-3 Hand Signal Devices for a Flagger.

Construction and Maintenance Zones 305

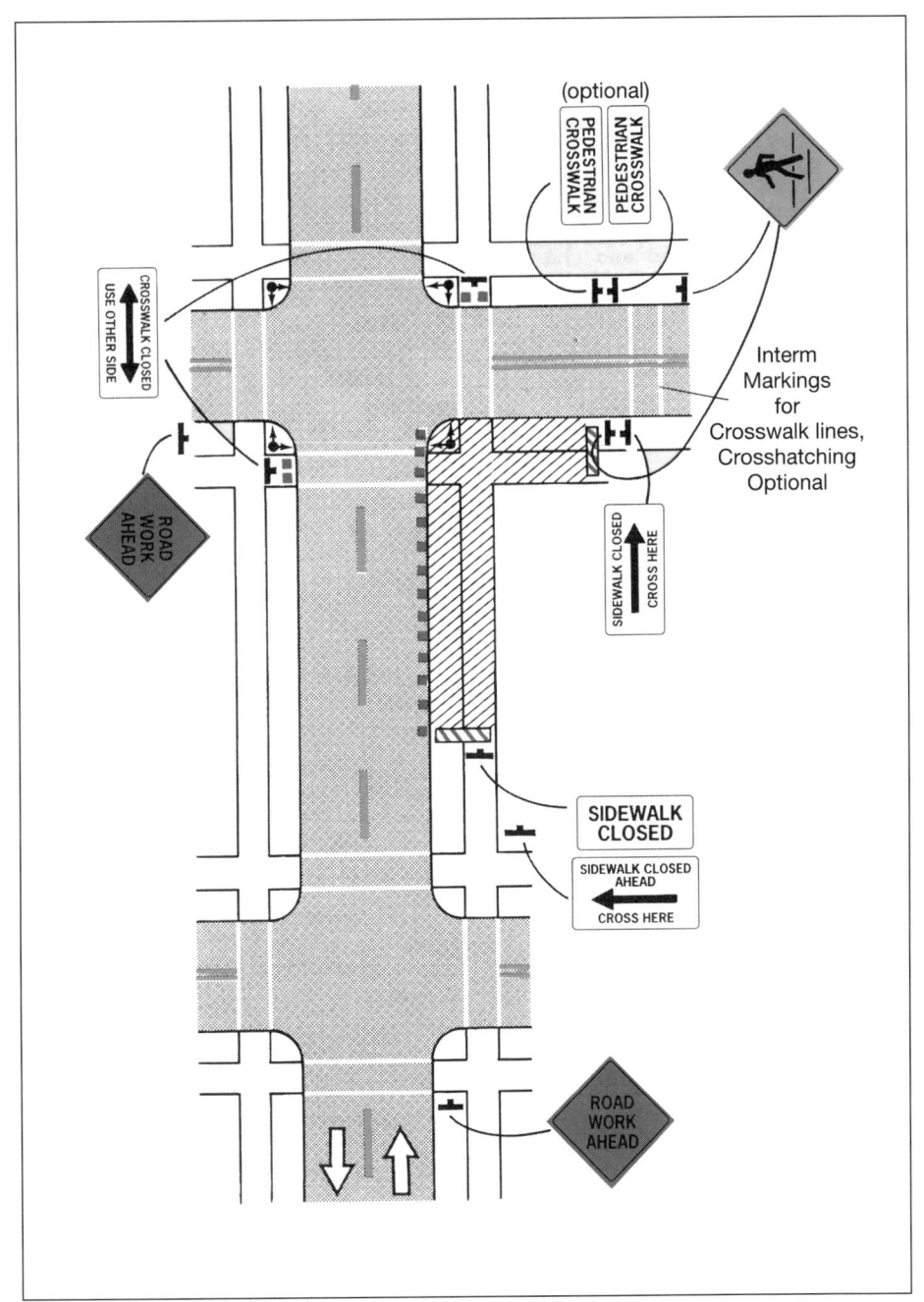

Figure 10-4 Typical Application for Crosswalk Closure and Pedestrian Detours.[2]

chain-link fences, and similar light-weight devices should not be substituted for crashworthy barriers, because they cannot prevent intrusion onto the sidewalk.

Worker Safety

Implementing all of the traffic controls to achieve the fundamental principles that command driver attention and provide positive guidance, are basic to workers safety. Other key elements to assure worker safety are:

1. *Training.* Workers should be trained to understand how to minimize vulnerability when working next to traffic.
2. *Clothing.* Workers exposed to traffic should wear bright, highly-visible clothing.
3. *Barriers.* Workers who work close to high-speed high-volume traffic for moderate to long periods should be protected by crashworthy barriers.
4. *Other Elements.* When worker exposure is particularly high, one or more of the following elements should be considered: speed control; police patrols; lighting for nighttime work areas; road closure; etc.

Typical Applications

Both the 1988 *MUTCD* and earlier versions have nine typical applications showing the sequences and spacing of traffic control devices in various types of construction and maintenance zones. These applications include detours, flagging operations, crossovers, and lane closures. Because these nine typical applications were insufficient for describing the best traffic control for all kinds of construction and maintenance operations, the 1993 *MUTCD* Revision[2] now includes 44 typical applications. Table 10-2, taken from the 1993 revision, lists the 44 typical application (TA) diagrams. Figures 10-4 through 10-8 show some of these typical application diagrams.

10.3 Traffic Control Devices

Traffic control devices for construction and maintenance zones include signs, signals, markings, channelizing devices, and any other standard devices placed on or adjacent to a street or highway to regulate, warn, or guide traffic. Many of these devices are the same as those used on normal roadways; other devices are unique to construction and maintenance zones. One particularly unique feature is the use of the color orange for warning signs and channelizing devices. The following discussion is a brief summary of the aspects of traffic control devices that are most important to safe operations in construction and maintenance zones.

Construction and Maintenance Zones

Table 10-2

Location Roadway Type Application	Duration of work	
	Stationary/ * short duration***	Mobile**
Roadside (outside of shoulder)		
All roadways		
Work beyond the shoulder	TA-1	
Blasting zone	TA-2	
Shoulder		
All roadways		
Work on shoulders	TA-3	
Mobile operation on shoulder		TA-4
Shoulder closed on freeway	TA-5	
Shoulder work with minor encroachment	TA-6	
Within traveled way		
Rural two-lane		
Road closed with on-site detour	TA-7	
Roads closed with off-site detour	TA-8	
Roads open and closed with detour	TA-9	
Lane closure on two-lane road using flaggers	TA-10	
Lane closure on low-volume, two-lane road	TA-11	
Lane closure on two-lane road using traffic signals	TA-12	
Temporary road closure	TA-13	
Haul road crossing	TA-14	
Work in center of low-volume roads	TA-15	
Surveying along centerline of low-volume road	TA-16	
Mobile operation on two-lane road		TA-17
Urban streets		
Lane closure on minor street	TA-18	
Detour for one travel direction	TA-19	
Detour for closed street	TA-20	
Intersections and walkways		
Lane closure near side of intersection	TA-21	
Right lane closure far side of intersection	TA-22	
Left lane closure far side of intersection	TA-23	
Half road closure far side of intersection	TA-24	
Multiple lane closures at intersection	TA-25	
Closure in center of intersection	TA-26	
Closure at side of intersection	TA-27	
Sidewalk closures and bypass walkway	TA-28	
Crosswalk closures and pedestrian detours	TA-29	
Multilane undivided		
Interior lane closure on multilane street	T-30	
Lane closure with uneven directional on streets volumes	TA-31	
Half road closure on multilane highway	TA-32	
Multilane divided		
Lane closure on divided highway	TA-33	
Lane closure with barrier	TA-34	
Mobile operation on multilane road		TA-35
Freeways		
Lane shift on freeway	TA-36	
Double lane closure on freeway	TA-37	
Interior lane closure on freeway	TA-38	
Median crossover on freeway	TA-39	
Median crossover for entrance ramp	TA-40	
Median crossover for exit ramp	TA-41	
Work in vicinity of exit ramp	TA-42	
Partial exit ramp closure	TA-43	
Work in vicinity of entrance ramp	TA-44	

*Long-term stationary: more than 3 days; Intermediate-term stationary: overnight up to 3 days; Short-term stationary: anytime, more than 60 minutes.
**Mobile: Intermittent and continuous moving.
***Short-duration: up to 60 minutes

308   Roadway Defects and Tort Liability

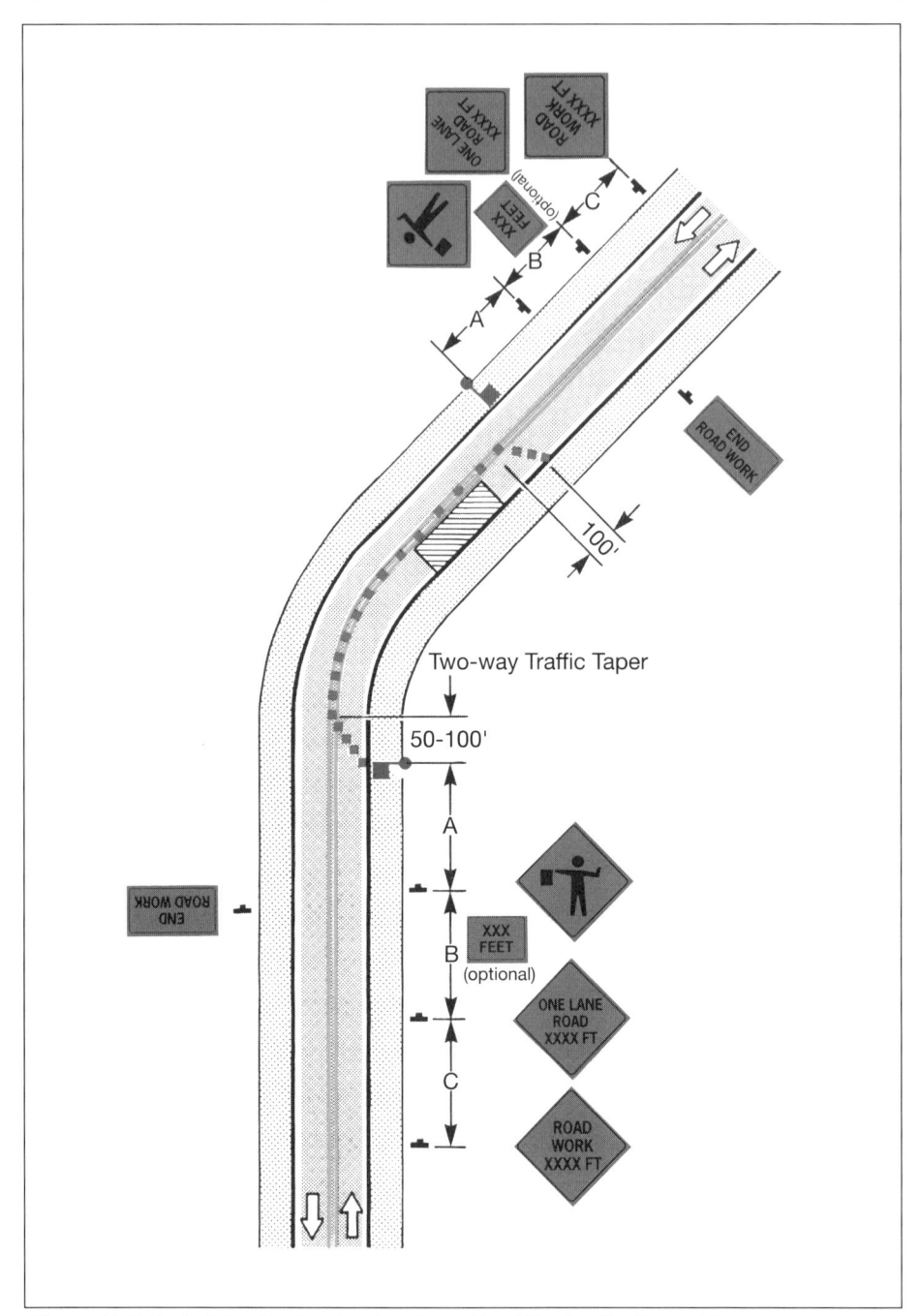

Figure 10-5 Typical Application for One-Lane Two-Way Traffic Control.[2]

Construction and Maintenance Zones 309

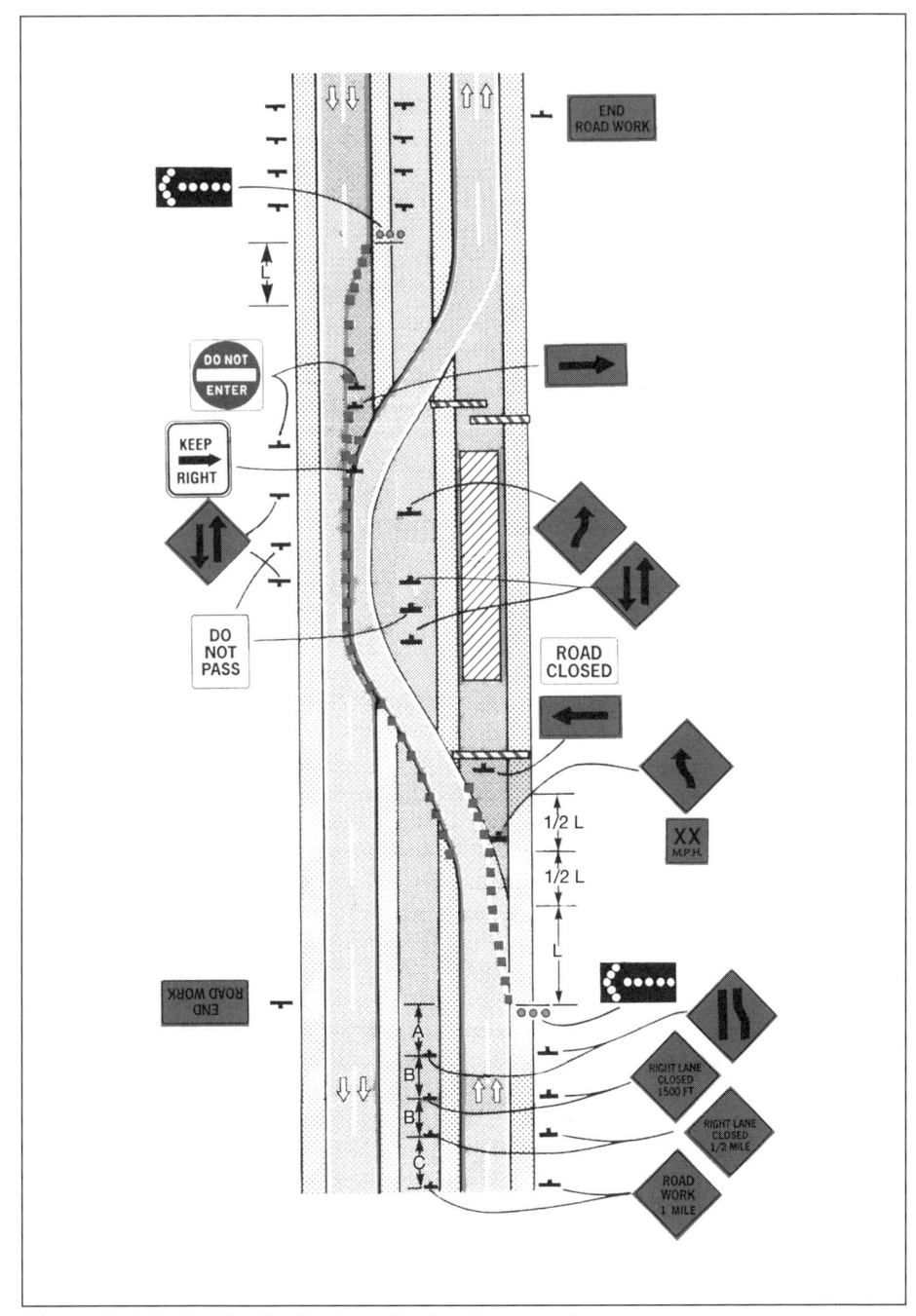

Figure 10-6 Typical Application for a Median Crossover.[2]

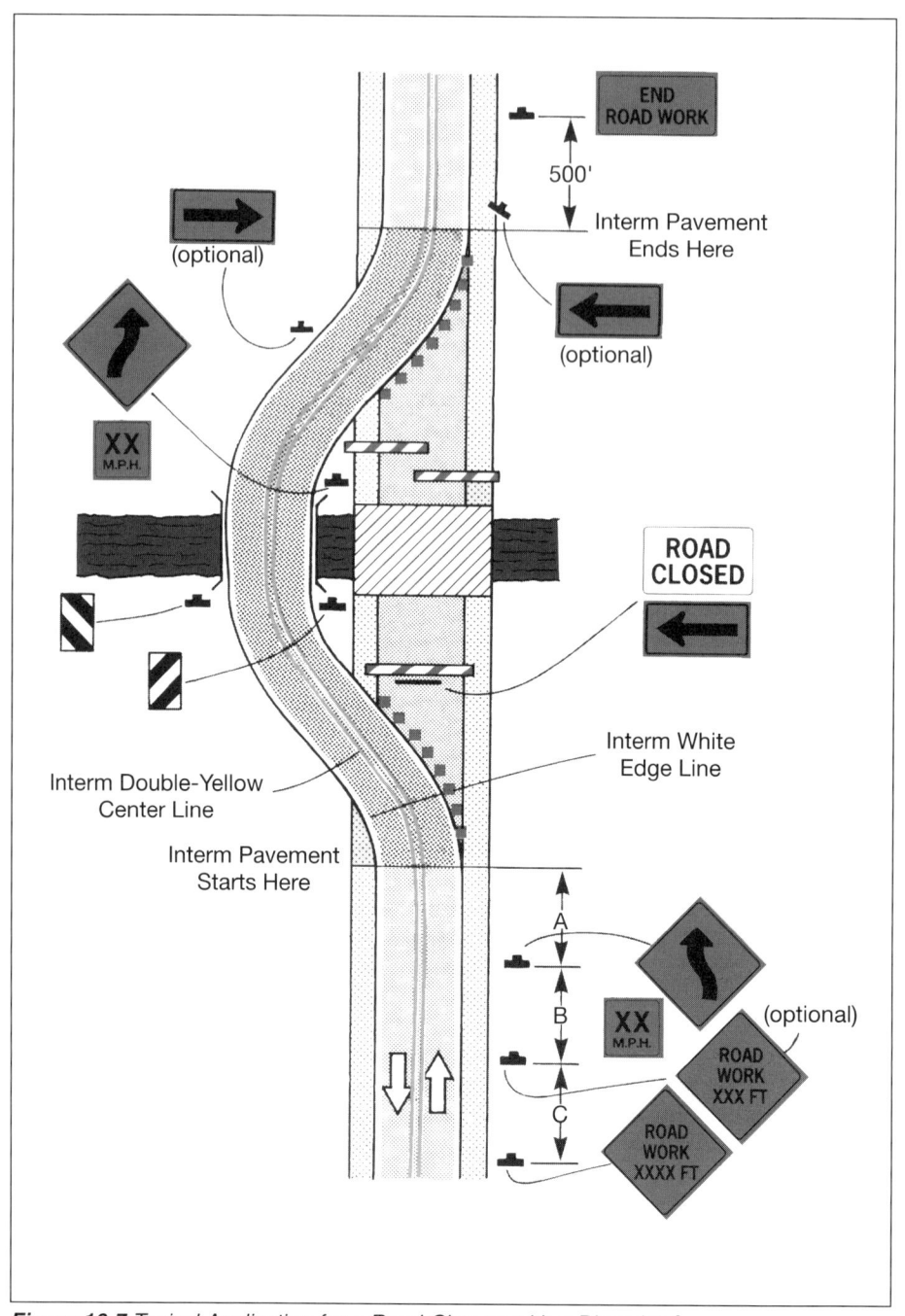

Figure 10-7 Typical Application for a Road Closure with a Diversion.[2]

Construction and Maintenance Zones 311

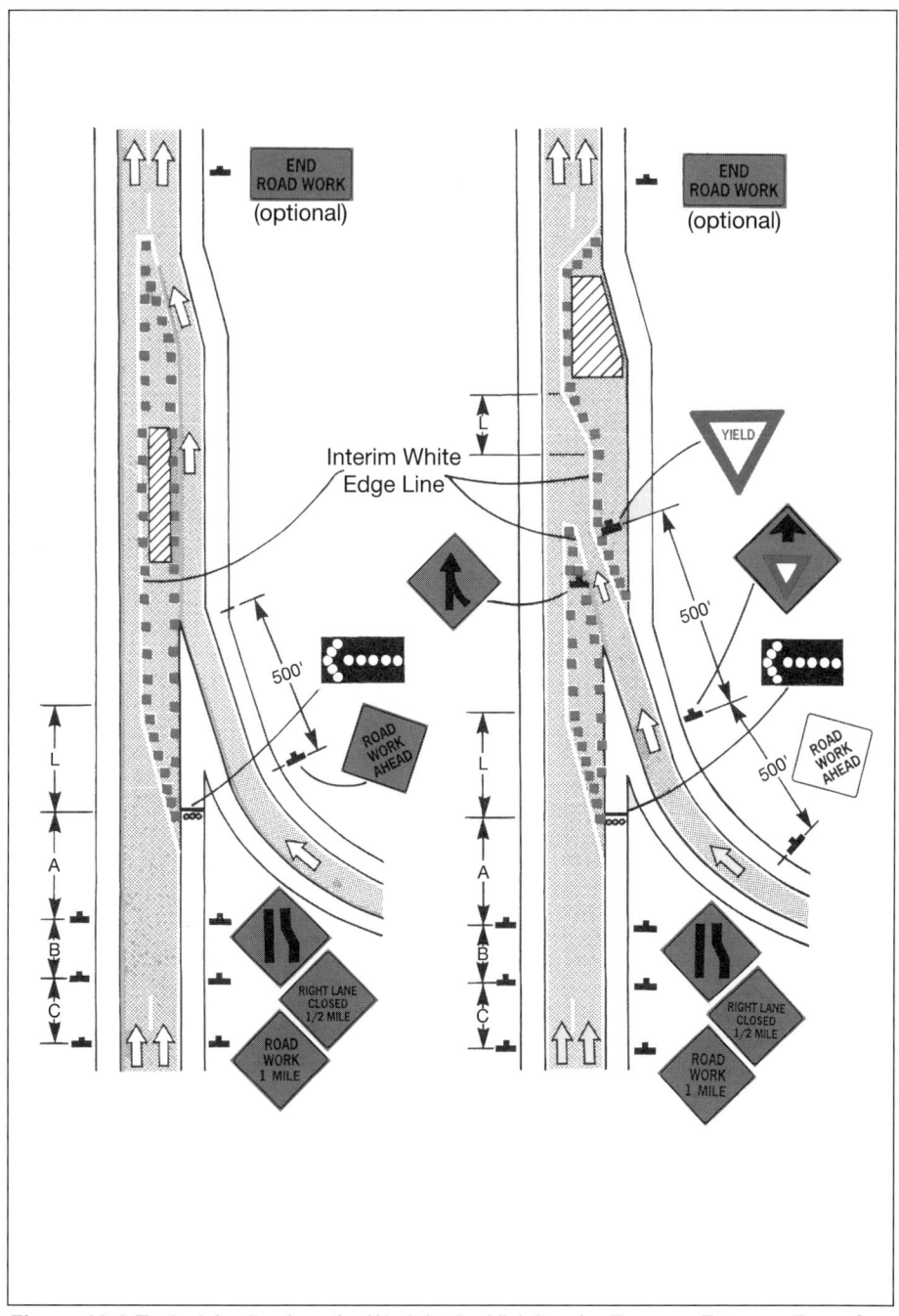

Figure 10-8 Typical Applications for Work in the Vicinity of a Freeway Entrance Ramp.[2]

Regulatory Signs

Figure 10-9 shows the commonly used regulatory signs. These signs impose legal restrictions and may not be used without permission from the authority having jurisdiction over the roadway. When they are used in construction or maintenance zones that requires regulatory measures different from those normally in effect, the existing permanent regulatory devices shall be temporarily removed or covered.

Road Closure Signs

The ROAD (STREET) CLOSED sign is usually used where the roadway is closed to all traffic except contractor's or other authorized vehicles. This sign is usually accompanied by appropriate detour signing. The ROAD CLOSED sign should be erected near the center of the roadway on a Type III barricade that closes the road. The words BRIDGE OUT or BRIDGE CLOSED may be substituted as appropriate. The ROAD CLOSED sign shall not be used where

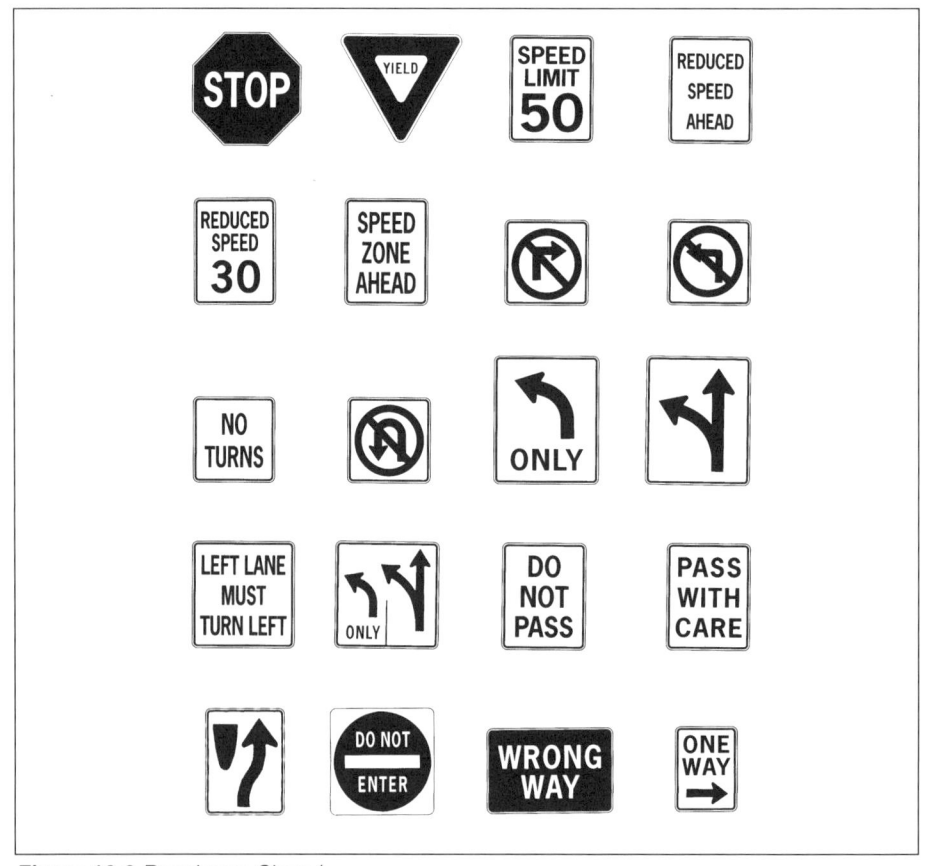

Figure 10-9 Regulatory Signs.[1]

Construction and Maintenance Zones 313

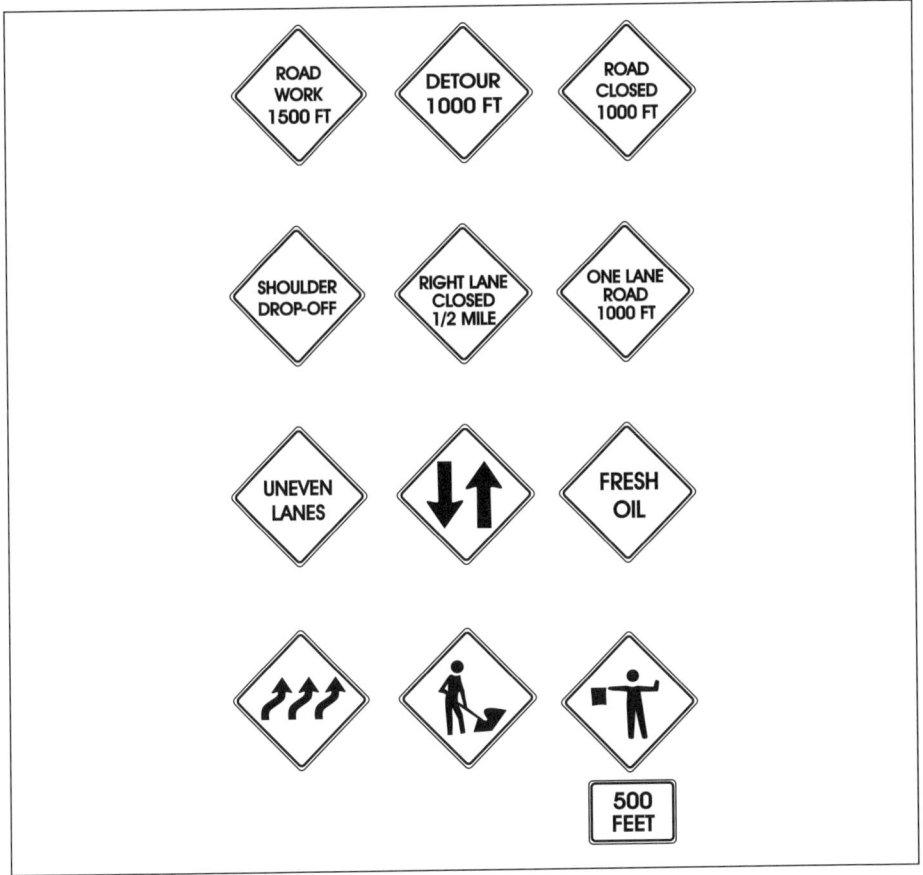

Figure 10-10 Warning Signs for Work Zones.[2]

traffic is maintained or where the physical closing is some distance beyond the sign.

The LOCAL TRAFFIC ONLY sign should be used where through traffic must detour to avoid a closing some distance beyond the sign, but where local traffic can move up to the point of closure.

Warning Signs

Warning signs give notice of conditions that are potentially hazardous. They should be used particularly where the danger is not obvious to or cannot be seen by the driver. They should not be overly restrictive or they will lose their credibility with drivers.

Where any part of the roadway is obstructed or closed, advance warning is required to alert drivers well in advance of these obstructions or restrictions. These signs may be used singly or in combination. Figure 10-10 is a sample of warning signs used in construction and maintenance zones.

Warning signs include the common approach signs such as; ROAD WORK 1500 ft; DETOUR 1000 ft; ROAD CLOSED 1000 ft; ONE LANE ROAD 1000 ft; and RIGHT LANE CLOSED 1/2 Mile. Other important and commonly used signs are the diagram signs for two-way traffic, flagger ahead, workers ahead, pavement edge drop, uneven lanes, and detours. In addition to these signs, of course, are all of the common warning signs for curves, stop ahead, truck crossing, narrow bridge, loose gravel, etc.

Detour Signs

Figure 10-11 shows the detour signing series. The Detour Arrow sign should be used where a detour route has been established because a road has been closed to through traffic. It should normally be mounted just below the ROAD CLOSED sign.

Each detour should be adequately marked with standard temporary route markers and destination signs. The DETOUR marker sign is mounted above the route marker assembly. The DETOUR sign is for unnumbered roadways. The END DETOUR sign is used to show the detour has ended.

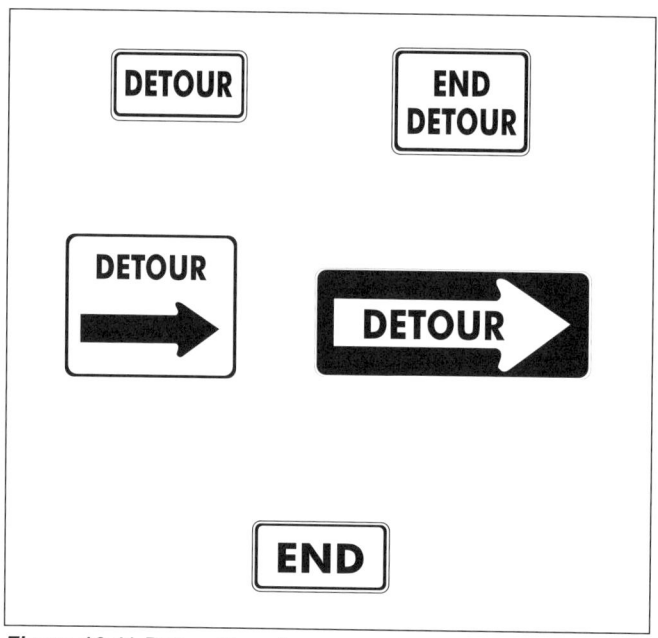

Figure 10-11 Detour Signs.[2]

Construction and Maintenance Zones

Channelizing Devices

These devices (shown in Figure 10-12) include but are not limited to cones, tubular markers, vertical panels, drums, barricades, and barriers (not shown). The major function of channelizing devices is to provide a clearly visible path to guide drivers safely past the hazards of the construction zone. They also serve the functions of guiding pedestrians and protecting workers.

Channelizing devices should provide for smooth and gradual traffic movement either from one lane to another, onto a detour or diversion, to laterally shift the travel lanes, or to reduce the width of the traveled way. They are also used to separate traffic from either the work space, pavement edge drops, pedestrian paths, or opposing lanes of traffic. Their spacing should not exceed a distance, in feet, equal to the speed limit, in mph, when used for a taper, and twice that distance when used along a tangent to separate traffic from the work space.

Channelizing devices are integral elements of the total system of traffic control devices used in construction and maintenance zones. These elements shall be preceded by a subsystem of warning devices that are adequate in size, number, and placement for the type of roadway. The following is a brief description of various channelizing devices:

Traffic cones are usually used for channelizing, delineation, and traffic separation for short-duration construction, maintenance, or utility activities. One of various recommended weighting methods should be used to ensure they will not be blown over.

Tubular Markers or Vertical Panels are used to divide opposing lanes, divide two lanes in the same direction, or delineate pavement edge drops where space restrictions do not allow for devices with more visible target area.

Drums, because of their size and high target value, command the respect of drivers. They are most commonly used to channelize or delineate traffic flow, particularly in construction zones where they will remain in place for a prolonged period. Drums should not be weighted with sand, water, or other material to the extent that they will be hazardous when impacted by a vehicle.

A *barricade* is a portable or fixed device having from one to three rails with appropriate markings. Type I or Type II barricades are used singly or in groups to mark a specific condition, or they are used in a series for channelizing traffic. Type I barricades are normally used on urban streets. Type II barricades have more target area and, therefore, are used more on high-speed roadways. Type III barricades are reserved for road closures.

The need for *portable barriers* should be determined by an engineering analysis of the protective needs of the location and should be designed accordingly. When serving the additional function of channelizing traffic, the

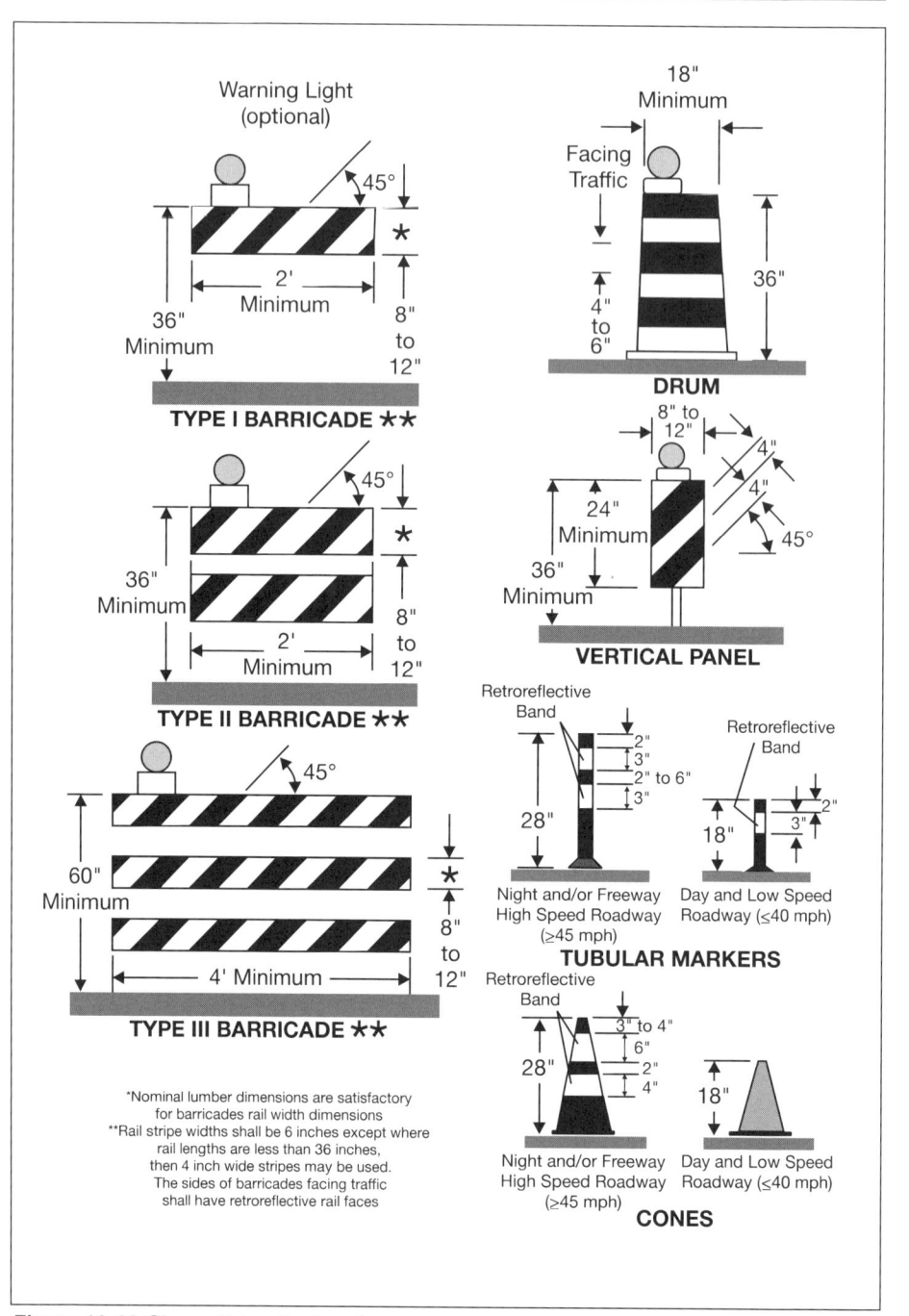

Figure 10-12 Channelizing Devices.[2]

channelizing taper shall be of standard length. The channelizing barrier shall be supplemented by standard delineators, channelizing devices, or pavement markings.

Warning Lights

These devices are yellow, lens-directed, enclosed, portable electric lights that can be used in a steady-burn or flashing mode. Flashing warning lights are most commonly mounted on barricades, drums, vertical panels, or advance warning signs to both periodically warn drivers of hazardous obstructions and to draw attention to warning messages. Flashers shall not be used for delineation, because a series of flashers tends to obscure the intended path. Where lights are needed to delineate the traveled way, the steady-burning mode shall be used.

Advance Warning Arrow Displays

These devices are large non-reflective black sign panels with a matrix of yellow lights capable of either flashing or sequential displays. They are not intended to solve difficult traffic problems by themselves, but they can be effective when used to reinforce signs and channelizing devices. If fact, the *MUTCD* recommends that necessary signs and channelizing devices be used along with Arrow Displays.

Arrow Displays are effective for encouraging drivers to leave a closed lane sooner. They provide additional advance warning and directional information where traffic must be shifted laterally along the roadway. Arrow Displays also assist in diverting and controlling traffic around construction or maintenance activities and give drivers positive guidance about a roadway path diversion that they might not otherwise expect.

Arrow Displays are used for both day and night lane closures, roadway diversions, and slow-moving maintenance activities on the traveled way. They are particularly effective for high-speed and high-density traffic conditions. At night, they are effective where other traffic control devices cannot provide adequate advance warning of a roadway path diversion. During daylight, Arrow Displays are effective under high-density traffic conditions that might otherwise block the driver's advanced view of construction or maintenance activities ahead.

Figure 10-13 shows the sizes, operating modes, and displays used for Arrow Displays. The panel shall be mounted on a vehicle, a trailer, or other suitable support. Ground-mounted displays should have a minimum height of 7 feet measured from the ground to the bottom of the panel. Vehicle-mounted displays should be as high as practical and have remote controls.

An Arrow Display set for the arrow mode or chevron mode may be used for stationary or moving lane closures. An Arrow Display in the caution mode shall

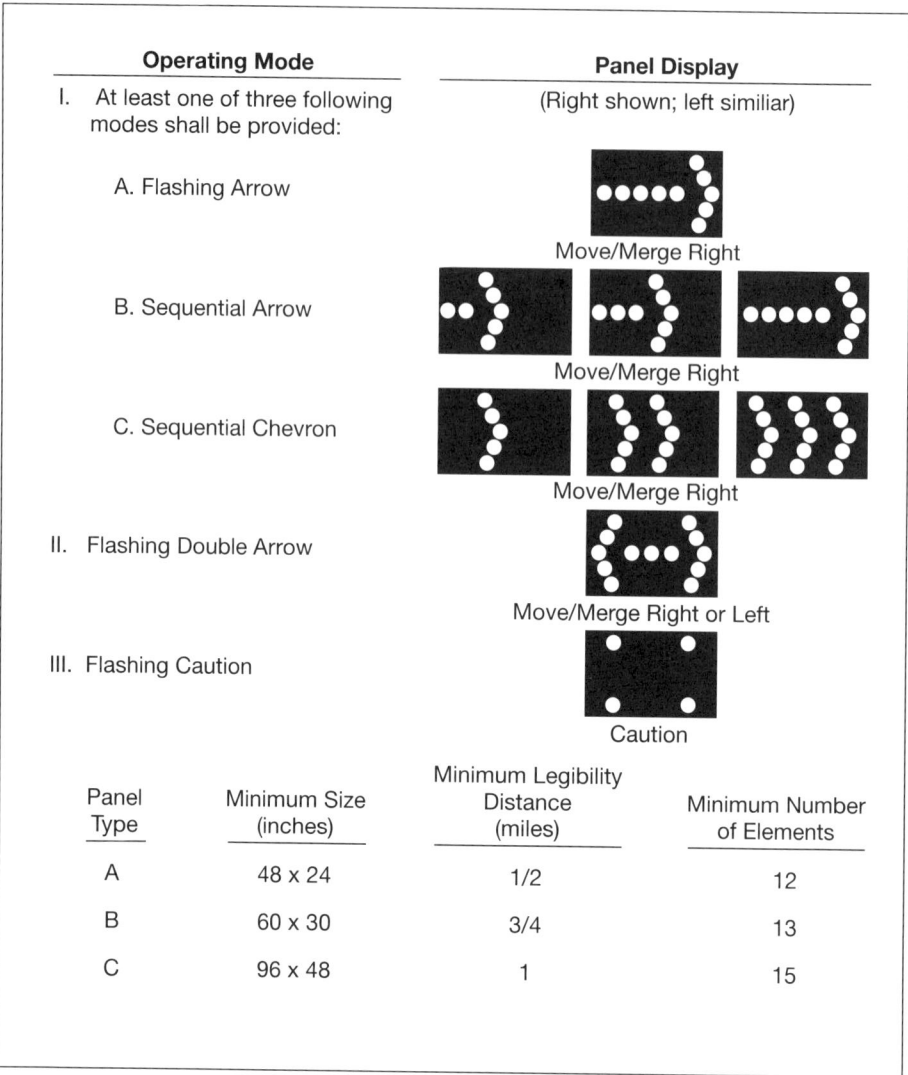

Operating Mode	Panel Display
I. At least one of three following modes shall be provided:	(Right shown; left similiar)
A. Flashing Arrow	Move/Merge Right
B. Sequential Arrow	Move/Merge Right
C. Sequential Chevron	Move/Merge Right
II. Flashing Double Arrow	Move/Merge Right or Left
III. Flashing Caution	Caution

Panel Type	Minimum Size (inches)	Minimum Legibility Distance (miles)	Minimum Number of Elements
A	48 x 24	1/2	12
B	60 x 30	3/4	13
C	96 x 48	1	15

Figure 10-13 Advance Warning Lighted Arrow Panels.[2]

be used only for shoulder work, blocking the shoulder, or roadside work near the shoulder.

For a stationary lane closure, the Arrow Display should be located on the shoulder at the beginning of the taper. Where the shoulder is narrow, the Arrow Display should be located in the closed lane. If Arrow Displays are used when multiple lanes are closed in tandem, the preferred position for additional Arrow

Construction and Maintenance Zones

Displays is in the closed lane at the start of the merge taper. Under various situations, such as for narrow shoulders, placement may be in the middle or at the end of the merge taper but always behind the channelizing devices. The display shall be located behind any channelizing devices used to transition traffic from a closed lane.

For moving maintenance operations where a lane is closed, the Arrow Display should be on a vehicle far enough behind the work operation to allow for appropriate reaction of approaching drivers. The vehicle with the Arrow Display shall be equipped with appropriate signing and/or lighting.

An Arrow Display shall not be used on a two-lane, two way roadway for temporary one-lane operation. Neither shall an Arrow Display be used on multi-lane roadways to laterally shift all lanes of traffic, because unnecessary lane changing may result.

Pavement Markings

The *MUTCD*[1] mandates that adequate pavement markings complying with *Part 3* be maintained in construction zones. It also requires that markings that are no longer applicable and might confuse drivers, be obliterated as soon as possible. Obliterated markings shall be unidentifiable as pavement markings under day or night, wet or dry conditions.

For short-term activities such as pavement resurfacing, temporary pavement markings can be substituted. These markings must comply with *Part 3* of the *MUTCD* with the following exceptions:

1. Broken-line temporary pavement markings (lane lines and centerlines) shall use the same cycle length as permanent markings and be at least four feet long.
2. For a duration of three days or less, signs may be substituted for no-passing stripes.

Temporary Traffic Signals

These devices are used in construction zones that have a temporary intersection with a haul road or equipment crossing and where alternate one-way traffic control is used, such as when reconstructing a bridge. All traffic signal and control equipment shall meet the standards of *Part 4* of the *MUTCD*. One-way traffic control requires an all-red interval of sufficient duration for traffic to clear the one-way area.

> **CONSTRUCTION ZONES**
> **Rule of Thumb**
> *If Something Seems Wrong With a Construction Zone, There Was Probably Some Violation of the MUTCD*

10.5 Technical Aspects of Construction or Maintenance Zone Defect Cases

Construction and maintenance zone cases have become more prevalent over the last few years because of increased Federal and State gasoline taxes, which have been used to accelerate both reconstruction and resurfacing of roadways. Although great effort has been devoted to training contractor and roadway agency employees about the fundamental principles of traffic safety, substandard, faulty, and confusing construction and maintenance zone layouts continue to be found.

A plaintiff, attempting to recover for injuries sustained from a motor-vehicle collision in an alleged hazardous construction or maintenance zone, needs to establish an evidence base to support that claim. Available sources of information about the zone layout and the traffic controls that were present at the time of collision may include: eye-witness testimony; photographs taken at the accident scene or a short time after; the police report; the traffic control plan (TCP); and construction diaries filled out by either the contractor or the roadway agency inspector or project engineer. Some of the more important bits of evidence might include the following:

a. Type, placement, number, size, and reflectivity of all traffic control devices.
b. Type, placement, number, size, and brightness of the warning devices on moving maintenance vehicles.
c. Length of the taper used to channel traffic.
d. Absence of mandated traffic control devices (e.g., temporary lane lines or centerlines).
e. Presence of non-standard traffic control devices.
f. Presence of wrong traffic control devices.
g. Absence of advance warning signs.

Construction and Maintenance Zones

h. Absence of warning signs for hazards that were not self-evident.
i. Improper spacing of channelizing devices.
j. Faulty or missing detour signs.
k. Presence of non-obliterated pavement markings that run contrary to the intended vehicle paths.
l. Improper placement of an Arrow Display.
m. Absence of a no-passing stripe (or sign).
n. Location, height, shape, and duration of a pavement edge drop.
o. Procedures used for flagging operations.
p. Improper use of positive barrier in place of a Type III barricade at a road closure.
q. Position of equipment, materials, and debris in a roadside clear zone.
r. Available sight distance preceding the point of impact.
s. Absence of warning signs on side roads or streets intersecting a construction or maintenance zone.
t. Absence of controlled paths for pedestrians and workers.
u. Procedures used for the monitoring of potential safety problems including regular drive-throughs and analysis of accident records.

The most common tort claim in construction and maintenance zones involves a *pavement edge drop*. Pavement edge drops occur on lane lines or at the edge of the travel way when resurfacing or excavating. Edge drops of 1.5 inches or greater within the travel way are clearly a hazard to avoid. Lower edge drop heights should be warned of with signs and eliminated as quickly as possible. Edge drops of 2 inches or more within or at the edge of the traveled way should be shielded with channelizing devices and preceded by advanced warning signs. Because edge drops of 6 inches or more are likely to cause vehicle undercarriage contact and subsequent vehicle rollover, positive barriers should be considered for these conditions. For a more complete discussion of pavement edge drops, please refer to Chapter 7 of this book.

Another common tort claim in construction and maintenance zones involves *faulty transition areas*. The specifics of these claims often include: (a) the taper was too short for a lane closure; (b) the median crossover had too low of a design speed; (c) the transition lacked adequate warning signs or channelizing devices; or (d) the lane closure was too close to a crossover.

Problems with traffic control devices at *road closures* are also fairly common to tort claims. Typically, advance signing is missing or ill-placed. Non-standard warning signs are sometimes used. Traffic barriers are often substituted for the recommended Type III barricade. Then too, the Type III barricade and/or ROAD CLOSED sign are sometimes missing at the point of closure.

Slow-moving maintenance operations are another major source of tort claims. Slow-moving maintenance vehicles present a special hazard to motorists on high-speed roadways because they violate driver expectancies, particularly when the sight distance or visibility is marginal or poor. Inadequately lighted snowplows that become hidden in a cloud of blowing snow create a rear-end hazard. A pavement sweeping vehicle operating in the left lane of a high-speed freeway without a trailing warning vehicle also is prone to being hit from the rear. Often these rear-end collisions happen with an automobile that is closely following a truck that sudden changes lanes to avoid the maintenance vehicle and leaves the automobile driver insufficient time to avoid collision.

10.6 Typical Defense Arguments

In defending a construction zone case, construction contractors usually argue that they only set up traffic control devices at the work site after direct instruction from the roadway agency people. On the other hand, the roadway agency will highlight that the responsibility for traffic control was vested in the contractor as spelled out in the contract, standard specifications, and the *MUTCD*. The definition will vary from State to State, but generally the courts hold that there is a shared responsibility where the contractor has the direct control and roadway agency must oversee the contractor.

Since most construction or maintenance zone cases allege a violation of the *MUTCD*, its language is often the major thrust of the defense arguments. With the exception of traffic control device colors, shapes, sizes, and reflectorization most provisions of *Part 6* of the *MUTCD* are either recommended (*shoulds*) or permissive (*mays*). For this reason, defendants will often argue that since no *MUTCD* mandates (*shalls*) were violated, that the traffic control applied in the subject construction or maintenance zone was discretionary and, therefore, not subject to claims of negligence. This argument may be particularly effective in those States that have tort claims acts that adhere to a strict interpretation of the *MUTCD*.

Other case-specific defense arguments allege either that the driver was the sole negligent party and/or the roadway condition had no contribution to the causation of the collision. Some of these arguments are:

1. Because the driver was within a construction or maintenance zone, the alleged hazards should have been self-evident.
2. The driver was negligent for exceeding the reduced speed limit imposed in the temporary traffic control zone.
3. The driver was negligent for driving while intoxicated.
4. The driver was negligent for running off the road.

Construction and Maintenance Zones

5. Direct evidence is lacking to indicate that the traffic control contribute to the subject collision.
6. The contractor or the roadway agency had no notice of the hazardous condition.

Endnotes

1. Federal Highway Administration, *Manual on Uniform Traffic Control Devices*, 1988. (available through Lawyers and Judges Publishing Co.)
2. Federal Highway Administration, *Part 6, Standards and Guides for Traffic Controls for Street and Highway Construction, Maintenance, Utility, and Incident Management Operations*, Revision 3 of the 1988 Edition of the Manual on Uniform Traffic Control Devices, September 3, 1993.
3. Federal Highway Administration, *Federal Aid Highway Program Manual*, Volume 6, Chapter 4, Section 2, Subsection 12, October 13, 1978, amended July 1, 1982.

References

Graham, Jerry I., Paulsen, Robert J., and Glennon, John C., *Accident Analysis of Highway Construction Zones*, Transportation Research Record 693, 1978.

Paulsen, Robert J., Harwood, Douglas W., Graham, Jerry L., and Glennon, John C., *State of Traffic Safety in Highway Construction Zones*, Transportation Research Record 693, 1978.

Graham, Jerry L., Migletz, James D., and Glennon, John C., *Guidelines for the Application of Arrowboards in Construction Zones*, Federal Highway Administration, 1979.

Graham, Jerry L., Paulsen, Robert J., and Glennon, John C., *Accident and Speed Studies in Construction Zones*, Federal Highway Administration, 1977.

Noel, E.C., et. al., *Work Zone Traffic Management Synthesis*, 5 volumes, Federal Highway Administration, 1989.

Federal Highway Administration, *Synthesis of Highway Safety Research Related to Traffic Control and Roadway Elements*, Volume 2, 1982.

Transportation Research Board, *Traffic Control for Freeway Maintenance*, NCHRP Synthesis Report 1, 1969.

Chapter 11

Traffic Control Devices

Traffic control devices include the signs, signals, and markings used on public roadways. They promote traffic safety by helping drivers with their guidance and navigation tasks (see Chapter 1). This purpose is achieved through the application of uniform standards for the design, placement, operation, and maintenance of traffic control devices. Uniforms standards for public roadways are presented in the *Manual on Uniform Traffic Control Devices*[1] *(MUTCD)*, which is part of Title 23 of the U.S. Code. In each State, the responsibility for the design, placement, operation, and maintenance of traffic control devices rests with various governmental bodies, who's action shall be in substantial conformance with the *MUTCD*.

In the *MUTCD*, the words *shall, should* and *may* are used to describe specific actions that are mandatory, advisory, or permissive. Figure 11-1 discusses the *definition* of these words. *Shall* is used about 900 times, *should* is used about 800 times, and *may* is used about 700 times. Anytime these three words are used in the text of this Chapter, the reader can interpret the text to be consistent with wording in the *MUTCD*.

SHALL	A *mandatory* condition. This word *requires* a specific design or placement of a sign, signal or marking.
SHOULD	An *advisory* condition. This word *recommends* a specific design or placement of a sign, signal, or marking.
MAY	A *permissive* condition. This word only *suggests* a possible design of placement of a sign, signal, or marking.

Figure 11-1 MUTCD Definitions.[1]

The *MUTCD*, in Part 1A-4, requires that an engineering study be done before installing a traffic control device on a public roadway. It states that qualified engineers are needed to exercise the engineering judgment to properly select and locate these devices. To help define what engineering studies are, the Federal Highway Administration has published the *Traffic Control Devices Handbook*[2] as an adjunct to the *MUTCD*. All regulatory devices also need to be backed by applicable laws, ordinances, or regulations. Effective traffic control also requires reasonable enforcement of regulations.

The uniform application of traffic control devices includes the selection of color codes to establish a general meaning for the devices. The *MUTCD* assigns the following meanings to colors:

YELLOW	- general warning
RED	- stop or prohibition
BLUE	- motorist service information
GREEN	- directional guidance
BROWN	- recreational/cultural guidance
ORANGE	- construction or maintenance warning
BLACK	- regulation
WHITE	- regulation

11.1 Traffic Control Device Requirements

The *MUTCD* describes basic principles for effective application of traffic control devices. These principles appear throughout the text in discussions of the various devices. To be effective, traffic control devices should meet five basic requirements:

1. Fulfill a need
2. Command the driver's attention
3. Convey a clear, simple message
4. Command the driver's respect
5. Give adequate time for proper response

To accomplish these basic requirements, five elements of application are important: design, placement, operation, maintenance, and uniformity.

Design

Traffic control devices use a combination of size, color, shape, and lighting or reflectorization to command the driver's attention. To convey a clear simple meaning, the features of shape, size, color, and simplicity of message are important. Legibility and size of lettering or symbols assist the driver in having adequate time for proper response.

Placement

Traffic control devices need to be placed within the cone of vision, so they will command the driver's attention. To convey a clear, simple meaning, they should be properly placed so they can be associate with the situation they are addressing. Devices also need to be placed so the driver has adequate time for proper response.

Operation or Application

Appropriate devices operated in a consistent manner are necessary to fulfill the needs of traffic at specific locations. Proper application not only elicits the proper response but also commands the respect of drivers, based on their previous exposure to similar traffic control situations.

Maintenance

Clear, legible, and properly-mounted traffic control devices that are in good working order command the respect of drivers. Also, devices need periodic adjustment or removal to account for changes in traffic.

Uniformity

In the *MUTCD*, uniformity means treating similar locations and traffic situations in the same way. Uniformity helps promote clear, simple messages because it aids drivers to recognize and understand roadway and traffic conditions. When standard devices are used incorrectly, they promote the disrespect of drivers.

11.2 Driver Information Needs

A driver needs information throughout the entire driving task. As described more fully in Chapter 1, these needs form a hierarchy of information handling complexity. At the *control level*, performance is relatively simple and is mostly subconscious for most drivers. At the *guidance* and *navigation* level, information handling becomes increasingly more complex, and drivers need more time to process decisions and respond safely to information.

Understanding this hierarchy is important in the placing traffic signs for drivers. For example, where drivers are busy with speed and path control, they should not be overloaded with directional signing. Likewise, placing too many signs in close succession can overload the driver's information-handling ability. The *Transportation and Traffic Engineering Handbook*[3] describes four basic qualities of traffic signs that aid drivers in information handling, as shown in Figure 11-2. These same principles apply to other traffic control devices as well.

TARGET VALUE	The quality that makes a sign stand out from its background (color, size, contrast, placement, lighting, and reflectance)
PRIORITY VALUE	The quality that makes it possible for the most important sign to be read first (size, contrast, and relative position)
LEGIBILITY	The quality of readability (character size and stroke width)
RECOGNITION	The quality of being easily recognized and understood (standard colors, shapes, legends, and symbols)

Figure 11-2 Qualities of Traffic Signs.[3]

11.3 Traffic Signs

The *MUTCD*, in *Part 2*, presents detailed standards for regulatory, warning, and guide signs. In doing so, the *MUTCD* states some basic principles as follows:

> Signs are essential where special regulations apply at specific places or specific time only, or where hazard are not self-evident...each standard sign shall be displayed only for the specific purpose prescribed for it in this Manual...Identical conditions should always be marked with the same type of sign...There is a need to establish an order of priority for sign installation. This is especially critical where space is limited for sign installation and there is a demand for several different types of signs. Overloading motorists with too much information can cause improper driving and impair safety...Signs, if used to excess, tend to lose their effectiveness.

The *MUTCD* sets forth several general specifications for the design and placement of signs. Some of the more important specifications are given in Figure 11-3. The rest of this section discusses special considerations for STOP signs, YIELD signs, warning signs, road closure signs, and wrong way traffic control.

STOP Signs

Because STOP signs cause inconvenience to motorists, they should only be used where warranted. The *MUTCD* lists four warrants as follows:

1. Intersections of less important roadways with a main roadway where application of the normal right-of-way rule is unduly hazardous.
2. Streets entering a through highway or street.
3. Unsignalized intersections in a signalized area.
4. Other intersections where a combination of high speed, restricted view, and serious accident record indicates a need for stop control.

Traffic Control Devices

Shapes
OCTAGON - Stop
INVERTED TRIANGLE - Yield
DIAMOND - Warning
RECTANGLE - Regulation
ROUND - Railroad Warning
X - Railroad crossbuck
PENNANT - No passing

Dimensions
The *MUTCD* prescribes standard dimensions. Larger signs are used on higher-speed roadways or where special emphasis is needed.

Symbols
The *MUTCD* prescribes many standard symbols. Symbols are preferred over words.

Word Messages
The *MUTCD* prescribes standard wordings.

Lettering
Sign lettering shall be upper case of a type approved by the Federal Highway Administration.

Illumination/Reflectorization
Regulatory and warning signs shall be illuminated or reflectorized.

Location
The General rule is to locate signs on the right side...Normally signs should be individually erected on separate posts...Signs should be placed so they are not hidden by other signs or objects.

Height
Rural roadside signs shall be mounted five feet above the near edge of the travel lanes.
Urban signs should have seven feet clearance under the sign.

Lateral Clearance
Signs should have the maximum lateral clearance possible. Clear-zone standards are applicable when signs are not breakaway.

Figure 11-3 Some MUTCD Sign Specifications.

Consideration should be given to using a YIELD sign where a full stop is not necessary all the time.

Although multi-way STOP signs are generally discouraged, they are useful as a safety measure under some circumstances where the traffic volumes on the intersecting roads are nearly the same. The *MUTCD* list the following warrants for multi-way STOP signs:

1. As an interim measure where traffic signals are urgently needed.
2. Where five or more reported accidents, susceptible to correction by multi-way STOP signs, occur in a 12-month period. These include right-angle and turning accidents.
3. Various traffic volume warrants list in Part 2B-6 of the *MUTCD*.

YIELD Signs

Vehicles controlled by a YIELD sign only need to stop when a conflicting vehicle, which has the right-of-way, is in close proximity to the intersection. The *MUTCD* list the following warrants for YIELD signs:

1. When the control of right-of-way is necessary where the approach speed exceeds 10 mph.
2. On the entrance ramp to an expressway where no acceleration lane is provided.
3. At intersections on divided roadways where the median width is greater than 30 feet.
4. At an intersection with a separate right-turn lane, without an adequate acceleration lane.
5. At any intersection where a special problem may be susceptible to correction by a YIELD sign.

Warning Signs

These yellow diamond-shaped signs are used whenever necessary to warn drivers of existing or potential hazards on or adjacent to the roadway. Warning signs put the driver on alert and may call for either a reduction in speed or a special maneuver to effect safety. To avoid breeding disrespect, warning signs should only be used where necessary.

Typical locations and hazards that may warrant the application of warning signs are:

1. Sharp changes in alignment
2. Intersections
3. Sight obstructed stop signs or signals
4. Converging traffic lanes
5. Narrow roadways
6. Changes in roadway cross section
7. Variable roadway surface conditions
8. Railroad crossings
9. Steep grades

Because warning signs are most beneficial to a driver who is either unacquainted or vaguely acquainted with the roadway, they should be carefully placed for maximum effectiveness. Warning signs should give adequate time for the driver to perceive, identify, decide, and perform any necessary speed or path control. Table 11-1 lists the *MUTCD* minimum sign placement distances for level roadways. These distances provide for a 3-second driver perception-reaction time, a 125-foot sign legibility distance, and a comfortable braking rate.

Traffic Control Devices

Table 11-1 A Guide for Advance Warning Sign Placement Distance[1]

Posted or 85 percentile speed (MPH)	Condition A High judgment needed	Condition B Stop Condition	\multicolumn{5}{c}{Condition C – Deceleration condition to listed advisory speed – MPH (or desired speed at condition)}
			0	10	20	30	40	50
20	175							
25	250							
30	325	100	150	100			
35	400	150	200	175			
40	475	225	275	250	175		
45	550	300	350	300	250		
50	625	375	425	400	325	225	
55	700	450	500	475	400	300	
60	775	550	575	550	500	400	300
65	850	650	650	625	575	500	375

Figure 11-4 shows a few of the *MUTCD* warning signs. Other warning signs have been discussed and illustrated in previous Chapters. Slippery Pavement warning signs are discussed in Chapter 6. Pavement edge drop warning signs are discussed in *Chapter 7*. Warning signs at rail-highway grade crossings are covered in *Chapter 8*. Roadway curve warning signs are discussed in *Chapter 9*. And, construction and maintenance zone warning signs are covered in *Chapter 10*. Some other warning signs (shown in Figure 11-4) used for particular purposes are described in the following paragraphs.

Intersection warning signs are used to give advance warning both of an intersection ahead and of its basic configuration. The *Cross Road* warning sign is used on a through roadway to warn of an obscure cross road intersection ahead. The *Side Road* warning sign, showing a side-road symbol either left or right of the through roadway and at an angle of either 90 or 45 degrees, is used under the same circumstances as the Cross Road sign. None of these signs are to be used on the same approach as a STOP sign.

The *T-Symbol or Y-Symbol* warning signs are used to warn drivers approaching a T or Y intersection from the roadway that forms the stem of the T or Y. It is not used where a STOP sign is on the same approach. A double-headed *Large Arrow* warning sign placed at the head of the T or fork of the Y, directly in line with approaching traffic, is a desirable supplement.

The STOP AHEAD or YIELD AHEAD warning signs are used on the approach to a STOP or YIELD sign that is not visible for a sufficient distance to allow the driver to bring his vehicle to a stop. A SIGNAL AHEAD sign is used on the approach to a signalized intersection where obstructions prevent a

Figure 11-4 Examples of Warning Signs.[1]

continuous view of at least two signal faces for the distance specified later in Figure 11-11. All of these signs may use either words or symbols.

The ROAD NARROWS warning sign is used on a two-lane roadway in advance of an abrupt reduction in the width such that two cars cannot pass safely without reducing speed. These locations can also be highlighted by delineators or object markers.

Traffic Control Devices

The NARROW BRIDGE warning sign is used in advance of a bridge or culvert either having a clear two-way width of 16 to 18 feet or having a clear two-way width less than the approach pavement. The ONE LANE BRIDGE warning sign is used in advance of a bridge or culvert either having a clear width less than 16 feet or having a clear width less than 18 feet for heavy commercial vehicle use and/or where approach alignment is poor.

The BUMP and DIP warning signs are used in advance of a sharp rise or depression in the roadway that is sufficient either to create a hazard, to cause considerable discomfort to motorists, to cause a shifting of cargo, or to deflect a vehicle from its true course at the normal driving speed for the roadway. Where appropriate, these signs are supplemented with advisory speed plates. The DIP sign shall not be used to indicate a section of depressed alignment.

Advance Crossing warning signs are used to alert drivers to unexpected entries into the roadway by pedestrians, trucks, bicyclists, animals, etc.. Where these crossings are confined to a single location, the Advanced Crossing sign can be supplemented with a distance plate. *Crossing* signs can be placed at a confined crossing to supplement the Advanced Crossing signs. These signs should only be placed adjacent to non-motorized crossings that are unusually hazardous or not readily apparent. These signs are the same as the Advance Crossing signs with the exception of added crossing lines.

Road Closure Signs

The ROAD CLOSED sign should be used on roadways that are closed to all traffic because of a temporary emergency. The words BRIDGE OUT may be substituted when appropriate. These signs should not be used where traffic is maintained or where traffic is detoured several miles in advance of the closure. Where the sign faces through traffic, it shall be preceded by an advance ROAD CLOSED AHEAD warning sign. The LOCAL TRAFFIC ONLY sign should be used where through traffic must detour for a temporary emergency, but where the roadway is open to traffic up to the point of closure.

Wrong Way Traffic Control

Wrong-way entries can occur because of driver confusion particularly at freeway exit ramps and at grade intersections between a crossroad and a divided roadway with a wide median. The signs that are used to dissuade wrong-way maneuvers are the DO NOT ENTER sign, the WRONG WAY sign, and the ONE WAY sign.

The DO NOT ENTER sign is placed in the most appropriate position at the end of an exit ramp or at the end of a one-way roadway. The sign is normally placed on the left side (right side for wrong-way drivers) of the one-way roadway, directly in the view of potential wrong-way driver.

The WRONG WAY sign can be used as a supplement to the DO NOT ENTER SIGN either where an exit ramp intersects a cross road or where a cross road intersects a divided roadway at grade. The WRONG WAY SIGN should be placed along the exit ramp or the divided roadway farther along the wrong-way direction than the DO NOT ENTER sign. Two particular locations where the application of wrong-way traffic controls are important are:

Grade Intersections on Divided Roadways - Wrong-way traffic control is used at these intersections where wrong-way use is experienced or where a wide median, a rural unlighted environment, or other factors indicate a likelihood of wrong-way use. Where the one-way roadways are separated by more than 30 feet, the intersections with the crossroad shall be signed as two separate intersections with ONE WAY signs visible on each crossroad approach. Turn prohibition, DO NOT ENTER, and WRONG WAY signs can be used to supplement the ONE WAY signs.

Freeway Exit Ramps - The following traffic control measures are to be implemented at exit ramps where wrong-way entries can be made:

1. ONE WAY signs shall be placed where the exit ramp intersects the crossroad.
2. DO NOT ENTER signs shall be placed near the end of the ramp in full view of a wrong-way driver.
3. At least one WRONG WAY sign shall be placed on the exit ramp.
4. On two-lane paved crossroads, double solid yellow centerlines should be painted for an adequate distance next to and both sides of an exit ramp.
5. Where needed for emphasis, a *Lane-Use Pavement Arrow* should be painted on the ramp near the crossroad.
6. FREEWAY ENTRANCE signs may be used at entrance ramps to reinforce proper maneuvers.

11.4 Traffic Markings

Traffic markings are a common component of the roadway system. Their primary purpose is to supply the visual information needed by the driver to steer his vehicle safely in a variety of situations.

As defined in the *MUTCD*, traffic markings are applied to guide and regulate the movement of traffic and to promote safety. Sometimes, they are used to

Traffic Control Devices

supplement other traffic control devices. In other places, they are the sole means of conveying regulations, warnings, and other information. Under these circumstances, they can provide clear, simple messages without diverting the driver's attention away from the roadway.

Pavement markings are classified as either longitudinal lines or transverse markings. They supply *positive* guidance by defining the limits of a driver's field of safe travel (lane lines, centerlines, edgelines, crosswalks, stop bars, etc.). They also convey *negative* guidance about where it is unsafe or not permissible to drive (no-passing zone markings, channelizing lines, edgelines, crosswalks, stop bars, etc.) Word or symbol markings on the pavement are another form of communication (directional arrows, STOP AHEAD, PED XING, RXR, etc.).

In addition to pavement markings, the *MUTCD* defines post-mounted delineators and object markers as roadway markings. Post-mounted delineators are used to outline the edge of the roadway, particularly on roadway curves (see Chapter 9). Object markers are placed on or next to obstructions within or near to the travel lanes. These obstructions include bridge piers and abutments, bridge rails, guardrails, culvert head walls, islands, etc.

Traffic markings must be easily recognized and understood. For this reason, the *MUTCD* promotes uniform standards for markings so they will convey exactly the same meaning whenever they are seen. The following paragraphs describe the uniform application of traffic markings.

Longitudinal Lines

In the U.S., the color of longitudinal markings shall be white, yellow or red. Meanings for these standard colors are established in the *MUTCD*, as shown in Figure 11-5. Longitudinal pavement markings include centerlines, no-passing

YELLOW LINES	Separate Opposing Traffic
	Mark the Left Edge on a One-Way Pavement
WHITE LINES	Separate Traffic Lanes in the Same Direction
	Mark the Right Edge of the Traveled Way
RED MARKINGS	Mark an Area of the Roadway that Shall Not be Entered
BROKEN LINES	Are Permissive
SOLID LINES	Are Restrictive
DOUBLE LINES	Indicates Maximum Restriction
LINE WIDTH	Indicates Degree of Emphasis
REFLECTORIZATION	Required by *MUTCD*

Figure 11-5 Uniform Meanings of Longitudinal Pavement Markings.[1]

zone markings, lane lines, edge lines, and channeling lines. The application of longitudinal pavement markings is described below.

Centerlines separate traffic traveling in opposite directions. They supply important guidance for drivers and should be used on most major paved roadways. On minor roadways where a continuous centerline is not used, short sections can be effective to help separate opposing traffic at roadway curves, hillcrests, intersection approaches, rail-highway crossings, and bridges.

Centerlines vary depending on the number of lanes and the availability of passing sight distance. A single broken-yellow centerline on two-lane, two-way roadways indicates that passing is permitted in either direction. A solid yellow line on either side of the broken yellow centerline indicates that passing is prohibited for the traffic adjacent to that solid line. Double solid yellow lines are used on two-way multilane roadways, on all roadways where the centerline is offset, and on two-lane two-way roadways where passing is prohibited in both directions.

No-Passing Zone Markings are warranted at two-lane roadways curves and hillcrests where the passing sight distance is less than the values shown in Table 11-2. Although the method used to establish these values in 1940[4] has been debated for years, recent studies by Glennon[5] and by Harwood and Glennon[6] have both corrected that method and at the same time validated the values. [Right values, accidentally derived with a wrong method]. The criteria for measuring these sight distances are a 3.5-foot eye height and a 3.5-foot object height measured along the centerline. The beginning of a no-passing zone is that point along the roadway where the passing sight distance first becomes less that the specified value. The end of the no-passing zone is that point where the passing sight distance is no longer less than the specified value.

Table 11.2 MUTCD Passing Sight Distances[1]

85th Percentile Speed (MPH)	Passing Sight Distance (Feet)
30	500
40	600
50	800
60	1000
70	1200

Lane lines are broken white lines that separate traffic traveling in the same direction. They shall be used on all Interstate highways, and should be used on all other multi-lane roadways. Solid lane lines can be used on approaches to intersections and in other areas where lane changes should be restricted. A dotted white line can be used to delineate the extension of a line not only through

Traffic Control Devices

intersections but also across deceleration lane openings where drivers may be otherwise confused.

Pavement Edge Lines (sometimes referred to as fog lines) supply guidance for the right edge of the traveled way. These solid white lines are particularly helpful at night and during adverse weather. They shall be painted or all Interstate highways and may be used on other roadways. Edge lines shall not be continued through intersections. For one-way roadways, the edge line on the left shall be yellow.

Channelizing Lines shall be wide or double solid white lines when they are used to form traffic islands where traffic in the same direction can travel on either side. Markings, such as cross hatching, inside these island areas shall be white. Two double solid yellow lines shall be used to form continuous median islands. Markings, such as cross hatching, inside these islands shall be yellow.

Approaches to Obstructions within the traveled way are a special cases where the *MUTCD* requires both channelizing lines and object markers (discussed later). Typical obstructions are bridge supports, bridge rails, and concrete gore abutments. For these conditions, the *MUTCD* requires that pavement markings shall be used to guide traffic around the obstruction. White channelizing lines shall be used when the obstruction is between lanes in the same direction. Yellow channelizing lines shall be used when the obstruction is between opposing lanes.

Obstruction approach markings shall consist of diagonal lines extending from the centerline or lane line to a point 1 to 2 feet on each sides of the approach end of the obstruction that is exposed to traffic. For roadways with a posted speed of 45 mph or greater, the taper length, L, in feet, of the marked area should be computed by multiplying the width of the offset, W, in feet, by the off-peak 85th-percentile speed, S, in mph. Where posted speeds are 40 mph or less, the formula $L = WS^2/60$ should be used. The minimum taper length shall be 100 feet in urban areas and 200 feet in rural areas.

Two-way Left-Turn Lanes are lanes reserved in the center of a roadway for left-turn vehicles from either direction. A two-way left-turn lane shall be marked on either side by a dashed yellow line inside and a solid yellow line outside to indicate the lane is reserved for left turns and cannot be used as a passing lane. Signs shall be used with the pavement markings.

Transverse Pavement Markings

These lines convey special messages to the driver, either alone or in combination with traffic signs. They include stop lines, pedestrian crosswalks, parking spaces, and pavement word and symbol markings as discussed below.

Stop Lines should be used at STOP signs and traffic signals to point out where the vehicle is required to stop. They are solid white lines, usually 12- to 24-inches wide, extending across all approach lanes. Where a marked crosswalk is in place, stop lines should be placed 4 feet in advance of the nearest crosswalk line. In the absence of a marked crosswalk, the stop line should be placed no more than 30 feet nor less than 4 feet from the intersecting roadway. For a STOP sign, the stop line should be placed in line with the sign.

Crosswalk Markings are used both to guide pedestrians in the proper path and to warn drivers of a pedestrian crossing. They should be marked at all intersections that have a substantial conflict between vehicle and pedestrian movements. They should also be marked at other places of pedestrian concentration, such as loading islands and mid-block locations. When necessary, they should be supplemented with advance warning signs and crosswalk signs.

Crosswalk lines shall be solid white lines, marking both edges of the crosswalk. They shall not be less than six inches wide nor spaced less than six feet apart.

Parking Space Markings encourage orderly use of parking spaces where parking turnover is substantial. They also discourage encroachment on fire hydrant zones, bus stops, approaches to corners, etc. Parking space markings shall be white.

Word and Symbol Markings include lane-use arrows, SCHOOL XING, PED XING, RXR, STOP, etc. Letters and numerals, 8 feet in height, should be placed so the message reads up (e.g., the first word of the message is nearest the approaching driver).

Object Markers

These devices are used to mark obstructions within or next to the roadway. They shall conform to one of three designs as shown in Figure 11-6. Type I object markers are either an 18-inch yellow reflectorized panel, or an 18-inch yellow or black panel with nine 3-inch yellow reflectors. Type II object markers are 6x12-inch yellow rectangular panels with either reflective material or three 3-inch yellow reflectors. Type III object markers are 12x36-inch rectangular reflective panels with yellow and black stripes angled down at 45° toward the side that traffic is to pass.

Objects in the Roadway shall be marked with either a Type I or a Type III object marker and with standard approach pavement markings (as discussed earlier). For a greater emphasis, large surfaces such as bridge piers may be painted with reflective diagonal stripes similar to the Type III design.

A structure with a low vertical clearance over the roadway is also considered an obstruction in the roadway. Where this vertical clearance is less than one foot

Traffic Control Devices

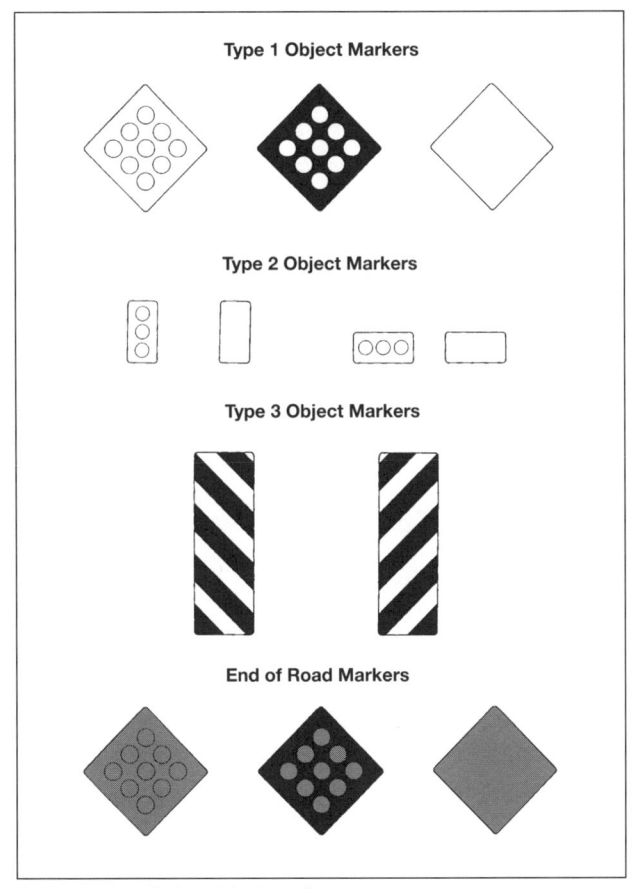

Figure 11-6 Object Markers.[1]

greater than the maximum legal height for trucks, the clearance height should be clearly marked, in feet and inches, on the overhead structure.

Objects Adjacent to the Roadway may be so close that they pose a hazard to traffic. Objects within a few feet of the travel lanes, such as bridge supports, bridge rails ends, and culvert headwalls should be marked with Type II or Type III object markers. The inside edge of the marker shall be in line with the inner edge of the object.

For *Ends of Roadways*, the MUTCD provides a special marker which is either a 18-inch red reflectorized diamond panel, or a 18-inch red or black diamond panel with nine 3-inch red reflectors (see Figure 11-6). One or more panels mounted at a 4-foot height may be used to mark the end where no alternative path is available.

Roadway Delineators

These are retro-reflective devices mounted in series at the side of the roadway to show the roadway alignment at night. They are considered as guidance devices rather than warning devices. The advantage of delineators over pavement markings is that they remain visible when the roadway is wet or snow covered.

Delineator reflector units shall reflect the light of the high beam on a standard automobile from 1000 feet. Like pavement edge lines, they shall be white except on the left side of a one-way roadway, where they shall be yellow.

Single delineators shall be installed along the right side of expressways and along at least one side of interchange ramps. These delineators are optional where continuous lighting is provided. Double or vertically elongated delineators should be provided at 100-foot intervals along acceleration and deceleration lanes. On one-way roadways, red delineators may be used on the back side of any regular delineator to help emphasize the incorrect path to wrong-way drivers.

Delineators of the appropriate color may be used to indicate a narrowing of the pavement. They are placed adjacent to the lane affected, along the full length of convergence.

Delineators shall be mounted so the reflector is about 4 feet above the near roadway edge. They shall be placed not less than 2 nor more than 6 feet outside the shoulder. Normally, delineators should be spaced from 200 to 528 feet along straight sections of roadway. Spacing should be adjusted on the approach and throughout a roadway curve so that several delineators are always visible to the driver. Table 11-3 shows the *MUTCD* suggested maximum spacing for delineators on roadway curves.

Table 11.3 Suggested Maximum Spacing for Delineators on Roadway Curves[1]

Radius of Curve (feet)	Spacing on Curve (feet)
150	30
250	40
400	55
600	70
800	80
1000	90
2000	130
3000	160

11.5 Traffic Signals

The *MUTCD* defines a traffic control signal as a type of signal by which traffic is alternately directed to stop and then permitted to proceed. By assigning right-of-way alternately to various traffic movements, traffic control signals exert a profound influence on traffic flow.

Traffic control signals, properly located and operated, usually have one or more of the following advantages:

1. Effect the orderly movement of traffic.
2. Increase the traffic-handling capacity of an intersection.
3. Reduce the rate of right-angle accidents.
4. Coordinate continuous traffic movement.
5. Interrupt heavy traffic at intervals to permit vehicular or pedestrian traffic to cross.

Traffic control signals, even though warranted by traffic and roadway conditions, can be ill-designed, poorly placed, poorly operated, or poorly maintained. The following disadvantages can result from improper or unwarranted traffic control signals:

1. Excessive delay.
2. Disobedience of the signal indications is encouraged.
3. Inadvertent diversion to less adequate routes.
4. Increase in the rate of rear-end accidents.

Other signal applications include: hazard identification beacons; intersection control beacons; lane-use control signals; freeway ramp metering signals; movable bridge signals; and railroad crossing signals. This discussion will be limited to traffic control signals, hazard beacons, and intersection control beacons. Railroad crossing signals are discussed in Chapter 8. The following paragraphs describe the uniform design, application, and operation of traffic signals.

Traffic Signal Warrants

Eleven warrants for traffic signals are presented in the *MUTCD* under the proviso that signals should not be installed unless at least one of these warrants is met. These warrants, which pertain to minimum traffic volumes, minimum pedestrian volumes, accident rates, signal spacing, and vehicular platooning, all require engineering studies to establish the justifications. If none of these warrants are met, a new traffic signal should not be put into operation nor should an existing signal be continued in operation.

Warrant 1, Minimum Vehicular Volume - This warrant applies where the volume of intersecting traffic establishes the need for a traffic signal. The warrant is met when for each of any 8 hours of a normal weekday, the traffic volumes shown in Table 11-4, exist on the major street and on the higher-volume minor-street approach to the intersection. These major-street and minor-street volumes are for the same 8 hours. During each hour, the higher approach volume on the minor street is used, regardless of direction.

When the 85-percentile speed of major-street traffic exceeds 40 mph, or when the intersection is in a built-up area of an isolated community with less than 10,000 people, Warrant 1 should be reduced to seventy percent of the Table 11-4 volumes.

Table 11-4 Minimum Vehicular Volumes for Signal Warrant 1[1]

Number of lanes for moving traffic on each approach		Vehicles per hour on major street (total of both approaches)	Vehicles per hour on higher-volume minor-street approach (one direction only)
Major Street	Minor Street		
1	1	500	150
2 or more	1	600	150
2 or more	2 or more	600	200
1	2 or more	500	200

Warrant 2, Interruption of Continuous Flow - This warrant applies where the traffic volume on the major street is so heavy that traffic on the minor street suffers excessive delay or hazard in entering or crossing the major street. The warrant is met when, for each of any 8 hours of a normal weekday, the traffic volumes shown in Table 11.5 exist on the major street and on the higher-volume minor-street approach to the intersection, and when the signal installation will not seriously disrupt progressive traffic flow. These major-street and minor-street volumes are for the same 8 hours. During each hour, the higher approach volume on the minor street is used, regardless of direction. As with Warrant 1, seventy percent of the Table 11.5 values can be used on higher-speed roadways or in smaller isolated communities.

Warrant 3, Minimum Pedestrian Volume - This warrant is met when, for a normal weekday, the pedestrian volume crossing the major street either at an intersection or at a mid-block point exceeds 100 for each of any 4 hours, or 190 during any one hour. These warranting volumes may be reduced as much as 50% when the predominant walking speed is less than 3.5 feet per second.

During the hours when the pedestrian volume warrant is met, traffic also shall exhibit less than 60 gaps per hour of adequate length for pedestrians to cross.

Traffic Control Devices

Table 11-5 Minimum Vehicular Volumes for Warrant 2[1]

Number of lanes for moving traffic on each approach		Vehicles per hour on major street (total of both approaches)	Vehicles per hour on higher-volume minor-street approach (one direction only)
Major Street	Minor Street		
1	1	750	75
2 or more	1	900	75
2 or more	2 or more	900	100
1	2 or more	750	100

When a divided street has a median where pedestrians are shielded, this gap requirement applies separately to each direction of traffic. Where coordinated traffic signals on each side of the location provide for platooned traffic that results in less than 60 gaps per hour of adequate length for pedestrians to cross, a traffic signal may not be warranted.

This warrant applies only where the nearest traffic signal on the major street is more the 300 feet away and where the new signal will not seriously disrupt progressive traffic flow.

Warrant 4, School Crossing - This warrant is met at an established school crossing when the number of adequate traffic gaps is less than the number of minutes that school children are using the crossing. When signals are installed entirely under this warrant, pedestrian signal indications should be added.

Warrant 5, Progressive Movement - Progressive movement control sometimes requires traffic signals at intersections where they would not otherwise be warranted, in order to maintain proper vehicle platooning and to control platoon speed. This warrant is met when:

1. On one-way streets or streets with predominant traffic in one direction, where adjacent signals are so far apart that they cannot effect the necessary degree of platooning and speed control.
2. On two-way streets, where adjacent signals do not provide the necessary degree of platooning and speed control, and where the proposed signal and adjacent signals could form a progressive signal system.
3. The new signal under this warrant will not be closer than 1000 feet to the nearest signal.

Warrant 6, Accident Experience - This warrant allows for the placement of a traffic signal based on accident experience at intersections that otherwise cannot quite satisfy other warrants. This warrant is met when:

1. Adequate trial of less restrictive remedies has failed to reduce the accident frequency;
2. Five or more reportable accidents, susceptible to correction by a traffic signal, have occurred within a 12-month period;
3. The vehicular and pedestrian volumes are not less than 80% of the requirements for Warrants 1, 2, or 3; and
4. The signal will not seriously disrupt progressive traffic flow.

Warrant 7, Systems Warrant - A signal may be warrant to encourage concentration and organization of traffic flow networks. This warrant is met when the intersection of two or more major routes has:

1. A total existing, or immediately projected, entering volume of 1000 vehicles during the peak hour of a normal weekday;
2. Five-year projected traffic volumes, which meet any of Warrants 1, 2, 8, 9, or 11; or
3. A total existing or immediately projected entering volume of 1000 vehicle for each of any five hours of a Saturday or Sunday.

Warrant 8, Combination of Warrants - In exceptional cases, signals may be justified where no single warrant is met but where traffic volumes exceed 80% of the Warrant 1 and 2 values. Adequate trial of other remedial measures should precede installation of signals under this warrant.

Warrant 9, Four Hour Volumes - This warrant is met when, for any four hours of a normal weekday, the hourly volumes plot above the curve in Figure 11-7(A) for the existing combination of approach lanes on the major and minor streets. When the 85th-percentile speed of major-street traffic exceeds 40 mph, or when the intersection is in a built-up area of an isolated community with less than 10,000 people, Figure 11-7(B) is used to meet Warrant 9.

Warrant 10, Peak Hour Delay - This warrant is met when the conditions given below exist for any four consecutive 15-minute periods during a normal weekday:

1. The total delay on one approach to a minor street STOP sign equals or exceeds four vehicle-hours.
2. The volume on the same approach equals or exceeds 100 vehicles per hour for one moving lane of traffic, or 150 vehicles per hour for two moving lanes; and
3. The total entering volume equals or exceeds 800 vehicles per hour for intersections with four or more approaches, or 650 vehicles per hour for intersections with three approaches.

Traffic Control Devices

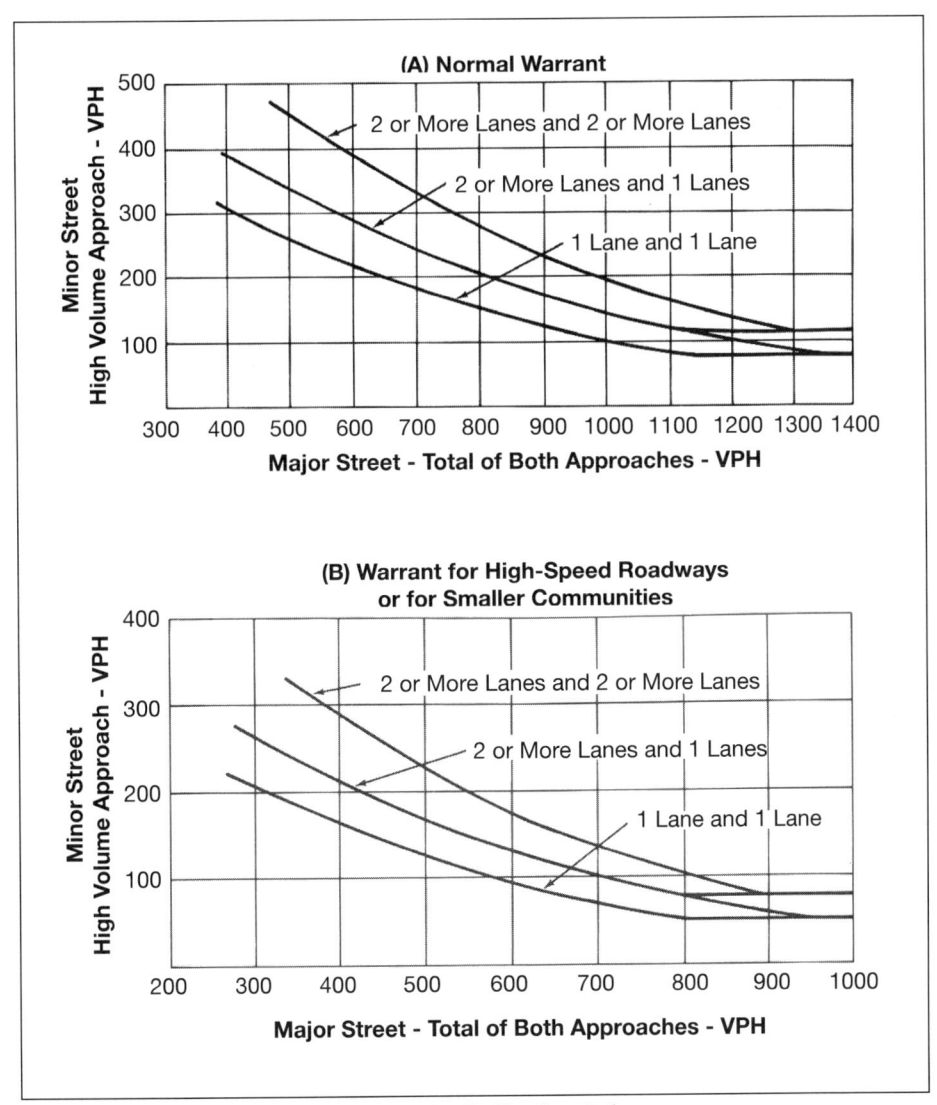

Figure 11-7 Four Hour Volume Warrant for Traffic Signals.[1]

Warrant 11, Peak Hour Volume - This warrant is met when, for a normal weekday, traffic volumes for any four consecutive 15-minute periods plot above the curve in Figure 11-8(A), for the existing combination of approach lanes on the major and minor streets. When the 85th-percentile speed of major-street traffic exceeds 40 mph, or when the intersection is in a built-up area of an isolated community with less than 10,000 people, Figure 11-8(B) is used to meet Warrant 11.

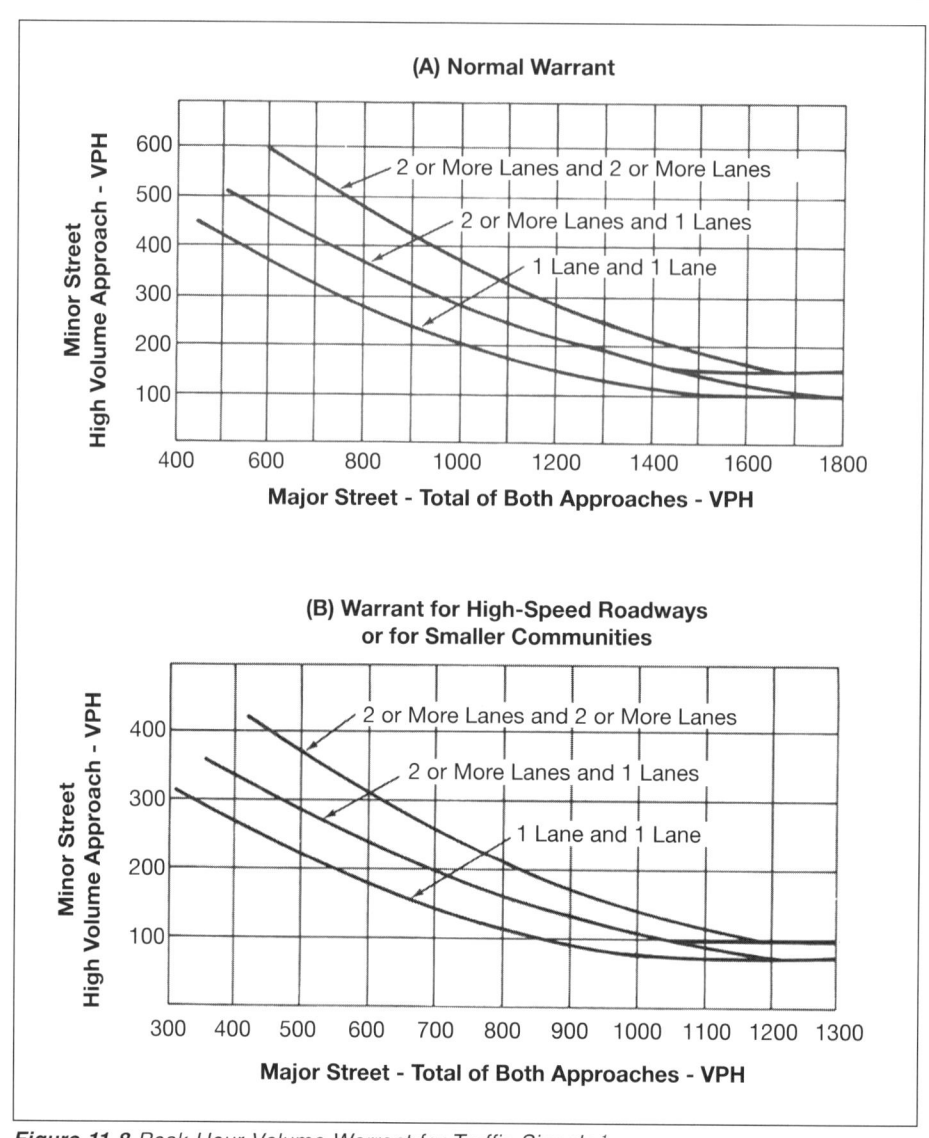

Figure 11-8 Peak Hour Volume Warrant for Traffic Signals.[1]

Number and Arrangement of Lenses per Signal Face

Traffic control signal lenses are Red (for stop), Yellow (for caution), Green (for go), and show either circular ball or arrow indications. With a few *MUTCD* exceptions, traffic signal faces shall have at least three lenses, but not more than five. The lenses shall be arranged vertically or horizontally as typified in Figure 11-9. The relative positions of lenses within the signal face shall be as shown in Table 11-6.

Traffic Control Devices 347

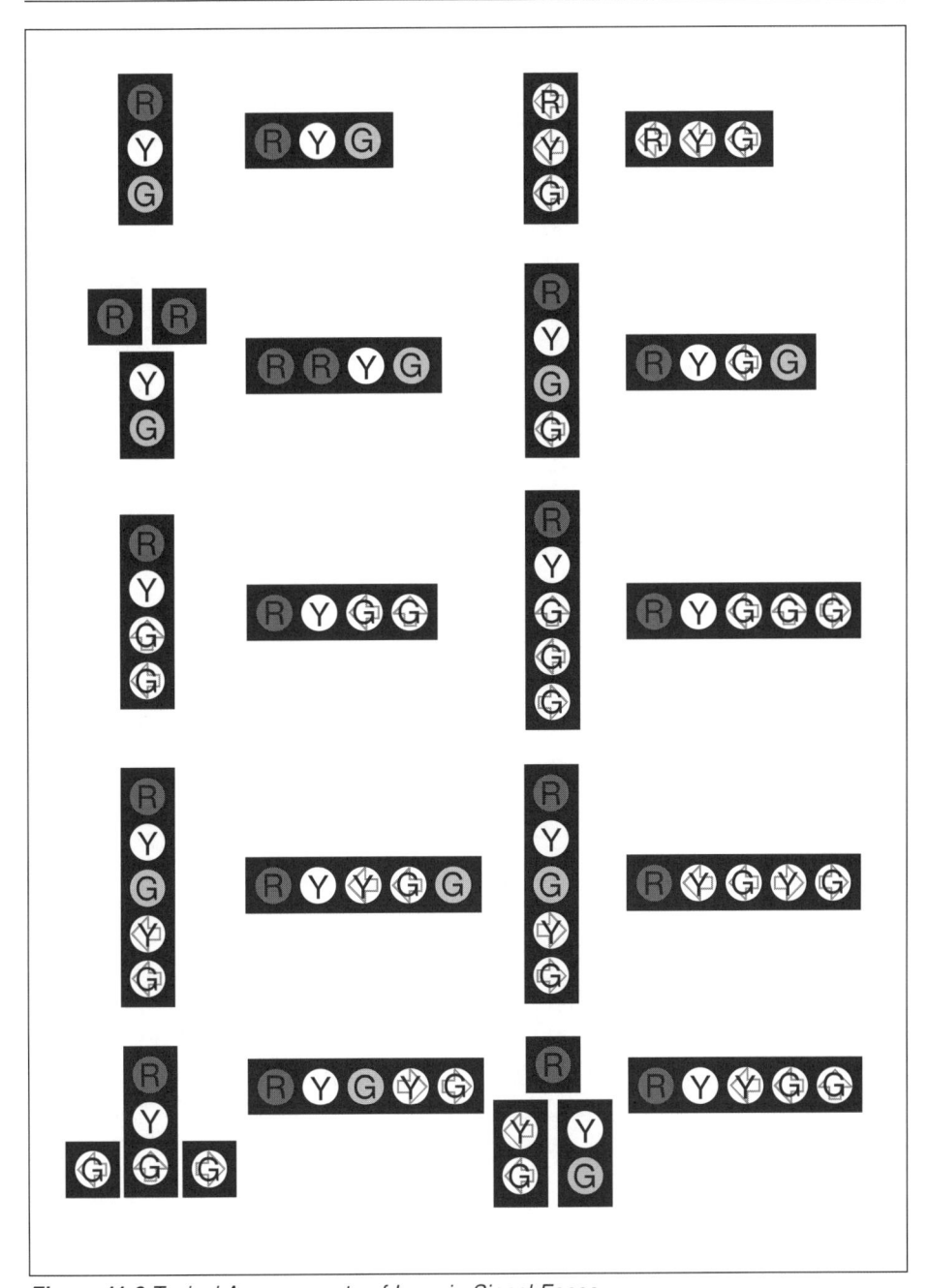

Figure 11-9 Typical Arrangements of Lens in Signal Faces.

Table 11.6 Allowable Signal Face Arrangements

Position of Lenses in a Vertical Signal Face (Top to Bottom)[1]

INDICATION	DIRECTION(S)
RED BALL	All
RED ARROW	Left Turn
RED ARROW	Right Turn
YELLOW BALL	All
GREEN BALL	All
GREEN ARROW	Straight Through
YELLOW ARROW	Left Turn
GREEN ARROW	Left Turn
YELLOW ARROW	Right Turn
GREEN ARROW	Right Turn

Position of Lenses in a Horizontal Signal Face (Left to Right)

INDICATION	DIRECTION(S)
RED BALL	All
RED ARROW	Left Turn
RED ARROW	Right Turn
YELLOW BALL	All
YELLOW ARROW	Left Turn
GREEN ARROW	Right Turn
GREEN BALL	All
GREEN ARROW	Straight Through
YELLOW ARROW	Right Turn
GREEN ARROW	Right Turn

The following combinations of signal indications shall not be displayed at the same time on any one signal face:
1. GREEN BALL with a YELLOW BALL
2. Straight-through GREEN ARROW with a RED BALL
3. RED BALL with a YELLOW BALL
4. GREEN BALL with a RED BALL

Straight-through RED ARROWS and YELLOW ARROWS shall not be allowed on any signal face.

Flashing Signals

When a red signal lens is flashing, drivers shall stop at the intersection. When a yellow signal lens is flashing, drivers may proceed through the intersection or past the signal only with caution.

Traffic Control Devices

When a traffic control signal is put on flashing operation, normally a yellow indication should be used for the major street and a red indication for other approaches. Yellow indications shall not be used for all approaches.

Size of Signal Lenses

All traffic control signal lenses shall be circular, with a diameter of either 8 inches or 12 inches. Different size lenses may be used in the same signal face except that no 8-inch RED indication shall be combined with a 12-inch GREEN BALL or a 12-inch YELLOW BALL. Figure 11-10 shows that the 12-inch lens has 2.25 times the target area of a 8-inch lens.

12-INCH LENS

12-Inch Lens has 2.25 times the area of 8-Inch Lens and is recommended where signal visability may be otherwise difficult.

8-INCH LENS

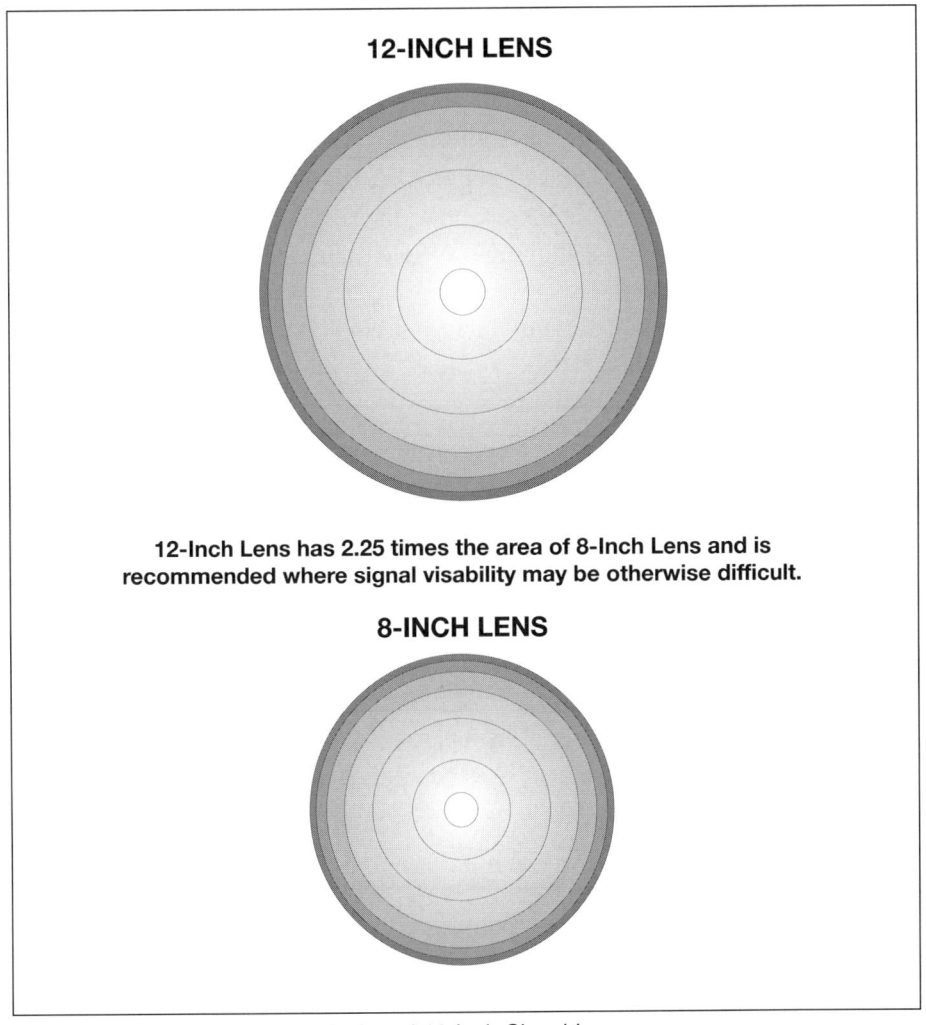

Figure 11-10 Comparison of 8-Inch and 12-Inch Signal Lens.

Twelve-inch lenses shall be used for the following conditions and applications:
1. For all arrow indications.
2. Where the nearest signal face is between 120 and 150 feet beyond the stop line, unless a supplemental near-side signal face is used.
3. For all signal faces beyond 150 feet from the stop line.
4. For all signal approaches where minimum visibility distance requirements cannot be met (see Figure 11-10).
5. For all signal approaches with 85th-percentile speeds above 40 mph.
6. For all unexpected signals.
7. Where drivers view both traffic control and lane-use control signals at the same time.
8. For all rural roadways where only post-mounted signals are used.

Locations and Visibility of Signal Faces

The primary factor in signal placement shall be visibility. Drivers shall be given a clear and unmistakable indication of their right-of-way assignment. For through traffic approaches, a minimum of two signal faces shall be presented. These signals shall be visible to approaching traffic not only from the minimum visibility distance shown in Figure 11-11 but also continuously up to the stop line. Where these visibility distances cannot be met, a SIGNAL AHEAD sign shall be erected to warn approaching traffic.

Except where a wide intersecting street precludes it, at least one and preferably both of the required signal faces shall be located not less than 40 feet nor more than 150 feet beyond the stop line (see Figure 11-11). Where both signal faces are post-mounted, they shall both be placed on the far side of the intersection, one on the right and one on the left. Where the nearest signal is 150 feet or more beyond the stop line, a supplemental near-side signal face shall be used.

Except where a wide intersecting street precludes it, at least one and preferably both signal faces shall be within the required cone of vision formed by 20° angles left and right of the center of the approach lanes at the stop line (as shown in Figure 11-11). This requirement is to be applied together with the maximum and minimum signal face distances described in the paragraph above and shown in Figure 11-11.

Required signal faces for through traffic on any one approach shall not be less than 8 feet apart, measured horizontally between the centers of the faces.

Height of Signal Faces

For post-mounted signal faces, the bottom of the housing shall be at least 8 feet but not more than 15 feet above the side walk or, if none, above the

Traffic Control Devices

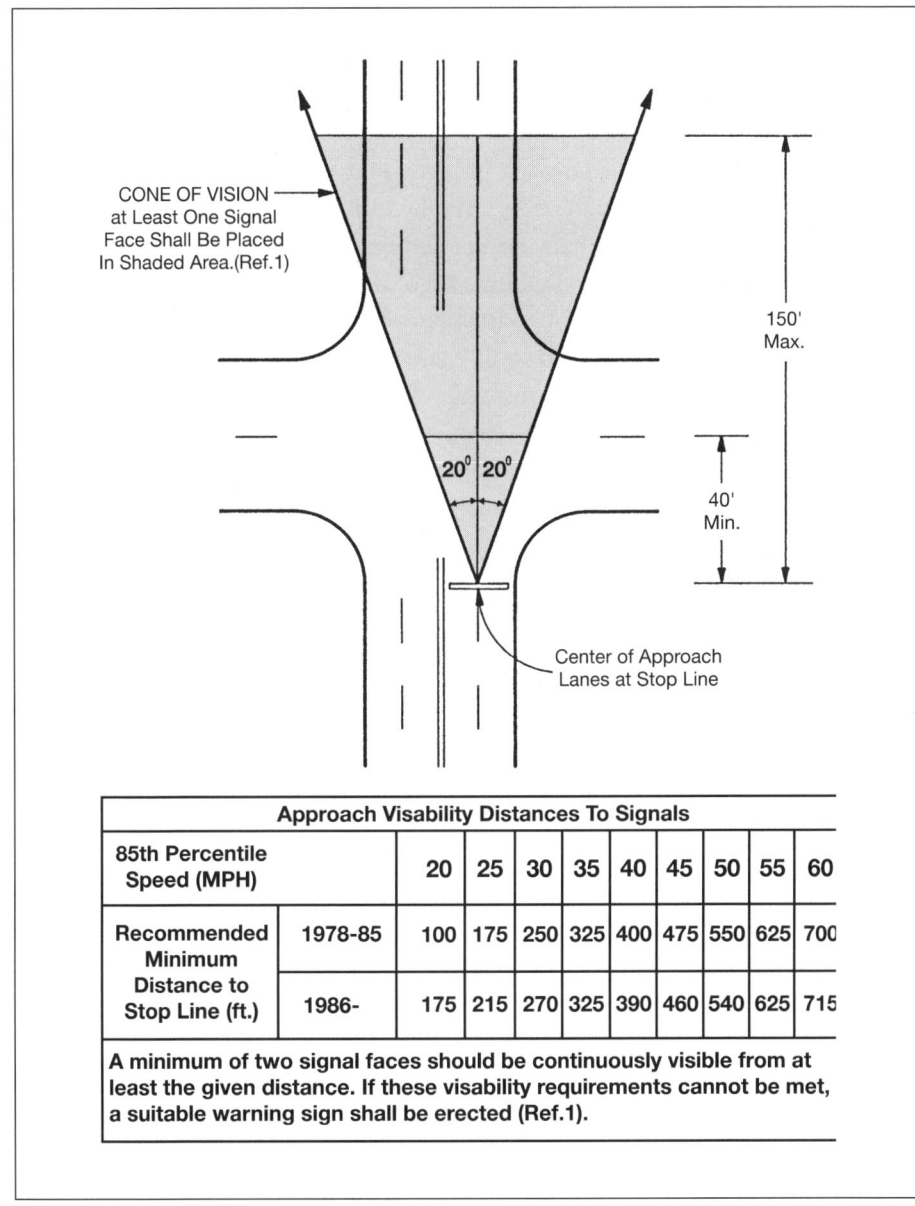

Figure 11-11 *Location and Visibility of Signal Faces.*[1]

Approach Visability Distances To Signals										
85th Percentile Speed (MPH)		20	25	30	35	40	45	50	55	60
Recommended Minimum Distance to Stop Line (ft.)	1978-85	100	175	250	325	400	475	550	625	700
	1986-	175	215	270	325	390	460	540	625	715

A minimum of two signal faces should be continuously visible from at least the given distance. If these visability requirements cannot be met, a suitable warning sign shall be erected (Ref.1).

pavement grade at the center of the roadway. One exception is for center median near-side signal faces, which may be mounted 4.5 feet above the median grade. For signal faces suspended over the travel lanes, the bottom of the housing shall be at least 15 feet but not more than 19 feet above the pavement grade at the center of the roadway. Within this required height range, optimum visibility distance and adequate overhead clearance should be the factors that set the signal height.

Shielding of Signal Faces

Each signal face shall be adjusted for maximum effectiveness for the intended approach traffic. Visors should be used in directing the signal indication toward approaching traffic and to reduce the *sun phantom* effect resulting from sunlight entering the lens. Back plates should be used to improve target value on signals that are viewed against bright or confusing backgrounds.

Signal faces should be aimed so that the optical axis of the signal lens intersects the roadway approach at the minimum visibility distance (see Figure 11-11) or greater. If the approach sight distance is restricted by roadway alignment, the signal faces shall be aimed where the signal first becomes visible.

Where irregular street design requires placing signal faces for approaches that intersect at small angles, each signal face shall, to the extent possible, be shielded so only the intended traffic can see it.

Signal Out of Operation

When a signal installation is not in operation, such as before placing it in service, during seasonal shutdowns, or when operating the signal is not desirable, the signal heads should be hooded, turned, or taken down to clearly indicate that the signal is not in operation.

Yellow Phase-Change Intervals

A yellow change interval shall be displayed following each GREEN BALL or GREEN ARROW indication. Its exclusive function shall be to warn drivers of the impending change in the right-of-way assignment.

Yellow change intervals should range from about 3 to 6 seconds depending on the approach speed. The yellow change interval may be followed by a all-red clearance interval (usually 1 second) that helps to clear the intersection before conflicting traffic movements are released. Table 11.7 shows suggested values[7] for both yellow and all-red clearance intervals.

Pedestrian's Signal Needs

Where pedestrian traffic regularly occurs at a signalized intersection, the design and operation of the signals should consider the needs of the pedestrian. Either the traffic control signal or a pedestrian signal should be properly visible to pedestrians on each approach. Where pedestrians experience excessive delay at traffic control signals, pedestrian actuation shall be installed.

Pedestrian signal indication are special traffic signals for the exclusive purpose of controlling pedestrian traffic. These indications are illuminated words or

Traffic Control Devices

Table 11.7 Traffic Signal Clearance Intervals[6]
(Courtesy of Criterion Press, Inc.)[7]

Approach Speed (mph)	Yellow Interval (seconds)	Minimum Yellow plus All-Red Interval Needed to Clear Intersection (seconds) For Cross-Street Widths (feet)			
		20	40	60	80
20	3.0	3.8	4.5	5.2	5.9
25	3.0	3.9	4.5	5.0	5.6
30	3.2	4.1	4.6	5.0	5.5
35	3.6	4.3	4.7	5.1	5.5
40	3.9	4.6	5.0	5.3	5.6
45	4.5	4.9	5.2	5.5	5.8
50	4.7	5.2	5.5	5.8	6.0
55	5.0	5.5	5.8	6.0	6.3

symbols showing WALK and DONT WALK. A steady DONT WALK means the pedestrian facing the signal shall not enter the roadway. A flashing DONT WALK facing pedestrians means both that they should not enter the roadway and that if they're already within the roadway they should proceed to a sidewalk or safety island. The steady WALK indication means that a pedestrian facing the signal may enter and proceed across the roadway.

Pedestrian signal indications shall be installed with a traffic control signal under any of the following conditions:

1. When a traffic signal is installed under either the Pedestrian Volume or School Crossing Warrant.
2. When an exclusive phase is provided for pedestrians with all vehicular traffic stopped.
3. Where vehicular indications are not visible or where their placement does not adequately serve pedestrians.
4. For school crossings at signalized intersections.

Hazard Identification Beacon

These beacons consists of one or more sections of a standard traffic signal head with a flashing YELLOW BALL indication in each section. They shall be used only to supplement an appropriate warning or regulatory sign or marker. Typical applications include:

1. Obstructions within or immediately adjacent to the travel lanes.
2. Supplementary to advance warning or regulatory signs (except STOP, YIELD, and DO NO ENTER)

3. At mid-block crosswalks.
4. At intersections where warning is required.

Intersection Control Beacon

These beacons consists of one or more sections of a standard traffic signal head, having a flashing YELLOW BALL or RED BALL indication for each face. They are used at intersections where traffic or physical conditions do not warrant conventional traffic control signals, but where high accident rates indicate a special hazard. Application of Intersection Control Beacons shall be limited to:

1. Yellow on the major road approaches and red for the remaining approaches; or
2. Red for all approaches.

A STOP sign should be used with a flashing red Intersection Control Beacon. Flashing yellow indications shall not face conflicting approaches.

Left-Turn Signal Displays

The three modes of left-turn signal control are: permitted; protected/prohibited; and protected/permitted. The simplest form of left-turn control is the *permitted* mode, where left-turning traffic moves on a GREEN BALL but must yield to opposing traffic.

Although a bit more complex, the *protected/prohibited* mode allows movement only with an exclusive phase using a left-turn arrow display, with or without a same direction GREEN BALL. In most cases, an exclusive left-turn lane is provided. A separate left-turn face must be used where the phasing does not provide simultaneous right-of-way to parallel through traffic. The clearance interval display may be either a left turn YELLOW ARROW or a YELLOW BALL. The red interval may use a RED ARROW only if a YELLOW ARROW is used; otherwise, a RED BALL is used. If a separate signal face is used, it should be placed in line with the left-turn movement. A LEFT TURN SIGNAL sign is required unless the signal face consists of arrows only, or is shielded to assure that conflicting YELLOW BALL or RED BALL indications are not visible to through drivers.

The *protected/permitted* mode is more complicated because it combines the other two modes. It has four distinct operational schemes as follows:

- Leading left turn with parallel through movement in one direction only.
- Simultaneous leading left turns in both directions with parallel through traffic stopped.

- Lagging left turn with a parallel through movement in one direction only.
- Simultaneous lagging left turns in both directions with parallel through traffic stopped.

11.6 The MUTCD at Court

[This section is an excerpt from a 1992 paper by Glennon[8]]

In matters of tort liability, the *Manual of Uniform Traffic Control Devices (MUTCD)*[1] is used daily in U.S. courts to argue about the safety of particular roadway locations, with reference to whether appropriate traffic control devices were present or needed. As such, the *MUTCD* serves as the underpinning of many judgments about the liability of Federal, State, and local jurisdictions across the U.S.

The *MUTCD* is both a legal document and an engineering standard for defining safe traffic control practices. However, its ambiguous language, its contradictions, and its lack of continuity create confusion when it is applied in the legal context of *negligence*. There are definite patterns of how the language of the *MUTCD* is used both to make claims of negligence against governmental jurisdictions or to defend such claims. Future *MUTCD* writers should be concerned with the format, the overall editing, the ways to promote continuity, and the ways to promote consistent safe practices.

MUTCD Word Meanings

In an attempt to say what is really meant, the *MUTCD* makes extensive use of three specifically defined words: *shall*, *should*, and *may* (see Figure 11-1). A major difficulty with the defined words is that, although they are sometimes used clearly to indicate the meaning intended, they are often used indiscriminately when they just happen to be the right word to fit the context. Looking closely at the entire context of the *MUTCD*, most traffic engineers would agree that a concise technical editing would change some *shalls* to *shoulds*, some *shoulds* to *shalls*, some *mays* to *shoulds*, etc.

Another difficulty in reading the *MUTCD* is the uncertainty of the authors in whether they want to use one of the defined words or whether they would rather avoid them. The result is a plethora of weasel words that either modify or take the place of the defined words. Phrases such as *should always*, *normally shall be*, *and may make it necessary*, frequently confuse the intent of the *MUTCD*. Table 11-8 shows additional examples of such phrases from every part of the *MUTCD*. This kind of language makes unclear whether the *MUTCD* is or is not

Table 11.8 Example MUTCD Phrases

Phrase	MUTCD Section
Basic principles that govern	1A-2
Conform to standards	1A-3
Should meet basic requirements	1A-2
Insure that requirements are met	1A-2
Engineering study required	1A-4
Signs are essential	2A-1
Is used	2A-6
Are desirable	2A-12
Is placed	2A-25
Intended for use	2B-4
Should ordinarily be used	2B-6
Shall be used when required	2B-29
Are used when deemed necessary	2C-1
Must serve to warn	3B-18
May be justified	4C-10
Meanings are	4D-1
Design requirements are	4D-4
Should be sufficient	4D-7
Should largely govern	4E-5
Should be realized	5A-1
Necessarily should be	5B-1
Is necessary	5B-2
A comprehensive guide	6A-1
A national standard	6A-1
Minimum desirable standards	6A-3
Must be provided	6A-3
Adherence to standards	6A-3
Standards should be adapted	6A-4
Shall be approximately	6B-6
Warrant special attention	6B-14
Prescribes standards	7A-4
In accordance with criteria	8B-2
Where conditions require	9A-4
Where desired	9B-22

recommending the practice, which can lead to unsafe practices in the field and confusion in the courtroom.

In considering the tort liability implications of the *MUTCD*, future authors should realize that words are the tools of a lawyer's trade and that indiscriminate use of words like *govern, necessary, requirement, always,* etc. can occasionally be incorrectly interpreted as requirements of the *MUTCD*. Similarly, phrases such as *is placed, is used,* and *intended for use* can occasionally be incorrectly interpreted as indicating no need for concern.

Traffic Control Devices 357

MUTCD Contradictions

The *MUTCD* has many contradictions that shadow its meaning and intent. Foremost is its discussion of legal implications. *Section 1A-4* states:

> It is the intent that the provisions of this Manual be standards for traffic control installation but not a legal requirement for installation.

What does this mean? Is the *MUTCD* just a technical guidebook that only needs to be referred to occasionally? Does it mean that roadway agencies can choose their own requirements, such as pink stop signs and lime green warning signs? Then how does this language comport with *Section 1A-3* that states:

> Traffic control devices placed and maintained by State and local officials are required by statute to conform to a State Manual which shall be in substantial conformance with this Manual?

What *Section IA-4* seems to be saying is that although the *MUTCD* should be followed as law in the field, it should not be used as law in the courtroom. This kind of ambivalence can only cause confusion and contempt for the document. Clearly the *MUTCD* has strong legal foundations that cannot be ignored or easily spoken away.

Another obvious contradiction of the *MUTCD* is the language of Section IA-4 titled *Engineering Study Required*, which infers a mandate. However, the text uses a *should* provision, which is a clear contradiction. Fortunately, the tort laws in most States are clear on how to interpret this conflict. Generally, laws say that when there is a conflict in language, the text outweighs the title. Therefore, the correct interpretation of *Section 1A-4* is that an engineering study is recommended (not required) for deciding on a particular traffic control device. Unfortunately, the *MUTCD* does very little to define what constitutes an engineering study, which makes compliance by non-engineering users particularly tenuous.

A particular characteristic of the *MUTCD* that promotes unsafe practices is its ambiguity between general and specific language within a Section. For example, in *Section 2-C*, the general language states that warnings are of great assistance to the vehicle operator and typical locations and hazards that may warrant signs are; (1) advance warning of control devices, (2) intersections...Yet when talking about warning signs for hidden STOP signs or for the end of a T intersection, the *MUTCD* simply says these signs are *intended for use*. This ambiguous language has been used in court to argue that a county roadway agency was not negligent for failing to sign an isolated rural T intersection where an unfamiliar driver on a dark night drove off the end of the T into a farm pond and drowned. Likewise, this language clearly promoted unsafe practices when a county roadway agency failed to warn of a STOP sign hidden over a sharp crest on a 55-mph roadway. Yet, the defense lawyer argued that the *MUTCD* did not *require* any signs for this condition.

Other contradictions are more direct. *Section 6C-9* of the *MUTCD* states:

> Type III barricades shall be erected at points of closure. They may extend completely across a roadway and its shoulders or from curb to curb.

In testing this text in a court case about a single 8-foot barricade used to close an on-ramp to an unopened Interstate highway, the plaintiff argued that the second sentence with its *may* provision simply clarified the first sentence by defining the only two allowable options. The defense argument was that the *may* indicated a permissive condition and, therefore, identified only two of several options.

Another example of a direct contradiction is found in *Part 8* on *Railroad Crossings*. In talking about post-mounted flashing signals in *Section 8C-2*, the *MUTCD* says that at crossings with traffic in both directions, back-to-back pairs of lights *shall* be placed on each side of the tracks. Yet when discussing cantilevered flashing signals, *Section 8C-3* not only fails to mention back-to-back pairs but also says flashing signals *should* usually be placed on the supporting post. Then to heighten the confusion on post-mounted signals, nearby *MUTCD* Figure 8-5 says post-mounted light units *may* be provided as conditions require. Therefore, in a span of 2 pages, the *MUTCD* uses the words *shall*, *should,* and *may* for the same consideration. This lack of clarity led one major city to justify using a non-uniform flashing traffic signal head hung on a guywire at a railroad crossing that had severely restricted sight triangles, a contiguous sharp curve, and other attendant hazards.

Lack of Continuity

The *MUTCD*, because of format and arrangement, may preclude the user (particularly the non-engineer) from identifying all of the traffic control devices available for a given roadway situation. Some applications of regulatory signs are found under the *Directional Sign* section and some devices that are sign-like are found in *Part 3* on markings. Then too, the user must look to several sections, often without any helpful clues, to coordinate signs, signals, and markings at any one location.

The *MUTCD* authors need to recognize the diversity of users by devoting more oversight in the preparation process to: (a) cross-referencing between Parts and Sections; (b) wider use of typical drawings; (c) better definition of what

Traffic Control Devices

constitutes an engineering study; and (d) indexing methods that lend quicker and more sure access to site-specific traffic control applications.

Some of the *MUTCD* features recommended above are currently treated well, particular in the stand-alone *Parts 6, 7, 8,* and *9*. A few examples of user difficulties in other parts, however, are:

1. For warning drivers of an **Object in The Road**, the *MUTCD* mandates both object markers and approach pavement markings. But, how does the unacquainted user ever find these devices? The object marker, which is sign-like, is one of a handful of devices in *Part 3* that is not a pavement marking. Although the section on object markers does cross-reference the section on approach pavement markings, the opposite is not true. Perhaps a simple drawing could help to show a standard layout.

2. The application of **Wrong-Way Signing** is adequately covered in several sections of the *MUTCD* with what would be good cross-referencing but for a typographical error. The major discussion, which seems hidden in *Section 2E* on *Guide Signs-Expressways*, is incorrectly cross-referenced both in the *Regulatory Sign* section and in the section on the general discussion of signs.

3. **Road Closure** sign and barricade application is another subject where all of the information is not covered in one place. The several *MUTCD* sections that cover this subject not only lack coordination but are generally inconsistent in using the defined words. Because a complete discussion of this subject is too detailed, the reader is referred to *Sections 2B-39, 2B-40, 2D-25, 3F-1, 6B-3 (Figs 6-2, 6-3, 6-4, 6-8), 6B-8, 6B-9, 6B-16, 6B-38, and 6C-9*. Although the reading of all these sections will eventually indicate all of the options for signing a road closure, the user will not be clear on what is mandated, what is recommended, and what is permissive.

4. In applying **Curve Warning Signs**, the *MUTCD* gives very little guidance to the non-engineering user when it says:

   > *The Curve sign may be used where engineering investigations of roadway geometric and operating conditions show the recommended speed on the curve to be greater than 30 mph and equal to or less than the speed limit.*

   What kind of engineering study is this passage talking about? What references should the users go to? Why not tell them about the *Traffic Control Devices Handbook*? What constitutes enough deviation between the safe speed and the speed limit to warrant the sign? Does *may be used* mean that the user can decide not to place any warning signs for a 31-mph roadway curve on a 55-mph roadway? The courts generally say *No*!

5. In applying **STOP Signs**, the *MUTCD* says:

> *A stop sign may be warranted at intersections where a combination of high speed, restricted view, and serious accident record indicates a need for control by the stop sign.*

Again, the guidance is not clear. Do all of these conditions have to be met simultaneously? How do you define *high speed, restricted view, and serious accident record?* Is the *MUTCD* giving blanket justification for the tens of thousands of low-volume rural uncontrolled intersections that have 55-mph speed limits, 25-mph or less (AASHTO) sight distance, but limited accident records? For these intersections, a pertinent question is how many accidents have to be saved to justify the cost of a STOP or YIELD sign? This question is being asked in the courts almost every day.

Plaintiff vs. Defendant

Clearly the *MUTCD* has language that will support arguments for both Plaintiffs and Defendants in court cases involving the use, misuse, or absence of traffic control devices. The following points contrast the wordsmithing of the *MUTCD* by plaintiff and defense lawyers:

1. Plaintiff lawyers use every *shall* provision that supports justification for a legal requirement. Defense lawyers argue that the *MUTCD* is not a legal document.
2. Defense lawyers point to the lack of any *shall* provisions that apply to the defendant. Plaintiff lawyers point to the violation of *should* provisions that are the commonly accepted good practices of the industry.
3. Defense lawyers use every *may* and *should* to point to the defendant's discretionary function. Plaintiff lawyers stress the *shoulds* as the appropriate standards of care.
4. Plaintiff lawyers make extensive use of the *MUTCD* discussions of basic concepts such as positive guidance, adequate time for proper response, target value, and uniformity. Defense lawyers argue that these concepts do not define a clear duty for the defendant.
5. Defense lawyers argue that the *MUTCD,* read literally, is the only way to judge the liability of traffic control device placement. Plaintiff lawyers use other prominent references, such as the *Traffic Control*

Device Handbook,[2] to add specificity to the *MUTCD's* general principles.
6. Plaintiff lawyers argue that the *MUTCD* requires an engineering study. Defense lawyers argue that the *MUTCD* does not require an engineering study.

Summary

The *MUTCD* because of its ambiguity, ambivalence, and ambage adds confusion to legal issues and, therefore, clouds the judgment of what is safe traffic control practice. Lawyers and judges, by their interpretations in the courtroom, often define what is and is not safe traffic control practice rather than the authors and users of the *MUTCD*. In fact, the efforts of the authors can become futile if the *MUTCD* is misused in the field, and that misuse is reinforced by the courts.

The *MUTCD* is an important document, but has many shortcomings that need to be corrected. The National Joint Committee needs to make a stronger commitment to eliminating the contradictions, the ambiguities, and the lack of continuity, so the liability of roadway agencies can be better defined, leading to a clearer communication of safe traffic control practices, particularly to the many non-engineering users. Surely, future editions of the *MUTCD* should strive for this objective. Rather than producing new editions that change a word, phrase, or paragraph here and there, major formatting, editing, and technical efforts are needed. No longer should Parts to be written as individual committee products without professional writing, formatting, and editing based on a well-defined consistency, continuity, and content of the entire document.

> **TRAFFIC CONTROL DEVICE DEFECTS**
> **Rule of Thumb**
> *Although Properly-Designed and Installed Traffic Control Devices Should Benefit All Drivers, They Should be Designed for the Unfamiliar Driver.*

11.7 Technical Aspects of Traffic Control Device Defect Cases

Thousands of roadway locations in the U.S. have missing, improperly placed, poorly maintained, or otherwise defective traffic control devices. These defects are often claimed as a proximate cause of traffic collisions in tort litigation against roadway agencies, particularly in local jurisdictions.

Some tort cases address the *lack* of traffic control devices. Common arguments are that the roadway agency was negligent for failure to warn of a defective roadway design feature. Often, because of the *design immunity* statute in most States, the plaintiff must first show that the roadway itself was defective but, second, and more important, show that the roadway agency not only had notice of the defect but also failed to properly warn of its hazard.

Arguments in other traffic control device defect cases focus on the direct defects of existing devices. Traffic signs can be too high, too low, too small, non-standard, not reflective, hidden behind trees, etc.. Traffic markings can be improperly placed, non-standard, or poorly maintained. And, traffic signals can have poor visibility, be hidden over a hillcrest, or have inadequate yellow clearance intervals.

Most tort cases about traffic control devices have a combination of defective elements that lead to stronger arguments about failure to warn, improper design, improper placement, failure to maintain, or any combination of these arguments. The following paragraphs describe some of the more common kinds of cases involving traffic control devices.

Hidden Traffic Signs

The STOP sign, probably more than any other sign, not only needs to be seen but needs to be seen for an adequate distance in order to be effective. STOP signs hidden over sharp hillcrests, around sharp roadway curves, or behind vegetation, other signs, or other obstructions can fail to give the driver adequate time for proper response. A typical collision under these circumstances involves a driver running through the STOP sign and hitting a vehicle on the intersecting roadway.

When a STOP sign is hidden over a sharp hillcrest, or around a sharp curve, the plaintiff has a viable argument if: (1) the STOP sign was not mounted as high as possible under the circumstances; and (2) a STOP AHEAD sign was not properly placed to give advance warning of the STOP condition. These defects are particularly noteworthy where the STOP sign is isolated, such that the driver has not experienced any stop condition for several miles previous to the subject location. Under these circumstance, the driver is falsely conditioned to not expect a stop condition. Figure 11-12 shows the application of a STOP AHEAD sign at a hillcrest that only allows about 150 feet of visibility to the STOP sign.

Traffic Control Devices   363

Figure 11-12 Sharp Hillcrest that Hides a Stop Sign.

Warning signs hidden by vegetation, other signs, parked cars, and other obstructions cannot effectively communicate their warnings. Figure 11-13 shows an advance railroad crossing sign partially hidden by weeds. Not only is this sign partially hidden, but the weeds convey a look of rail abandonment.

Figure 11-13 Traffic Signs Obscured by Weeds.

Illusive Traffic Signs

Another common difficulty with STOP signs in rural areas is where a roadway route is realigned and the old route is maintained for local access from the point where it intersects with the new route. Because the two route alignments usually cross at a very flat angle, a new roadway curve is usually built at the end of the old route to form a 90° intersection angle with the new route. A STOP sign is then placed on the old route where it intersects the new route.

This kind of intersection has at least two associated problems. First, the STOP sign placed perpendicular to the adjacent roadway cannot be viewed adequately for a very great distance because of the curvature. This can be particularly difficult at night, because a retroreflective sign has a limited cone of reflectance, and, therefore, may not be brightly presented to the driver for an adequate distance. The second associated problem is that steering the roadway curve can command most of the driver's attention to where he does not see the STOP sign at all. Any other driver distractions or sight obstructions under this kind of intersection environment can further degrade the driver's ability to see and respond to the STOP sign.

The same considerations described above pertain to an isolated rural STOP sign. If on a particular route, drivers can travel several miles without stopping before reaching an isolated single normal-size STOP sign, they either may be surprised by the STOP sign, or they may think it doesn't apply to them, or worse, because of their false expectancy pattern, they may not see the STOP sign at all.

The problems with these intersection environments are exacerbated by STOP signs that are too low, improperly aimed, or inadequately reflectorized. These problems can all be minimized with properly-placed STOP AHEAD signs and oversized STOP signs on both sides of the approach roadway.

Non-Reflective Signs

All traffic signs are required by the *MUTCD* to either be illuminated or reflectorized. The vast majority of traffic signs have retro-reflective material.

Traffic signs in place need to be checked regularly to ensure their nighttime reflectivity is effective. The two key factors that affect this reflectivity are grime and fading. Signs not only need to be routinely cleaned but also need to be replaced periodically as their reflectivity degrades to less than 70% of the new condition.

Because STOP, STOP AHEAD, curving warning, and construction zone warning signs need their reflectivity to properly function at night, serious accidents can occur when they lose their reflectivity. These conditions often become the subject of tort claims. Figure 11-14 shows a faded reverse-curve warning sign that has lost both its color and reflectivity.

Traffic Control Devices 365

Figure 11-14 Non-reflective Reverse Curve Warning Sign.

Failure to Warn

Even though the application of many warning signs is not mandated by the *MUTCD*, roadway conditions exist where the lack of a proper warning sign can be judged as negligent in many States. On high-speed roadways some conditions that are hazardous without any warning signs are:
1. At very sharp roadway curves.
2. On the stem approach to T or Y intersections without STOP or YIELD signs.
3. Where a bridge is significantly narrower than the approach roadway.
4. Where a roadway narrows significantly.
5. Where a roadway changes from paved to unpaved.
6. Where a significant pavement edge drop exists.
7. Where the pavement on a high-speed roadway is very slippery when wet.
8. Where a significant bump or dip on a high-speed roadway can cause loss of control.

Wrong Sign for Location

Untrained roadway personnel often install signs that may be confusing to motorists. Each *MUTCD* sign has a specific application and should only be used for that application. For example, TURN signs should not be used to indicate where the predominant flow turns through a three- or four-way 90° intersection, as shown in Figure 11-15.

Figure 11-15 Improper Use of a Traffic Sign.

Unclear Messages

Warning signs should not be used to convey non-traffic messages or messages that are otherwise unfamiliar to drivers. Figure 11-16 shows a sign with the message, UNDERMINED AREA. What this means is anyone's guess. Frequent use of unclear messages in fact "undermines" the respect of drivers who desire and need simple, clear messages to operate safely.

Non-Standard Traffic Signs

A non-standard warning sign that shows notice of a hazardous condition, but fails to communicate that information to a driver who has an accident involving the hazard, can clearly be the subject of a tort claim. Figure 11-17 shows a confusing and wordy warning message on a wrong color, wrong shape, and poorly maintained sign.

Traffic Control Devices 367

Figure 11-16 Traffic Sign with Obscure Message.

Figure 11-17 Non-standard Use of Traffic Sign.

Damaged Traffic Signs

Traffic signs occasionally get knocked down, stolen, or shot full of holes (Figure 11-18). Vigilance by the roadway agency is required to minimize the hazard associated with these conditions. Roadway agencies normally use the combined efforts of routine maintenance surveillance, local police, and citizen call-ins to correct these conditions. When these conditions occur and are not corrected for several weeks, they can become the source of a tort claim.

Figure 11-18 Stop Sign Shot Full of Holes.

Warning Sign in Only One Direction

Particularly on two-lane roadways, many conditions that warrant warning signs need those signs for both directions. Curve warning signs are the best example of this situation. Sharp roadway curves usually present the same hazard in both directions. If warning signs are placed for one direction but not the other, the roadway agency demonstrates not only that they had notice of the dangerous condition but also that they were negligent in their failure to warn half of the traffic. Figure 11-19 shows a roadway location where these warning signs were not duplicated over the hillcrest for the opposite direction of traffic. [Also notice the improper use of two warning signs on the same post].

Traffic Control Devices 369

Figure 11-19 Traffic Signs Placed for One Direction without Identical Signs for the Other Direction.

Partially Obscured Signs

The *MUTCD* prohibits both the use of advertising on the face of traffic signs and the placement of other kinds of signs on traffic sign posts. Figure 11-20 shows a rail-highway crossing warning sign with a election campaign sticker on the sign and a large yard sale sign on the same post. These kinds of violations, taken to the extreme, can seriously degrade the effectiveness of the traffic sign.

Signs that Directly Indicate Negligence

Figure 11-21 shows a non-standard warning sign with the legend *Dangerous Intersection*. When the danger is not apparent to the driver and/or when other standard warning signs are proper to warn of the danger, the roadway agency may only be highlighting its own negligence.

Missing End of Road Markers

For a high-speed roadway or any other roadway that appears to be a through roadway, End of Road markers (Figure 11-6) should be used to inform drivers

Figure 11-20 Traffic Sign Obscured by Election Sticker on its Face and a Yard Sale Sign on its Post.

Figure 11-21 Traffic Sign that may Highlight Roadway Agency Negligence.

that the road is discontinued. Figure 11-22 shows a end-of-road situation that was the subject of a tort claim. This rural road, which had neither speed limit signs nor advance warning signs, was alleged as a negligent condition. The

Traffic Control Devices 371

Figure 11-22 End of a Roadway with No Markers to Warn Drivers.

Figure 11-23 Obstructions in the Roadway.

vehicle, traveling at 40 mph, ran off the end of the road, collided with a large tree, and seriously or fatally injured several occupants. Properly placed End of Road markers could have prevented this tragedy.

Obstructions in the Roadway

The *MUTCD* mandates both object markers and pavement channelizing lines for obstructions in the roadway. When either of these is missing, a potential tort claim seems evident when a collision occurs with the obstruction. Figure 11-23 shows a railroad underpass where several obstructions exist. Although the bridge columns do have object markers and channelizing lines, the object markers are badly deteriorated and have non-standard white stripes instead of standard yellow stripes.

Missing or Faded Pavement Markings

When painted centerlines, edgelines, or channelizing lines are not properly maintained, they lose their effectiveness. The fact that they were painted means that the roadway agency thought they were needed. When they are allowed to lose their effectiveness, they could become an alleged proximate cause to an accident.

Isolated Traffic Signals

As high-speed (50-mph or higher) rural highways enter an urban area, drivers can be presented with a traffic signal after many miles of no control. These kinds of signals are notorious for having serious accidents involving high-speed approach vehicles. For this reason, these locations can be judged as negligently operated unless extraordinary traffic control measures are taken. Figure 11-24 shows a warning system used as an effective countermeasure at a previously high-accident intersection.

Signal Visibility Defects

The visibility of traffic signals is important for their safe operation. The *MUTCD* mandates that each intersection approach have two signal faces, that at least one signal be within a defined area, and, for speed limits of 45-mph or above, that 12-inch signal lenses be used. Violation of any of these mandates could be judged as a proximate cause in a signalized intersection accident.

11.8 Typical Defense Arguments

The most common defense argument, which is supported by tort claims legislation in many States, is that the placement of traffic control devices is a discretionary function that is protected from liability. The major exception is when the traffic control defect violated a mandate (*shall*) of the *MUTCD*. Another exception granted by the courts in some States is in accordance with Sections 2A-1 of the *MUTCD* which says:

> *Signs are essential where hazards are not self-evident.*

Traffic Control Devices 373

Figure 11-24 Isolated Traffic Signal with Active Advance Warnings.

Since most traffic control defect cases allege a violation of the *MUTCD*, its language is often the major thrust of the defense arguments. With the exception of traffic control device colors, shapes, sizes, and reflectorization, most provisions of the *MUTCD* are either recommended (*should*s) or permissive (*may*s). For this reason, defendants will often argue that, since no *MUTCD* mandates (*shall*s) were violated, the traffic controls placed or not placed were discretionary and, therefore, not subject to claims of negligence.

Other case-specific defense arguments allege either that the driver was the sole negligent party or that the roadway condition did not contribute to cause the collision. Some of these arguments are:

1. The roadway geometry and/or contiguous development gave fair advance warning to the driver.
2. The driver was negligent for speeding.
3. The driver was negligent for driving while intoxicated.
4. The driver did not exercise proper lookout.
5. The driver did not drive with the highest degree of care.
6. Direct evidence is lacking to indicate that the traffic control devices, either placed or not placed, contributed to the accident.
7. The roadway agency had no notice of the defective traffic control condition.
8. If we put up signs, they will only be stolen or shot full of holes.

Endnotes

1. Federal Highway Administration, *Manual on Uniform Traffic Control Devices*, 1988. (available through Lawyers and Judges Publishing Co.)
2. Federal Roadway Administration, *Traffic Control Devices Handbook*, 1983.
3. Institute of Transportation Engineers, *Transportation and Traffic Engineering Handbook*, Second Edition, 1982.
4. American Association of State Highway Officials, *A Policy on Sight Distance for Highways*, 1940.
5. Glennon, John C., *A New and Improved Model of Passing Sight Distance on Two-Lane Highways*, Transportation Research Record 1195, 1988.
6. Harwood, Douglas W., and Glennon, John C., *Passing Sight Distance for Passenger Cars and Trucks*, Transportation Research Record 1208, 1989.
7. Kell, James H., and Fullerton, Iris J., *Manual of Traffic Signal Design*, Institute of Transportation Engineers, 1982.
8. Glennon, John C., *The MUTCD at Court*, Proceedings, Institute of Transportation Engineers, 1992.

References

Glennon, John C., *Design and Traffic Control Guidelines for Low-Volume Rural Roads*, Transportation Research Board, NCHRP Report 214, 1977.

Alexander, G.J. and Lunenfeld, H., *Positive Guidance in Traffic Control*, Federal Highway Administration, 1975.

Alexander, G.J., and Lunenfeld, H., *Driver Expectancy in Highway Design and Engineering*, Federal Highway Administration, 1986.

Fullerton, I.J., *Roadway Delineation Practice Handbook*, Federal Highway Administration, 1981.

McGee, H.W., and Mace, D.L., *Retroreflectivity of Roadway Signs for Adequate Visibility: A Guide*, Federal Highway Administration, Report No. FHWA-5A-88-050, 1988.

Chapter 12

Roadway Maintenance

Roadway maintenance is an activity aimed at minimizing or correcting the effects of aging, material failure, weather, vegetation growth, traffic wear, damage, and vandalism. The American Association of State Highway and Transportation Officials in its 1987 publication, *AASHTO Maintenance Manual*,[1] defines highway maintenance as follows:

> *Highway Maintenance – A program to preserve and repair a system of roadways with its elements to its designed or accepted configuration. System elements include travelway surfaces, shoulders, roadsides, drainage facilities, bridges, tunnels, signs, markings, lighting fixtures, truck weighing and inspection facilities, etc. Included in the program are such traffic services as lighting and signal operation, snow and ice removal, and operation of roadside rest areas, movable span bridges, etc.*

Another major source of information about maintenance problems and practices is the *Street and Highway Maintenance Manual*[2] published by the American Public Works Association.

12.1 Unpaved Surfaces Maintenance

An unpaved surface is composed of an earth and aggregate mixture without any binder such as asphalt or portland cement. This type of surface accounts for over 60% of the total U.S. road mileage. These roadways, however, only carry about 10% of the total vehicle-miles of U.S. roadway travel.

Unpaved surfaces, usually crushed rock, gravel, or other natural materials, form a crust when the correct blend of aggregate, fines, and water is used. When the water dries, this hard crust forms the wearing surface. The strength of this surface depends largely on the strength of the base material because the surface material has no bending strength.

The primary objective in maintaining an unpaved roadway is a usable and safe surface for anticipated traffic. Vehicles using the roadway should not encounter holes, bumps, oversize material, or loose material that will cause loss of control. Unpaved surfaces are smoothed by blading to break the crust. To form a new crust, the blend of earth and aggregate must have the correct moisture content.

Frequent inspections and regularly scheduled maintenance are more essential to this type of roadway than to paved surfaces because of their tendency to deteriorate rapidly under traffic or adverse weather. A variety of factors can deteriorate an unpaved roadway. The crown of the roadway will decrease under normal traffic operation. Vehicle braking causes corrugations (washboard). Dry weather can bring on extreme dust problems. And, unusually dry or wet weather can break the hard crust causing rutting, soft spots, and holes.

Maintenance of unpaved roadways includes smoothing, crown reshaping, dry weather blading, adjusting the aggregate/fine mix, and dust prevention. Rehabilitation of unpaved roadways includes base stabilization using either portland cement, lime, or asphalt.

Roadway surfaces are maintained by blading to: (1) pull loose surface material that has been whipped onto the shoulder by traffic back onto the travel surface; (2) cut down high spots and fill low areas to remove corrugations; and (3) reshape the traveled surface to provide a crown of one half to one inch per foot sloping out from the center so the surface will drain. Motorgraders are used for blading (see Figure 12-1). Grading speed should not be excessive because this action can cause corrugations. Blading should be done as soon as practical after a rain or when the surface is moist. Dry blading may become necessary to remove excessive surface material and to fill holes during dry weather.

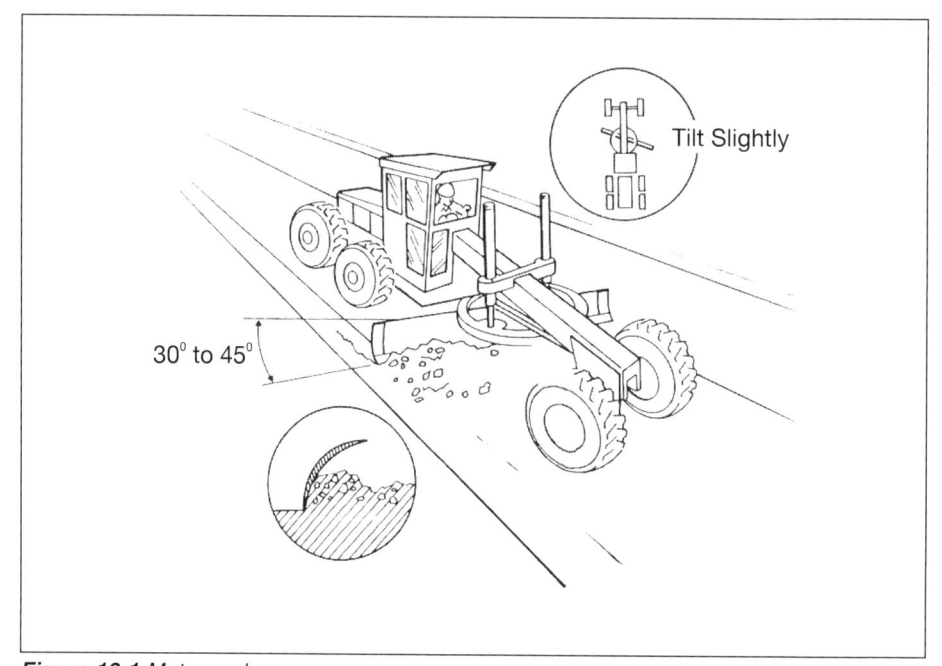

Figure 12-1 Motorgrader.

Roadway Maintenance

Windrows of aggregate should be held to a minimum size sufficient for maintaining the surface and placed as far from the travel surface as possible to avoid vehicles losing control by contact with the windrows. This is particularly important at hilltops and roadway curves on narrow roadways. On roadway curves, the windrow should always be placed on the outside of the curve.

Stabilization of earth-aggregate surfaces offers the following advantages at a reasonable cost: (1) a firm load-bearing surface; (2) resistance to raveling and loose aggregate; and (3) resistance to water and frost action. Stabilization is usually done by pulverizing the soil and road-mixing it with a material such as asphalt, portland cement, calcium chloride, sodium chloride, or lime. The mix is then shaped and compacted with a roller.

The principal problems with unpaved roadways that require maintenance are described below.

Corrugations

These are surface distortions that resemble the pattern of washboard (see Figure 12-2). The progressive growth of corrugations on unpaved roadways is usually caused by the braking action of wheels. The gradation of the surface aggregate will also contribute to corrugation when rounded coarse particles are bound with a fine binder. Once started, corrugations usually worsen and spread.

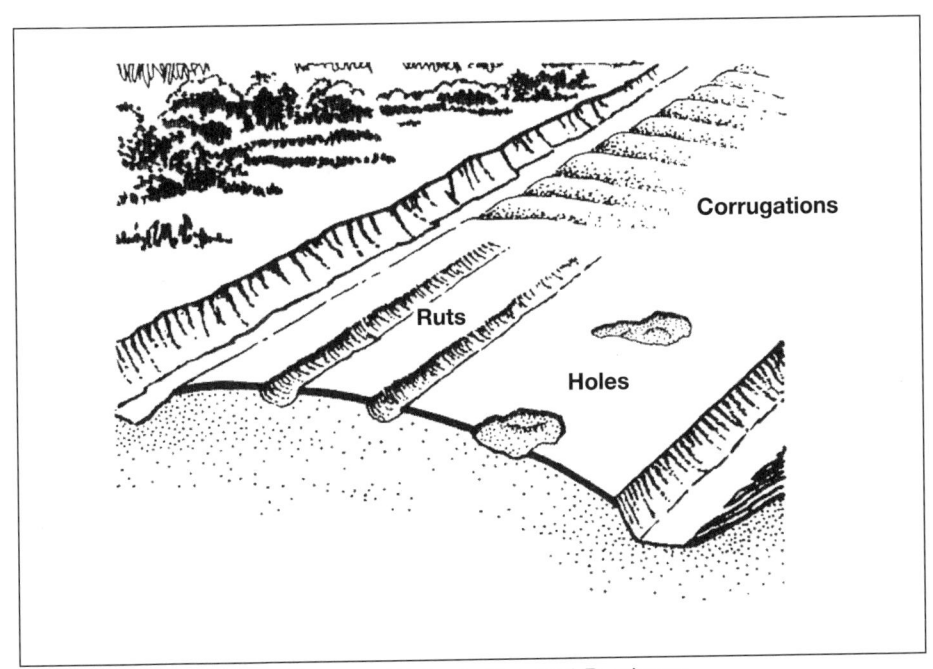

Figure 12-2 Ruts, Corrugations, and Holes in Unpaved Roadways.

Prompt action is needed to mitigate the progressive growth of corrugations. The best way to repair corrugations is to scarify, pulverize, reshape, and compact the surface. The best time is after a soaking rain. Also, stabilizing material mixed with the pulverized material can substantially resist future corrugations. Dragging or blading can repair corrugations as a temporary measure.

Wheel Ruts

Wheel ruts can occur in loose or wet unpaved surfaces. Heavy trucks are often the cause. The tendency of ruts to redirect the front wheels of a vehicle makes them a potential hazard. Filled with water, ruts can create splash and icing problems. Dragging and blading while the road is wet, followed by compaction, is the most common way to repair ruts.

Soft Spots

Localized settlement of unpaved roadways often occurs because of low areas or the lack of crown that leads to ponding. Other soil or drainage problems are caused by a high water table, hydrostatic pressure, capillary action, or ground water seepage. Saturated base material can be removed and replaced with coarse-graded material. An alternative remedy is to place subdrains. When settlement is caused by traffic loads, filling, compacting, and regrading should provide an adequate repair.

Dust

During dry weather, dust can become a serious visibility hazard on unpaved roadways, particularly on narrow roadways where head-on conflicts are relatively hazardous even without dust problems. Dust palliatives such as calcium chloride and sodium chloride can be effective to minimize dust.

Potholes

The repair of potholes in unpaved surfaces is a continuous process. Materials similar to the existing surface should be used for temporary repair. More permanent repairs to include scarifying, mixing, and compacting, should be done when moisture is available.

12.2 Asphalt Pavement Maintenance

Bituminous concrete (asphalt) pavements consist of a mixture of graded aggregate and bituminous materials. Asphalt is refined petroleum and must meet standard specifications. It is produced in a variety of types and grades. The three major types are asphalt cements, liquid asphalts, and asphalt emulsions. Liquid asphalt is prepared by blending asphalt cements with petroleum distillates, and asphalt emulsions are prepared by homogenizing asphalt cement with water.

Asphalt surfaces are machine laid on a substantial base coarse of either rigid or non-rigid design. The thickness of the surface wearing coarse is usually 1.25 inches or more.

Asphalt surfaces are subject to several different forms of distress as indicated in Table 12.1. Unattended, most of these pavement irregularities become extreme and can contribute to vehicular loss of control on high-speed roadways, particularly at roadway curves. For this reason and also to maintain the integrity of the pavement, surveillance and regular maintenance are important for asphalt pavements. Some of the more typical maintenance problems and practices are described below.

Corrugations and Shoving

Corrugations are surface undulations perpendicular to the roadway with closely-spaced valleys and crests. These irregularities are caused by traffic action on an unstable pavement or base. Corrugations become hazardous as the vertical deviations increase, particularly on roadway curves.

Shoving is a permanent displacement of the pavement parallel to the direction of traffic. It usually occurs at intersections or at hills where vehicles are accelerating or braking. The first indications of shoving are crescent shaped slippage cracks with the apex pointing in the direction of the shove. To correct shoving, the section of pavement should be removed down to a firm base and replaced.

Potholes

Potholes are steep-sided holes where localized pavement disintegration has been caused by poor drainage and aggravated by traffic loads. As shown in Figure 12-3, potholes are repaired by careful patching that includes cleaning out and sealing the hole, filling it with asphalt material, and compacting that material to the level of the surrounding pavement.

Research[3] indicates that potholes are usually not a major contributor to loss of control for passenger cars unless the size of the hole is large. However, when vehicles demand side friction such as on roadway curves, impact with a pothole could trigger instability. Then too the control of motorcycles, mopeds, and bicycles is quite sensitive to pavement irregularities such as potholes.

Upheavals and Settlements

Excessive moisture in the base coarse caused by capillary action, swelling soils, unrepaired surface cracks, or poor pavement drainage, can causing heaving or settlement of the pavement. Frost heave is one of the major actions causing pavement upheaval.

Table 12.1 Asphalt Pavement Distress

Alligator Cracking	Interconnecting cracks caused by fatigue of the surface under repeated loads.
Bumps and Sags	Small localized, upward displacements of the pavement surface caused by shrinkage and daily temperature cycling of hardened asphalt. Not load connected.
Corrugations (Washboarding)	A series of closely spaced ridges occurring at regular intervals caused by traffic action on an unstable pavement or base.
Depressions	Localized pavement areas that are slightly lower than the surrounding areas caused by foundation soil settlement.
Edge Cracking	Cracks parallel and close to the outer edge of pavement caused by the lack of shoulder support.
Flushing (Bleeding)	A film of asphalt on the pavement surface caused by excessive asphalt or insufficient voids in the mix.
Joint Reflection Cracking	Cracks in the asphalt surface over the joints in a portland cement concrete slab.
Longitudinal and Transverse Cracking	Non-reflective cracks caused by poorly constructed lane joints, shrinking of the surface, or hardening of the asphalt.
Pavement Edge Drops	Difference in elevation between the pavement and the shoulder caused by shoulder settlement or the erosion of shoulder material.
Polished Aggregate	Smooth aggregate caused by the repeated action of traffic.
Pot Holes	Small steep-sided depressions in the pavement surface caused by poor mixes or weak spots in the subgrade, aggravated by traffic.
Rutting	A surface depression in the wheel path due to the deformation of the pavement or subgrade caused by repeated and heavy wheel loads.
Shoving	Permanent longitudinal displacement of a localized area of pavement surface caused by braking and accelerating.
Slippage Cracking	Crescent shaped cracks where the ends point away from the direction of traffic. Occur when braking wheels cause the pavement to slide or deform.
Swelling	Upward bulge in pavement surface caused by swelling soil or frost action.
Weathering and Raveling	Wearing away of the pavement caused by loss of binder and then aggregate.

Roadway Maintenance

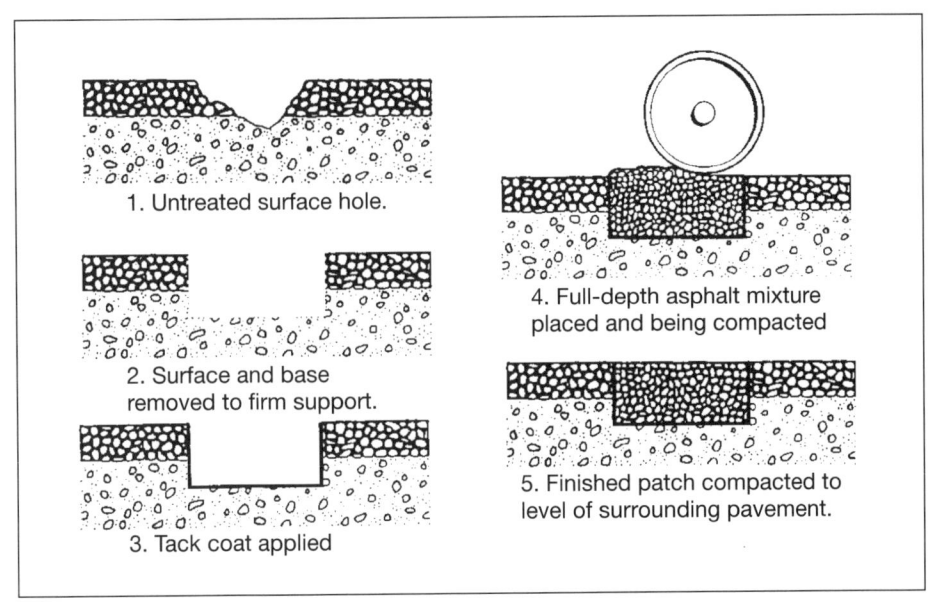

Figure 12-3 Pothole Repair Regimen.

1. Untreated surface hole.
2. Surface and base removed to firm support.
3. Tack coat applied
4. Full-depth asphalt mixture placed and being compacted
5. Finished patch compacted to level of surrounding pavement.

In the extreme, these bumps and dips can cause vehicular loss of control. Unstable material should be removed and replaced with material that has good drainage qualities. Subdrains may be necessary. Repairs to the subgrade should be well compacted and both the base and surface coarse replaced. Mudjacking is sometimes effective for localized distress.

Flushing

This condition, sometimes called bleeding, is a concentration of asphalt on the surface caused by either too much asphalt in the mix, an unstable mix, or too much asphalt in applied patches. Flushing, which occurs in hot weather, can create a slippery pavement if not treated. Slight flushing can sometimes be corrected by blotting with sand or porous aggregate. Permanent repair of heavy flushing requires overlaying with a new surface having a suitable asphalt content.

Cracks

Cracks result from distress in the structure underlying the pavement surface. *Reflection cracks*, either parallel or perpendicular to the roadway, appear in an asphalt overlay caused by joints and cracks in a concrete base coarse. *Edge cracks* occur near and parallel to the outer edge of the pavement because of inadequate shoulder support. *Alligator cracks* are interconnected cracks, forming a series of small polygons, exhibiting pavement fatigue caused by traffic loads and inadequate base support. *Shrinkage cracks* are random interconnected cracks exhibiting a shrinkage of the pavement volume due to improper materials or proportions and/or aging of the asphalt mix.

Crack filling and sealing is good maintenance practice. Several crack filling methods are used. The practice, however, ranges from filling all cracks, to only filling large cracks, to not filling any cracks. The permeability of the base is a major determinant. When cracking is caused by poor surface materials, overlaying the surface may be the only remedy.

Pavement Edge Drops

These irregularities often occur when a pavement is overlaid and no shoulder material is added. Where shoulders are unpaved, edge drops are created by wheel ruts, by material being blown away by large trucks, and by natural erosion. Edge drops of 1.5 to 2 inches or more can significantly contributed to vehicular loss of control. For a complete discussion, the reader should refer to Chapter 7, which is dedicated to this topic.

Wheel Ruts

Wheel ruts in asphalt pavements are the result of frequent and/or heavy traffic. They are formed both through traffic wear and by the displacement and compaction of either the surface or base course by heavy vehicles. Other causes include inadequate compacting or improper mix design during construction. Ruts can be repaired by leveling with an asphalt mixture or by milling or planing.

Many roadway agencies repair wheel ruts that are one-half inch deep or more. These depths will hold enough water to cause slippery conditions. During warmer weather, water in wheel ruts can cause hydroplaning under some circumstances (see Chapter 6). In the winter, during freeze and thaw cycles, wheel ruts can collect water during a daytime thaw, which can then freeze in the wheel track as temperatures drop.

Aggregate Polishing

All aggregates polish with traffic and weather action. Although some aggregates are very hard and are highly resistant to polishing, the majority of aggregate available in the U.S. today are not highly resistant. Smooth rounded aggregate or soft limestones may polish to the point where the surface becomes slippery when wet. A polished surface should be corrected with an open-graded asphalt surface coarse, a chip seal, or a slurry seal. For further discussion of slippery pavements, refer to Chapter 6.

Debris on the Pavement

Surfaces, gutters, medians, bridge floors and walks, and other paved areas should be kept clear of dirt, gravel, sand, and other foreign matter that can be a hazard to traffic. Cleaning paved areas is done by flushing with water, by sweeping with a motorized sweeper, or by sweeping by hand.

Sand and gravel on the travel lanes can seriously reduce the effective friction, particularly for motorcycles. When vehicles from a gravel road at an intersection enter and kick gravel onto the paved roadway, that gravel should be routinely sweep from the pavement. If the condition persists and/or is particularly hazardous, an asphalt apron about 30 feet long should be paved along the unpaved roadway. Where gravel is routinely kicked from a gravel shoulder on a narrow paved road, a two-foot width of the shoulder should be paved with asphalt. This may particularly be a problem on roadway curves where vehicles may more frequently ride onto the shoulder, and also, where motorcycles require more friction than on straight roadway sections.

12.3 Concrete Pavement Maintenance

Portland cement concrete pavements provide excellent service but, like any other structure, they require maintenance to provide that service. The most common forms of concrete pavement distress are summarized in Table 12-2. Unattended, most of these pavement irregularities become extreme and can contribute to vehicular loss of control on high-speed roadways, particularly on roadway curves.

Surface Texture Distress

Map cracking, scaling, and aggregate polishing of portland cement concrete are the typical surface failures. *Map cracking* appears as fine interconnected lines resembling a road map. They are usually shallow. If they are wide enough to allow water to enter, additional damage will result because of freeze-thaw cycles or because of winter maintenance chlorides that reach the reinforcing steel.

Scaling

This can result from the freeze-thaw cycles of water that has entered small surface cracks. It leaves the surface rough with exposed aggregate. Scaling is usually a result of inadequate finishing, inadequate curing, improper air entrainment, or faulty concrete mix.

Small areas of cracking or scaling can be repaired with an epoxy surface patch. Large areas of scaling or aggregate polishing can be corrected with an asphaltic concrete or slurry seal overlay.

Crack and Joint Distress

Portland cement concrete pavements will expand with heat and contract with cold. The extreme temperature variations between summer and winter in some areas can cause severe cracking in concrete slabs. Cracking is usually controlled

Table 12.2 Types of Distress in Portland Cement Concrete Pavements

Blow Up	A localized upward thrust of the slab edge usually due to insufficient openings to accommodate joint movement.
Corner Break	A crack that intersects a joint less than half the slab length that extends vertically through the entire slab usually caused by loss of support and repetitive loads.
Shattered Slab	The slab is divided into four or more pieces due to overloading and/or inadequate support.
Durability Cracking	Usually a pattern of cracks running parallel and close to a joint caused by break down of large aggregate due to freeze-thaw expansion.
Faulting	This is a difference in elevation across a joint caused by settlement, pumping, or curling of slab edges due to temperature and moisture changes.
Joint Seal Damage	Any damage to joint material that allows intrusion of foreign material.
Linear Cracking	Cracks that divide the slab in two or three pieces caused by repeated loading, thermal curling, and moisture.
Pavement Edge Drops	A difference in elevation between the lane and the shoulder caused by shoulder erosion or settlement.
Polished Aggregate	Aggregate that is so smooth that the tire adhesion is considerably reduced due to abrasion of repeated traffic application.
Popouts	A void left by the loss of a piece of aggregate that has been dislodged through expansion caused by freeze-thaw action.
Punchouts	The localized breaking into pieces of the slab caused by heavy loads, inadequate thickness, loss of support, or concrete deficiency.
Pumping	The ejection of material under a slab caused by the deflection of the slab acting on the water under the pavement.
Sealing, Mapcracking, and Crazing	A network of hairline cracks through the upper pavement surface caused by overfinishing, action of salts, and improper construction.
Shrinkage Cracks	Cracks that do not extend through the entire slab or completely across it, usually caused by improper curing.
Spalling Corners	The breaking off of the corner within 2 ft. (0.6m) of the corner.

by building in expansion and contraction joints. Contraction joints are sawed into the pavement very early after the concrete is placed. They are sawed to a sufficient depth to ensure that the slab will crack at that joint. Expansion joints are designed into the pavement structure and extend for the full depth.

Both expansion and contraction joints are filled with a material that should expand during cold weather and compress during hot weather to seal the joint against the entry of water. If the joint material is inadequate for this purpose, damage to the base or subgrade will likely result from the intrusion of water.

Some cracks result from horizontal stresses caused by an improper joint or an inadequate number of joints. Other cracks result from unusual vertical stresses on the slab caused by a weak base or subgrade support or by improper mixing or curing of the concrete. Cracks caused by inadequate base or subgrade support usually first occurs on the corners of the slab. To repair this defect, the broken area should be removed, the base properly stabilized, and the corner replaced. Other cracks should be checked and/or sealed regularly.

Variable temperatures between the top and bottom of a concrete pavement can cause the slab to curl. This action can loosen and eject solid joint seals. Defective, missing, or inadequate joint filler material should be removed or replaced.

Spalling

This expansion distress in concrete pavements appears as chipping or splintering of a localized area. Expansion results from either heat or from the corrosion of the reinforcing steel caused by water and/or chlorides. Spalls should be routinely repaired with joint sealants. For larger spalls, high-density concrete or other high-quality repair materials should be applied.

Blowups

This distress appears as buckling or shattering at a transverse crack or joint. Blowups occur during hot weather when inadequate space between slabs causes high longitudinal compressive stresses. On high-speed roadways, blowups over one inch are dangerous and should be repaired as soon as possible, and blowups over two inches should be repaired immediately. On lower-speed roadways, blowups over two inches are dangerous and should be repaired as soon as possible, and blowups over three inches should be repaired immediately.

A blowup usually begins with a slight spalling at the joint. The condition may either progressively worsen or the blowup may occur suddenly. The repair of a blowup often requires cutting off the end of the slab to permit it to fall back in position. The widening of the joint should be kept as narrow as possible. If the slab has been shattered, complete removal and replacement of the damaged section is necessary.

Frost Heaves

This distress appears as differential upward displacements of adjoining concrete slabs due to the freezing of water under the joint. Repairs require that frost susceptible material be removed below the joint and replaced with frost free material. Subdrains are sometimes a less costly remedy. Also joint sealants should be added or upgraded when necessary.

Pumping

This distress appears as the ejection of mud along joints, cracks, and pavement edges. Pumping is activated by the downward movement of the slab caused by the passage of heavy axle loads over the pavement when free water has accumulated under the pavement. Excessive pumping should be corrected as soon as possible by mudjacking the pavement and sealing cracks and joints.

Pavement Edge Drops

Because a concrete pavement has a formed vertical edge with a 90° corner, edge drops of 1.5 or more can be quite dangerous for an errant driver. Edge drops on unpaved shoulders are created by wheel ruts, by shoulder material being blown away by large trucks, and by natural erosion. Edge drops, on asphalt-paved shoulders adjacent to concrete pavements, occur when a differential settlement occurs. Chapter 7 provides a complete discussion of the hazard associated with pavement edge drops.

Debris on the Pavement

Surfaces, gutters, medians, bridge floors and walks, and other paved areas should be kept clear of dirt, gravel, sand, and other foreign matter that can be a hazard to traffic. Cleaning paved areas is done by flushing with water, by sweeping with a motorized sweeper, or by sweeping by hand.

Sand and gravel on the travel lanes can seriously reduce the effective friction, particularly for motorcycles. When vehicles from a gravel road at an intersection enter onto and kick gravel onto the paved roadway, that gravel should be routinely sweep from the pavement. If the condition persists and/or is particularly hazardous, an asphalt apron about 30 feet long should be paved along the unpaved roadway. Where gravel is routinely kicked from a gravel shoulder on a narrow paved road, a two-foot width of the shoulder should be paved with asphalt. This may particularly be a problem on roadway curves where vehicles may more frequently ride onto the shoulder, and also, where motorcycles require more friction than on straight roadway sections.

12.4 Shoulder Maintenance

Shoulders are those areas parallel to the travel lanes that extend to the outer edge of the roadbed. They provide a secondary recovery area for roadside encroachments, extra width for collision avoidance, and lateral support for the pavement edge.

Shoulders should be maintained so they are reasonably smooth, free of holes and ruts, and sloped away from the pavement edge for surface drainage. They should be kept level and flush with the travel surface. Shoulder width should be maintained as constructed unless widening is desirable during normal ditching operations. Shoulder types are discussed in Table 12.3.

Shoulder cross slopes in areas without curbs are normally from 0.03 to 0.05 feet per foot on paved shoulders, 0.04 to 0.07 feet per foot on gravel or crushed stone shoulders, and 0.08 feet per foot on turf shoulders. Newly constructed shoulders should be checked frequently during wet weather for damage, particularly for settlement that may produce a dangerous edge drop and affect the lateral support of the pavement. Shoulders that settle or sink should be checked for drainage problems.

The shoulders on the inside of roadway curves should be better stabilized than other shoulder areas because of vehicles cutting the curve. At least the first few feet of earth, sod, and aggregate shoulders are sometimes paved on the inside of roadway curves to prevent the formation of dangerous pavement edge drops.

Routine maintenance for shoulders should be part of a regular schedule. The applied maintenance measures and their frequency is a function of traffic volumes, traffic loads, type and width of the travel surface, type of shoulder, and time of year.

The major types of shoulders, their problems, and their maintenance needs are described below.

Earth Shoulders

These shoulders require extensive maintenance because they do not have any great amount of turf. Every effort should be made to establish turf and to protect any turf present, in order to prevent erosion. Where turf is establishing itself, ruts, holes, and settlements should be corrected by adding soil rather than by blading. If erosion, rutting, and settling are extensive, the shoulder should be reshaped by blading at a slope of one-half inch per foot down from the pavement. Precaution is necessary during blading to keep dirt off the travel lanes.

Table 12.3 Shoulder Types

Earth Shoulder	Turf has not been established. Requires extensive maintenance. To prevent erosion, turf is a necessary improvement.
Sod Shoulder	Solid turf has been established. Build up of turf with time can develop drainage problems and must be cut down.
Aggregate Surfaced Shoulder	Thickness of gravel will affect amount of maintenance. Where surfacing is light, turf will provide additional stability. Where surfacing is thicker, aggregate should be bladed periodically to keep shoulder flush with pavement.
Stabilized Aggregate Shoulder	Aggregate shoulder with stabilizing agents mixed in.
Asphalt Surface Treated Shoulder	Seal coat or prime coat placed on prepared flexible base. A wedge of asphaltic concrete may be placed at the pavement edge.
Asphalt Paved Shoulder	Asphalt mat on a prepared base. May be extension of the roadway surface paving.
Concrete Paved Shoulder	Used in urban areas where conditions warrant this type.

Sod Shoulders

These are earth shoulders on which a solid turf has been established. Sod shoulders normally do not require much maintenance, particularly if they have enough granular material for stability. Holes, ruts, and settlements are usually repaired by hand labor with shovels using sod or stabilized material. Light blading at frequent intervals is necessary to keep the turf from building up to where it will retain water on the travel lanes.

Aggregate Surfaced Shoulders

These shoulders vary in thickness, which affects the amount of maintenance needed. Where the surfacing is light, turf will establish to some extent and should be encouraged to provide additional stability and to prevent erosion. On these shoulders, blading should be light and spot repairs should be made by adding aggregate. Where the surfacing is thicker, periodic blading is needed to keep the surface level and flush with the pavement edge. Blading should be done after a soaking rain to prevent segregation of the larger aggregate. The shoulders should be rolled after the blading.

Surface Treated Shoulders

These shoulders have a seal coat or prime coat placed on a prepared flexible base. They require frequent attention and should be spot sealed, spot patched, and base failures repaired with methods similar to roadway surface repairs. For more extreme deterioration of the surface, continuous seal coats may be needed.

Paved Shoulders

Asphalt and concrete shoulders should receive the same maintenance that is applied to roadway surfaces.

Maintenance of Pavement Edge Drops

This distress is a continual maintenance problem or earth, sod, and aggregate shoulders, particularly where travel surfaces are less than 24 feet in width. They are caused by the sweeping and vacuum action of large vehicles, by vehicles running partially or totally off the roadway surface, and by erosion. In addition, settlement of these shoulders is caused by the difference in stability between the roadway pavement and the shoulder.

Edge ruts can be helped by filling the rut with stabilized material and compacting it thoroughly. Repeated applications of an asphalt and aggregate mix has proved effective in providing extra stability when repairing edge ruts

12.5 Roadside Drainage and Vegetation Maintenance

Roadside maintenance includes those activities necessary to keep the roadside clean, attractive, and safe, and to facilitate the drainage of runoff water away from the roadway structure. For purposes of this discussion the roadside is considered as that area between the outside edges of the shoulders and the right-of-way lines.

Several aspects of the roadside can effect either the frequency or severity of vehicular collisions. Culverts, ditches, and inlets, which are clogged, can redirect water across the pavement creating a hydroplaning (see Chapter 6) potential. These drainage problems can also cause the pavement and shoulder distresses discussed in the previous sections of this Chapter. Uncut vegetation can hide important traffic control devices and can severely restrict both stopping sight distance (see Chapter 4) and intersection sight distance (see Chapter 5). Trees within the clear zone (see Chapter 2) that are allowed to grow beyond a four-inch diameter are a serious threat to errant vehicles. The following discussion briefly covers some of the maintenance problems and practices associated with these aspects of the roadside.

Drainage Facilities

Water is the enemy of roadway maintenance. If water is not removed quickly under strict control, damage to the pavement base and subgrade and eventually the pavement can occur. Then too, drainage facilities that cannot handle normal runoff will spill water across the pavement with attendant hazards to traffic.

Roadway drainage systems include ditches, culverts, underdrains, storm sewers, gutters, and inlets. A consistent program of inspection and maintenance is necessary for roadway drainage system to operate efficiently. Proper maintenance of drainage facilities includes keeping ditches free of accumulations of debris and vegetation, keeping gutters, inlets, culverts, and storm sewers free of silt, sand, and debris, and repairing settlements and breaks in drainage structures.

Vegetation Control

The driver needs to see conflicting vehicles, intersections, roadway curves, and other potential hazards far enough ahead of time to avoid collision. The driver also needs to see important traffic control devices in time to properly respond. During the growing season, grass, weeds, brush, and trees often grow to where they restrict a driver's view to less than the desired sight distance. Maintenance crews need to be trained both to be sensitive to sight distance restrictions and to know how much sight distance is necessary for various speeds on various parts of the roadway system. Without this training, vegetation control for sight distance will be minimal in most jurisdictions.

Roadside Hazard Control

Chapter 2 provides a complete discussion of roadside safety considerations. The most important concept is of the forgiving roadside, one that has gentle slopes and is free of large rigid objects. Although proper roadway design is important in providing forgiving roadsides, roadway maintenance is important to keeping the roadside safe. Repair of slope erosion, turf maintenance, mowing, removal of large loose rocks and other debris, and the removal of large trees within the clear zone are all important maintenance functions. In addition, vegetation control should include removing small volunteer trees that can grow to be large trees.

12.6 Traffic Barrier Maintenance

Traffic barriers are safety devices used to compensate for an otherwise unsafe roadside. They include guardrails, bridge rails, median barriers, and crush cushions. Each of these devices supplies its own function through proper design, placement, installation, and maintenance. If elements of a barrier are improperly installed, missing, or damaged, the barrier will probably not function to minimize the occupant injury severity for colliding vehicles.

Because barriers are safety devices, any unsafe aspect or element that could contribute to a less than acceptable impact should be considered for upgrading or repair as a normal maintenance activity. A brief summary of traffic barrier deficiencies and maintain practices is given below. Chapter 3 gives a complete description of traffic barrier systems and their application.

Guardrails

Guardrails are installed at steep embankments, at ends of bridge rails, around bridge supports, and at large sign supports. Guardrail should be routinely inspected for lack or loss of functional integrity. Table 12.4 gives a checklist for guardrail maintenance.

Table 12.4 Guardrail Maintenance Checklist

— Is the guardrail actually needed here? If not, remove it!
— Is the guardrail a current design? Check standard drawings and upgrade obsolete installations where possible.
— Is the guardrail long enough to effectively shield the hazard? Consider extending the guardrail!
— Should the guardrail be joined to another nearby guardrail to close up short gaps (less than 200 feet)?
— Is the guardrail located on a too steep slope or in a ditch? Can regrading correct the deficiency?
— Is the guardrail behind a curb, dike, sidewalk, or drop inlet that would cause vehicle ramping? If so, can the guardrail be repositioned?
— Is snow packed in front of the guardrail that could cause vehicle ramping? Is additional plowing necessary?
— Is there enough clearance between the barrier and the hazard for the expected deflection? If not, can the guardrail be stiffened?
— Is there sufficient soil behind the posts to prevent them from being pushed out on impact? If not, can the slope be modified?
— Is the guardrail height correct? If not, can the posts be jacked up or driven in?
— Are splices lapped in the right direction? If not, reattach correctly!
— Are all rail/post connections installed and tight? If not, correct!
— Are the rails, cables, or posts damaged or corroded? If so, replace deficient elements!
— Is there vegetation, rough ground, or debris in front of the barrier? Correct by cleaning up, regrading, or applying weed killer.
— Is the end treatment a current and acceptable crashworthy design? If not, consider replacement!
— Is the end treatment hardware correct?
— For a 3-strand cable guardrail, is it properly anchored and tensioned?
— Is the end treatment on a 10:1 or flatter slope? If not, consider modifications!
— Are transitions to rigid barriers and bridge rails adequately anchored and stiffened? If not, consider modifications!
— Have all damaged guardrail sections been properly repaired?

Bridge Rails

Most bridge rails designed before 1964 have rails either with insufficient strength and/or with an open-faced design that will cause vehicle snagging or impalement. These rails are also subject to collision damage or corrosion that makes them completely dysfunctional. Table 12.5 gives a checklist for bridge rail maintenance.

Table 12.5 Bridge Rail Maintenance Checklist

— Is the bridge rail an acceptable design? If not, can it be retrofitted with W-beam or Thrie-beam rails to stiffen the system and provide a smooth face?
— Is there a curb in front of the bridgerail that would cause vehicle ramping? If so, can the curb be removed or can a secondary rail be positioned flush with the curb face?
— Is snow packed in front of the bridge rail that could cause vehicle ramping? Is additional plowing necessary?
— Are the ends of the bridge rail shielded with guardrails? If not, consider adding guardrails!
— Are transitions between approach guardrails and bridge rails properly anchored and stiffened? If not, consider modifications!
— Have all damaged bridge rail sections been properly repaired?

Median Barriers

Median barriers are of flexible (cable), semi-rigid (post and beam), or rigid (concrete) design. Flexible barriers are for wide medians. Semi-rigid barriers are for moderate width medians. And rigid barriers are for narrow medians.

Concrete median barriers require very little maintenance. Collision damage usually amounts to minor spalling that can be repaired with epoxy cement grout. Pre-cast concrete barriers, however, are sometimes subject to displacement. A crane may be necessary to realign section where large displacements occur. Semi-rigid and flexible median barriers do need maintenance to ensure their functional integrity, particularly after a collision. Table 12.6 gives a checklist for maintenance of these median barriers.

Crash Cushions

These devices are used to reduce the severity of a collision with a fixed object that cannot be removed or adequately protected by other kinds of barriers. Crush cushions are usually placed in off-ramp gores, at bridge supports, and at large sign supports.

Most maintenance of crash cushions involves removing debris from collisions, replacing damaged elements, and reconfiguring the system. Because crash cushions elements are usually used up after a collision, crash cushions should normally be restored to their original design as quickly as possible. Table 12.7 is a checklist of additional maintenance considerations for crash cushions.

Roadway Maintenance

Table 12.6 Median Barrier Maintenance Checklist

— Is the median barrier an acceptable design? Upgrade obsolete installations where possible.
— Is the median barrier in a median between separate alignments? Does the barrier need to be staggered to function separately for each direction?
— Is the median barrier behind a curb, dike, or drop inlet that would cause vehicle ramping? If so can the curb, dike, or drop inlet be removed or repositioned?
— Is snow packed in front of the median barrier that could cause vehicle ramping? Is additional plowing necessary?
— Are splices lapped in the right direction? If not, reattach correctly!
— Are all splice bolts installed? If not, add necessary bolts!
— Are all rail/post connections installed and tight? If not, correct!
— Are the rail, cable, or posts damaged or corroded? If so, replace deficient elements!
— Will the median barrier deflect too much in a narrow median?
— Is the end treatment a current and acceptable crashworthy design? If not, consider replacement!
— For a 3-Strand cable barrier, is it properly anchored and tensioned?
— Are transitions proper between rigid and semi-rigid barriers? If not, consider modifications?
— Have all damaged median barrier sections been properly repaired?

Table 12.7 Crash Cushion Maintenance Checklist

Sand Barrel Crash Cushions

— Are modules tipped, split, or losing sand?
— Are any modules damaged and needing replacement?
— Are any lids missing?
— Is there debris or snow packed around the modules that could cause ramping?
— Is the sand at the proper level in each module?
— Are the modules placed in the proper sequence of size?
— Have the modules walked out of their original positions because of road vibrations?
— Is the sand caked or hardened? The sand needs to be loose to function on impact.

Compression Crash Cushions

— Are cartridge and diaphragm covers or lids properly shut?
— Are fender panels closed?
— Are fender panels level and not sagging?
— Are HI-DRO cell diaphragms vertical?
— Are all nuts and bolts tight?
— Are shear pins in place?
— Are tension cables tight?
— Are pull-out cable clamps tight?
— Is all cartridge or diaphragm hardware workable?
— Is there debris or snow packed around the barrier that could ramp a vehicle?

12.7 Traffic Control Device Maintenance

Maintenance management has been influenced by the recognition of traffic engineering as a special discipline. In many roadway agencies, some aspects of traffic engineering have been joined with maintenance activities in an Operations Department. In other agencies, these functions are separated with a Maintenance Department performing the road work and the Traffic department supplying technical guidance. A third type of agency structure puts the maintenance function under the Traffic Department.

Traffic control devices include roadway signs, pavement markings, delineators, object markers, traffic islands, traffic signals, and roadway lighting. The design, placement, installation, and maintenance of traffic control devices in the U.S. in specified in the *Manual on Uniform Traffic Control Devices*.[4] Many of the maintenance aspects of traffic control devices are discussed in the *Traffic Control Devices Handbook*.[5] The reader is referred to Chapter 11 of this book for a complete discussion of traffic control devices.

The safe operation of the roadway system relies heavily on sign, signals, and markings. To ensure continuing dependable operation and performance of these devices, good maintenance practices are required. Repairing damaged devices, cleaning dirty devices, and replacing malfunctioning or deteriorated devices should be regularly scheduled maintenance activities. The following discussion summarizes the maintenance aspects and activities for various devices.

Traffic Signs

These devices should be kept in proper position, clean, and legible. Damaged signs should be repaired or replaced as soon as possible. Vegetation, construction materials, grime, mud, and snow should not be allowed to obscure a sign face. Illegal signs, non-standard signs, or signs that are no longer applicable should be removed.

Signs should be inspected at least twice a year by trained personnel looking for damage, poor legibility, and other problems. Night inspections of sign reflectivity should also be done at least twice a year. Periodic cleaning of reflective materials may be necessary to maintain nighttime legibility. Signs obscured by dust, fungus, pollen, road film, spray paint, or cosmetics should be cleaned as needed. Washing should be done with a detergent mix and a soft bristle brush.

Sign replacement is required when legibility or reflectivity has been degraded by age or damage. Severely damaged or missing signs should be replaced promptly. Reflectorized signs should be replaced when a night inspection indicates that the reflectivity is no longer effective.

Some signs have breakaway supports (see Chapter 2). To function properly, these supports are fastened to their slip bases by bolts and nuts that are torqued to precise levels. This torque should be checked periodically for conformance with specifications.

Traffic Markings

These devices include pavement markings, curb markings, object markers, and delineators. Their function is to convey regulations, warnings, and guidance to the driver without diverting his attention from the roadway. Maintenance of markings includes the following activities:

1. *Cleaning* - to remove grime that impairs visibility.
2. *Repainting or Re-applying* - to improve visibility caused by wear or damage.
3. *Repairing or Replacing* - to restore damaged markings to their original configuration.
4. *Marking Removal* - to avoid the confusion of obsolete markings. The removal method should not leave a trace line or message on the pavement.

Pavement markings are usually applied by machine methods. Paint should not be applied to pavements that are dirty, damp, or cold (less than 40°). Unless rapid drying paint is used, newly painted markings should be protected with traffic cones until the paint is dry.

Pavement markings that become obsolete because of the rechanneling of traffic can either falsely direct drivers or confuse the meaning of the replacement markings. These obsolete markings should be removed immediately using water jets, sand blasting, grinding, chemicals, or heat.

Traffic Signals

Because of potentially critical consequences, maintenance personnel should respond to signal failures as quickly as possible. Traffic signals should not be left dark when inoperable, but should be either set on flashing operation or hooded.

Periodic inspections to check traffic signal operations and timing should be conducted about every three months. Lamps in traffic signals should be replaced by a group replacement program with the period calculated considering the rated life and operating frequency. Group replacement at 60% -75% of rated lamp life is recommended in the *AASHTO Maintenance Manual*.[1]

Roadway Lighting

Illumination is provided to improve nighttime visibility and to promote safer and more efficient use of roadway ramps, intersections, and other potentially hazardous areas. Lighting failure should be given special attention whenever critical areas are affected. A preventive maintenance schedule should be established to relamp each luminaire on a regular basis. Typically, fluorescent and halide lamps are replaced every two years. Sodium and mercury vapor luminaires should be relamped every five years. Luminaires, such as in tunnels, that are exposed to traffic splash and road grime should be cleaned as needed.

12.8 Snow and Ice Control

The impact of snow and ice storms has always been a concern of travelers. Before the advent of motor vehicles, travel during and after snow and ice storms was usually very limited. Eventually with motor vehicles and all weather roadways widely available, government agencies responded to the public needs with improved ice and snow control methods. In the U.S. today, ice and snow only effects the safety and mobility of roadway transportation for short periods during the winter. The handling of snow and ice before, during, and right after a storm is not only a great concern for safety but also is a frequent subject of tort litigation.

Our modern economy depends heavily on all-weather roadways. When roadways are closed or only partly functioning due to snow or ice, commerce and public safety are adversely affected. Collisions that result from slippery pavements or other dangerous conditions created by winter maintenance activities claim an expensive toll of life, injury, and property damage. Therefore, the goal of any effective snow and ice control program should be to keep the roadway system open and safe as much as possible within the constraints of the weather and the available winter maintenance manpower and equipment.

A significant amount of manpower and money is spent by State, County, and Municipal roadway agencies on snow and ice control. These agencies spend over one billion dollars annually on these activities.[6] Resources allocated to snow and ice control will vary from agency to agency depending on the annual snowfall, the topography, the size and population of the jurisdiction, and the sensitivity of local drivers to snow and ice.

Winter Maintenance Policies and Plans

Snow and ice removal doesn't just happen. A plan must be developed and deployed to ensure safe roadways. Like a successful football team, roadway agencies must devise, and try to comply with a game plan. Regardless of size, all agencies should develop a plan that predetermines basic policy. The basic plan of attack should be finalized before the first snowflake falls. It should set a chain of command, the relative priority of different roadways, and the trade-off between level of service and cost. Snow and ice control plans should be based on

Table 12.8 Typical Winter Maintenance Priority Scheme

Classification	Policy	Level of Service	Priorities
Class A Routes Interstates, freeways, expressways, and other roadways over 2,500 vehicles per day.	To maintain as near normal surface conditions on the traveled way as capabilities and weather conditions allow. Ramps and interchanges will be treated the same as the traveled way. Chemicals, abrasives and/or chemical-abrasive mixtures may be used, as required, to prevent ice and snow pack conditions.	Maintain 24-Hr. coverage until near normal surface conditions are restored. Coverage will be performed at a rate of 2 to 4 cycles/shift, as conditions and resources allow. Other pavement surfaces and shoulder clean-up operations will usually be performed during regular working hours.	1. Aid for emergency vehicles responding to calls (if requested) 2. Interstate 3. Urban commuter 4. Rural commuter 5. Freeways and expressways 6. Higher volume roadways, or as storm conditions dictate
Class B Routes 750 to 2500 vehicles per day.	To maintain a near normal or at least a snow-packed surface for two-way traffic as conditions allow. During the storm, a mixture of chemicals and abrasives may be used as required. After the storm and with proper weather conditions, the use of chemicals, abrasives, and/or chemical-abrasive mixtures are authorized to aid removal of the existing ice and snow pack.	Maintain at least 16-Hr. coverage or until near normal surface conditions are restored or the ice and snow-pack condition has been treated. Coverage will be performed at a rate of 2 to 4 cycles/shift, as conditions and resources allow. Plow and continue treating ice and snow-pack conditions during regular working hours as required. All clean-up operations will usually be performed during regular working hours.	1. Aid for emergency vehicles responding to calls (if requested) 2. Known problem areas (hills, curves, bridges, stop signs, etc.) 3. Higher volume roadways, or as storm conditions dictate
Class C Routes Under 750 vehicles per day.	To maintain a serviceable traveled way for one- or two-way traffic, as conditions allow. Chemicals, abrasives and/or chemical-abrasive mixtures may be used, as needed, at specific locations.	Maintain at least 12-Hr. coverage or until at least a one-way traveled way is established. All clean-up operations will usually be performed during regular working hours.	1. Aid for emergency vehicles responding to calls (if requested) 2. Known problem areas (hills, curves, bridges, stop signs, etc.) 3. Higher volume roadways, or as storm conditions dictate

past experiences and updated annually. The intent of a snow removal and ice control policy should be to remove snow and provide skid protection as rapidly as possible with available manpower and equipment. Operations should begin as soon as weather conditions demand and continue until all primary routes are open to two-way traffic and/or all critical skid locations are treated.

Although the highest level of service on all roadway is a desirable goal, the budget of most roadway agencies will not allow this. Therefore, most agencies develop a priority scheme for allocating resources. Four criteria typically used are roadway class, traffic volume, special hazardous locations, and the level of service desired. Table 12.8 shows a typical priority scheme employed by a State roadway agency. Note the emphasis on higher traffic volumes, aid for emergency vehicles, and attending to known problem areas such as bridges, hills, curves, and stop signs.

Storm Watch

Weather information is essential to give advance notice for snow and ice control operations. Many agencies contract with a meteorological service to provide detailed weather forecasts 24 hours a day, seven day a week, throughout the season. Reports should forecast the expected conditions, whether it be frost, snow, ice, sleet, or freezing rain. They also should provide information on: how much precipitation to expect and how long it will last; temperatures before, during, and after the storm; wind velocity and direction; and anticipated conditions 24 to 48 hours after the storm.

The objective of a storm watch is to minimize the time lapse between the recognition of need and the initiation of snow and ice control operations. This objective is met by readying enough personnel, equipment, and supplies for prompt action as the weather conditions arrive. In a maintenance operation as critical and temporal as snow and ice control, the work cannot be left to a normal daytime work force. Some degree of shifting or a great amount of overtime may be needed. Also, special attention should be given to scheduling personnel to avoid having an employee assigned to two consecutive 8-hour shifts.

Training

No effective snow and ice control plan can be implemented without adequate equipment and materials. But the best facilities will be useless without trained manpower to supervise and operate the equipment. Emphasis should be directed to training equipment operators. New operators should be tested for their manual abilities and understanding of maps, procedures, and communications equipment.

Basic truck driving skills should be mastered before any attempt is made to teach the extra skills needed for snow plowing. These extra skills include how to operate the truck with a 9- or 10-foot blade extended along the surface in an area not visible to the operator. This difficulty is compounded by the poor visibility created by both a snow storm and the snow blowing off of the plow.

General Maintenance Procedures

The plan of attack for any winter storm depends on (1) the type and intensity of precipitation, (2) the air temperature and whether it is rising or falling, and (3) the temperature and conditions of the roadway surface. Example procedures for the four common combinations of conditions are given below. For those routes designated, continuous treatment should be applied throughout the storm. All other routes should be treated at danger spots only. The example procedures are as follows:

- *Rising Temperatures of 25°F or Above, with Snowfall, Sleet, or Freezing Rain, and a Wet Pavement* – The immediate treatment is to apply salt. If snowfall or sleet continues to accumulate to one-half inch or more, begin plowing and repeat salting if necessary. If rain continues to freeze, reapply salt. When the storm stops, continue operations until the pavement is clear. A mixture of sand and salt may be substituted for the straight salt.
- *Falling Temperatures of 32°F or Below, with Dry Snowfall, and a Dry Pavement* - The immediate treatment is to apply a mixture of three parts salt to one part calcium chloride. If snowfall or sleet continues, plow and repeat the chemical mixture. If freezing rain continues, reapply the chemical mixture. When the storm stops, continue operations until the pavement is clear.
- *Temperatures of 10°F or Below, with Snowfall, Sleet, or Freezing Rain, and an Accumulation of Packed Snow or Ice* - The immediate treatment is to plow snow or sleet and apply abrasives. For ice, only abrasives are applied. For subsequent treatments, apply a mixture of sand and chemicals.

Snow Plowing

This operation should begin after an accumulation of one-half inch or more and/or before snow becomes packed. Packed snow can be hazardous and is difficult to remove by plowing.

Many types of equipment are available for snow plowing. The equipment chosen should fit the job to be done under most circumstances. A plow blade should be long enough to plow the average lane width, yet, it cannot be too large for the pushing vehicle. The following kinds of plows are available:

1. *One-way Plow* - These plows can be ordered to throw snow either to the right or to the left. The right-hand one-way plow is the most common. The left-hand one-way plow is useful only on expressways and one-way streets. The moldboards of these plows are usually curved with the top forward.
2. *Reversible Plow* – These plows are designed to plow either right or left. They are usually power reversible. The moldboard is curved and usually of a more vertical configuration than a one-way plow.
3. *V Plow* – These plows push snow both directions from the center of the pushing vehicle. The advantage of these plows is for deep snow because the maximum forward push can be obtained from the vehicle.
4. *Underbody Plows* – These plows are mounted under the truck. While the moldboard isn't very tall, its location enables the operator to apply downward hydraulic pressure on the blade. This makes the plow superior for removing ice or packed snow. This plow, however, is not well-suited for deep snow.
5. *Leveling Wing* – This added plow is mounted on the side of the plowing vehicle. This wing is used in combination with the front plow to increase the plowed width by three to five feet.
6. *Motor Grader* – The blade under the motor grader has the advantage of the under body snow plow, making it very effective for removing ice or packed snow, particularly when coupled with a serrated blade. The extra traction and steering ability of a motor grader makes it particularly effective.
7. *Rotary Plows* – These plows are mounted on trucks or front-end loaders. They couple multiple augurs to move snow to the center of the unit where a large fan blows the snow great distances through a chute directed toward open areas.

Plowing is continued throughout a storm as needed to keep the accumulation less than the service level. Heavy storms will require occasional passes on the shoulders or parking lanes to provide enough room to store the snow. Operations are repeated until the storm stops, and the roadway surfaces are clear.

Normally, expressway plow trucks operate in pairs with one truck discharging left and one right. Reduced speeds should be used when plowing overpasses to prevent the discharged snow from falling onto the roadway below.

On major streets, tandem plowing has several advantages. The lead operator only has to set the left edge of his plow on the centerline, while the second operator watches out for parked cars. With this procedure, there is usually no need to return later to widen the plowed area.

Every advantage should be taken of the temperature rise, which usually occurs between 11:00 a.m. and 3:00 p.m., to clear surfaces of melting snow or slush. Trucks should be operated at moderate speeds when removing snow, particularly when slush exists, when plowing shoulders, and when meeting traffic. Plowing slush at high speeds deposits the slush on signs and on other vehicles. Plowing snow at high speeds causes excessive snow clouds making the roadway unsafe for nearby motorists.

Chemical and Abrasive Application

The removal of ice or packed snow by plowing is practically impossible under certain temperature conditions where the only effective treatments are either removal using chemicals or skid control using abrasives. Localized spots of ice or packed snow are particularly hazardous and should be treated. Caution is necessary in using chemicals during periods of sharp temperature drops to prevent refreezing of the brine.

Spreading chemicals and abrasives the full width of the roadway is recommended for urban and congested areas and danger spots where fast melting or full skid control is essential for the entire area. On rural two-lane roadways, chemicals and abrasives may be spread on the middle third of the roadway. The concentrated material quickly melts the snow or ice over the middle section and works out to each side with very little waste. When abrasives are spread this way, some skid control is provided in both lanes immediately.

ROADWAY MAINTENANCE DEFECTS

Rule of Thumb

Regular Maintenance Activities are Important to Prevent the Sequential Degradation of the Functional Integrity of Many Roadway Elements that Affect the Safety of the Motoring Public.

12.9 Technical Aspects of Maintenance Defect Cases

Maintenance defects are leading topics of roadway tort claims involving motor-vehicle collisions. Potholes, washboard surfaces, windrows of gravel, pavement edge drops, rutted pavements, missing traffic barrier hardware, faded signs, and faulty ice and snow control are a few of these alleged defects. Clearly the most importantly technical aspect of these tort claims relates to whether the roadway agency adhered to some reasonable bases of maintenance surveillance, scheduling, and application.

Roadway agencies have a clear duty to perform the roadway maintenance needed to keep roadways safe. Roadway maintenance is not considered a discretionary function, but a ministerial function. The only discretion applies when technical uncertainty exists about the degree of hazard associate with the degree of defect. For example, a 3-inch vertical pavement edge drop is clearly a hazardous roadway defect, but a 1.5- inch rounded pavement edge drop may be questionable. For the higher edge drop, roadway maintenance should have been done well before it became so severe. For the lower edge drop, maintenance is indicated but not so urgently.

In considering the negligence of a roadway agency for various maintenance defects, the descriptions of eye witnesses at the time of collision and the roadway agency's recorded dates of the last inspection or last repair may be important. If eye witnesses indicate that a defect was present, but roadway agency records indicate the condition was not defective a few days earlier, there may be no cause of action. If records indicate that the condition had been repaired shortly before the subject collision, a cause of action will be tenuous. On the other hand, if eye witnesses indicate that a serious defect was present, and that it had existed for some time, a roadway agency can often be found negligent.

With the exceptions of surface irregularities and the winter maintenance problems discussed below, most other maintenance defects have already been discussed in another chapter of this book as follows:

Chapter 2 – Roadside Safety
Chapter 3 – Traffic Barriers
Chapters 4 and 5 – Sight Distances
Chapter 6 – Slippery Pavements and Hydroplaning
Chapter 7 – Pavement Edge Drops
Chapter 11 – Traffic Control Devices

Other kinds of maintenance defects are discussed below.

Washboard, Wheel Ruts, and Holes on Unpaved Roadways

This group of maintenance defects is probably the most common topic of tort claims on unpaved roadways. Unattended, these defects worsen to where they can cause loss of vehicular control, particularly on roadway curves. Typical issues are how long the condition had existed and whether citizen complaints had been received. These defects are also particularly hazardous when they are isolated, hidden over the crest of a hill, or otherwise unexpected by the driver.

Windrows of Gravel on Unpaved Roadways

Windrows of gravel left on narrow roadways create hazards for opposing traffic. Windrows can be a problem almost anywhere on a narrow roadway. They are most hazardous at sharp roadway curves and sharp hillcrests. When windrows are left on a narrow roadway curve, drivers can very easily get caught up in the gravel and lose control. When windrows narrow the roadway to less than 18 feet at a sharp hillcrest with very limited sight distance, drivers on a head-on collision course may only be able to see each other for 2-3 seconds or less. Under these circumstances, drivers have almost no time for collision avoidance and may collide because neither driver was able to do any more than brake just before collision. If drivers attempt to steer right, they will plow through the windrow and usually either lose control or be redirected into the head-on collision.

Dust on Unpaved Roadways

Dust is a problem on unpaved roadways during hot dry periods and particularly after the roadway has been graded. Unless dust palliatives have been applied, dust can sometime completely restrict a driver's visibility. The typical tort claim involving a motor-vehicle collision is where one vehicle is following another vehicle, which causes a blinding dust cloud. Usually, the following vehicle will collide head-on with an oncoming vehicle, with neither driver seeing the other.

Roughness, Potholes, and Bumps on Paved Roadways

Most people are aware of extreme cases of roughness, potholes, or bumps on paved roadways. As stated by Wambold, et. al.[7]:

> *The effect of these irregularities on safety is widely recognized. The hazard of a washboard road where a driver can lose control at high speeds, is readily understood. However, the transition of a roadway from a smooth surface to a rough one, which may give the driver difficulty in controlling his vehicle, can be more subtle in its influence on safety. By violating driver expectancy, a road that is differentially rough may be less safe than a uniformly rough road.*

With regard to potholes, Baker[8] offers the following observations:

> *Holes in the pavement have to be a foot or more long and wider than a tire to be hazardous. If a driver claims his vehicle was thrown out of control by a small hole, treat this statement with suspicion and look for driver actions which may be contributing factors, such as cutting back into lane after over-taking...A vehicle can be turned over by hitting a chuck hole without signs on either the tire or the hole, especially when the edges of the hole are rounded.*

In 1977, Klein, et. al.,[9] reported on the answers on about 400 returned questionnaires in which potholes ranked third out of 13 identified disturbances in terms of a driver's perception of hazard. Potholes are clearly perceived to be a significant threat to public safety. For this reason, they are often the topic of a tort claim for injuries in a motor-vehicle collision.

In an attempt to verify the correctness of this public opinion, Zimmer and Ivey[3] conducted a series of controlled impact tests with potholes. Their conclusions were:

> *It is apparent that a hole must be relatively large to constitute a significant safety influence when rim or tire damage are the guiding criteria. At common highway speeds, in excess of 40 mph, a hole must be in excess of 60 inches long and three inches deep to constitute a threat to the smallest automobiles. On urban streets, with traffic speeds as low as 20 mph, holes must still be over 30 inches in length and over three inches deep to have the potential of damaging tires and/or rims.*
>
> *Damage to tires and rims, with the associated potential for an air-out, is the only significant safety related influence of holes identified in this study...*
>
> *Problems can arise if a driver reacts to the hole inappropriately. For example, it is counter-productive to react with braking or extreme cornering to a hole in the vehicle's path...Losses of control can occur if extreme braking is produced at highway speeds. Extreme cornering can have two results. First, if a driver reacts with a large steering input to avoid a hole he may produce a loss of control on a low friction surface. Second, he may put his vehicle in a hazardous position with respect to other traffic...*
>
> *The influence of holes encountered when cornering deserves further attention. A cornering (turning) vehicle transfers weight from the wheels on the inside of the turn to the outside wheels. The springs on the heavily loaded side are compressed. When one of these tires encounters a hole, it goes down faster due to the acceleration of the higher spring force. Thus, it gets in*

position to be damaged somewhat more quickly...A second and potentially more hazardous situation is if a tire is moving laterally and encounters the side of a hole.

Although the Zimmer and Ivey[3] study generally discounts potholes as a roadway hazard, it disagrees with the accounts of drivers who have had accidents where they believe a pothole was the reason for their loss of control.

Because of conflicting conclusions in the literature, the effect of potholes on loss of control in most collisions is difficult to prove or disprove based on a purely technical argument, with two exceptions. First, most technical literature agrees that potholes or almost any kind of pavement irregularity can be a major factor in loss of control for motorcycles. And, second, most technical literature agrees that potholes or almost any kind of pavement irregularity can be a major factor in the loss of control on roadway curves, particularly for sharp curves at higher speeds.

Gravel on the Pavement

Gravel on an asphalt or concrete surface can seriously degrade the available tire-pavement friction. This condition is particularly hazardous for motorcycles.

Gravel on pavements is particularly common at the following three types of locations:

1. *Narrow Roadway with Gravel Shoulders.* Gravel is kicked onto the pavement when vehicles, particularly wider trucks, run onto gravel shoulders where pavements are less than 22 feet wide. The hazard is exacerbated on roadway curves where not only do vehicles run onto the shoulder more often but also vehicles on the pavement have higher demands for friction. If the pavement cannot regularly be cleaned by maintenance crews, this condition can be minimized by paving 2-3 feet of the shoulder next to the pavement, particularly on the inside of roadway curves.

2. *Intersection of a Paved Roadway with a Gravel Roadway.* For these locations, vehicles entering the intersection from the gravel roadway will kick gravel onto the paved surface creating an isolated hazard for through vehicles traveling on the paved roadway. If the pavement cannot regularly be cleaned by maintenance crews, an asphalt apron should be paved for at least 30 feet along each gravel roadway approach to the intersection.

3. *Change in Roadway Surface From Paved to Unpaved.* For these locations, vehicle traveling from the unpaved surface to the paved surface will kick gravel onto the paved surface creating a isolated hazard for vehicles traveling on the paved surface. This pavement

needs to be regularly cleaned by maintenance crews and also have warning signs indicating this change in surface and perhaps an appropriate advisory speed.

Flooded Pavements

When culverts become plugged or ditches become blocked, rain water during heavy rainstorms will often fill the ditches to where the water starts flowing across the pavement. This condition, of course, creates the potential for hydroplaning. When this condition occurs because of a long-term lack of maintenance, a roadway agency can often be found negligent.

Soft Shoulders

Shoulders that are not designed for stability are a hazard when wet to even the casual encroachments of vehicles. This condition violates the purposes of shoulders in providing (1) a secondary recovery area, (2) a refuge for disabled vehicles, and (3) an area for evasive maneuvers. A vehicle that traverses onto a soft shoulder at higher speeds will encounter most of the same difficulties associated with pavement edge drops (see Chapter 7).

Icy Bridges

Ice and frost at localized spots on otherwise clear traveled ways are universally recognized as creating a particularly hazardous condition for unsuspecting or inattentive drivers. The formation of ice and frost on bridge decks while the rest of the traveled way remains ice- or frost-free is a well-known phenomenon. A 1978 estimate is that 25,000 accidents costing $80 million dollars occur annually in the U.S. due to ice and frost on bridge decks.[10] This phenomenon is treated in various ways and to various degrees depending on roadway agency policy and location. Clearly each agency should recognize the problem and implement a maintenance strategy that as a minimum has:

1. Warning signs at bridges, where necessary, during winter months.
2. A method in place for early detection of frost.
3. A readiness plan for quick response of manpower and equipment.
4. A priority response strategy for placement of chemicals and /or abrasives.

Sight Distance Obstruction Created by Snow Plowing

Not uncommon, where deep snowfalls occur on two-lane roadways, is for a snow plow to only initially plow a 10-12 foot swath. This method creates a head-on collision hazard where stopping sight distance is otherwise deficient, such as at sharp roadway curves and at sharp hillcrests. Every effort should be taken to plow wider paths at these hazardous locations.

Roadway Maintenance

Another common problem created by snow plowing is when deep snow is piled up along the edge of the roadway. Sometimes, this snow can seriously degrade the sight distance at uncontrolled or stop-controlled intersections or on sharp roadway curves. Actually where the sight distance is otherwise minimally acceptable because of an embankment, a small snow pile can sometimes completely cut off the available sight distance. Every effort should be taken to minimize or eliminate snow piles at locations where safe sight distance can be critically diminished.

Snow Plow Cloud

Vehicles running into snow plows or other vehicles are a common winter occurrence, particularly where the snow plow leaves a blinding snow cloud. Every effort should be taken to enhance snow plow visibility through the use of strobe lights and flashers. Also, the snow plow speed should be tempered to account for the hazard associated with snow clouds.

Hydroplaning Sections in the Winter

Those locations that typically have hydroplaning vehicles during rainy weather (see Chapter 6) can also experience hydroplaning during winter weather. The usual occurrence involves alternate melting and freezing of snow and ice. Usually ice and snow on the pavement, shoulder, and/or roadside melts during midday either collecting water in wheel ruts and sag vertical curves with no cross slope or causing sheet flow across the ends of roadway curves. Not only can the collection of water create a hydroplaning hazard, but when this water freezes toward the end of the day, isolated black ice can form, creating an extremely slippery pavement to the unsuspecting driver.

12.10 Typical Defense Arguments

When defending claims that allege negligent maintenance, roadway agencies may have technical arguments both about whether a defect was actually present or about the degree of hazard, but they generally only have one legal defense—*notice*. The burden is usually on the plaintiff to prove that the roadway agency had either actual or constructive notice of the defect. Most courts hold that the roadway agency must have sufficient advance notice of the defect to have a reasonable opportunity to either correct the defect or warn of its hazard. Under the concept of constructive notice, however, actual notice may not be necessary if the condition was a direct result of the agency's own negligence.

Other case-specific arguments are that either the driver was the sole negligent party and/or the roadway condition had no contribution to the causation of the collision. Some of these arguments are:

1. The defect should have been self-evident, and the driver should have controlled his speed accordingly.
2. The driver was negligent for exceeding the speed limit.
3. The driver was negligent for driving while intoxicated.
4. The driver should have expected the maintenance defect because of the generally substandard nature of the roadway.
5. The driver was negligent for oversteering the vehicle.
6. The driver did not exercise proper lookout.
7. The driver did not drive with the highest degree of care.
8. There is no official standard on maintenance practices.
9. Physical evidence is lacking that the maintenance defect caused the collision.
10. The degree of the maintenance defect was not sufficient to warrant a repair.
11. The roadway agency has limited funds, which prevent it from fixing every maintenance defect.
12. The subject section of roadway has been selected for future rehabilitation, therefore, no maintenance is being performed.

Endnotes

1. American Association of State Highway and Transportation Officials, *AASHTO Maintenance Manual*, 1987.
2. American Public Works Association, *Street and Highway Maintenance Manual*, 1985.
3. Zimmer, R.A., and Ivey, D.L., *The Influence of Roadway Surface Holes and the Potential for Vehicle Loss of Control*, Texas Transportation Institute, Research Report 328-2F, 1983.
4. Federal Highway Administration, *Manual on Uniform Traffic Control Devices*, 1988. (available through Lawyers and Judges Publishing Co.)
5. Federal Highway Administration, *Traffic Control Devices Handbook*, 1983.
6. Migletz, J., et. al., *Safety Restoration During Snow Removal – Guidelines*, Federal Highway Administration, 1987.
7. Wambold, James C., Zimmer, Richard A., Ross, Hayes E., Jr., and Ivey, Don L. *Roughness, Holes, and Bumps*, Chapter 3, The Influence of Roadway Surface Discontinuities on Safety, Transportation Research Board, 1984.
8. Baker. J.S., *Traffic Accident Investigation's Manual*, 4th Edition, Traffic Institute, Northwestern University, 1975.
9. Klein, R.H., Johnson, W.A., and Szostak, H.T., *Influence of Roadway Disturbances on Vehicle Handling*, National Highway Traffic Safety Administration, 1977.
10. Blackburn, R.R., Glennon, J.C., Glauz, W.D., and St. John, A.D., *Economic Evaluation of Ice and Frost on Bridge Decks*, Transportation Research Board, NCHRP Report 182, 1978.

References

Federal Highway Administration, *W-Beam Guardrail Repair and Maintenance*, Report FHWA-RT-90-001, 1990.

Federal Highway Administration, *Vegetation Control for Safety*, Report FHWA-RT-90-003, 1990.

Glennon, John C., *Effects of Pavement/Shoulder Drop-Offs on Highway Safety*, Transportation Research Board, 1984.

Federal Highway Administration, *Maintenance and Highway Safety Handbook*, 1977.

Appendix

Parts of Title 23 U.S. Code Related to Highway Safety
April 1, 1995

Section	Subject	Title 23 page number
	Explanation	v.
Part 460	Public Road Mileage for Apportionment of Federal Highway Safety Funds	121
Part 625	Design Standards for Highways	158
Part 645	Utilities	228
Part 646	Railroads	245
Part 650	Bridges, Structures, and Hydraulics	255
Part 652	Pedestrian and Bicycle Accommodations and Projects	268
Part 655	Traffic Operations	271
Part 924	Highway Safety improvement Programs	456
Part 1200	Uniform Procedures for State Highway Safety Programs	460

Explanation

The Code of Federal Regulations is a codification of the general and permanent rules published in the Federal Register by the Executive departments and agencies of the Federal Government. The Code is divided into 50 titles which represent broad areas subject to Federal regulation. Each title is divided into chapters which usually bear the name of the issuing agency. Each chapter is further subdivided into parts covering specific regulatory areas.

Each volume of the Code is revised at least once each calendar year and issued on a quarterly basis approximately as follows:

Title 1 through Title 16..as of January 1
Title 17 through Title 27..as of April 1
Title 28 through Title 41..as of July 1
Title 42 through Title 50..as of October 1

The appropriate revision date is printed on the cover of each volume.

LEGAL STATUS

The contents of the Federal Register are required to be judicially noticed (44 U.S.C. 1507). The Code of Federal Regulations is prima facie evidence of the text of the original documents (44 U.S.C. 1510).

HOW TO USE THE CODE OF FEDERAL REGULATIONS

The Code of Federal Regulations is kept up to date by the individual issues of the Federal Register. These two publications must be used together to determine the latest version of any given rule.

To determine whether a Code volume has been amended since its revision date (in this case, April 1, 1995), consult the "List of CFR Sections Affected (LSA)," which is issued monthly, and the "Cumulative List of Parts Affected," which appears in the Reader Aids section of the daily Federal Register. These two lists will identify the Federal Register page number of the latest amendment of any given rule.

EFFECTIVE AND EXPIRATION DATES

Each volume of the Code contains amendments published in the Federal Register since the last revision of that volume of the Code. Source citations for the regulations are referred to by volume number and page number of the Federal Register and date of publication. Publication dates and effective dates are usually not the same and care must be exercised by the user in determining the actual effective date. In instances where the effective date is beyond the cut-off date for the Code a note has been inserted to reflect the future effective date. In those instances where a regulation published in the Federal Register states a date certain for expiration, an appropriate note will be inserted following the text.

OMB CONTROL NUMBERS

The Paperwork Reduction Act of 1980 (Pub. L. 96-511) requires Federal agencies to display an OMB control number with their information collection request. Many agencies have begun publishing numerous OMB control numbers as amend-

Appendix – Parts of Title 23 U.S. Code 413

ments to existing regulations in the CFR. These OMB numbers are placed as close as possible to the applicable recordkeeping or reporting requirements.

OBSOLETE PROVISIONS

Provisions that become obsolete before the revision date stated on the cover of each volume are not carried. Code users may find the text of provisions in effect on a given date in the past by using the appropriate numerical list of sections affected. For the period before January 1, 1986, consult either the List of CFR Sections Affected, 1949–1963, 1964–1972, or 1973–1985, published in seven separate volumes. For the period beginning January 1, 1986, a "List of CFR Sections Affected" is published at the end of each CFR volume.

INCORPORATION BY REFERENCE

What is incorporation by reference? Incorporation by reference was established by statute and allows Federal agencies to meet the requirement to publish regulations in the Federal Register by referring to materials already published elsewhere. For an incorporation to be valid, the Director of the Federal Register must approve it. The legal effect of incorporation by reference is that the material is treated as if it were published in full in the Federal Register (5 U.S.C. 552(a)). This material, like any other properly issued regulation, has the force of law.

What is a proper incorporation by reference? The Director of the Federal Register will approve an incorporation by reference only when the requirements of 1 CFR part 51 are met. Some of the elements on which approval is based are:

(a) The incorporation will substantially reduce the volume of material published in the Federal Register.

(b) The matter incorporated is in fact available to the extent necessary to afford fairness and uniformity in the administrative process.

(c) The incorporating document is drafted and submitted for publication in accordance with 1 CFR part 51.

Properly approved incorporations by reference in this volume are listed in the Finding Aids at the end of this volume.

What if the material incorporated by reference cannot be found? If you have any problem locating or obtaining a copy of material listed in the Finding Aids of this volume as an approved incorporation by reference, please contact the agency that issued the regulation containing that incorporation. If, after contacting the agency, you find the material is not available, please notify the Director of the Federal Register, National Archives and Records Administration, Washington DC 20408, or call (202) 523–4534.

CFR INDEXES AND TABULAR GUIDES

A subject index to the Code of Federal Regulations is contained in a separate volume, revised annually as of January 1, entitled CFR INDEX AND FINDING AIDS. This volume contains the Parallel Table of Statutory Authorities and Agency Rules (Table I), and Acts Requiring Publication in the Federal Register (Table II). A list of CFR titles, chapters, and parts and an alphabetical list of agencies publishing in the CFR are also included in this volume.

An index to the text of "Title 3—The President" is carried within that volume.

The Federal Register Index is issued monthly in cumulative form. This index is based on a consolidation of the "Contents" entries in the daily Federal Register.

A List of CFR Sections Affected (LSA) is published monthly, keyed to the revision dates of the 50 CFR titles.

REPUBLICATION OF MATERIAL

There are no restrictions on the republication of material appearing in the Code of Federal Regulations.

INQUIRIES AND SALES

For a summary, legal interpretation, or other explanation of any regulation in this volume, contact the issuing agency. Inquiries concerning editing procedures and reference assistance with respect to the Code of Federal Regulations may be addressed to the Director, Office of the Federal Register, National Archives and Records Administration, Washington, DC 20408 (telephone 202–523–3517). All mail order sales are handled exclusively by the Superintendent of Documents, Attn: New Orders, P.O. Box 371954, Pittsburgh, PA 15250–7954. Charge orders may be telephoned to the Government Printing Office order desk at 202–512–1800.

MARTHA L. GIRARD,
Director,
Office of the Federal Register.

April 1, 1995.

Federal Highway Administration, DOT

far such strategies can go in eliminating the need for additional SOV capacity in the corridor. If the analysis demonstrates that additional SOV capacity is warranted, then all reasonable strategies to manage the facility effectively (or to facilitate its management in the future) shall be incorporated into the proposed facility. Other travel demand reduction and operational management strategies appropriate for the corridor, but not appropriate for incorporation into the SOV facility itself must be committed to by the State and the MPO for implementation in a timely manner but no later than completion of construction of the SOV facility. If the area does not already have a traffic management and carpool/vanpool program, the establishment of such programs must be a part of the commitment.

(3) In TMAs that are nonattainment for carbon monoxide and/or ozone, the MPO, a State and/or transit operator may not advance a project utilizing Federal funds that provides a significant capacity increase for SOVs (adding general purpose lanes, with the exception of safety improvements or the elimination of bottlenecks, or a new highway on a new location) beyond the NEPA process unless an interim CMS is in place that meets the criteria in paragraphs (b)(1) and (b)(2) of this section and the project results from this interim CMS.

(4) Projects that are part of or consistent with a State mandated congestion management system/plan are not subject to the requirements in paragraphs (b)(1) and (b)(2) of this section.

(5) Projects advanced beyond the NEPA process as of April 6, 1992 and which are being implemented, e.g., right-of-way acquisition has been approved, will be deemed to be programmed and not subject to this requirement.

(6) At such time as a final CMS is fully operational the provisions of § 450.320(b) apply.

§ 460.2

PART 460—PUBLIC ROAD MILEAGE FOR APPORTIONMENT OF HIGHWAY SAFETY FUNDS

Sec.
460.1 Purpose.
460.2 Definitions.
460.3 Procedures.

AUTHORITY: 23 U.S.C. 315, 402(c); 49 CFR 1.48.

SOURCE: 40 FR 44322, Sept. 26, 1975, unless otherwise noted.

§ 460.1 Purpose.

The purpose of this part is to prescribe the policies and procedures followed in identifying and reporting public road mileage for utilization in the statutory formula for the apportionment of highway safety funds under 23 U.S.C. 402(c).

§ 460.2 Definitions.

As used in this part:

(a) *Public road* means any road under the jurisdiction of and maintained by a public authority and open to public travel.

(b) *Public authority* means a Federal, State, county, town, or township, Indian tribe, municipal or other local government or instrumentality thereof, with authority to finance, build, operate or maintain toll or toll-free highway facilities.

(c) *Open to public travel* means that the road section is available, except during scheduled periods, extreme weather or emergency conditions, passable by four-wheel standard passenger cars, and open to the general public for use without restrictive gates, prohibitive signs, or regulation other than restrictions based on size, weight, or class of registration. Toll plazas of public toll roads are not considered restrictive gates.

(d) *Maintenance* means the preservation of the entire highway, including surfaces, shoulders, roadsides, structures, and such traffic control devices as are necessary for its safe and efficient utilization.

§ 460.3

(e) *State* means any one of the 50 States, the District of Columbia, Puerto Rico, the Virgin Islands, Guam, and American Samoa. For the purpose of the application of 23 U.S.C. 402 on Indian reservations, *State* and *Governor of a State* include the Secretary of the Interior.

§ 460.3 Procedures.

(a) *General requirements.* 23 U.S.C. 402(c) provides that funds authorized to carry out section 402 shall be apportioned according to a formula based on population and public road mileage of each State. Public road mileage shall be determined as of the end of the calendar year preceding the year in which the funds are apportioned and shall be certified to by the Governor of the State or his designee and subject to the approval of the Federal Highway Administrator.

(b) *State public road mileage.* Each State must annually submit a certification of public road mileage within the State to the Federal Highway Administration Division Administrator by the date specified by the Division Administrator. Public road mileage on Indian reservations within the State shall be identified and included in the State mileage and in computing the State's apportionment.

(c) *Indian reservation public road mileage.* The Secretary of the Interior or his designee will submit a certification of public road mileage within Indian reservations to the Federal Highway Administrator by June 1 of each year.

(d) *Action by the Federal Highway Administrator.* (1) The certification of Indian reservation public road mileage, and the State certifications of public road mileage together with comments thereon, will be reviewed by the Federal Highway Administrator. He will make a final determination of the public road mileage to be used as the basis for apportionment of funds under 23 U.S.C. 402(c). In any instance in which the Administrator's final determination differs from the public road mileage certified by a State or the Secretary of the Interior, the Administrator will advise the State or the Secretary of the Interior of his final determination and the reasons therefor.

(2) If a State fails to submit a certification of public road mileage as required by this part, the Federal Highway Administrator may make a determination of the State's public road mileage for the purpose of apportioning funds under 23 U.S.C. 402(c). The State's public road mileage determined by the Administrator under this subparagraph may not exceed 90 percent of the State's public road mileage utilized in determining the most recent apportionment of funds under 23 U.S.C. 402(c).

PART 470—HIGHWAY SYSTEMS

Subpart A—Federal-Aid Highway Systems

Sec.
470.101 Purpose.
470.103 Definitions.
470.105 System classification.
470.107 General procedures.
470.109 Specific systems procedures.
470.111 Reclassifications, deletions, and reinstatements.
470.113 Proposals for system actions.
470.115 Approval authority.
470.117 Realignment schedule.

APPENDICES TO SUBPART A

APPENDIX A—FLORIDA (NATIONAL SYSTEM OF INTERSTATE AND DEFENSE HIGHWAYS)
APPENDIX B—PRIMARY FEDERAL-AID SYSTEM
APPENDIX C—URBANIZED FEDERAL-AID URBAN SYSTEM

Subpart B-C—(Reserved)

Subpart A—Federal-Aid Highway Systems

AUTHORITY: 23 U.S.C. 103(b)(2), 103(c)(2), 103(d)(2), 103(e)(1), 103(e)(3), 103(f), and 315; 49 CFR 1.48(b)(2) and (b)(35), unless otherwise noted.

SOURCE: 40 FR 42344, Sept. 12, 1975, unless otherwise noted. Redesignated at 41 FR 51396, Nov. 22, 1976.

§ 470.101 Purpose.

This regulation sets forth policies and procedures relating to the designation of the National System of Interstate and Defense Highways, the Federal-aid primary system, the Federal-aid secondary system, and the Federal-aid urban system after June 30, 1976.

§ 625.1

proper operation of the Federal-aid highway, the State shall take immediate action to restore such facility to its jurisdiction without cost to Federal-aid highway funds.

(3) If it is found at any time that a relinquished frontage road or portion thereof or any part of the right-of-way therefor has been abandoned by local governmental authority and a showing cannot be made that such abandoned facility is no longer required as a public road, it is to be understood that the Federal Highway Administrator may cause to be withheld from Federal-aid highway funds due to the State an amount equal to the Federal-aid participation in the abandoned facility.

(4) In no case shall any relinquishment include any portion of the right-of-way within the access control lines as shown on the plans for a Federal-aid project approved by the FHWA, without the prior approval of the Federal Highway Administrator.

(5) There cannot be additional Federal-aid participation in future construction or reconstruction on any relinquished "off the Federal-aid system" facility unless the underlying reason for such future work is caused by future improvement of the associated Federal-aid highway.

(g) In the event that a State desires to apply for approval by the Federal Highway Administrator for the relinquishment of a facility such as described in paragraph (d) (1) and (2) of this section, the facts pertinent to such proposal are to be presented to the division engineer of the FHWA. The division engineer shall have appropriate review made of such presentation and forward the material presented by the State together with his findings thereon through the Regional Federal Highway Administrator for consideration by the Federal Highway Administrator and determination of action to be taken.

(h) No change may be made in control of access, without the joint determination and approval of the SHA and FHWA. This would not prevent the relinquishment of title, without prior approval of the FHWA, of a segment of the right-of-way provided there is an abandonment of a section of highway inclusive of such segment.

(i) Relinquishments must be justified by the State's finding concurred in by the FHWA, that:

(1) The subject land will not be needed for Federal-aid highway purposes in the foreseeable future;

(2) That the right-of-way being retained is adequate under present day standards for the facility involved;

(3) That the release will not adversely affect the Federal-aid highway facility or the traffic thereon;

(4) That the lands to be relinquished are not suitable for retention in order to restore, preserve, or improve the scenic beauty adjacent to the highway consonant with the intent of 23 U.S.C. 319 and Pub. L. 89–285, Title III, sections 302–305 (Highway Beautification Act of 1965).

(j) If a relinquishment is to a Federal, State, or local governmental agency for highway purposes, there need not be a charge to the said agency, nor in such event any credit to Federal funds. If for any reason there is a charge to the said agency and Federal funds participated in the cost of the right-of-way, there shall be a credit to Federal funds on the same basis Federal funds participated in the cost of the right-of-way.

PART 625—DESIGN STANDARDS FOR HIGHWAYS

Sec.
625.1 Purpose.
625.2 Policy.
625.3 Application.
625.4 Standards, policies, and standard specifications.
625.5 Guides and references.
APPENDIX A—DOCUMENTS CITED IN PART 625

AUTHORITY: 23 U.S.C. 109, 315, and 402; Sec. 1073 of Pub. L. 102–240, 105 Stat. 1914, 2012; 49 CFR 1.48(b) and (n).

EDITORIAL NOTE: For documents cited by bracketed numbers, see appendix A.

§ 625.1 Purpose.

To designate those standards, specifications, policies, guides and references that are acceptable to the Federal Highway Administration (FHWA) for application in the geometric and structural design of highways.

[51 FR 16832, May 7, 1986]

Federal Highway Administration, DOT

§ 625.2 Policy.

(a) Plans and specifications for proposed Federal-aid highway projects shall provide for a facility that will—

(1) Adequately meet the existing and probable future traffic needs and conditions in a manner conducive to safety, durability, and economy of maintenance; and

(2) Be designed and constructed in accordance with standards best suited to accomplish the foregoing objectives and to conform to the particular needs of each locality.

(b) The development and overall management of highway facilities must be considered as a continuing program. This process of highway management commences with planning and extends through design, construction, maintenance, and operation. To assure a continuing acceptable level of safe traffic service, it is essential to provide for adequate maintenance and periodic resurfacing, restoration, and rehabilitation (RRR) throughout the life of the highway. The RRR work is defined as work undertaken to extend the service life of an existing highway and enhance highway safety. This includes placement of additional surface material and/or other work necessary to return an existing roadway, including shoulders, bridges, the roadside, and appurtenances to a condition of structural or functional adequacy. The RRR work may include upgrading of geometric features, such as minor roadway widening, flattening curves, or improving sight distances. The RRR work is an essential part of any highway program, and each State and local agency should provide for these types of improvements in each annual highway program.

(c) An important goal of the FHWA is to provide the highest practical and feasible level of safety for people and property associated with the Nation's highway transportation systems and to reduce highway hazards and the resulting number and severity of accidents on all the Nation's highways. Accordingly, the only constraint on the application of Federal-aid funds to RRR work is that they must be used to provide a facility that adequately meets existing and probable future traffic needs and conditions in a manner conducive to safety, durability, and economy of maintenance, and acceptable levels of community and environmental impact. The RRR projects shall be designed and constructed in a manner that will enhance highway safety and accomplish the foregoing objectives according to the particular needs of each State and locality.

[51 FR 16832, May 7, 1986, as amended at 54 FR 281, Jan. 5, 1989]

§ 625.3 Application.

(a) The standards, policies, and standard specifications contain specific criteria and controls for the design of highways. Deviations from specific minimum values therein are to be handled in accordance with procedures in paragraph (f) of this section. If there is a conflict between criteria in the documents enumerated in § 625.4 of this part, the latest listed standard, policy, or standard specification will govern.

(b) The guides and references (handbooks, reports, etc.) include information and general controls that are valuable in attaining good design and in promoting uniformity. They are intended to provide general program direction. Project-by-project deviations from the criteria in guides and references do not require handling as exceptions under paragraph (f) of this section.

(c) Application of FHWA regulations, although cited in § 625.4 of this part as standards, policies, and standard specifications, shall be as set forth therein.

(d) This regulation does not establish Federal standards for work that is not federally funded; however, the safety related criteria of the referenced documents are established as goals for developing State and local safety programs for all public highways as required by Highway Safety Program Standard 12, 23 CFR 1204.4.

(e) The Division Administrator shall determine the applicability of the roadway geometric design standards to traffic engineering and safety projects for signing, marking, signal installation, and traffic barriers which include very minor or no roadway work. Formal findings of applicability are expected only as needed to resolve controversies.

159

§ 625.4

(f) *Exceptions.* (1) Approval within the delegated authority provided by FHWA Order 1-1 [2] may be given on a project basis to designs which do not conform to the minimum criteria as set forth in the standards, policies, and standard specifications for:

(i) Experimental features on projects; and

(ii) Projects where conditions warrant that exceptions be made.

(2) The determination to approve a project design that does not conform to the minimum criteria is to be made only after due consideration is given to all project conditions such as maximum service and safety benefits for the dollar invested, compatibility with adjacent sections of roadway and the probable time before reconstruction of the section due to increased traffic demands or changed conditions.

[51 FR 16832, May 7, 1986]

§ 625.4 Standards, policies, and standard specifications.

The documents listed in this section are incorporated by reference in accordance with 5 U.S.C. 552(a) and are on file at the Office of the Federal Register in Washington, DC. They are available for inspection as noted in footnote numbers 1 and 2 in the appendix to this part.

(a) *Roadway and appurtenances.* (1) A Policy on Geometric Design of Highways and Streets, AASHTO 1990. [3]

(2) A Policy on Design Standards—Interstate System, AASHTO, 1991. [3]

(3) Interim Selected Metric Values for Geometric Design, AASHTO, 1993. [3]

(4) The geometric design standards for resurfacing, restoration, and rehabilitation (RRR) projects on highways other than freeways shall be the procedures and the design or design criteria established for individual projects, groups of projects, or all nonfreeway RRR projects in a State, and as approved by the FHWA. The other geometric design standards in this section do not apply to RRR projects on highways other than freeways, except as adopted on an individual State basis. The RRR design standards shall reflect the consideration of the traffic, safety, economic, physical, community, and environmental needs of the projects.

(5) A Policy on U-Turn Median Openings on Freeways, AASHO 1960. [3]

(6) A Policy on Access Between Adjacent Railroads and Interstate Highways, AASHO 1960. [3]

(7) Erosion and Sediment Control on Highway Construction Projects, FHWA, 23 CFR part 650, subpart B.

(8) Location and Hydraulic Design of Encroachments on Flood Plains, FHWA, 23 CFR part 650, subpart A.

(9) Water Supply and Sewage Treatment at Safety Rest Areas, FHWA, 23 CFR part 650, subpart E.

(10) Procedures for Abatement of Highway Traffic Noise and Construction Noise, FHWA, 23 CFR part 772.

(11) Accommodation of Utilities, FHWA, 23 CFR part 645, subpart B.

(12) Pavement Design, FHWA, 23 CFR part 626.

(b) *Bridges and structures.* (1) Standard Specifications for Highway Bridges, Thirteenth Edition, AASHTO 1983, and Interim Specifications, Bridges, AASHTO, issued annually, 1984 through 1988. [3]

(2) Standard Specifications for Movable Highway Bridges, AASHTO 1988. [3]

(3) Bridge Welding Code (D1.5–88), AASHTO/AWS 1988. [3, 4]

(4) Reinforcing Steel Welding Code AWS D 1.4–79. [4]

(5) Standard Specifications for Structural Supports for Highway Signs, Luminaires and Traffic Signals, AASHTO 1985, and Interim Specifications, Bridges, AASHTO 1986 through 1988. For use on Federal-aid highway projects, the requirement for maximum change in velocity in Section 7, Breakaway Supports, may be 16 fps in lieu of the 15 fps contained in the AASHTO specifications. [3]

(6) Navigational Clearances for Bridges, FHWA, 23 CFR part 650, subpart H.

(c) *Materials.* (1) General Materials Requirements, FHWA, 23 CFR part 635, subpart D.

(2) Standard Specifications for Transportation Materials and Methods of Sampling and Testing, parts I and II, AASHTO 1986. [3]

Federal Highway Administration, DOT   **Pt. 625, App. A**

(3) Sampling and Testing of Materials and Construction, FHWA, 23 CFR part 637, subpart B.

[51 FR 16832, May 7, 1986, as amended at 52 FR 36247, Sept. 28, 1987; 54 FR 281, Jan. 5, 1989; 58 FR 25939, Apr. 29, 1993; 58 FR 25943, Apr. 29, 1993; 58 FR 64897, Dec. 10, 1993]

§ 625.5 **Guides and references.**

The following are citations to publications which are primarily informational or guidance in character and serve to assist the public in knowing those materials which are considered by FHWA to provide valuable information in attaining good design.

(a) *Roadway and appurtenances.* (1) An Informational Guide for Roadway Lighting, AASHTO 1984. [3]

(2) Highway Design and Operational Practices Related to Highway Safety, Report of the Special AASHTO Traffic Safety Committee, AASHTO 1974. [3]

(3) Roadside Design Guide, AASHTO 1989. [3]

(4) An Informational Guide on Fencing Controlled Access Highways, AASHO 1967. [3]

(5) An Informational Guide for Preparing Private Driveway Regulations for Major Highways, AASHO 1960. [3]

(6) Highway Capacity Special Report 209, Highway Capacity Manual, Transportation Research Board 1985. [5]

(7) Guidelines on Pavement Management, AASHTO 1985. [3]

(8) Handbook of Highway Safety Design and Operating Practices, FHWA 1978. [2]

(9) Skid Accident Reduction Program, T 5040.17, FHWA December 23, 1980. [2]

(10) Guidelines for Skid Resistant Pavement Design, AASHTO 1976. [3]

(11) Special Report 214, Designing Safer Roads, Practices for Resurfacing, Restoration, and Rehabilitation, TRB 1987. [5]

(12) AASHTO Guide for Design of Pavement Structures, AASHTO 1986. [3]

(13) National Cooperative Highway Research Program Report 350, Recommended Procedures for the Safety Performance Evaluation of Highway Features, TRB 1993. [5]

(b) *Bridges and structures.* (1) A Guide for Protective Screening of Overpass Structures, AASHO 1969. [3]

(2) Manual for Maintenance Inspection of Bridges, AASHTO 1983, and Interim Specifications, Bridges, AASHTO 1986 through 1988. [3]

(3) Guide Specifications for Fracture Critical Non-Redundant Steel Bridge Members, AASHTO 1978, and amending Interim Specifications, Bridges, AASHTO 1981 through 1987. [3]

(4) Guide Specifications for Horizontally Curved Highway Bridges, AASHTO 1980, and amending Interim Specifications, Bridges, AASHTO 1981 through 1988. [3]

(5) Guide Specifications for Seismic Design of Highway Bridges, AASHTO 1983, and amending Interim Specifications, Bridges, AASHTO 1985 through 1988. [3]

(6) Highway Drainage Guidelines, Volumes I through VIII, AASHTO 1987. [3]

(7) Manual on Subsurface Investigations, AASHTO, 1988. [3]

(8) Guide Specifications for Bridge Railings, AASHTO 1989. [3]

(c) *Other.* (1) Transportation Glossary, AASHTO 1983. [3]

(2) A Guide for Erecting Mailboxes on Highways, AASHTO 1984. [3]

(3) A Guide on Safety Rest Areas for the National System of Interstate and Defense Highways, AASHO 1968. [3]

(4) Guide for Development of New Bicycle Facilities, AASHTO 1981. [3]

(5) Guide Specifications for Highway Construction AASHTO 1985. [3]

[51 FR 16832, May 7, 1986, as amended at 53 FR 15671, May 3, 1988; 54 FR 281, Jan. 3, 1989; 54 FR 1844, Jan. 17, 1989; 54 FR 25117, June 13, 1989; 55 FR 25828, June 25, 1990; 58 FR 38298, July 16, 1993]

APPENDIX A—DOCUMENTS CITED IN PART 625

1. The design standards listed in § 625.4 are incorporated by reference and are on file at the Office of the Federal Register, 800 North Capitol Street, NW., suite 700, Washington, DC.

2. These documents may be reviewed at the Department of Transportation Library, 400 7th Street, SW., Washington, DC 20590 in Room 2200. These documents are also available for inspection and copying as provided in 49 CFR part 7, appendix D.

3. American Association of State Highway and Transportation Officials, Suite 225, 444 North Capitol Street, NW., Washington, DC 20001.

Federal Highway Administration, DOT

§ 645.105

Subpart A—Utility Relocations, Adjustments, and Reimbursement

SOURCE: 50 FR 20345, May 15, 1985, unless otherwise noted.

§ 645.101 Purpose.

To prescribe the policies, procedures, and reimbursement provisions for the adjustment and relocation of utility facilities on Federal-aid and direct Federal projects.

§ 645.103 Applicability.

(a) The provisions of this regulation apply to reimbursement claimed by a State highway agency (SHA) for costs incurred under an approved and properly executed highway agency (HA)/utility agreement and for payment of costs incurred under all Federal Highway Administration (FHWA)/utility agreements.

(b) Procedures on the accommodation of utilities are set forth in 23 CFR part 645, subpart B, Accommodation of Utilities.

(c) When the lines or facilities to be relocated or adjusted due to highway construction are privately owned, located on the owner's land, devoted exclusively to private use and not directly or indirectly serving the public, the provisions of the FHWA's right-of-way procedures in 23 CFR chapter I, subchapter H, Right-of-Way and Environment, apply. When applicable, under the foregoing conditions, the provisions of this regulation may be used as a guide to establish a cost-to-cure.

(d) The FHWA's reimbursement to the SHA will be governed by State law (or State regulation) or the provisions of this regulation, whichever is more restrictive. When State law or regulation differs from this regulation, a determination shall be made by the SHA subject to the concurrence of the FHWA as to which standards will govern, and the record documented accordingly, for each relocation encountered.

(e) For direct Federal projects, all references herein to the SHA or HA are inapplicable, and it is intended that the FHWA be considered in the relative position of the SHA or HA.

§ 645.105 Definitions.

For the purposes of this regulation, the following definitions shall apply:

(a) *Authorization*—for Federal-aid projects authorization to the SHA by the FHWA, or for direct Federal projects authorization to the utility by the FHWA, to proceed with any phase of a project. The date of authorization establishes the date of eligibility for Federal funds to participate in the costs incurred on that phase of work.

(b) *Betterment*—any upgrading of the facility being relocated that is not attributable to the highway construction and is made solely for the benefit of and at the election of the utility.

(c) *Cost of relocation*—the entire amount paid by or on behalf of the utility properly attributable to the relocation after deducting from that amount any increase in value of the new facility, and any salvage derived from the old facility.

(d) *Cost of Removal*—the amount expended to remove utility property including the cost of demolishing, dismantling, removing, transporting, or otherwise disposing of utility property and of cleaning up to leave the site in a neat and presentable condition.

(e) *Cost of salvage*—the amount expended to restore salvaged utility property to usable condition after its removal.

(f) *Direct Federal projects*—highway projects such as projects under the Federal Lands Highways Program which are under the direct administration of the FHWA.

(g) *Highway agency (HA)*—that department, commission, board, or official of any State or political subdivison thereof, charged by its law with the responsibility for highway administration.

(h) *Indirect or overhead costs*—those costs which are not readily identifiable with one specific task, job, or work order. Such costs may include indirect labor, social security taxes, insurance, stores expense, and general office expenses. Costs of this nature generally are distributed or allocated to the applicable job or work orders, other accounts and other functions to which they relate. Distribution and allocation is made on a uniform basis which is reasonable, equitable, and in accord-

229

§ 645.107

ance with generally accepted cost accounting practices.

(i) *Relocation*—the adjustment of utility facilities required by the highway project. It includes removing and reinstalling the facility, including necessary temporary facilities, acquiring necessary right-of-way on the new location, moving, rearranging or changing the type of existing facilities and taking any necessary safety and protective measures. It shall also mean constructing a replacement facility that is both functionally equivalent to the existing facility and necessary for continuous operation of the utility service, the project economy, or sequence of highway construction.

(j) *Salvage value*—the amount received from the sale of utility property that has been removed or the amount at which the recovered material is charged to the utility's accounts, if retained for reuse.

(k) *State highway agency*—the highway agency of one of the 50 States, the District of Columbia, or Puerto Rico.

(l) *Use and occupancy agreement*—the document (written agreement or permit) by which the HA approves the use and occupancy of highway right-of-way by utility facilities or private lines.

(m) *Utility*—a privately, publicly, or cooperatively owned line, facility or system for producing, transmitting, or distributing communications, cable television, power, electricity, light, heat, gas, oil, crude products, water, steam, waste, storm water not connected with highway drainage, or any other similar commodity, including any fire or police signal system or street lighting system, which directly or indirectly serves the public. The term utility shall also mean the utility company inclusive of any wholly owned or controlled subsidiary.

(n) *Work order system*—a procedure for accumulating and recording into separate accounts of a utility all costs to the utility in connection with any change in its system or plant.

§ 645.107 Eligibility.

(a) When requested by the SHA, Federal funds may participate, subject to the provisions of § 645.103(d) of this part and at the pro rata share applicable, in an amount actually paid by an HA for the costs of utility relocations. Federal funds may participate in safety corrective measures made under the provisions of § 645.107(k) of this part. Federal funds may also participate for relocations necessitated by the actual construction of highway project made under one or more of the following conditions when:

(1) The SHA certifies that the utility has the right of occupancy in its existing location because it holds the fee, an easement, or other real property interest, the damaging or taking of which is compensable in eminent domain,

(2) The utility occupies privately or publicly owned land, including public road or street right-of-way, and the SHA certifies that the payment by the HA is made pursuant to a law authorizing such payment in conformance with the provisions of 23 U.S.C. 123, and/or

(3) The utility occupies publicy owned land, including public road and street right-of-way, and is owned by a public agency or political subdivision of the State, and is not required by law or agreement to move at its own expense, and the SHA certifies that the HA has the legal authority or obligation to make such payments.

(b) On projects which the SHA has the authority to participate in project costs, Federal funds may not participate in payments made by a political subdivision for relocation of utility facilities, other than those proposed under the provisions of § 645.107(k) of this part, when State law prohibits the SHA from making payment for relocation of utility facilities.

(c) On projects which the SHA does not have the authority to participate in project costs, Federal funds may participate in payments made by a political subdivision for relocation of utility facilities necessitated by the actual construction of a highway project when the SHA certifies that such payment is based upon the provisions of § 645.107(a) of this part and does not violate the terms of a use and occupancy agreement, or legal contract, between the utility and the HA or for utility safety corrective measures under the provisions of § 645.107(k) of this part.

(d) Federal funds are not eligible to participate in any costs for which the

Federal Highway Administration, DOT §645.109

utility contributes or repays the HA, except for utilities owned by the political subdivision on projects which qualify under the provisions of §645.107(c) of this part in which case the costs of the utility are considered to be costs of the HA.

(e) The FHWA may deny Federal fund participation in any payments made by a HA for the relocation of utility facilities when such payments do not constitute a suitable basis for Federal fund participation under the provisions of title 23 U.S.C.

(f) The rights of any public agency or political subdivision of a State under contract, franchise, or other instrument or agreement with the utility, pertaining to the utility's use and occupancy of publicly owned land, including public road and street right-of-way, shall be considered the rights of the SHA in the absence of State law to the contrary.

(g) In lieu of the individual certifications required by §645.107(a) and (c), the SHA may file a statement with the FHWA setting forth the conditions under which the SHA will make payments for the relocation of utility facilities. The FHWA may approve Federal fund participation in utility relocations proposed by the SHA under the conditions of the statement when the FHWA has made an affirmative finding that such statement and conditions form a suitable basis for Federal fund participation under the provisions of 23 U.S.C. 123.

(h) Federal funds may not participate in the cost of relocations of utility facilities made solely for the benefit or convenience of a utility, its contractor, or a highway contractor.

(i) When the advance installation of new utility facilities crossing or otherwise occupying the proposed right-of-way of a planned highway project is underway, or scheduled to be underway, prior to the time such right-of-way is purchased by or under control of the HA, arrangements should be made for such facilities to be installed in a manner that will meet the requirements of the planned highway project. Federal funds are eligible to participate in the additional cost incurred by the utility that are attributable to, and in accommodation of, the highway project provided such costs are incurred subsequent to authorization of the work by the FHWA. Subject to the other provisions of this regulation, Federal participation may be approved under the foregoing circumstances when it is demonstrated that the action taken is necessary to protect the public interest and the adjustment of the facility is necessary by reason of the actual construction of the highway project.

(j) Federal funds are eligible to participate in the costs of preliminary engineering and allied services for utilities, the acquisition of replacement right-of-way for utilities, and the physical construction work associated with utility relocations. Such costs must be incurred by or on behalf of a utility after the date the work is included in an approved program and after the FHWA has authorized the SHA to proceed in accordance with 23 CFR part 630, subpart A, Federal-Aid Programs Approval and Project Authorization.

(k) Federal funds may participate in projects solely for the purpose of implementing safety corrective measures to reduce the roadside hazards of utility facilities to the highway user. Safety corrective measures should be developed in accordance with the provisions of 23 CFR 645.209(k).

(Information collection requirements in paragraph (g) were approved by the Office of Management and Budget under control number 2125-0515)

[50 FR 20345, May 15, 1985, as amended at 53 FR 24932, July 1, 1988]

§645.109 Preliminary engineering.

(a) As mutually agreed to by the HA and utility, and subject to the provisions of paragraph (b) of this section, preliminary engineering activities associated with utility relocation work may be done by:

(1) The HA's or utility's engineering forces;

(2) An engineering consultant selected by the HA, after consultation with the utility, the contract to be administered by the HA; or,

(3) An engineering consultant selected by the utility, with the approval of the HA, the contract to be administered by the utility.

§ 645.111

(b) When a utility is not adequately staffed to pursue the necessary preliminary engineering and related work for the utility relocation, Federal funds may participate in the amount paid to engineers, architects, and others for required engineering and allied services provided such amounts are not based on a percentage of the cost of relocation. When Federal participation is requested by the SHA in the cost of such services, the utility and its consultant shall agree in writing as to the services to be provided and the fees and arrangements for the services. Federal funds may participate in the cost of such services performed under existing written continuing contracts when it is demonstrated that such work is performed regularly for the utility in its own work and that the costs are reasonable. Prior approval by the FHWA of consulting services is necessary, except the FHWA may forgo preaward review and/or approval of any proposed consultant contract which is not expected to exceed $10,000.

(c) The procedures in 23 CFR part 172, Administration of Negotiated Contracts, may be used as a guide for reviewing proposed consultant contracts.

§ 645.111 Right-of-way.

(a) Federal participation may be approved for the cost of replacement right-of-way provided:

(1) The utility has the right of occupancy in its existing location beause it holds the fee, an easement, or another real property interest, the damaging or taking of which is compensable in eminent domain, or the acquisition is made in the interest of project economy or is necessary to meet the requirements of the highway project, and

(2) There will be no charge to the project for that portion of the utility's existing right-of-way being transferred to the HA for highway purposes.

(b) The utility shall determine and make a written valuation of the replacement right-of-way that it acquires in order to justify amounts paid for such right-of-way. This written valuation shall be accomplished prior to negotiation for acquisition.

(c) Acquisition of replacement right-of-way by the HA on behalf of a utility or acquisition of nonoperating real property from a utility shall be in accordance with the Uniform Relocation Assistance and Real Property Acquisition Policies Act of 1970 (42 U.S.C. 4601 et seq.) and applicable right-of-way procedures in 23 CFR chapter I, subchapter H, Right-of-Way and Environment.

(d) When the utility has the right-of-occupancy in its existing location because it holds the fee, an easement, or another real property interest, and it is not necessary by reason of the highway construction to adjust or replace the facilities located thereon, the taking of and damage to the utility's real property, including the disposal or removal of such facilities, may be considered a right-of-way transaction in accordance with provisions of the applicable right-of-way procedures in 23 CFR chapter I, subchapter H, Right-of-Way and Environment.

§ 645.113 Agreements and authorizations.

(a) On Federal-aid and direct Federal projects involving utility relocations, the utility and the HA shall agree in writing on their separate responsibilities for financing and accomplishing the relocation work. When Federal participation is requested, the agreement shall incorporate this regulation by reference and designate the method to be used for performing the work (by contract or force account) and for developing relocation costs. The method proposed by the utility for developing relocation costs must be acceptable to both the HA and the FHWA. The preferred method for the development of relocation costs by a utility is on the basis of actual direct and related indirect costs accumulated in accordance with a work order accounting procedure prescribed by the applicable Federal or State regulatory body.

(b) When applicable, the written agreement shall specify the terms and amounts of any contribution or repayments made or to be made by the utility to the HA in connection with payments by the HA to the utility under the provisions of § 645.107 of this regulation.

(c) The agreement shall be supported by plans, specifications when required, and itemized cost estimates of the work agreed upon, including appro-

Federal Highway Administration, DOT

priate credits to the project, and shall be sufficiently informative and complete to provide the HA and the FHWA with a clear description of the work required.

(d) When the relocation involves both work to be done at the HA's expense and work to be done at the expense of the utility, the written agreement shall state the share to be borne by each party.

(e) In the event there are changes in the scope of work, extra work or major changes in the planned work covered by the approved agreement, plans, and estimates, Federal participation shall be limited to costs covered by a modification of the agreement, a written change, or extra work order approved by the HA and the FHWA.

(f) When the estimated cost to the HA of proposed utility relocation work on a project for a specific utility company is $25,000 or less, the FHWA may approve an agreement between the HA and the utility for a lump-sum payment without later confirmation by audit of actual costs. Lump-sum agreements in excess of $25,000 may be approved when the FHWA finds that this method of developing costs would be in the best interest of the public.

(g) Except as otherwise provided by § 645.113(h), authorization by the FHWA to the SHA to proceed with the physical relocation of a utility's facilities may be given after:

(1) The utility relocation work, or the right-of-way, or physical construction phase of the highway construction work is included in an approved program,

(2) The appropriate environmental evaluation and public hearing procedures required by 23 CFR part 771, Environmental Impact and Related Procedures, have been satisfied.

(3) The FHWA has reviewed and approved the plans, estimates, and proposed or executed agreements for the utility work and is furnished a schedule for accomplishing the work.

(h) The FHWA may authorize the physical relocation of utility facilities before the requirements of § 645.113(g)(2) are satisfied when the relocation or adjustment of utility facilities meets the requirements of § 645.107(i) of this regulation.

§ 645.115

(i) Whenever the FHWA has authorized right-of-way acquisition under the hardship and protective buying provisions of 23 CFR part 712, the Acquisition Functions, the FHWA may authorize the physical relocation of utility facilities located in whole or in part on such right-of-way.

(j) When all efforts by the HA and utility fail to bring about written agreement of their separate responsibilities under the provisions of this regulation, the SHA shall submit its proposal and a full report of the circumstances to the FHWA. Conditional authorizations for the relocation work to proceed may be given by the FHWA to the SHA with the understanding that Federal funds will not be paid for work done by the utility until the SHA proposal has been approved by the FHWA.

(k) The FHWA will consider for approval any special procedure under State law, or appropriate administrative or judicial order, or under blanket master agreements with the utilities, that will fully accomplish all of the foregoing objectives and accelerate the advancement of the construction and completion of projects.

§ 645.115 Construction.

(a) Part 635, subpart B, of this title, Force Account Construction (justification required for force account work), states that it is cost-effective for certain utility adjustments to be performed by a utility with its own forces and equipment, provided the utility is qualified to perform the work in a satisfactory manner. This cost-effectiveness finding covers minor work on the utility's existing facilities routinely performed by the utility with its own forces. When the utility is not adequately staffed and equipped to perform such work with its own forces and equipment at a time convenient to and in coordination with the associated highway construction, such work may be done by:

(1) A contract awarded by the HA or utility to the lowest qualified bidder based on appropriate solicitation,

(2) Inclusion as part of the HA's highway construction contract let by the HA as agreed to by the utility,

233

§ 645.117

(3) An existing continuing contract, provided the costs are reasonable, or

(4) A contract for low-cost incidental work, such as tree trimming and the like, awarded by the HA or utility without competitive bidding, provided the costs are reasonable.

(b) When it has been determined under part 635, subpart B, that the force account method is not the most cost-effective means for accomplishing the utility adjustment, such work is to be done under competitive bid contracts as described in § 645.115(a) (1) and (2) or under an existing continuing contract provided it can be demonstrated this is the most cost-effective method.

(c) Costs for labor, materials, equipment, and other services furnished by the utility shall be billed by the utility directly to the HA. The special provisions of contracts let by the utility or the HA shall be explicit in this respect. The costs of force account work performed for the utility by the HA and of contract work performed for the utility under a contract let by the HA shall be reported separately from the costs of other force account and contract items on the highway project.

§ 645.117 Cost development and reimbursement.

(a) *Developing and recording costs.* (1) All utility relocation costs shall be recorded by means of work orders in accordance with an approved work order system except when another method of developing and recording costs, such as lump-sum agreement, has been approved by the HA and the FHWA. Except for work done under contracts, the individual and total costs properly reported and recorded in the utility's accounts in accordance with the approved method for developing such costs, or the lump-sum agreement, shall constitute the maximum amount on which Federal participation may be based.

(2) Each utility shall keep its work order system or other approved accounting procedure in such a manner as to show the nature of each addition to or retirement from a facility, the total costs thereof, and the source or sources of cost. Separate work orders may be issued for additions and retirements. Retirements, however, may be included with the construction work order provided that all items relating to retirements shall be kept separately from those relating to construction.

(b) *Direct labor costs.* (1) Salaries and wages, at actual or average rates, and related expenses paid by the utility to individuals for the time worked on the project are reimbursable when supported by adequate records. This includes labor associated with preliminary engineering, construction engineering, right-of-way, and force account construction.

(2) Salaries and expenses paid to individuals who are normally part of the overhead organization of the utility may be reimbursed for the time worked directly on the project when supported by adequate records and when the work performed by such individuals is essential to the project and could not have been accomplished as economically by employees outside the overhead organization.

(3) Amounts paid to engineers, architects and others for services directly related to projects may be reimbursed.

(c) *Labor surcharges.* (1) Labor surcharges include worker compensation insurance, public liability and property damage insurance, and such fringe benefits as the utility has established for the benefit of its employees. The cost of labor surcharges will be reimbursed at actual cost to the utility, or, at the option of the utility, average rates which are representative of actual costs may be used in lieu of actual costs if approved by the SHA and the FHWA. These average rates should be adjusted at least once annually to take into account known anticipated changes and correction for any over or under applied costs for the preceding period.

(2) When the utility is a self-insurer, there may be reimbursement at experience rates properly developed from actual costs. The rates cannot exceed the rates of a regular insurance company for the class of employment covered.

(d) *Overhead and indirect construction costs.* (1) Overhead and indirect construction costs not charged directly to work order or construction accounts may be allocated to the relocation provided the allocation is made on an equitable basis. All costs included in the allocation shall be eligible for Federal

234

Federal Highway Administration, DOT §645.117

reimbursement, reasonable, and actually incurred by the utility.

(2) Costs not eligible for Federal reimbursement include, but are not limited to, the costs associated with advertising, sales promotion, interest on borrowings, the issuance of stock, bad debts, uncollectible accounts receivable, contributions, donations, entertainment, fines, penalties, lobbying, and research programs.

(3) The records supporting the entries for overhead and indirect construction costs shall show the total amount, rate, and allocation basis for each additive, and are subject to audit by representatives of the State and Federal Government.

(e) *Material and supply costs.* (1) Materials and supplies, if available, are to be furnished from company stock except that they may be obtained from other sources near the project site when available at a lower cost. When not available from company stock, they may be purchased either under competitive bids or existing continuing contracts under which the lowest available prices are developed. Minor quantities of materials and supplies and proprietary products routinely used in the utility's operation and essential for the maintenance of system compatibility may be excluded from these requirements. The utility shall not be required to change its existing standards for materials used in permanent changes to its facilities. Costs shall be determined as follows:

(i) Materials and supplies furnished from company stock shall be billed at the current stock prices for such new or used materials at time of issue.

(ii) Materials and supplies not furnished from company stock shall be billed at actual costs to the utility delivered to the project site.

(iii) A reasonable cost for plant inspection and testing may be included in the costs of materials and supplies when such expense has been incurred. The computation of actual costs of materials and supplies shall include the deduction of all offered discounts, rebates, and allowances.

(iv) The cost of rehabilitating rather than replacing existing utility facilities to meet the requirements of a project is reimbursable, provided this cost does not exceed replacement costs.

(2) Materials recovered from temporary use and accepted for reuse by the utility shall be credited to the project at prices charged to the job, less a considertion for loss in service life at 10 percent. Materials recovered from the permanent facility of the utility that are accepted by the utility for return to stock shall be credited to the project at the current stock prices of such used materials. Materials recovered and not accepted for reuse by the utility, if determined to have a net sale value, shall be sold to the highest bidder by the HA or utility following an opportunity for HA inspection and appropriate solicitation for bids. If the utility practices a system of periodic disposal by sale, credit to the project shall be at the going prices supported by records of the utility.

(3) Federal participation may be approved for the total cost of removal when either such removal is required by the highway construction or the existing facilities cannot be abandoned in place for aesthetic or safety reasons. When the utility facilities can be abandoned in place but the utility or highway constructor elects to remove and recover the materials, Federal funds shall not participate in removal costs which exceed the value of the materials recovered.

(4) The actual and direct costs of handling and loading materials and supplies at company stores or material yards, and of unloading and handling recovered materials accepted by the utility at its stores or material yards are reimbursable. In lieu of actual costs, average rates which are representative of actual costs may be used if approved by the SHA and the FHWA. These average rates should be adjusted at least once annually to take into account known anticipated changes and correction for any over or under applied costs for the preceding period. At the option of the utility, 5 percent of the amounts billed for the materials and supplies issued from company stores and material yards or the value of recovered materials will be reimbursed in lieu of actual or average costs for handling.

§ 645.119

(f) *Equipment costs.* The average or actual costs of operation, minor maintenance, and depreciation of utility-owned equipment may be reimbursed. Reimbursement for utility-owned vehicles may be made at average or actual costs. When utility-owned equipment is not available, reimbursement will be limited to the amount of rental paid (1) to the lowest qualified bidder, (2) under existing continuing contracts at reasonable costs, or (3) as an exception by negotiation when paragraph (f) (1) and (2) of this section are impractical due to project location or schedule.

(g) *Transportation costs.* (1) The utility's cost, consistent with its overall policy, of necessary employee transportation and subsistence directly attributable to the project is reimbursable.

(2) Reasonable cost for the movement of materials, supplies, and equipment to the project and necessary return to storage including the associated cost of loading and unloading equipment is reimbursable.

(h) *Credits.* (1) Credit to the highway project will be required for the cost of any betterments to the facility being replaced or adjusted, and for the salvage value of the materials removed.

(2) Credit to the highway project will be required for the accrued depreciation of a utility facility being replaced, such as a building, pumping station, filtration plant, power plant, substation, or any other similar operational unit. Such accrued depreciation is that amount based on the ratio between the period of actual length of service and total life expectancy applied to the original cost. Credit for accrued depreciation shall not be required for a segment of the utility's service, distribution, or transmission lines.

(3) No betterment credit is required for additions or improvements which are:

(i) Required by the highway project,

(ii) Replacement devices or materials that are of equivalent standards although not identical,

(iii) Replacement of devices or materials no longer regularly manufactured with next highest grade or size,

(iv) Required by law under governmental and appropriate regulatory commission code, or

(v) Required by current design practices regularly followed by the company in its own work, and there is a direct benefit to the highway project.

(4) When the facilities, including equipment and operating facilities, described in § 645.117(h)(2) are not being replaced, but are being rehabilitated and/or moved, as necessitated by the highway project, no credit for accrued depreciation is needed.

(5) In no event will the total of all credits required under the provisions of this regulation exceed the total costs of adjustment exclusive of the cost of additions or improvements necessitated by the highway construction.

(i) *Billings.* (1) After the executed HA/utility agreement has been approved by the FHWA, the utility may be reimbursed through the SHA by progress billings for costs incurred. Cost for materials stockpiled at the project site or specifically purchased and delivered to the utility for use on the project may also be reimbursed on progress billings following approval of the executed HA/utility agreement.

(2) The utility shall provide one final and complete billing of all costs incurred, or of the agreed-to lump-sum, at the earliest practicable date. The final billing to the FHWA shall include a certification by the SHA that the work is complete, acceptable, and in accordance with the terms of the agreement.

(3) All utility cost records and accounts relating to the project are subject to audit by representatives of the State and Federal Government for a period of 3 years from the date final payment has been received by the utility.

(4) Reimbursement for a final utility billing shall not be approved until the HA furnishes evidence that it has paid the utility from its own funds.

(Information collection requirements in paragraph (i) were approved by the Office of Management and Budget under control number 2125–0159)

§ 645.119 Alternate procedure.

(a) This alternate procedure is provided to simplify the processing of utility relocations or adjustments under the provisions of this regulation. Under this procedure, except as otherwise

Federal Highway Administration, DOT § 645.201

provided in paragraph (b) of this section, the SHA is to act in the relative position of the FHWA for reviewing and approving the arrangements, fees, estimates, plans, agreements, and other related matters required by this regulation as prerequisites for authorizing the utility to proceed with and complete the work.

(b) The scope of the SHA's approval authority under the alternate procedure includes all actions necessary to advance and complete all types of utility work under the provisions of this regulation except in the following instances:

(1) Utility relocations and adjustments involving major transfer, production, and storage facilities such as generating plants, power feed stations, pumping stations and reservoirs.

(2) Utility relocations falling within the scope of § 645.113 (h), (i), and (j), and § 645.107(i) of this regulation.

(c) Each SHA is encouraged to adopt the alternate procedure and file a formal application for approval by the FHWA. The application must include the following:

(1) The SHA's written policies and procedures for administering and processing Federal-aid utility adjustments. Those policies and procedures must make adequate provisions with respect to the following:

(i) Compliance with the requirements of this regulation, except as otherwise provided by § 645.119(b), and the provisions of 23 CFR part 645, subpart B, Accommodation of Utilities.

(ii) Advance utility liaison, planning, and coordination measures for providing adequate lead time and early scheduling of utility relocation to minimize interference with the planned highway construction.

(iii) Appropriate administrative, legal, and engineering review and coordination procedures as needed to establish the legal basis of the HA's payment; the extent of eligibility of the work under State and Federal laws and regulations; the more restrictive payment standards under § 645.103(d) of this regulation; the necessity of the proposed utility work and its compatibility with proposed highway improvements; and the uniform treatment of all utility matters and actions, consistent with sound management practices.

(iv) Documentation of actions taken in compliance with SHA policies and the provisions of this regulation, shall be retained by the SHA.

(2) A statement signed by the chief administrative officer of the SHA certifying that:

(i) Federal-aid utility relocations will be processed in accordance with the applicable provisions of this regulation, and the SHA's utility policies and procedures submitted under § 645.119(c)(1).

(ii) Reimbursement will be requested only for those costs properly attributable to the proposed highway construction and eligible for participation under the provisions of this regulation.

(d) The SHA's application and any changes to it will be submitted to the FHWA for review and approval.

(e) After the alternate procedure has been approved, the FHWA may authorize the SHA to proceed with utility relocation on a project in accordance with the certification, subject to the following conditions:

(1) The utility work must be included in an approved program.

(2) The SHA must submit a request in writing for such authorization. The request shall include a list of the utility relocations to be processed under the alternate procedure, along with the best available estimate of the total costs involved.

(f) The FHWA may suspend approval of the alternate procedure when any FHWA review discloses noncompliance with the certification. Federal funds will not participate in relocation costs incurred that do not comply with the requirements under § 645.119(c)(1).

(Information collection requirements in paragraph (c) were approved by the Office of Management and Budget under control number 2125–0533)

Subpart B—Accommodation of Utilities

SOURCE: 50 FR 20354, May 15, 1985, unless otherwise noted.

§ 645.201 Purpose.

To prescribe policies and procedures for accommodating utility facilities

§ 645.203

and private lines on the right-of-way of Federal-aid or direct Federal highway projects.

§ 645.203 Applicability.

This subpart applies to:

(a) New utility installations within the right-of-way of Federal-aid or direct Federal highway projects,

(b) Existing utility facilities which are to be retained, relocated, or adjusted within the right-of-way of active projects under development or construction when Federal-aid or direct Federal highway funds are either being or have been used on the involved highway facility. When existing utility installations are to remain in place without adjustments on such projects the highway agency and utility are to enter into an appropriate agreement as discussed in § 645.213 of this part,

(c) Existing utility facilities which are to be adjusted or relocated under the provisions of § 645.209(k), and

(d) Private lines which may be permitted to cross the right-of-way of a Federal-aid or direct Federal highway project pursuant to State law and regulations and the provisions of this subpart. Longitudinal use of such right-of-way by private lines is to be handled under the provisions of 23 CFR 1.23(c).

§ 645.205 Policy.

(a) Pursuant to the provisions of 23 CFR 1.23, it is in the public interest for utility facilities to be accommodated on the right-of-way of a Federal-aid or direct Federal highway project when such use and occupancy of the highway right-of-way do not adversely affect highway or traffic safety, or otherwise impair the highway or its aesthetic quality, and do not conflict with the provisions of Federal, State or local laws or regulations.

(b) Since by tradition and practice highway and utility facilities frequently coexist within common right-of-way or along the same transportation corridors, it is essential in such situations that these public service facilities be compatibly designed and operated. In the design of new highway facilities consideration should be given to utility service needs of the area traversed if such service is to be provided from utility facilities on or near the highway. Similarly the potential impact on the highway and its users should be considered in the design and location of utility facilities on or along highway right-of-way. Efficient, effective and safe joint highway and utility development of transportation corridors is important along high speed and high volume roads, such as major arterials and freeways, particularly those approaching metropolitan areas where space is increasingly limited. Joint highway and utility planning and development efforts are encouraged on Federal-aid highway projects.

(c) The manner is which utilities cross or otherwise occupy the right-of-way of a direct Federal or Federal-aid highway project can materially affect the highway, its safe operation, aesthetic quality, and maintenance. Therefore, it is necessary that such use and occupancy, where authorized, be regulated by highway agencies in a manner which preserves the operational safety and the functional and aesthetic quality of the highway facility. This subpart shall not be construed to alter the basic legal authority of utilities to install their facilities on public highways pursuant to law or franchise and reasonable regulation by highway agencies with respect to location and manner of installation.

(d) When utilities cross or otherwise occupy the right-of-way of a direct Federal or Federal-aid highway project on Federal lands, and when the right-of-way grant is for highway purposes only, the utility must also obtain and comply with the terms of a right-of-way or other occupancy permit for the Federal agency having jurisdiction over the underlying land.

[50 FR 20354, May 15, 1985, as amended at 53 FR 2833, Feb. 2, 1988]

§ 645.207 Definitions.

For the purpose of this regulation, the following definitions shall apply:

(a) *Aesthetic quality*—those desirable characteristics in the appearance of the highway and its environment, such as harmony between or blending of natural and manufactured objects in the environment, continuity of visual form without distracting interruptions, and simplicity of designs which are desir-

Federal Highway Administration, DOT

§ 645.207

ably functional in shape but without clutter.

(b) *Clear recovery area*—that portion of the roadside, within the highway right-of-way as established by the highway agency, free of nontraversable hazards and fixed objects. The purpose of such areas is to provide drivers of errant vehicles which leave the traveled portion of the roadway a reasonable opportunity to stop safely or otherwise regain control of the vehicle. The clear recovery area may vary with the type of highway, terrain traversed, and road geometric and operating conditions. The American Association of State Highway and Transportation Officials (AASHTO) "Roadside Design Guide," 1989, should be used as a guide for establishing clear recovery areas for various types of highways and operating conditions. It is available for inspection from the FHWA Washington Headquarters and all FHWA Division and Regional Offices as prescribed in 49 CFR part 7, appendix D. Copies of current AASHTO publications are available for purchase from the American Association of State Highway and Transportation Officials, Suite 225, 444 North Capitol Street, NW., Washington, DC 20001.

(c) *Clear roadside policy*—that policy employed by a highway agency to provide a clear recovery area in order to increase safety, improve traffic operations, and enhance the aesthetic quality of highways by designing, constructing and maintaining highway roadsides as wide, flat, and rounded as practical and as free as practical from natural or manufactured hazards such as trees, drainage structures, nonyielding sign supports, highway lighting supports, and utility poles and other ground-mounted structures. The policy should address the removal of roadside obstacles which are likely to be associated with accident or injury to the highway user, or when such obstacles are essential, the policy should provide for appropriate countermeasures to reduce hazards. Countermeasures include placing utility facilities at locations which protect out-of-control vehicles, using breakaway features, using impact attenuation devices, or shielding. In all cases full consideration shall be given to sound engineering principles and economic factors.

(d) *Direct Federal highway projects*—those active or completed highway projects such as projects under the Federal Lands Highways Program which are under the direct administration of the Federal Highway Administration (FHWA)

(e) *Federal-aid highway projects*—those active or completed highway projects administered by or through a State highway agency which involve or have involved the use of Federal-aid highway funds for the development, acquisition of right-of-way, construction or improvement of the highway or related facilities, including highway beautification projects under 23 U.S.C. 319, Landscaping and Scenic Enhancement.

(f) *Freeway*—a divided arterial highway with full control of access.

(g) *Highway agency*—that department, agency, commission, board, or official of any State or political subdivision thereof, charged by its law with the responsibility for highway administration.

(h) *Highway*—any public way for vehicular travel, including the entire area within the right-of-way and related facilities constructed or improved in whole or in part with Federal-aid or direct Federal highway funds.

(i) *Private lines*—privately owned facilities which convey or transmit the commodities outlined in paragraph (m) of this section, but devoted exclusively to private use.

(j) *Right-of-way*—real property, or interests therein, acquired, dedicated or reserved for the construction, operation, and maintenance of a highway in which Federal-aid or direct Federal highway funds are or have been involved in any stage of development. Lands acquired under 23 U.S.C. 319 shall be considered to be highway right-of-way.

(k) *State highway agency*—the highway agency of one of the 50 States, the District of Columbia, or Puerto Rico.

(l) *Use and occupancy agreement*—the document (written agreement or permit) by which the highway agency approves the use and occupancy of highway right-of-way by utility facilities or private lines.

239

§ 645.209

(m) *Utility facility*—privately, publicly or cooperatively owned line, facility, or system for producing, transmitting, or distributing communications, cable television, power, electricity, light, heat, gas, oil, crude products, water, steam, waste, storm water not connected with highway drainage, or any other similar commodity, including any fire or police signal system or street lighting system, which directly or indirectly serves the public. The term utility shall also mean the utility company inclusive of any substantially owned or controlled subsidiary. For the purposes of this part, the term includes those utility-type facilities which are owned or leased by a government agency for its own use, or otherwise dedicated solely to governmental use. The term utility includes those facilities used solely by the utility which are a part of its operating plant.

[50 FR 20345, May 15, 1985, as amended at 51 FR 16834, May 7, 1986; 53 FR 2833, Feb. 2, 1988; 55 FR 25828, June 25, 1990]

§ 645.209 General requirements.

(a) *Safety.* Highway safety and traffic safety are of paramount, but not of sole, importance when accommodating utility facilities within highway right-of-way. Utilities provide an essential public service to the general public. Traditionally, as a matter of sound economic public policy and law, utilities have used public road right-of-way for transmitting and distributing their services. However, due to the nature and volume of highway traffic, the effect of such joint use on the traveling public must be carefully considered by highway agencies before approval of utility use of the right-of-way of Federal-aid or direct Federal highway projects is given. Adjustments in the operating characteristics of the utility or the highway or other special efforts may be necessary to increase the compatibility of utility-highway joint use. The possibility of this joint use should be a consideration in establishing right-of-way requirements for highway projects. In any event, the design, location, and manner in which utilities use and occupy the right-of-way of Federal-aid or direct Federal highway projects must conform to the clear roadside policies for the highway involved and otherwise provide for a safe traveling environment as required by 23 U.S.C. 109(1)(1).

(b) *New above ground installations.* On Federal-aid or direct Federal highway projects, new above ground utility installations, where permitted, shall be located as far from the traveled way as possible, preferably along the right-of-way line. No new above ground utility installations are to be allowed within the established clear recovery of the highway unless a determination has been made by the highway agency that placement underground is not technically feasible or is unreasonably costly and there are no feasible alternate locations. In exceptional situations when it is essential to locate such above ground utility facilities within the established clear recovery area of the highway, appropriate countermeasures to reduce hazards shall be used. Countermeasures include placing utility facilities at locations which protect or minimize exposure to out-of-control vehicles, using breakaway features, using impact attenuation devices, using delineation, or shielding.

(c) *Installations within freeways.* (1) Each State highway agency shall submit an accommodation plan in accordance with §§ 645.211 and 645.215 which addresses how the State highway agency will consider applications for longitudinal utility installations within the access control lines of a freeway. This includes utility installations within interchange areas which must be constructed or serviced by direct access from the main lanes or ramps. If a State highway agency elects to permit such use, the plan must address how the State highway agency will oversee such use consistent with this subpart, Title 23 U.S.C., and the safe and efficient use of the highways.

(2) Any accommodation plan shall assure that installations satisfy the following criteria:

(i) The effects utility installations will have on highway and traffic safety will be ascertained, since in no case shall any use be permitted which would adversely affect safety.

(ii) The direct and indirect environmental and economic effects of any loss of productive agricultural land or any productivity of any agricultural

Federal Highway Administration, DOT

§ 645.209

land which would result from the disapproval of the use of such right-of-way for accommodation of such utility facility will be evaluated.

(iii) These environmental and economic effects together with any interference with or impairment of the use of the highway in such right-of-way which would result from the use of such right-of-way for the accommodation of such utility facility will be considered.

(iv) [Reserved]

(v) A utility strip will be established along the outer edge of the right-of-way by locating a utility access control line between the proposed utility installation and the through roadway and ramps. Existing fences should be retained and, except along sections of freeways having frontage roads, planned fences should be located at the freeway right-of-way line. The State or political subdivision is to retain control of the utility strip right-of-way including its use by utility facilities. Service connections to adjacent properties shall not be permitted from within the utility strip.

(3) Nothing in this part shall be construed as prohibiting a highway agency from adopting a more restrictive policy than that contained herein with regard to longitudinal utility installations along freeway right-of-way and access for constructing and/or for servicing such installations.

(d) *Uniform policies and procedures.* For a highway agency to fulfill its responsibilities to control utility use of Federal-aid highway right-of-way within the State and its political subdivisions, it must exercise or cause to be exercised, adequate regulation over such use and occupancy through the establishment and enforcement of reasonably uniform policies and procedures for utility accommodation.

(e) *Private lines.* Because there are circumstances when private lines may be allowed to cross or otherwise occupy the right-of-way of Federal-aid projects, highway agencies shall establish uniform policies for properly controlling such permitted use. When permitted, private lines must conform to the provisions of this part and the provisions of 23 CFR 1.23(c) for longitudinal installations.

(f) *Direct Federal highway projects.* On direct Federal highway projects, the FHWA will apply, or cause to be applied, utility and private line accommodation policies similar to those required on Federal-aid highway projects. When appropriate, agreements will be entered into between the FHWA and the highway agency or other government agencies to ensure adequate control and regulation of use by utilities and private lines of the right-of-way on direct Federal highway projects.

(g) *Projects where state lacks authority.* On Federal-aid highway projects where the State highway agency does not have legal authority to regulate highway use by utilities and private lines, the State highway agency must enter into formal agreements with those local officials who have such authority. The agreements must provide for a degree of protection to the highway at least equal to the protection provided by the State highway agency's utility accommodation policy approved under the provisions of § 645.215(b) of this part. The project agreement between the State highway agency and the FHWA on all such Federal-aid highway projects shall contain a special provision incorporating the formal agreements with the responsible local officials.

(h) *Scenic areas.* New utility installations, including those needed for highway purposes, such as for highway lighting or to serve a weigh station, rest area or recreation area, are not permitted on highway right-of-way or other lands which are acquired or improved with Federal-aid or direct Federal highway funds and are located within or adjacent to areas of scenic enhancement and natural beauty. Such areas include public park and recreational lands, wildlife and waterfowl refuges, historic sites as described in 23 U.S.C. 138, scenic strips, overlooks, rest areas and landscaped areas. The State highway agency may permit exceptions provided the following conditions are met:

(1) New underground or aerial installations may be permitted only when they do not require extensive removal or alteration of trees or terrain features visible to the highway user or im-

§ 645.209

pair the aesthetic quality of the lands being traversed.

(2) Aerial installations may be permitted only when:

(i) Other locations are not available or are unusually difficult and costly, or are less desirable from the standpoint of aesthetic quality,

(ii) Placement underground is not technically feasible or is unreasonably costly, and

(iii) The proposed installation will be made at a location, and will employ suitable designs and materials, which give the greatest weight to the aesthetic qualities of the area being traversed. Suitable designs include, but are not limited to, self-supporting armless, single-pole construction with vertical configuration of conductors and cable.

(3) For new utility installations within freeways, the provisions of paragraph (c) of this section must also be satisfied.

(i) *Joint use agreements.* When the utility has a compensable interest in the land occupied by its facilities and such land is to be jointly occupied and used for highway and utility purposes, the highway agency and utility shall agree in writing as to the obligations and responsibilities of each party. Such joint-use agreements shall incorporate the conditions of occupancy for each party, including the rights vested in the highway agency and the rights and privileges retained by the utility. In any event, the interest to be acquired by or vested in the highway agency in any portion of the right-of-way of a Federal-aid or direct Federal highway project to be vacated, used or occupied by utilities or private lines, shall be adequate for the construction, safe operation, and maintenance of the highway project.

(j) *Traffic control plan.* Whenever a utility installation, adjustment or maintenance activity will affect the movement of traffic or traffic safety, the utility shall implement a traffic control plan and utilize traffic control devices as necessary to ensure the safe and expeditious movement of traffic around the work site and the safety of the utility work force in accordance with procedures established by the highway agency. The traffic control plan and the application of traffic control devices shall conform to the standards set forth in the "Manual on Uniform Traffic Control Devices" (MUTCD) and 23 CFR part 630, subpart J. (This publication is incorporated by reference and is on file at the Office of the Federal Register in Washington, DC. It is available for inspection and copying from the FHWA Washington Headquarters and all FHWA Division and Regional Offices as prescribed in 49 CFR part 7, appendix D.)

(k) *Corrective measures.* When the highway agency determines that existing utility facilities are likely to be associated with injury or accident to the highway user, as indicated by accident history or safety studies, the highway agency shall initiate or cause to be initiated in consultation with the affected utilities, corrective measures to provide for a safer traffic environment. The corrective measures may include changes to utility or highway facilities and should be prioritized to maximum safety benefits in the most cost-effective manner. The scheduling of utility safety improvements should take into consideration planned utility replacement or upgrading schedules, accident potential, and the availability of resources. It is expected that the requirements of this paragraph will result in an orderly and positive process to address the identified utility hazard problems in a timely and reasonable manner with due regard to the effect of the corrective measures on both the utility consumer and the road user. The type of corrective measures are not prescribed. Any requests received involving Federal participation in the cost of adjusting or relocating utility facilities pursuant to this paragrpah shall be subject to the provisions of 23 CFR part 645, subpart A, Utility Relocations, Adjustments and Reimbursement, and 23 CFR part 924, Highway Safety Improvement Program.

(l) *Wetlands.* The installation of privately owned lines or conduits on the right-of-way of Federal-aid or direct Federal highway projects for the purpose of draining adjacent wetlands onto the highway right-of-way is considered to be inconsistent with Executive Order 11990, Protection of Wet-

Federal Highway Administration, DOT § 645.211

lands, dated May 24, 1977, and shall be prohibited.

[50 FR 20354, May 15, 1985, as amended at 53 FR 2833, Feb. 2, 1988]

§ 645.211 State highway agency accommodation policies.

The FHWA shall use the AASHTO publication, "A Guide for Accommodating Utilities Within Highway Right-of-Way," 1981, and should use the AASHTO "Roadside Design Guide," 1989, to assist in the evaluation of adequacy of State highway agency utility accommodation policies. The "Guide for Accommodating Utilities" is incorporated by reference with the approval of the Director of the Federal Register in accordance with 5 U.S.C. 552(a) and 1 CFR part 51, and is on file at the Office of the Federal Register in Washington, DC. It is available for inspection from the FHWA Washington Headquarters and all FHWA Division and Regional Offices as prescribed in 49 CFR part 7, appendix D. Copies of current AASHTO publications are available for purchase from the American Association of State Highway and Transportation Officials, Suite 225, 444 North Capitol Street, NW., Washington, DC 20001. As a minimum, such policies shall make adequate provisions with respect to the following:

(a) Utilities must be accommodated and maintained in a manner which will not impair the highway or adversely affect highway or traffic safety. Uniform procedures controlling the manner, nature and extent of such utility use shall be established.

(b) Consideration shall be given to the effect of utility installations in regard to safety, aesthetic quality, and the costs or difficulty of highway and utility construction and maintenance.

(c) The State highway agency's standards for regulating the use and occupancy of highway right-of-way by utilities must include, but are not limited to, the following:

(1) The horizontal and vertical location requirements and clearances for the various types of utilities must be clearly stated. These must be adequate to ensure compliance with the clear roadside policies for the particular highway involved.

(2) The applicable provisions of government or industry codes required by law or regulation must be set forth or appropriately referenced, including highway design standards or other measures which the State highway agency deems necessary to provide adequate protection to the highway, its safe operation, aesthetic quality, and maintenance.

(3) Specifications for and methods of installation; requirements for preservation and restoration of highway facilities, appurtenances, and natural features and vegetation on the right-of-way; and limitations on the utility's activities within the right-of-way including installation within areas set forth by § 645.209(h) of this part should be prescribed as necessary to protect highway interests.

(4) Measures necessary to protect traffic and its safe operation during and after installation of facilities, including control-of-access restrictions, provisions for rerouting or detouring traffic, traffic control measures to be employed, procedures for utility traffic control plans, limitations on vehicle parking and materials storage, protection of open excavations, and the like must be provided.

(5) A State highway agency may deny a utility's request to occupy highway right-of-way based on State law, regulation, or ordinances or the State highway agency's policy. However, in any case where the provisions of this part are to be cited as the basis for disapproving a utility's request to use and occupy highway right-of-way, measures must be provided to evaluate the direct and indirect environmental and economic effects of any loss of productive agricultural land or any impairment of the productivity of any agricultural land that would result from the disapproval. The environmental and economic effects on productive agricultural land together with the possible interference with or impairment of the use of the highway and the effect on highway safety must be considered in the decision to disapprove any proposal by a utility to use such highway right-of-way.

(d) Compliance with applicable State laws and approved State highway agency utility accommodation policies

§ 645.213

must be assured. The responsible State highway agency's file must contain evidence of the written arrangements which set forth the terms under which utility facilities are to cross or otherwise occupy highway right-of-way. All utility installations made on highway right-of-way shall be subject to written approval by the State highway agency. However, such approval will not be required where so provided in the use and occupancy agreement for such matters as utility facility maintenance, installation of service connections on highways other than freeways, or emergency operations.

(e) The State highway agency shall set forth in its utility accommodation plan detailed procedures, criteria, and standards it will use to evaluate and approve individual applications of utilities on freeways under the provisions of § 645.209(c) of this part. The State highway agency also may develop such procedures, criteria and standards by class of utility. In defining utility classes, consideration may be given to distinguishing utility services by type, nature or function and their potential impact on the highway and its user.

(f) The means and authority for enforcing the control of access restrictions applicable to utility use of controlled access highway facilities should be clearly set forth in the State highway agency plan.

(Information collection requirements in paragraphs (a), (b) and (c) were approved under control number 2125-0522, and paragraph (d) under control number 2125-0514)

[50 FR 20354, May 15, 1985, as amended at 53 FR 2834, Feb. 2, 1988; 55 FR 25828, June 25, 1990]

§ 645.213 Use and occupancy agreements (permits).

The written arrangements, generally in the form of use and occupancy agreements setting forth the terms under which the utility is to cross or otherwise occupy the highway right-of-way, must include or incorporate by reference:

(a) The highway agency standards for accommodating utilities. Since all of the standards will not be applicable to each individual utility installation, the use and occupancy agreement must, as a minimum, describe the requirements for location, construction, protection of traffic, maintenance, access restriction, and any special conditions applicable to each installation.

(b) A general description of the size, type, nature, and extent of the utility facilities being located within the highway right-of-way.

(c) Adequate drawings or sketches showing the existing and/or proposed location of the utility facilities within the highway right-of-way with respect to the existing and/or planned highway improvements, the traveled way, the right-of-way lines and, where applicable, the control of access lines and approved access points.

(d) The extent of liability and responsibilities associated with future adjustment of the utilities to accommodate highway improvements.

(e) The action to be taken in case of noncompliance with the highway agency's requirements.

(f) Other provisions as deemed necessary to comply with laws and regulations.

(Approved by the Office of Management and Budget under control number 2125-0522)

§ 645.215 Approvals.

(a) Each State highway agency shall submit a statement to the FHWA on the authority of utilities to use and occupy the right-of-way of State highways, the State highway agency's power to regulate such use, and the policies the State highway agency employs or proposes to employ for accommodating utilities within the right-of-way Federal-aid highways under its jurisdiction. Statements previously submitted and approved by the FHWA need not be resubmitted provided the statement adequately addresses the requirements of this part. When revisions are deemed necessary the changes to the previously approved statement may be submitted separately to the FHWA for approval. The State highway agency shall include similar information on the use and occupancy of such highways by private lines where permitted. The State shall identify those areas, if any, of the Federal-aid highway system within its borders where the State highway agency is without legal authority to regulate use by utilities. The statement shall address the

Federal Highway Administration, DOT

§ 646.103

nature of the formal agreements with local officials required by § 645.209(g) of this part. It is expected that the statements required by this part or necessary revisions to previously submitted and approved statements will be submitted to FHWA within 1 year of the effective date of this regulation.

(b) Upon determination by the FHWA that a State highway agency's policies satisfy the provisions of 23 U.S.C. 109, 111, and 116, and 23 CFR 1.23 and 1.27, and meet the requirements of this regulation, the FHWA will approve their use on Federal-aid highway projects in that State

(c) Any changes, additions or deletions the State highway agency proposes to the approved policies are subject to FHWA approval.

(d) When a utility files a notice or makes an individual application or request to a State highway agency to use or occupy the right-of-way of a Federal-aid highway project, the State highway agency is not required to submit the matter to the FHWA for prior concurrence, except under the following circumstances:

(1) The proposed installation is not in accordance with this regulation or the State highway agency's utility accommodation policy approved by the FHWA for use on Federal-aid highway projects.

(2) Longitudinal installations of private lines.

(e) The State highway agency's practices under the policies or agreements approved under § 645.215(b) of this part shall be periodically reviewed by the FHWA.

(Information collection requirements in paragraph (a) were approved by the Office of Management and Budget under control number 2125-0514)

[50 FR 20354, May 15, 1985, as amended at 53 FR 2834, Feb. 2, 1988]

PART 646—RAILROADS

Subpart A—Railroad-Highway Insurance Protection

Sec.
646.101 Purpose.
646.103 Application.
646.105 Contractor's public liability and property damage insurance.
646.107 Railroad protective insurance.
646.109 Types of coverage.
646.111 Amount of coverage.

Subpart B—Railroad-Highway Projects

646.200 Purpose and applicability.
646.202 Authority.
646.204 Definitions.
646.206 Types of projects.
646.208 Funding.
646.210 Classification of projects and railroad share of the cost.
646.212 Federal share.
646.214 Design.
646.216 General procedures.
646.218 Simplified procedure for accelerating grade crossing improvements.
646.220 Alternate Federal-State procedure.

APPENDIX TO SUBPART B—HORIZONTAL AND VERTICAL CLEARANCE PROVISIONS FOR OVERPASS AND UNDERPASS STRUCTURES

AUTHORITY: 23 U.S.C. 109(e), 120(d), 130, and 315; 49 CFR 1.48(b).

Subpart A—Railroad-Highway Insurance Protection

SOURCE: 39 FR 36474, Oct. 10, 1974, unless otherwise noted.

§ 646.101 Purpose.

The purpose of this part is to prescribe provisions under which Federal funds may be applied to the costs of public liability and property damage insurance obtained by contractors (a) for their own operations, and (b) on behalf of railroads on or about whose right-of-way the contractors are required to work in the construction of highway projects financed in whole or in part with Federal funds.

§ 646.103 Application.

(a) This part applies:

(1) To a contractors' legal liability for bodily injury to, or death of, persons and for injury to, or destruction of, property.

(2) To the liability which may attach to railroads for bodily injury to, or death of, persons and for injury to, or destruction of, property.

(3) To damage to property owned by or in the care, custody or control of the railroads, both as such liability or damage may arise out of the contractor's operations, or may result from work performed by railroads at or about railroad rights-of-way in connec-

§ 646.105

tion with projects financed in whole or in part with Federal funds.

(b) Where the highway construction is under the direct supervision of the Federal Highway Administration (FHWA), all references herein to the State shall be considered as references to the FHWA.

§ 646.105 Contractor's public liability and property damage insurance.

(a) Contractors may be subject to liability with respect to bodily injury to or death of persons, and injury to, or destruction of property, which may be suffered by persons other than their own employees as a result of their operations in connection with construction of highway projects located in whole or in part within railroad right-of-way and financed in whole or in part with Federal funds. Protection to cover such liability of contractors shall be furnished under regular contractors' public liability and property damage insurance policies issued in the names of the contractors. Such policies shall be so written as to furnish protection to contractors respecting their operations in performing work covered by their contract.

(b) Where a contractor sublets a part of the work on any project to a subcontractor, the contractor shall be required to secure insurance protection in his own behalf under contractor's public liability and property damage insurance policies to cover any liability imposed on him by law for damages because of bodily injury to, or death of, persons and injury to, or destruction of, property as a result of work undertaken by such subcontractors. In addition, the contractor shall provide for and on behalf of any such subcontractors protection to cover like liability imposed upon the latter as a result of their operations by means of separate and individual contractor's public liability and property damage policies; or, in the alternative, each subcontractor shall provide satisfactory insurance on his own behalf to cover his individual operations.

(c) The contractor shall furnish to the State highway department evidence satisfactory to such department and to the FHWA that the insurance coverages required herein have been provided. The contractor shall also furnish a copy of such evidence to the railroad or railroads involved. The insurance specified shall be kept in force until all work required to be performed shall have been satisfactorily completed and accepted in accordance with the contract under which the construction work is undertaken.

§ 646.107 Railroad protective insurance.

In connection with highway projects for the elimination of hazards of railroad-highway crossings and other highway construction projects located in whole or in part within railroad right-of-way, railroad protective liability insurance shall be purchased on behalf of the railroad by the contractor. The standards for railroad protective insurance established by §§ 646.109 through 646.111 shall be adhered to insofar as the insurance laws of the State will permit.

[39 FR 36474, Oct. 10, 1974, as amended at 47 FR 33955, Aug. 5, 1982]

§ 646.109 Types of coverage.

(a) Coverage shall be limited to damage suffered by the railroad on account of occurrences arising out of the work of the contractor on or about the railroad right-of-way, independent of the railroad's general supervision or control, except as noted in § 646.109(b)(4).

(b) Coverage shall include:

(1) Death of or bodily injury to passengers of the railroad and employees of the railroad not covered by State workmen's compensation laws;

(2) Personal property owned by or in the care, custody or control of the railroads;

(3) The contractor, or any of his agents or employees who suffer bodily injury or death as the result of acts of the railroad or its agents, regardless of the negligence of the railroad;

(4) Negligence of only the following classes of railroad employees:

(i) Any supervisory employee of the railroad at the job site;

(ii) Any employee of the railroad while operating, attached to, or engaged on, work trains or other railroad equipment at the job site which are assigned exclusively to the contractor; or

Federal Highway Administration, DOT

(iii) Any employee of the railroad not within (b)(4) (i) or (ii) who is specifically loaned or assigned to the work of the contractor for prevention of accidents or protection of property, the cost of whose services is borne specifically by the contractor or governmental authority.

§ 646.111 Amount of coverage.

(a) The maximum dollar amounts of coverage to be reimbursed from Federal funds with respect to bodily injury, death and property damage is limited to a combined amount of $2 million per occurrence with an aggregate of $6 million applying separately to each annual period except as provided in paragraph (b) of this section.

(b) In cases involving real and demonstrable danger of appreciably higher risks, higher dollar amounts of coverage for which premiums will be reimbursable from Federal funds shall be allowed. These larger amounts will depend on circumstances and shall be written for the individual project in accordance with standard underwriting practices upon approval of the FHWA.

[39 FR 36474, Oct. 10, 1974, as amended at 47 FR 33955, Aug. 5, 1982]

Subpart B—Railroad-Highway Projects

SOURCE: 40 FR 16059, Apr. 9, 1975, unless otherwise noted.

§ 646.200 Purpose and applicability.

(a) The purpose of this subpart is to prescribe policies and procedures for advancing Federal-aid projects involving railroad facilities.

(b) This subpart, and all references hereinafter made to *projects*, applies to Federal-aid projects involving railroad facilities, including projects for the elimination of hazards of railroad-highway crossings, and other projects which use railroad properties or which involve adjustments required by highway construction to either railroad facilities or facilities that are jointly owned or used by railroad and utility companies.

(c) Additional instructions for projects involving the elimination of hazards of railroad-highway grade crossings pursuant to 23 U.S.C. 405 and Section 203 of the Highway Safety Act of 1973 are set forth in 23 CFR part 924.

(d) Procedures on reimbursement for projects undertaken pursuant to this subpart are set forth in 23 CFR part 140, subpart I.

(e) Procedures on insurance required of contractors working on or about railroad right-of-way are set forth in 23 CFR part 646, subpart A.

(f) Audit requirements for work undertaken pursuant to this subpart which is not accomplished under competitive bidding procedures are set forth in 23 CFR part 170.

[40 FR 16059, Apr. 9, 1975, as amended at 45 FR 20795, Mar. 31, 1980]

§ 646.202 Authority.

This subpart is issued under authority of 23 U.S.C. 109(e), 120(d), 130, 315, and 405, Section 203 of the Highway Safety Act of 1973, and 49 CFR 1.48.

§ 646.204 Definitions.

For the purposes of this subpart, the following definitions apply:

(a) *Railroad* shall mean all rail carriers, publicly-owned, private, and common carriers, including line haul freight and passenger railroads, switching and terminal railroads and passenger carrying railroads such as rapid transit, commuter and street railroads.

(b) *Utility* shall mean the lines and facilities for producing, transmitting or distributing communications, power, electricity, light, heat, gas, oil, water, steam, sewer and similar commodities.

(c) *Company* shall mean any railroad or utility company including any wholly owned or controlled subsidiary thereof.

(d) *'G' funds* signify those Federal-aid highway funds which, pursuant to 23 U.S.C. 120(d), may be used for projects for the elimination of hazards of railroad-highway crossings not to exceed 10 percent of Federal-aid funds apportioned in accordance with 23 U.S.C. 104.

(e) *Preliminary engineering* shall mean the work necessary to produce construction plans, specifications, and estimates to the degree of completeness required for undertaking construction thereunder, including locating, surveying, designing, and related work.

§ 646.206

(f) *Construction* shall mean the actual physical construction to improve or eliminate a railroad-highway grade crossing or accomplish other railroad involved work.

(g) *A diagnostic team* means a group of knowledgeable representatives of the parties of interest in a railroad-highway crossing or a group of crossings.

(h) *Main line railroad track* means a track of a principal line of a railroad, including extensions through yards, upon which trains are operated by timetable or train order or both, or the use of which is governed by block signals or by centralized traffic control.

(i) *Passive warning devices* means those types of traffic control devices, including signs, markings and other devices, located at or in advance of grade crossings to indicate the presence of a crossing but which do not change aspect upon the approach or presence of a train.

(j) *Active warning devices* means those traffic control devices activated by the approach or presence of a train, such as flashing light signals, automatic gates and similar devices, as well as manually operated devices and crossing watchmen, all of which display to motorists positive warning of the approach or presence of a train.

§ 646.206 Types of projects.

(a) Projects for the elimination of hazards, to both vehicles and pedestrians, of railroad-highway crossings may include but are not limited to:

(1) Grade crossing elimination;
(2) Reconstruction of existing grade separations; and
(3) Grade crossing improvements.

(b) Other railroad-highway projects are those which use railroad properties or involve adjustments to railroad facilities required by highway construction but do not involve the elimination of hazards of railroad-highway crossings. Also included are adjustments to facilities that are jointly owned or used by railroad and utility companies.

§ 646.208 Funding.

(a) Federal-aid funding for projects which involve the elimination of hazards of railroad-highway crossings may, at the option of the State, be provided through one of the following alternative methods, within the qualifications prescribed for each:

(1) "G" funding, as provided by 23 U.S.C. 120(d) and 130;
(2) Regular pro rata sharing as provided by 23 U.S.C. 120(a) and 120(c); and
(3) Funding authorized by 23 U.S.C. 405 and section 203 of the Highway Safety Act of 1973.

(b) The adjustment of railroad facilities which does not involve the elimination of hazards of railroad-highway crossings may be funded through regular pro rata sharing, as provided by 23 U.S.C. 120 (a) and (c).

§ 646.210 Classification of projects and railroad share of the cost.

(a) State laws requiring railroads to share in the cost of work for the elimination of hazards at railroad-highway crossings shall not apply to Federal-aid projects.

(b) Pursuant to 23 U.S.C. 130(b), and 49 CFR 1.48:

(1) Projects for grade crossing improvements are deemed to be of no ascertainable net benefit to the railroads and there shall be no required railroad share of the costs.

(2) Projects for the reconstruction of existing grade separations are deemed to generally be of no ascertainable net benefit to the railroad and there shall be no required railroad share of the costs, unless the railroad has a specific contractual obligation with the State or its political subdivision to share in the costs.

(3) On projects for the elimination of existing grade crossings at which active warning devices are in place or ordered to be installed by a State regulatory agency, the railroad share of the project costs shall be 5 percent.

(4) On projects for the elimination of existing grade crossings at which active warning devices are not in place and have not been ordered installed by a State regulatory agency, or on projects which do not eliminate an existing crossing, there shall be no required railroad share of the project cost.

(c) The required railroad share of the cost under § 646.210(b)(3) shall be based on the costs for preliminary engineering, right-of-way and construction within the limits described below:

Federal Highway Administration, DOT

(1) Where a grade crossing is eliminated by grade separation, the structure and approaches required to transition to a theoretical highway profile which would have been constructed if there were no railroad present, for the number of lanes on the existing highway and in accordance with the current design standards of the State highway agency.

(2) Where another facility, such as a highway or waterway, requiring a bridge structure is located within the limits of a grade separation project, the estimated cost of a theoretical structure and approaches as described in § 646.210(c)(1) to eliminate the railroad-highway grade crossing without considering the presence of the waterway or other highway.

(3) Where a grade crossing is eliminated by railroad or highway relocation, the actual cost of the relocation project, the estimated cost of the relocation project, or the estimated cost of a structure and approaches as described in § 646.210(c)(1), whichever is less.

(d) Railroads may voluntarily contribute a greater share of project costs than is required. Also, other parties may voluntarily assume the railroad's share.

§ 646.212 Federal share.

(a) *General.* (1) Federal funds are not eligible to participate in costs incurred solely for the benefit of the railroad.

(2) At grade separations Federal funds are eligible to participate in costs to provide space for more tracks than are in place when the railroad establishes to the satisfaction of the State highway agency and FHWA that it has a definite demand and plans for installation of the additional tracks within a reasonable time.

(3) The Federal share of the cost of a grade separation project shall be based on the cost to provide horizontal and/or vertical clearances used by the railroad in its normal practice subject to limitations as shown in the appendix or as required by a State regulatory agency.

(b) *"G" Funds.* (1) The Federal share of the cost of a "G" funded project may be up to 100 percent of the cost of preliminary engineering and construction and 75 percent of the cost of right-of-way and property damage, except that the Federal share shall be reduced by the amount of any required railroad share of the cost.

(2) Projects for the elimination of hazards of railroad-highway crossings, either by crossing elimination, improvement, or the reconstruction of existing grade separations, as described in § 646.206(a) are eligible for "G" funding subject to the following limitations:

(i) For a new or reconstructed grade separation, the entire structure or structures and necessary highway and railroad approaches to accommodate both vehicular and pedestrian traffic.

(ii) Where another facility, such as a highway or waterway requiring a bridge structure, is located within the limits of a grade separation project, the estimated cost and limits of work for a theoretical structure and necessary approaches as in § 646.212 (b)(2)(i) without considering the presence of the waterway or other highway.

(iii) For railroad or highway relocation the actual cost of the relocation project or the estimated cost of a theoretical structure and necessary approaches to eliminate the grade crossing(s) as in § 646.212(b)(2)(i), whichever is less.

(iv) Grade crossing improvements in the vicinity of the crossing and related work, including construction or reconstruction of the approaches as necessary to provide an acceptable transition to existing or improved highway gradients and alignments, and advance warning devices.

[40 FR 16059, Apr. 9, 1975, as amended at 47 FR 33955, Aug. 5, 1982; 53 FR 32218, Aug. 24, 1988]

§ 646.214 Design.

(a) *General.* (1) Facilities that are the responsibility of the railroad for maintenance and operation shall conform to the specifications and design standards used by the railroad in its normal practice, subject to approval by the State highway agency and FHWA.

(2) Facilities that are the responsibility of the highway agency for maintenance and operation shall conform to the specifications and design standards and guides used by the highway agency

§ 646.216

in its normal practice, subject to approval by FHWA.

(b) *Grade crossing improvements.* (1) All traffic control devices proposed shall comply with the latest edition of the Manual on Uniform Traffic Control Devices for Streets and Highways supplemented to the extent applicable by State standards.

(2) Pursuant to 23 U.S.C. 109(e), where a railroad-highway grade crossing is located within the limits of or near the terminus of a Federal-aid highway project for construction of a new highway or improvement of the existing roadway, the crossing shall not be opened for unrestricted use by traffic or the project accepted by FHWA until adequate warning devices for the crossing are installed and functioning properly.

(3)(i) *Adequate warning devices*, under § 646.214(b)(2) or on any project where Federal-aid funds participate in the installation of the devices are to include automatic gates with flashing light signals when one or more of the following conditions exist:

(A) Multiple main line railroad tracks.

(B) Multiple tracks at or in the vicinity of the crossing which may be occupied by a train or locomotive so as to obscure the movement of another train approaching the crossing.

(C) High Speed train operation combined with limited sight distance at either single or multiple track crossings.

(D) A combination of high speeds and moderately high volumes of highway and railroad traffic.

(E) Either a high volume of vehicular traffic, high number of train movements, substantial numbers of schoolbuses or trucks carrying hazardous materials, unusually restricted sight distance, continuing accident occurrences, or any combination of these conditions.

(F) A diagnostic team recommends them.

(ii) In individual cases where a diagnostic team justifies that gates are not appropriate, FHWA may find that the above requirements are not applicable.

(4) For crossings where the requirements of § 646.214(b)(3) are not applicable, the type of warning device to be installed, whether the determination is made by a State regulatory agency, State highway agency, and/or the railroad, is subject to the approval of FHWA.

(c) *Grade crossing elimination.* All crossings of railroads and highways at grade shall be eliminated where there is full control of access on the highway (a freeway) regardless of the volume of railroad or highway traffic.

[40 FR 16059, Apr. 9, 1975, as amended at 47 FR 33955, Aug. 5, 1982]

§ 646.216 General procedures.

(a) *General.* Unless specifically modified herein, applicable Federal-aid procedures govern projects undertaken pursuant to this subpart.

(b) *Preliminary engineering and engineering services.* (1) As mutually agreed to by the State highway agency and railroad, and subject to the provisions of § 646.216(b)(2), preliminary engineering work on railroad-highway projects may be accomplished by one of the following methods:

(i) The State or railroad's engineering forces;

(ii) An engineering consultant selected by the State after consultation with the railroad, and with the State administering the contract; or

(iii) An engineering consultant selected by the railroad, with the approval of the State and with the railroad administering the contract.

(2) Where a railroad is not adequately staffed, Federal-aid funds may participate in the amounts paid to engineering consultants and others for required services, provided such amounts are not based on a percentage of the cost of construction, either under contracts for individual projects or under existing written continuing contracts where such work is regularly performed for the railroad in its own work under such contracts at reasonable costs.

(c) *Rights-of-way.* (1) Acquisition of right-of-way by a State highway agency on behalf of a railroad or acquisition of nonoperating real property from a railroad shall be in accordance with the Uniform Relocation Assistance and Real Property Acquisition Policies Act of 1970 (42 U.S.C. 4601 *et seq.*) and applicable FHWA right-of-way procedures in 23 CFR, chapter I, subchapter H. On projects for the elimination of hazards

Federal Highway Administration, DOT § 646.216

of railroad-highway crossings by the relocation of railroads, acquisition or replacement right-of-way by a railroad shall be in accordance with 42 U.S.C. 4601 et seq.

(2) Where buildings and other depreciable structures of the railroad (such as signal towers, passenger stations, depots, and other buildings, and equipment housings) which are integral to operation of railroad traffic are wholly or partly affected by a highway project, the costs of work necessary to functionally restore such facilities are eligible for participation. However, when replacement of such facilities is necessary, credits shall be made to the cost of the project for:

(i) Accrued depreciation, which is that amount based on the ratio between the period of actual length of service and total life expectancy applied to the original cost.

(ii) Additions or improvements which provide higher quality or increased service capability of the facility and which are provided solely for the benefit of the railroad.

(iii) Actual salvage value of the material recovered from the facility being replaced. Total credits to a project shall not be required in excess of the replacement cost of the facility.

(3) Where Federal funds participate in the cost of replacement right-of-way, there will be no charge to the project for the railroad's existing right-of-way being transferred to the State highway agency except when the value of the right-of-way being taken exceeds the value of the replacement right-of-way.

(d) *State-railroad agreements.* (1) Where construction of a Federal-aid project requires use of railroad properties or adjustments to railroad facilities, there shall be an agreement in writing between the State highway agency and the railroad company.

(2) The written agreement between the State and the railroad shall, as a minimum include the following, where applicable:

(i) The provisions of this subpart and of 23 CFR part 140, subpart I, incorporated by reference.

(ii) A detailed statement of the work to be performed by each party.

(iii) Method of payment (either actual cost or lump sum),

(iv) For projects which are not for the elimination of hazards of railroad-highway crossings, the extent to which the railroad is obligated to move or adjust its facilities at its own expense,

(v) The railroad's share of the project cost,

(vi) An itemized estimate of the cost of the work to be performed by the railroad,

(vii) Method to be used for performing the work, either by railroad forces or by contract,

(viii) Maintenance responsibility,

(ix) Form, duration, and amounts of any needed insurance,

(x) Appropriate reference to or identification of plans and specifications,

(xi) Statements defining the conditions under which the railroad will provide or require protective services during performance of the work, the type of protective services and the method of reimbursement to the railroad, and

(xii) Provisions regarding inspection of any recovered materials.

(3) On work to be performed by the railroad with its own forces and where the State highway agency and railroad agree, subject to approval by FHWA, an agreement providing for a lump sum payment in lieu of later determination of actual costs may be used for any of the following:

(i) Installation or improvement of grade crossing warning devices and/or grade crossing surfaces, regardless of cost, or

(ii) Any other eligible work where the estimated cost to the State of the proposed railroad work does not exceed $25,000 or

(iii) Where FHWA finds that the circumstances are such that this method of developing costs would be in the best interest of the public.

(4) Where the lump sum method of payment is used, periodic reviews and analyses of the railroad's methods and cost data used to develop lump sum estimates will be made.

(5) Master agreements between a State and a railroad on an areawide or statewide basis may be used. These agreements would contain the specifications, regulations, and provisions required in conjunction with work per-

251

§ 646.216

formed on all projects. Supporting data for each project or group of projects must, when combined with the master agreement by reference, satisfy the provisions of § 646.216(d)(2).

(6) Official orders issued by regulatory agencies will be accepted in lieu of State-railroad agreements only where, together with supplementary written understandings between the State and the railroad, they include the items required by § 646.216(d)(2).

(7) In extraordinary cases where FHWA finds that the circumstances are such that requiring such agreement or order would not be in the best interest of the public, projects may be approved for construction with the aid of Federal funds, provided satisfactory commitments have been made with respect to construction, maintenance and the railroad share of project costs.

(e) *Authorizations.* (1) The costs of preliminary engineering, right-of-way acquisition, and construction incurred after the date each phase of the work is included in an approved program and authorized by FHWA are eligible for Federal-aid participation. Preliminary engineering and right-of-way acquisition costs which are otherwise eligible, but incurred by a railroad prior to authorization by FHWA, although not reimbursable, may be included as part of the railroad share of project cost where such a share is required.

(2) Prior to issuance of authorization by FHWA either to advertise the physical construction for bids or to proceed with force account construction for railroad work or for other construction affected by railroad work, the following must be accomplished:

(i) The plans, specifications and estimates must be approved by FHWA.

(ii) A proposed agreement between the State and railroad must be found satisfactory by FHWA. Before Federal funds may be used to reimburse the State for railroad costs the executed agreement must be approved by FHWA. However, cost for materials stockpiled at the project site or specifically purchased and delivered to the company for use on the project may be reimbursed on progress billings prior to the approval of the executed State-Railroad Agreement in accordance with 23 CFR 140.922(a) and § 646.218 of this part.

(iii) Adequate provisions must be made for any needed easements, right-of-way, temporary crossings for construction purposes or other property interests.

(iv) The pertinent portions of the State-railroad agreement applicable to any protective services required during performance of the work must be included in the project specifications and special provisions for any construction contract.

(3) In unusual cases, pending compliance with § 646.216(e)(2)(ii), (iii) and (iv), authorization may be given by FHWA to advertise for bids for highway construction under conditions where a railroad grants a right-of-entry to its property as necessary to prosecute the physical construction.

(f) *Construction.* (1) Construction may be accomplished by:

(i) Railroad force account,

(ii) Contracting with the lowest qualified bidder based on appropriate solicitation,

(iii) Existing continuing contracts at reasonable costs, or

(iv) Contract without competitive bidding, for minor work, at reasonable costs.

(2) Reimbursement will not be made for any increased costs due to changes in plans:

(i) For the convenience of the contractor, or

(ii) Not approved by the State and FHWA.

(3) The State and FHWA shall be afforded a reasonable opportunity to inspect materials recovered by the railroad prior to disposal by sale or scrap. This requirement will be satisfied by the railroad giving written notice, or oral notice with prompt written confirmation, to the State of the time and place where the materials will be available for inspection. The giving of notice is the responsibility of the railroad, and it may be held accountable for full value of materials disposed of without notice.

(4) In addition to normal construction costs, the following construction costs are eligible for participation with Federal-aid funds when approved by the State and FHWA:

(i) The cost of maintaining temporary facilities of a railroad company

Federal Highway Administration, DOT § 646.220

required by and during the highway construction to the extent that such costs exceed the documented normal cost of maintaining the permanent facilities.

(ii) The cost of stage or extended construction involving grade corrections and/or slope stabilization for permanent tracks of a railroad which are required to be relocated on new grade by the highway construction. Stage or extended construction will be approved by FHWA only when documentation submitted by the State establishes the proposed method of construction to be the only practical method and that the cost of the extended construction within the period specified is estimated to be less than the cost of any practicable alternate procedure.

(iii) The cost of restoring the company's service by adustments of existing facilities away from the project site, in lieu of and not to exceed the cost of replacing, adjusting or relocating facilities at the project site.

(iv) The cost of an addition or improvement to an existing railroad facility which is required by the highway construction.

[40 FR 16059, Apr. 9, 1975, as amended at 40 FR 29712, July 15, 1975; 47 FR 33956, Aug. 5, 1982]

§ 646.218 Simplified procedure for accelerating grade crossing improvements.

(a) The procedure set forth in this section is encouraged for use in simplifying and accelerating the processing of single or multiple grade crossing improvements.

(b) Eligible preliminary engineering costs may include those incurred in selecting crossings to be improved, determining the type of improvement for each crossing, estimating the cost and preparing the required agreement.

(c) The written agreement between a State and a railroad shall contain as a minimum:

(1) Identification of each crossing location.

(2) Description of improvement and estimate of cost for each crossing location.

(3) Estimated schedule for completion of work at each location.

(d) Following programming, authorization and approval of the agreement under § 646.218(c), FHWA may authorize construction, including acquisition of warning device materials, with the condition that work at any particular location will not be undertaken until the proposed or executed State-railroad agreement under § 646.216(d)(2) is found satisfactory by FHWA and the final plans, specifications, and estimates are approved and with the condition that only material actually incorporated into the project will be eligible for Federal participation.

(e) Work programmed and authorized under this simplified procedure should include only that which can reasonably be expected to reach the construction stage within one year and be completed within two years after the initial authorization date.

§ 646.220 Alternate Federal-State procedure.

(a) On other than Interstate projects, an alternate procedure may be used, at the election of the State, for processing certain types of railroad-highway work. Under this procedure, the State highway agency will act in the relative position of FHWA for reviewing and approving projects.

(b) The scope of the State's approval authority under the alternate procedure includes all actions necessary to advance and complete the following types of railroad-highway work:

(1) All types of grade crossing improvements under § 646.206(a)(3).

(2) Minor adjustments to railroad facilities under § 646.206(b).

(c) The following types of work are to be reviewed and approved in the normal manner, as prescribed elsewhere in this subpart.

(1) All projects under § 646.206(a) (1) and (2).

(2) Major adjustments to railroad facilities under § 646.206(b).

(d) Any State wishing to adopt the alternate procedure may file a formal application for approval by FHWA. The application must include the following:

(1) The State's written policies and procedures for administering and processing Federal-aid railroad-highway work, which make adequate provisions with respect to all of the following:

253

Pt. 646, Subpt. B, App.

(i) Compliance with the provisions of title 23 U.S.C., title 23 CFR, and other applicable Federal laws and Executive Orders.

(ii) Compliance with this subpart and 23 CFR part 140, subpart I and 23 CFR part 172.

(iii) For grade crossing safety improvements, compliance with the requirements of 23 CFR part 924.

(2) A statement signed by the Chief Administrative Officer of the State highway agency certifying that:

(i) The work will be done in accordance with the applicable provisions of the State's policies and procedures submitted under § 646.220(d)(1), and

(ii) Reimbursement will be requested in only those costs properly attributable to the highway construction and eligible for Federal fund participation.

(e) When FHWA has approved the alternate procedure, it may authorize the State to proceed in accordance with the State's certification, subject to the following conditions:

(1) The work has been programmed.

(2) The State submits in writing a request for such authorization which shall include a list of the improvements or adjustments to be processed under the alternate procedure, along with the best available estimate of cost.

(f) The FHWA Regional Administrator may suspend approval of the certified procedure, where FHWA reviews disclose noncompliance with the certification. Federal-aid funds will not be eligible to participate in costs that do not qualify under § 646.220(d)(1).

[40 FR 16059, Apr. 9, 1975; 40 FR 29712, July 15, 1975; 40 FR 31211, July 25, 1975; 42 FR 30835, June 17, 1977, as amended at 45 FR 20795, Mar. 31, 1980]

APPENDIX TO SUBPART B—HORIZONTAL AND VERTICAL CLEARANCE PROVISIONS FOR OVERPASS AND UNDERPASS STRUCTURES

The following implements provisions of 23 CFR 646.212(a)(3).

a. *Lateral Geometrics*

A cross section with a horizontal distance of 20 feet, measured at right angles from the centerline of track at the top of rails, to the face of the embankment slope, may be approved. The 20-foot distance may be increased at individual structure locations as appropriate to provide for drainage if justified by a hydraulic analysis or to allow adequate room to accommodate special conditions, such as where heavy and drifting snow is a problem. The railroad must demonstrate that this is its normal practice to address these special conditions in the manner proposed. Additionally, this distance may also be increased up to 8 feet as may be necessary for off-track maintenance equipment, provided adequate horizontal clearance is not available in adjacent spans and where justified by the presence of an existing maintenance road or by evidence of future need for such equipment. All piers should be placed at least 9 feet horizontally from the centerline of the track and preferably beyond the drainage ditch. For multiple track facilities, all dimensions apply to the centerline of the outside track.

Any increase above the 20-foot horizontal clearance distance must be required by specific site conditions and be justified by the railroad to the satisfaction of the State highway agency (SHA) and the FHWA.

b. *Vertical Clearance*

A vertical clearance of 23 feet above the top of rails, which includes an allowance for future ballasting of the railroad tracks, may be approved. Vertical clearance greater than 23 feet may be approved when the State regulatory agency having jurisdiction over such matters requires a vertical clearance in excess of 23 feet or on a site by site basis where justified by the railroad to the satisfaction of the SHA and the FHWA. A railroad's justification for increased vertical clearance should be based on an analysis of engineering, operational and/or economic conditions at a specific structure location.

Federal-aid highway funds are also eligible to participate in the cost of providing vertical clearance greater than 23 feet where a railroad establishes to the satisfaction of a SHA and the FHWA that it has a definite formal plan for electrification of its rail system where the proposed grade separation project is located. The plan must cover a logical independent segment of the rail system and be approved by the railroad's corporate headquarters. For 25 kv line, a vertical clearance of 24 feet 3 inches may be approved. For 50 kv line, a vertical clearance of 26 feet may be approved.

A railroad's justification to support its plan for electrification shall include maps and plans or drawings showing those lines to be electrified; actions taken by its corporate headquarters committing it to electrification including a proposed schedule; and actions initiated or completed to date implementing its electrification plan such as a showing of the amounts of funds and identification of structures, if any, where the railroad has expended its own funds to provide added clearance for the proposed electrification. If available, the railroad's justification should include information on its con-

Federal Highway Administration, DOT

templated treatment of existing grade separations along the section of its rail system proposed for electrification.

The cost of reconstructing or modifying any existing railroad-highway grade separation structures solely to accommodate electrification will not be eligible for Federal-aid highway fund participation.

c. *Railroad Structure Width*

Nine feet of structure width outside of the centerline of the outside tracks may be approved for a structure carrying railroad tracks. Greater structure width may be approved when in accordance with standards established and used by the affected railroad in its normal practice.

In order to maintain continuity of off-track equipment roadways at structures carrying tracks over limited access highways, consideration should be given at the preliminary design stage to the feasibility of using public road crossings for this purpose. Where not feasible, an additional structure width of 8 feet may be approved if designed for off-track equipment only.

[53 FR 32218, Aug. 24, 1988]

PART 650—BRIDGES, STRUCTURES, AND HYDRAULICS

Subpart A—Location and Hydraulic Design of Encroachments on Flood Plains

Sec.
650.101 Purpose.
650.103 Policy.
650.105 Definitions.
650.107 Applicability.
650.109 Public involvement.
650.111 Location hydraulic studies.
650.113 Only practicable alternative finding.
650.115 Design standards.
650.117 Content of design studies.

Subpart B—Erosion and Sediment Control on Highway Construction Projects

650.201 Purpose.
650.203 Policy.
650.205 Definitions.
650.207 Plans, specifications, and estimates.
650.209 Construction.
650.211 Guidelines.

Subpart C—National Bridge Inspection Standards

650.301 Application of standards.
650.303 Inspection procedures.
650.305 Frequency of inspections.
650.307 Qualifications of personnel.
650.309 Inspection report.
650.311 Inventory.

§ 650.103

Subpart D—Highway Bridge Replacement and Rehabilitation Program

650.401 Purpose.
650.403 Definition of terms.
650.405 Eligible projects.
650.407 Application for bridge replacement or rehabilitation.
650.409 Evaluation of bridge inventory.
650.411 Procedures for bridge replacement and rehabilitation projects.
650.413 Funding.
650.415 Reports.

Subparts E–F—(Reserved)

Subpart G—Discretionary Bridge Candidate Rating Factor

650.701 Purpose.
650.703 Eligible projects.
650.705 Application for discretionary bridge funds.
650.707 Rating factor.
650.709 Special considerations.

Subpart H—Navigational Clearances for Bridges

650.801 Purpose.
650.803 Policy.
650.805 Bridges not requiring a USCG permit.
650.807 Bridges requiring a USCG permit.
650.809 Movable span bridges.

AUTHORITY: 23 U.S.C. 109 (a) and (h), 144, 151, 315, and 319; 23 CFR 1.32; 49 CFR 1.48(b); E.O. 11988 (3 CFR, 1977 Comp., p. 117); Department of Transportation Order 5650.2 dated April 23, 1979 (44 FR 24678); § 161 of Public Law 97–424, 96 Stat. 2097, 3135; § 4(b) of Public Law 97–134, 95 Stat. 1699; 33 U.S.C. 401, 491 et seq., 511 et seq.; and § 1057 of Public Law 102–240, 105 Stat. 2002.

Subpart A—Location and Hydraulic Design of Encroachments on Flood Plains

SOURCE: 44 FR 67580, Nov. 26, 1979, unless otherwise noted.

§ 650.101 **Purpose.**

To prescribe Federal Highway Administration (FHWA) policies and procedures for the location and hydraulic design of highway encroachments on flood plains, including direct Federal highway projects administered by the FHWA.

§ 650.103 **Policy.**

It is the policy of the FHWA:

Roadway Defects and Tort Liability

§ 650.309

Training Manual,"[2] which has been developed by a joint Federal-State task force, and subsequent additions to the manual.[3]

(b) An individual in charge of a bridge inspection team shall possess the following minimum qualifications:

(1) Have the qualifications specified in paragraph (a) of this section; or

(2) Have a minimum of 5 years experience in bridge inspection assignments in a responsible capacity and have completed a comprehensive training course based on the "Bridge Inspector's Training Manual," which has been developed by a joint Federal-State task force.

(3) Current certification as a Level III or IV Bridge Safety Inspector under the National Society of Professional Engineer's program for National Certification in Engineering Technologies (NICET)[4] is an alternate acceptable means for establishing that a bridge inspection team leader is qualified.

[36 FR 7851, Apr. 27, 1971. Redesignated at 39 FR 10430, Mar. 20, 1974, and amended at 44 FR 25435, May 1, 1979; 53 FR 32616, Aug. 26, 1988]

§ 650.309 **Inspection report.**

The findings and results of bridge inspections shall be recorded on standard forms. The data required to complete the forms and the functions which must be performed to compile the data are contained in section 3 of the AASHTO Manual.

[39 FR 29590, Aug. 16, 1974]

---

[2] The "Bridge Inspector's Training Manual" may be purchased from the Superintendent of Documents, U.S. Government Printing Office, Washington, DC 20402.

[3] The following publications are supplements to the "Bridge Inspector's Training Manual": "Bridge Inspector's Manual for Movable Bridges," 1977, GPO Stock No. 050-002-00103-5; "Culvert Inspector's Training Manual," July 1986, GPO Stock No. 050-001-0030-7; and "Inspection of Fracture Critical Bridge Members," 1986, GPO Stock No. 050-001-00302-3.

[4] For information on NICET program certification contact: National Institute for Certification in Engineering Technologies, 1420 King Street, Alexandria, Virginia 22314, Attention: John D. Antrim, P.E., Phone (703) 684-2835.

23 CFR Ch. I (4-1-95 Edition)

§ 650.311 **Inventory.**

(a) Each State shall prepare and maintain an inventory of all bridge structures subject to the Standards. Under these Standards, certain structure inventory and appraisal data must be collected and retained within the various departments of the State organization for collection by the Federal Highway Administration as needed. A tabulation of this data is contained in the structure inventory and appraisal sheet distributed by the Federal Highway Administration as part of the Recording and Coding Guide for the Structure Inventory and Appraisal of the Nation's Bridges (Coding Guide) in January of 1979. Reporting procedures have been developed by the Federal Highway Administration.

(b) Newly completed structures, modification of existing structures which would alter previously recorded data on the inventory forms or placement of load restriction signs on the approaches to or at the structure itself shall be entered in the State's inspection reports and the computer inventory file as promptly as practical, but no later than 90 days after the change in the status of the structure for bridges directly under the State's jurisdiction and no later than 180 days after the change in status of the structure for all other bridges on public roads within the State.

[44 FR 25435, May 1, 1979, as amended at 53 FR 32617, Aug. 26, 1988]

Subpart D—Highway Bridge Replacement and Rehabilitation Program

SOURCE: 44 FR 15665, Mar. 15, 1979, unless otherwise noted.

§ 650.401 **Purpose.**

The purpose of this regulation is to prescribe policies and outline procedures for administering the Highway Bridge Replacement and Rehabilitation Program in accordance with 23 U.S.C. 144.

§ 650.403 **Definition of terms.**

As used in this regulation:

(a) *Bridge.* A structure, including supports, erected over a depression or an

Federal Highway Administration, DOT § 650.409

obstruction, such as water, a highway, or a railway, having a track or passageway for carrying traffic or other moving loads, and having an opening measured along the center of the roadway of more than 20 feet between undercopings of abutments or spring lines of arches, or extreme ends of the openings for multiple boxes; it may include multiple pipes where the clear distance between openings is less than half of the smaller contiguous opening.

(b) *Sufficiency rating.* The numerical rating of a bridge based on its structural adequacy and safety, essentiality for public use, and its serviceability and functional obsolescence.

(c) *Rehabilitation.* The major work required to restore the structural integrity of a bridge as well as work necessary to correct major safety defects.

§ 650.405 Eligible projects.

(a) *General.* Deficient highway bridges on all public roads may be eligible for replacement or rehabilitation.

(b) *Types of projects which are eligible.* The following types of work are eligible for participation in the Highway Bridge Replacement and Rehabilitation Program (HBRRP), hereinafter known as the bridge program.

(1) *Replacement.* Total replacement of a structurally deficient or functionally obsolete bridge with a new facility constructed in the same general traffic corridor. A nominal amount of approach work, sufficient to connect the new facility to the existing roadway or to return the gradeline to an attainable touchdown point in accordance with good design practice is also eligible. The replacement structure must meet the current geometric, construction and structural standards required for the types and volume of projected traffic on the facility over its design life.

(2) *Rehabilitation.* The project requirements necessary to perform the major work required to restore the structural integrity of a bridge as well as work necessary to correct major safety defects are eligible except as noted under ineligible work. Bridges to be rehabilitated both on or off the F-A System shall, as a minimum, conform with the provisions of 23 CFR part 625, Design Standards for Federal-aid Highways, for the class of highway on which the bridge is a part.

(c) *Ineligible work.* Except as otherwise prescribed by the Administrator, the costs of long approach fills, causeways, connecting roadways, interchanges, ramps, and other extensive earth structures, when constructed beyond the attainable touchdown point, are not eligible under the bridge program.

§ 650.407 Application for bridge replacement or rehabilitation.

(a) Agencies participate in the bridge program by conducting bridge inspections and submitting Structure Inventory and Appraisal (SI&A) sheet inspection data. Federal and local governments supply SI&A sheet data to the State agency for review and processing. The State is responsible for submitting the six computer card format or tapes containing all public road SI&A sheet bridge information through the Division Administrator of the Federal Highway Administration (FHWA) for processing. These requirements are prescribed in 23 CFR 650.309 and 650.311, the National Bridge Inspection Standards.

(b) Inventory data may be submitted as available and shall be submitted at such additional times as the FHWA may request.

(c) Inventory data on bridges that have been strengthened or repaired to eliminate deficiencies, or those that have been replaced or rehabilitated using bridge replacement and/or other funds, must be revised in the inventory through data submission.

(d) The Secretary may, at the request of a State, inventory bridges, on and off the Federal-aid system, for historic significance.

[44 FR 15665, Mar. 15, 1979, as amended at 44 FR 72112, Dec. 13, 1979]

§ 650.409 Evaluation of bridge inventory.

(a) *Sufficiency rating of bridges.* Upon receipt and evaluation of the bridge inventory, a sufficiency rating will be assigned to each bridge by the Secretary in accordance with the approved

§ 650.411

AASHTO[1] sufficiency rating formula. The sufficiency rating will be used as a basis for establishing eligibility and priority for replacement or rehabilitation of bridges; in general the lower the rating, the higher the priority.

(b) *Selection of bridges for inclusion in State program.* After evaluation of the inventory and assignment of sufficiency ratings, the Secretary will provide the State with a selection list of bridges within the State that are eligible for the bridge program. From that list or from previously furnished selection lists, the State may select bridge projects.

§ 650.411 Procedures for bridge replacement and rehabilitation projects.

(a) Consideration shall be given to projects which will remove from service highway bridges most in danger of failure.

(b) *Submission and approval of projects.* (1) Bridge replacement or rehabilitation projects shall be submitted by the State to the Secretary in accordance with 23 CFR part 630, subpart A Federal-Aid Programs, Approval and Authorization.

(2) Funds apportioned to a State shall be made available throughout each State on a fair and equitable basis.

(c)(1) Each approved project will be designed, constructed, and inspected for acceptance in the same manner as other projects on the system on which the project is located. It shall be the responsibility of the State agency to properly maintain, or cause to be properly maintained, any project constructed under this bridge program. The State highway agency shall enter into a formal agreement for maintenance with appropriate local government officials in cases where an eligible project is located within and is under the legal authority of such a local government.

(2) Whenever a deficient bridge is replaced or its deficiency alleviated by a new bridge under the bridge program, the deficient bridge shall either be dismantled or demolished or its use limited to the type and volume of traffic the structure can safely service over its remaining life. For example, if the only deficiency of the existing structure is inadequate roadway width and the combination of the new and existing structure can be made to meet current standards for the volume of traffic the facility will carry over its design life, the existing bridge may remain in place and be incorporated into the system.

[44 FR 15665, Mar. 15, 1979, as amended at 44 FR 72112, Dec. 13, 1979]

§ 650.413 Funding.

(a) Funds authorized for carrying out the Highway Bridge Replacement and Rehabilitation Program are available for obligation at the beginning of the fiscal year for which authorized and remain available for expenditure for the same period as funds apportioned for projects on the Federal-aid primary system.

(b) The Federal share payable on account of any project carried out under 23 U.S.C. 144 shall be 80 percent of the eligible cost.

(c) Not less than 15 percent nor more than 35 percent of the apportioned funds shall be expended for projects located on public roads, other than those on a Federal-aid system. The Secretary after consultation with State and local officials may, with respect to a State, reduce the requirement for expenditure for bridges not on a Federal-aid system when he determines that such State has inadequate needs to justify such expenditure.

§ 650.415 Reports.

The Secretary must report annually to the Congress on projects approved and current inventories together with recommendations for further improvements.

Subparts E–F—(Reserved)

Subpart G—Discretionary Bridge Candidate Rating Factor

SOURCE: 48 FR 52296, Nov. 17, 1983, unless otherwise noted.

---

[1] American Association of State Highway and Transportation Officials, Suite 225, 444 North Capitol Street, NW, Washington, DC 20001.

Appendix – Parts of Title 23 U.S. Code

Federal Highway Administration, DOT

§ 650.701 Purpose.

The purpose of this regulation is to describe a rating factor used as part of a selection process of allocation of discretionary bridge funds made available to the Secretary of Transportation under 23 U.S.C. 144.

§ 650.703 Eligible projects.

(a) Deficient highway bridges on Federal-aid highway system roads may be eligible for allocation of discretionary bridge funds to the same extent as they are for bridge funds apportioned under 23 U.S.C. 144, provided that the total project cost for a discretionary bridge candidate is at least $10 million or twice the amount of 23 U.S.C. 144 funds apportioned to the State during the fiscal year for which funding for the candidate bridge is requested.

(b) After the effective date of this regulation for the discretionary bridge candidate rating factor, only candidate bridges not previously selected with a computed rating factor of 100 or less will be eligible for consideration.

§ 650.705 Application for discretionary bridge funds.

Each year through its field offices, the FHWA will issue an annual call for discretionary bridge candidate submittals including updates of previously submitted but not selected projects. Each State is responsible for submitting such data as required for candidate bridges. Data requested will include structure number, funds needed by fiscal year, total project cost, current average daily truck traffic and a narrative describing the existing bridge, the proposed new or rehabilitated bridge and other relevant factors which the State believes may warrant special consideration.

§ 650.707 Rating factor.

(a) The following formula is to be used in the selection process for ranking discretionary bridge candidates:

$$\text{Rating Factor (RF)} = \frac{SR}{D} \times \frac{TPC}{ADT'} \times \left[1 + \frac{\text{Unobligated HBRRP Balance}}{\text{Total HBRRP Funds Received}} \right]$$

The lower the rating factor, the higher the priority for selection and funding.

(b) The terms in the rating factor are defined as follows:

SR is Sufficiency Rating computed as illustrated in appendix A of the Recording and Coding Guide for the Structure Inventory and Appraisal of the Nation's Bridges, USDOT/FHWA (latest edition); (If SR is less than 1.0, use SR=1.0);

ADT is Average Daily Traffic in thousands taking the most current value from the national bridge inventory data;

ADTT is Average Daily Truck Traffic in thousands (Pick up trucks and light delivery trucks not included);

For load posted bridges, the ADTT furnished should be that which would use the bridge if traffic were not restricted.

The ADTT should be the annual average volume, not peak or seasonal.

D is Defense Highway System Status

D=1 if not on defense highway

D=1.5 if bridge carries a designated defense highway

The last term of the rating factor expression includes the State's unobligated balance of funds received under 23 U.S.C. 144 as of June 30 preceding the date of calculation, and the total funds received under 23 U.S.C. 144 for the last four fiscal years ending with the most recent fiscal year of the FHWA's annual call for discretionary bridge candidate submittals; (if unobligated HBRRP balance is less than $10 million, use zero balance);

TPC is Total Project Cost in millions of dollars;

HBRRP is Highway Bridge Replacement and Rehabilitation Program;

ADT' is ADT plus ADTT.

(c) In order to balance the relative importance of candidate bridges with very low (less than one) sufficiency ratings and very low ADT's against candidate bridges with high ADT's, the minimum sufficiency rating used will be 1.0. If the computed sufficiency rating for a candidate bridge is less than 1.0, use 1.0 in the rating factor formula.

§ 650.709

(d) If the unobligated balance of HBRRP funds for the State is less than $10 million, the HBRRP modifier is 1.0. This will limit the effect of the modifier on those States with small apportionments or those who may be accumulating funds to finance a major bridge.

[48 FR 52296, Nov. 17, 1983; 48 FR 53407, Nov. 28, 1983]

§ 650.709 Special considerations.

(a) The selection process for new discretionary bridge projects will be based upon the rating factor priority ranking. However, although not specifically included in the rating factor formula, special consideration will be given to bridges that are closed to all traffic or that have a load restriction of less than 10 tons. Consideration will also be given to bridges with other unique situations, and to bridge candidates in States which have not previously been allocated discretionary bridge funds.

(b) The need to administer the program from a balanced national perspective requires that the special cases set forth in paragraph (a) of this section and other unique situations be considered in the discretionary bridge candidate evaluation process.

(c) Priority consideration will be given to the continuation and completion of bridge projects previously begun with discretionary bridge funds.

Subpart H—Navigational Clearances for Bridges

SOURCE: 52 FR 28139, July 28, 1987, unless otherwise noted.

§ 650.801 Purpose.

The purpose of this regulation is to establish policy and to set forth coordination procedures for Federal-aid highway bridges which require navigational clearances.

§ 650.803 Policy.

It is the policy of FHWA:
(a) To provide clearances which meet the reasonable needs of navigation and provide for cost-effective highway operations,
(b) To provide fixed bridges wherever practicable, and

(c) To consider appropriate pier protection and vehicular protective and warning systems on bridges subject to ship collisions.

§ 650.805 Bridges not requiring a USCG permit.

(a) The FHWA has the responsibility under 23 U.S.C. 144(h) to determine that a USCG permit is not required for bridge construction. This determination shall be made at an early stage of project development so that any necessary coordination can be accomplished during environmental processing.

(b) A USCG permit shall not be required if the FHWA determines that the proposed construction, reconstruction, rehabilitation, or replacement of the federally aided or assisted bridge is over waters (1) which are not used or are not susceptible to use in their natural condition or by reasonable improvement as a means to transport interstate or foreign commerce and (2) which are (i) not tidal, or (ii) if tidal, used only by recreational boating, fishing, and other small vessels less than 21 feet in length.

(c) The highway agency (HA) shall assess the need for a USCG permit or navigation lights or signals for proposed bridges. The HA shall consult the appropriate District Offices of the U.S. Army Corps of Engineers if the susceptibility to improvement for navigation of the water of concern is unknown and shall consult the USCG if the types of vessels using the waterway are unknown.

(d) For bridge crossings of waterways with navigational traffic where the HA believes that a USCG permit may not be required, the HA shall provide supporting information early in the environmental analysis stage of project development to enable the FHWA to make a determination that a USCG permit is not required and that proposed navigational clearances are reasonable.

(e) Since construction in waters exempt from a USCG permit may be subject to other USCG authorizations, such as approval of navigation lights and signals and timely notice to local mariners of waterway changes, the USCG should be notified whenever the

Federal Highway Administration, DOT § 650.807

proposed action may substantially affect local navigation.

§ 650.807 Bridges requiring a USCG permit.

(a) The USCG has the responsibility (1) to determine whether a USCG permit is required for the improvement or construction of a bridge over navigable waters except for the exemption exercised by FHWA in § 650.805 and (2) to approve the bridge location, alignment and appropriate navigational clearances in all bridge permit applications.

(b) A USCG permit shall be required when a bridge crosses waters which are: (1) tidal and used by recreational boating, fishing, and other small vessels 21 feet or greater in length or (2) used or susceptible to use in their natural condition or by reasonable improvement as a means to transport interstate or foreign commerce. If it is determined that a USCG permit is required, the project shall be processed in accordance with the following procedures.

(c) The HA shall initiate coordination with the USCG at an early stage of project development and provide opportunity for the USCG to be involved throughout the environmental review process in accordance with 23 CFR part 771. The FHWA and Coast Guard have developed internal guidelines which set forth coordination procedures that both agencies have found useful in streamlining and expediting the permit approval process. These guidelines include (1) USCG/FHWA Procedures for Handling Projects which Require a USCG Permit[1] and (2) the USCG/FHWA Memorandum of Understanding on Coordinating The Preparation and Processing of Environmental Projects.[2]

(d) The HA shall accomplish sufficient preliminary design and consultation during the environmental phase of project development to investigate bridge concepts, including the feasibility of any proposed movable bridges, the horizontal and vertical clearances that may be required, and other location considerations which may affect navigation. At least one fixed bridge alternative shall be included with any proposal for a movable bridge to provide a comparative analysis of engineering, social, economic and environmental benefit and impacts.

(e) The HA shall consider hydraulic, safety, environmental and navigational needs along with highway costs when designing a proposed navigable waterway crossing.

(f) For bridges where the risk of ship collision is significant, HA's shall consider, in addition to USCG requirements, the need for pier protection and warning systems as outlined in FHWA Technical Advisory 5140.19, Pier Protection and Warning Systems for Bridges Subject to Ship Collisions, dated February 11, 1983.

(g) Special navigational clearances shall normally not be provided for accommodation of floating construction equipment of any type that is not required for navigation channel maintenance. If the navigational clearances are influenced by the needs of such equipment, the USCG should be consulted to determine the appropriate clearances to be provided.

(h) For projects which require FHWA approval of plans, specifications and estimates, preliminary bridge plans shall be approved at the appropriate level by FHWA for structural concepts, hydraulics, and navigational clearances prior to submission of the permit application.

(i) If the HA bid plans contain alternative designs for the same configuration (fixed or movable), the permit application shall be prepared in sufficient detail so that all alternatives can be evaluated by the USCG. If appropriate, the USCG will issue a permit for all alternatives. Within 30 days after award of the construction contract, the USCG shall be notified by the HA of the alternate which was selected. The USCG procedure for evaluating permit applications which contain alternates is presented in its Bridge Administration

---

[1] This document is an internal directive in the USCG Bridge Administration Manual, Enclosure 1a, COMDT INST M16590.5, change 2 dated Dec. 1, 1983. It is available for inspection and copying from the U.S. Coast Guard or the Federal Highway Administration as prescribed in 49 CFR part 7, appendices B and D.

[2] FHWA Notice 6640.22 dated July 17, 1981, is available for inspection and copying as prescribed in 49 CFR part 7, appendix D.

267

§ 650.809

Manual (COMDT INST M16590.5).[3] The FHWA policy on alternates, Alternate Design for Bridges; Policy Statement, was published at 48 FR 21409 on May 12, 1983.

§ 650.809 Movable span bridges.

A fixed bridge shall be selected wherever practicable. If there are social, economic, environmental or engineering reasons which favor the selection of a movable bridge, a cost benefit analysis to support the need for the movable bridge shall be prepared as a part of the preliminary plans.

PART 652—PEDESTRIAN AND BICYCLE ACCOMMODATIONS AND PROJECTS

Sec.
652.1 Purpose.
652.3 Definitions.
652.5 Policy.
652.7 Eligibility.
652.9 Federal participation.
652.11 Planning.
652.13 Design and construction criteria.

AUTHORITY: 23 U.S.C. 109, 217, 315, 402(b)(1)(F); 49 CFR 1.48(b).

SOURCE: 49 FR 10662, Mar. 22, 1984, unless otherwise noted.

§ 652.1 Purpose.

To provide policies and procedures relating to the provision of pedestrian and bicycle accommodations on Federal-aid projects, and Federal participation in the cost of these accommodations and projects.

§ 652.3 Definitions.

(a) *Bicycle.* A vehicle having two tandem wheels, propelled solely by human power, upon which any person or persons may ride.

(b) *Bikeway.* Any road, path, or way which in some manner is specifically designated as being open to bicycle travel, regardless of whether such facilities are designated for the exclusive use of bicycles or are to be shared with other transportation modes.

---

[3] United States Coast Guard internal directives are available for inspection and copying as prescribed in 49 CFR part 7, appendix B.

(c) *Bicycle Path (Bike Path).* A bikeway physically separated from motorized vehicular traffic by an open space or barrier and either within the highway right-of-way or within an independent right-of-way.

(d) *Bicycle Lane (Bike Lane).* A portion of a roadway which has been designated by striping, signing and pavement markings for the preferential or exclusive use of bicyclists.

(e) *Bicycle Route (Bike Route).* A segment of a system of bikeways designated by the jurisdiction having authority with appropriate directional and informational markers, with or without a specific bicycle route number.

(f) *Shared Roadway.* Any roadway upon which a bicycle lane is not designated and which may be legally used by bicycles regardless of whether such facility is specifically designated as a bikeway.

(g) *Pedestrian Walkway or Walkway.* A continuous way designated for pedestrians and separated from the through lanes for motor vehicles by space or barrier.

(h) *Highway Construction Project.* A project financed in whole or in part with Federal-aid or Federal funds for the construction, reconstruction or improvement of a highway or portions thereof, including bridges and tunnels.

(i) *Independent Bicycle Construction Project (Independent Bicycle Project).* A project designation used to distinguish a bicycle facility constructed independently and primarily for use by bicyclists from an improvement included as an incidental part of a highway construction project.

(j) *Independent Pedestrian Walkway Construction Project (Independent Walkway Project).* A project designation used to distinguish a walkway constructed independently and solely as a pedestrian walkway project from a pedestrian improvement included as an incidental part of a highway construction project.

(k) *Incidental Bicycle or Pedestrian Walkway Construction Project (Incidental Feature).* One constructed as an incidental part of a highway construction project.

(l) *Nonconstruction Bicycle Project.* A bicycle project not involving physical

Federal Highway Administration, DOT

construction which enhances the safe use of bicycles for transportation purposes.

(m) *Snowmobile.* A motorized vehicle solely designed to operate on snow or ice.

§ 652.5 Policy.

The safe accommodation of pedestrians and bicyclists should be given full consideration during the development of Federal-aid highway projects, and during the construction of such projects. The special needs for the elderly and the handicapped shall be considered in all Federal-aid projects that include pedestrian facilities. Where current or anticipated pedestrian and/or bicycle traffic presents a potential conflict with motor vehicle traffic, every effort shall be made to minimize the detrimental effects on all highway users who share the facility. On highways without full control of access where a bridge deck is being replaced or rehabilitated, and where bicycles are permitted to operate at each end, the bridge shall be reconstructed so that bicycles can be safely accommodated when it can be done at a reasonable cost. Consultation with local groups of organized bicyclists is to be encouraged in the development of bicycle projects.

§ 652.7 Eligibility.

(a) Independent bicycle projects, incidental bicycle projects, and nonconstruction bicycle projects must be principally for transportation rather than recreational use and must meet the project conditions for authorization where applicable.

(b) The implementation of pedestrian and bicycle accommodations may be authorized for Federal-aid participation as either incidental features of highways or as independent projects where all of the following conditions are satisfied.

(1) The safety of the motorist, bicyclist, and/or pedestrian will be enhanced by the project.

(2) The project is initiated or supported by the appropriate State highway agency(ies) and/or the Federal land management agency. Projects are to be located and designed pursuant to an overall plan, which provides due consideration for safety and contiguous routes.

(3) A public agency has formally agreed to:

(i) Accept the responsibility for the operation and maintenance of the facility.

(ii) Ban all motorized vehicles other than maintenance vehicles, or snowmobiles where permitted by State or local regulations, from pedestrian walkways and bicycle paths, and

(iii) Ban parking, except in the case of emergency, from bicycle lanes that are contiguous to traffic lanes.

(4) The estimated cost of the project is consistent with the anticipated benefits to the community.

(5) The project will be designed in substantial conformity with the latest official design criteria. (See § 652.13.)

[49 FR 10662, Mar. 22, 1984; 49 FR 14729, Apr. 13, 1984]

§ 652.9 Federal participation.

(a) Independent walkway projects, independent bicycle projects and nonconstruction bicycle projects shall be financed with 100 percent Federal-aid primary, secondary or urban highway funds, provided the total amount obligated for all such projects in any one State in any fiscal year does not exceed $4.5 million of Federal-aid funds or a lesser amount apportioned by the Federal Highway Administrator to avoid exceeding the annual $45 million cost limitation on these projects for all States in a fiscal year. The Federal Highway Administrator may, upon application, waive this limitation for a State for any fiscal year. This limitation also applies to projects funded under § 652.9(d). This limitation does not apply to projects of the type described in § 652.9(c). The FHWA Offices of Direct Federal Programs and Engineering will coordinate projects of the type described in § 652.9(d) to ensure that the annual cost limitations will not be exceeded.

(b) Specific eligibility requirements for Federal-aid participation in independent and nonconstruction projects are:

(1) An independent walkway project must be constructed on highway right-of-way or easement, or right-of-way acquired for this purpose. Independent

§ 652.11

walkway projects may be constructed separately or in conjunction with a Federal-aid highway construction project. Where an independent walkway project is located away from the Federal-aid highway right-of-way, it must serve pedestrians who would normally desire to use the Federal-aid route.

(2) An independent bicycle project may include the acquisition of land needed for the facility, or such projects may be constructed on existing highway right-of-way or easement acquired for this purpose. Independent bicycle projects may include construction of bicycle lanes, paths, shelters, bicycle parking facilities and other roadway and bridge work necessary to accommodate bicyclists.

(3) Nonconstruction bicycle projects must be related to the safe use of bicycles for transportation, and may include safety educational material and route maps for safe bicycle transportation purposes. Nonconstruction bicycle projects shall not include salaries for administration, law enforcement, maintenance and similar items required to operate transportation networks and programs, but may include cost of staff or consultants for development of specific nonconstruction projects.

(c) Bicycle and pedestrian accommodations may also be constructed as incidental features of highway construction projects. These incidental features may be financed with the same type of Federal-aid funds, including funds of the type described in § 652.9(d) (except Interstate construction funds) and at the same Federal share payable as a basic highway project. These accommodations are not subject to the funding limitations for independent walkway, independent bicycle and nonconstruction bicycle projects. In the case of the Interstate construction projects, Federal-aid Interstate construction funds may only be used to replace existing facilities that would be interrupted by construction of the project, or to mitigate specific environmental impacts. Interstate 4R funds provided by 23 U.S.C. 104(b)(5)(B) may be used only for incidental features. As incidental features, these accommodations must be part of a highway improvement and must be located within the right-of-way of the highway, including land acquired under 23 U.S.C. 319 (Scenic Enhancement Program).

(d) Funds authorized for Federal lands highways (forest highways, public lands highways, park roads, parkways, and Indian reservation roads which are public roads), forest development roads and trails (i.e., roads or trails under the jurisdiction of the Forest Service), and public lands development roads and trails (i.e., roads or trails which the Secretary of the Interior determines are of primary importance for the development, protection, administration, and utilization of public lands and resources under his/her control), may be used for independent bicycle routes and independent walkway projects. These funds may not be used for nonconstruction bicycle projects.

§ 652.11 Planning.

Federally aided bicycle and pedestrian projects implemented within urbanized areas must be included in the transportation improvement program/annual (or biennial) element unless excluded by agreement between the State and the metropolitan planning organization.

§ 652.13 Design and construction criteria.

(a) The American Association of State Highway and Transportation Officials' "Guide for Development of New Bicycle Facilities, 1981" (AASHTO Guide) or equivalent guides developed in cooperation with State or local officials and acceptable to the division office of the FHWA, shall be used as standards for the construction and design of bicycle routes. Copies of the AASHTO Guide may be obtained from the American Association of State Highway and Transportation Officials, 444 North Capitol Street, NW., Suite 225, Washington, DC 20001.

(b) Curb cuts and other provisions as may be appropriate for the handicapped are required on all Federal and Federal-aid projects involving the provision of curbs or sidewalks at all pedestrian crosswalks.

Appendix – Parts of Title 23 U.S. Code

Federal Highway Administration, DOT

PART 655—TRAFFIC OPERATIONS

Subparts A–C—(Reserved)

Subpart D—Traffic Surveillance and Control

Sec.
655.401 Purpose.
655.403 Traffic surveillance and control systems.
655.405 Policy.
655.407 Eligibility.
655.409 Traffic engineering analysis.
655.411 Project administration.

Subpart E—(Reserved)

Subpart F—Traffic Control Devices on Federal-Aid and Other Streets and Highways

655.601 Purpose.
655.602 Definitions.
655.603 Standards.
655.604 Achieving basic uniformity.
655.605 Project procedures.
655.606 Higher cost materials.
655.607 Funding.
APPENDIX TO SUBPART F—ALTERNATE METHOD OF DETERMINING THE COLOR OF RETROREFLECTIVE SIGN MATERIALS

Subpart G—(Reserved)

AUTHORITY: 23 U.S.C. 101(a), 104, 105, 109(d), 114(a), 135, 217, 307, 315, and 402(a); 23 CFR 1.32 and 1204.4; and 48 CFR 1.48(b).

Subparts A–C—(Reserved)

Subpart D—Traffic Surveillance and Control

SOURCE: 49 FR 8436, Mar. 7, 1984, unless otherwise noted.

§ 655.401 Purpose.

The purpose of this regulation is to provide policies and procedures relating to Federal-aid requirements of traffic surveillance and control system projects.

§ 655.403 Traffic surveillance and control systems.

(a) A traffic surveillance and control system is an array of human, institutional, hardware and software components designed to monitor and control traffic, and to manage transportation on streets and highways and thereby improve transportation performance, safety, and fuel efficiency.

(b) Systems may have various degrees of sophistication. Examples include, but are not limited to, the following systems: traffic signal control, freeway surveillance and control, and highway advisory radio, reversible lane control, tunnel and bridge control, adverse weather advisory, remote control of movable bridges, and priority lane control.

(c) Systems start-up is the process necessary to assure the surveillance and control project operates effectively. The start-up process is accomplished in a limited time period immediately after the system is functioning and consists of activities to achieve optimal performance. These activities include evaluation of the hardware, software and system performance on traffic; completion and updating of basic data needed to operate the system; and any modifications or corrections needed to improve system performance.

§ 655.405 Policy.

Implementation and efficient utilization of traffic surveillance and control systems are essential to optimize transporation systems efficiency, fuel conservation, safety, and environmental quality.

§ 655.407 Eligibility.

Traffic surveillance and control system projects are an integral part of Federal-aid highway construction and all phases of these projects are eligible for funding with appropriate Federal-aid highway funds. The degree of sophistication of any system must be in scale with needs and with the availability of personnel and budget resources to operate and maintain the system.

§ 655.409 Traffic engineering analysis.

Traffic surveillance and control system projects shall be based on a traffic engineering analysis. The analysis should be on a scale commensurate with the project scope. The basic elements of the analysis are:

(a) *Preliminary analysis.* The Preliminary Traffic Engineering Analysis should determine: The area to be controlled; transportation characteristics; objectives of the system; existing sys-

§ 655.411

tems resources (including communications); existing personnel and budget resources for the maintenance and operation of the system.

(b) *Alternative systems analysis.* Alternative systems should be analyzed as applicable. For the alternatives considered, the analysis should encompass incremental initial costs; required maintenance and operating budget and personnel resources; and expected benefits. Improved use of existing resources, as applicable, should be considered also.

(c) *Procurement and system start-up analysis.* Procurement and system start-up methods should be considered in the analysis. Federal-aid laws, regulations, policies, and procedures provide considerable flexibility to accommodate the special needs of systems procurement.

(d) *Special features analysis.* Unique or special features including special components and functions (such as emergency vehicle priority control, redundant hardware, closed circuit television, etc.) should be specifically evaluated in relation to the objectives of the system and incremental initial costs, operating costs, and resource requirements.

(e) *Analysis of laws and ordinances.* Existing traffic laws, ordinances, and regulations relevant to the effective operation of the proposed system shall be reviewed to ensure compatibility.

(f) *Implementation plan.* The final element in the traffic engineering analysis shall be an implementation plan. It shall include needed legislation, systems design, procurement methods, construction management procedures including acceptance testing, system start-up plan, operation and maintenance plan. It shall include necessary institutional arrangements and the dedication of needed personnel and budget resources required for the proposed system.

(Approved by the Office of Management and Budget under control number 2125–0512)

[49 FR 8436, Mar. 7, 1984, as amended at 59 FR 33910, July 1, 1994]

§ 655.411 Project administration.

(a) Prior to authorization of Federal-aid highway funds for construction, there should be a commitment to the operations plan (see § 655.409 (f)).

(b) The plans, specifications and estimates submittal shall include a total system acceptance plan.

(c) Project approval actions are delegated to the Division Administrator. Approval actions for traffic surveillance and control system projects costing over $1,000,000 are subject to review by the Regional Administrator prior to approval of plans, specifications, and estimates.

(d) System start-up is an integral part of a surveillance and control project.

(1) Costs for system start-up, over and above those attributable to routine maintenance and operation, are eligible for Federal-aid funding.

(2) Final project acceptance should not occur until after completion of the start-up phase.

Subpart E—[Reserved]

Subpart F—Traffic Control Devices on Federal-Aid and Other Streets and Highways

SOURCE: 48 FR 46776, Oct. 14, 1983, unless otherwise noted.

§ 655.601 Purpose.

To prescribe the policies and procedures of the Federal Highway Administration (FHWA) to obtain basic uniformity of traffic control devices on all streets and highways in accordance with the following references that are approved by the FHWA for application on Federal-aid projects:

(a) Manual on Uniform Traffic Control Devices for Streets and Highways (MUTCD), FHWA, 1988, including Revision No. 1 dated January 17, 1990, Revision No. 2 dated March 17, 1992, Revision No. 3 dated September 3, 1993, and Revision No. 4 dated November 1, 1994. This publication is incorporated by reference in accordance with 5 U.S.C. 552(a) and 1 CFR part 51 and is on file at the Office of the Federal Register in Washington, DC. The 1988 MUTCD Stock No. 050–001–00308–2 and "1988 MUTCD Revision 3," dated September 3, 1993 (Stock No. 050–001–00316–3), may be purchased from the Superintendent of Documents, U.S. Government Print-

Federal Highway Administration, DOT § 655.603

ing Office (GPO), Washington, DC 20402. The amendments to the MUTCD, titled "1988 MUTCD Revision 1," dated January 17, 1990, "1988 MUTCD Revision 2," dated March 17, 1992, and "1988 MUTCD Revision 4, dated November 1, 1994," are available from the Federal Highway Administration, Office of Highway Safety, HHS-21, 400 Seventh Street SW., Washington, DC 20590. These documents are available for inspection and copying as prescribed in 49 CFR part 7, appendix D.

(b) Standard Alphabets for Highway Signs, FHWA, 1966 Edition, Reprinted May 1972. (This publication is incorporated by reference and is on file at the Office of the Federal Register in Washington, DC. This document is available for inspection and copying as provided in 49 CFR part 7, appendix D).

(c) Traffic Surveillance and Control, FHWA, 23 CFR part 655, subpart D.

(d) Motorist Aid Systems, FHWA, 23 CFR part 655, subpart G.

(e) Pavement Marking Demonstration Program, FHWA, 23 CFR part 920.

[51 FR 16834, May 7, 1986, as amended at 57 FR 53030, Nov. 6, 1992; 60 FR 365, Jan. 4, 1995]

§ 655.602 Definitions.

The terms used herein are defined in accordance with definitions and usages contained in the MUTCD and 23 U.S.C. 101(a).

§ 655.603 Standards.

(a) *National MUTCD.* The MUTCD approved by the Federal Highway Administrator is the national standard for all traffic control devices installed on any street, highway, or bicycle trail open to public travel in accordance with 23 U.S.C 109(d) and 402(a). The national MUTCD is specifically approved by the FHWA for application on any highway project in which Federal highway funds participate and on projects in federally administered areas where a Federal department or agency controls the highway or supervises the traffic operations.

(b) *State or other Federal MUTCD.* (1) Where State or other Federal agency MUTCDs or supplements are required, they shall be in substantial conformance with the national MUTCD. Changes to the national MUTCD issued by the FHWA shall be adopted by the States or other Federal agencies within 2 years of issuance. The FHWA Regional Administrator has been delegated the authority to approve State MUTCDs and supplements.

(2) The Direct Federal Program Administrator has been delegated the authority to approve other Federal agency MUTCDs with the concurrence of the Office of Traffic Operations. States and other Federal agencies are encouraged to adopt the national MUTCD as their official Manual on Uniform Traffic Control Devices.

(c) *Color specifications.* Color determinations and specifications of sign and pavement marking materials shall conform to requirements of the FHWA Color Tolerance Charts.[2] An alternate method of determining the color of retroreflective sign material is provided in the appendix.

(d) *Compliance*—(1) *Existing highways.* Each State, in cooperation with its political subdivisions, and Federal agencies shall have a program as required by Highway Safety Program Standard Number 13, Traffic Engineering Services (23 CFR 1204.4) which shall include provisions for the systematic upgrading of substandard traffic control devices and for the installation of needed devices to achieve conformity with the MUTCD.

(2) *New or reconstructed highways.* Federal-aid projects for the construction, reconstruction, resurfacing, restoration, or rehabilitation of streets and highways shall not be opened to the public for unrestricted use until all appropriate traffic control devices, either temporary or permanent, are installed and functioning properly. Both temporary and permanent devices shall conform to the MUTCD.

(3) *Construction area activities.* All traffic control devices installed in construction areas using Federal-aid funds shall conform to the MUTCD. Traffic control plans for handling traffic and pedestrians in construction zones and for protection of workers shall conform to the requirements of 23 CFR part 630,

---

[2] Available for inspection from the Office of Traffic Operations, Federal Highway Administration, 400 Seventh Street, SW., Washington, DC 20590.

273

§ 655.604

subpart J, Traffic Safety in Highway and Street Work Zones.

(4) *MUTCD changes.* The FHWA may establish target dates for achieving compliance with changes to specific devices in the MUTCD.

(e) *Specific information signs.* Standards for specific information signs are contained in the MUTCD.

[48 FR 46776, Oct. 14, 1983, as amended at 51 FR 16834, May 7, 1986]

§ 655.604 **Achieving basic uniformity.**

(a) *Programs.* Programs for the orderly and systematic upgrading of existing traffic control devices or the installation of needed traffic control devices on or off the Federal-aid system should be based on inventories made in accordance with 23 CFR 1204.4, Highway Safety Program Standards. These inventories provide the information necessary for programming traffic control device upgrading projects.

(b) *Inventory.* An inventory of all traffic control devices is required by Highway Safety Program Standard Number 13, Traffic Engineering Services (23 CFR 1204.4). Highway planning and research funds and highway related safety grant program funds may be used in statewide or systemwide studies or inventories. Also, metropolitan planning (PL) funds may be used in urbanized areas provided the activity is included in an approved unified work program.

§ 655.605 **Project procedures.**

(a) *Federal-aid highways.* Federal-aid projects involving the installation of traffic control devices shall follow procedures as established in 23 CFR part 630, subpart A, Federal-Aid Programs Approval and Project Authorization. Simplified and timesaving procedures are to be used to the extent permitted by existing policy.

(b) *Off-system highways.* Certain federally funded programs are available for installation of traffic control devices on streets and highways that are not on the Federal-aid system. The procedures used in these programs may vary from project to project but, essentially, the guidelines set forth herein should be used.

23 CFR Ch. I (4-1-95 Edition)

§ 655.606 **Higher cost materials.**

The use of signing, pavement marking, and signal materials (or equipment) having distinctive performance characteristics, but costing more than other materials (or equipment) commonly used may be approved by the FHWA Division Administrator when the specific use proposed is considered to be in the public interest.

§ 655.607 **Funding.**

(a) *Federal-aid highways.* (1) Funds apportioned or allocated under 23 U.S.C. 104(b) are eligible to participate in projects to install traffic control devices in accordance with the MUTCD on newly constructed, reconstructed, resurfaced, restored, or rehabilitated highways, or on existing highways when this work is classified as construction in accordance with 23 U.S.C. 101(a). Federal-aid highway funds for eligible pavement markings and traffic control signalization may amount to 100 percent of the construction cost. Federal-aid highway funds apportioned or allocated under other sections of 23 U.S.C. are eligible for participation in improvements conforming to the MUTCD in accordance with the provisions of applicable program regulations and directives.

(2) Traffic control devices are eligible, in keeping with paragraph (a)(1) of this section, provided that the work is classified as construction in accordance with 23 U.S.C. 101(a) and the State or local agency has a policy acceptable to the FHWA Division Administrator for selecting traffic control devices material or equipment based on items such as cost, traffic volumes, safety, and expected service life. The State's policy should provide for cost-effective selection of materials which will provide for substantial service life taking into account expected and necessary routine maintenance. For these purposes, effectiveness would normally be measured in terms of durability, service life and/or performance of the material. Specific projects including material or equipment selection shall be developed in accordance with this policy. Proposed work may be approved on a project-by-project basis when the work is (i) clearly warranted, (ii) on a Federal-aid system, (iii) clearly identified

Federal Highway Administration, DOT

Pt. 655, Subpt. F, App.

by site, (iv) substantial in nature, and (v) of sufficient magnitude at any given location to warrant Federal-aid participation as a construction item.

(3) The method of accomplishing the work will be in accordance with 23 CFR part 635, subpart A, Contract Procedures.

(b) *Off-system highways.* Certain Federal-aid highway funds are eligible to participate in traffic control device improvement projects on off-system highways. In addition, Federal-aid highway funds apportioned or allocated in 23 U.S.C. are eligible for the installation of traffic control devices on any public road not on the Federal-aid system when the installation is directly related to a traffic improvement project on a Federal-aid system route.

APPENDIX TO SUBPART F—ALTERNATE METHOD OF DETERMINING THE COLOR OF RETROREFLECTIVE SIGN MATERIALS

1. The FHWA Color Tolerance Charts provide that conventional color measuring instruments such as spectrophotometers and tristimulus photoelectric colorimeters should not be used for measurement of retroreflective material colors and that such materials should be evaluated visually using the Color Tolerance Charts and paying strict attention to prescribed illumination and viewing conditions.

2. As an alternate to visual testing, the diffuse day color of retroreflective sign material may be determined in accordance with ASTM E 97, "Standard Method of Test for 45-Degree, 0-Degree Directional Reflectance of Opaque Specimens by Filter Photometry." Geometric characteristics must be confined to illumination incident within 10 degrees of, and centered about, a direction of 45 degrees from the perpendicular to the test surface; viewing is within 15 degrees of, and centered about, the perpendicular to the test surface. Conditions of illumination and observation must not be interchanged.

3. Standards to be used for reference are the Munsell Papers designated in Table 1 or Table II, attached. The papers must be recently calibrated on a spectrophotometer. Acceptable test instruments are:

a. Gardner Multipurpose Reflectometer or Model XL 20 Color Difference Meter,

b. Gardner Model Ac-2a or XL 30 Color Difference Meter,

c. Meeco Model V Colormaster,

d. Hunter lab D25 Color Difference Meter, or

e. Approved equal.

4. Average performance sheeting is identified as Types I and II sheeting and high performance sheeting is identified as Types III and IV sheeting in Standard Specifications for Construction of Roads and Bridges on Federal Highway Projects[3] (FP-79, Section 633).

TABLE I—COLOR SPECIFICATION LIMITS AND REFERENCE STANDARDS, TYPES I AND II SHEETING

Color	Chromaticity coordinates[1] (corner points)								Reflectance limits (percent Y) Y		Reference[3] standard (munsell papers)
	1		2		3		4		Minium	Maximum	
	x	y	x	y	x	y	x	y			
White[2]	.305	.290	.350	.342	.321	.361	.276	.308	35	—	6.3Gy 6.77/0.8.
Red	.602	.317	.664	.336	.644	.356	.575	.356	8	12	8.2R 3.78/14.0.
Orange	.535	.375	.607	.393	.582	.417	.535	.399	18	30	2.5YR 5.5/14.0
Brown	.445	.353	.604	.396	.556	.443	.445	.386	4	9	5.0YR 3/6.
Yellow	.482	.450	.532	.465	.505	.494	.475	.485	29	45	1.25Y 6/12.
Green	.130	.369	.180	.391	.155	.460	.107	.439	3.5	9	0.65BG 2.84/8.45.
Blue	.147	.075	.176	.091	.176	.151	.106	.113	1.0	4	5 8PB 1.32/6.8.

[1] The four pairs of chromaticity coordinates determine the acceptable color in terms of the CIE 1931 standard colorimetric system measured with standard illumination source C.

[2] Silver white is an acceptable color designation.

[3] Available from Munsell Color Company. 2441 Calvert Street, Baltimore. Maryland 21218.

[3]This document is available for inspection and copying as prescribed in 49 CFR part 7,

§ 924.1

SUBCHAPTER J—HIGHWAY SAFETY

PART 924—HIGHWAY SAFETY IMPROVEMENT PROGRAM

Sec.
924.1 Purpose.
924.3 Definitions.
924.5 Policy.
924.7 Program structure.
924.9 Planning.
924.11 Implementation.
924.13 Evaluation.
924.15 Reporting.

AUTHORITY: 23 U.S.C. 105(f), 152, 315, and 402; sec. 203 of the Highway Safety Act of 1973, as amended; 49 CFR 1.48(b).

SOURCE: 44 FR 11544, Mar. 1, 1979, unless otherwise noted.

§ 924.1 Purpose.

The purpose of this regulation is to set forth policy for the development and implementation of a comprehensive highway safety improvement program in each State.

§ 924.3 Definitions.

(a) The term *highway*, as used in this regulation, includes in addition to those items listed in 23 U.S.C. 101(a), those facilities specifically provided for the accommodation and protection of pedestrians and bicyclists.

(b) The term *State*, as used in this regulation, means any one of the 50 States, the District of Columbia, Puerto Rico, the Virgin Islands, Guam, American Samoa, and the Commonwealth of the Northern Mariana Islands except that, for the purpose of implementing section 203 of the Highway Safety Act of 1973, as amended, *State* means any one of the 50 States, the District of Columbia, and Puerto Rico.

§ 924.5 Policy.

Each State shall develop and implement, on a continuing basis, a highway safety improvement program which has the overall objective of reducing the number and severity of accidents and decreasing the potential for accidents on all highways.

§ 924.7 Program structure.

The highway safety improvement program in each State shall consist of components for planning, implementation, and evaluation of safety programs and projects. These components shall be comprised of processes developed by the States and approved by the Federal Highway Administration (FHWA). Where appropriate, the processes shall be developed cooperatively with officials of the various units of local governments. The processes may incorporate a range of alternate procedures appropriate for the administration of an effective highway safety improvement program on individual highway systems, portions of highway systems and in local political subdivisions, but combined shall cover all public roads in the State.

[48 FR 44066, Sept. 26, 1983]

§ 924.9 Planning.

(a) The planning component of the highway safety improvement program shall incorporate:

(1) A process for collecting and maintaining a record of accident, traffic, and highway data, including, for railroad-highway grade crossings, the characteristics of both highway and train traffic;

(2) A process for analyzing available data to identify highway locations, sections and elements determined to be hazardous on the basis of accident experience or accident potential;

(3) A process for conducting engineering studies of hazardous locations, sections, and elements to develop highway safety improvement projects as defined in 23 U.S.C. 101(a); and

(4) A process for establishing priorities for implementing highway safety improvement, projects, considering:

(i) The potential reduction in the number and/or severity of accidents,

(ii) The cost of the projects and the resources available,

(iii) The relative hazard of public railroad-highway grade crossings based on a hazard index formula,

(iv) Onsite inspection of public grade crossings,

(v) The potential danger to large numbers of people at public grade crossings used on a regular basis by

Federal Highway Administration, DOT §924.13

passenger trains, school buses, transit buses, pedestrians, bicyclists, or by trains and/or motor vehicles carrying hazardous materials, and

(vi) Other criteria as appropriate in each State.

(b) In planning a program of safety improvement projects at railroad-highway grade crossings, special emphasis shall be given to the legislative requirement that all public crossings be provided with standard signing.

(c) The planning component of the highway safety improvement program may be financed with funds made available through 23 U.S.C. 402, 307(c), and, where applicable, 104(f).

§924.11 Implementation.

(a) The implementation component of the highway safety improvement program in each State shall include a process for scheduling and implementing safety improvement projects in accordance with (1) the procedures set forth in 23 CFR part 630, subpart A (Federal-Aid Program Approval and Project Authorization) and (2) the priorities developed in accordance with §924.9. The States are encouraged to utilize the timesaving procedures incorporated in FHWA directives for the minor type of work normal to highway safety improvement projects.

(b) Funds apportioned under 23 U.S.C. 152, Hazard Elimination Program, are to be used to implement highway safety improvement projects on any public road other than Interstate.

(c) Funds apportioned under section 203(b) of the Highway Safety Act of 1973, as amended, Rail-Highway Crossings, are to be used to implement railroad-highway grade crossing safety projects on any public road. At least 50 percent of the funds apportioned under section 203(b) must be made available for the installation of grade crossing protective devices. The railroad share, if any, of the cost of grade crossing improvements shall be determined in accordance with 23 CFR part 646, subpart B (Railroad-Highway Projects).

(d) Highway safety improvement projects may be implemented on the Federal-aid system with funds apportioned under 23 U.S.C. 104(b), and with funds apportioned under section 104(b)(1) of the Federal-Aid Highway Act of 1978 and section 103(a) of the Highway Improvement Act of 1982, if excess to Interstate System needs.

(e) Funds apportioned under 23 U.S.C. 219, Safer Off-System Roads, may be used to implement highway safety improvement projects on public roads which are not on a Federal-aid system.

(f) Major safety defects on bridges, including related approach improvements, may be corrected as part of a bridge rehabilitation project on any public road with funds apportioned under 23 U.S.C. 144, if such project is considered eligible under 23 CFR part 650, subpart D (Special Bridge Replacement Program).

(g) Award of contracts for highway safety improvement programs shall be in accordance with 23 CFR part 635.

[48 FR 44066, Sept. 26, 1983]

§924.13 Evaluation.

(a) The evaluation component of the highway safety improvement progam in each State shall include a process for determining the effect that highway safety improvement projects have in reducing the number and severity of accidents and potential accidents, including:

(1) The cost of, and the safety benefits derived from the various means and methods used to mitigate or eliminate hazards,

(2) A record of accident experience before and after the implementation of a highway safety improvement, project, and

(3) A comparison of accident numbers, rates, and severity observed after the implementation of a highway safety improvement project with the accident numbers, rates, and severity expected if the improvement had not been made.

(b) The findings resulting from the evaluation process shall be incorporated as basic source data in the planning process outlined in §924.9(a).

(c) The evaluation component may be financed with funds made available through 23 U.S.C. 402, 307(c), and, where applicable, 104(f). In addition, when highway safety improvement projects are undertaken with funds apportioned under 23 U.S.C. 152 or section 203 of the Highway Safety Act of 1973, as amend-

§ 924.15

ed, these funds may also be used to evaluate the improvements.

§ 924.15 Reporting.

(a) Each State shall submit to the FHWA Division Administrator no later than August 31 of each year a report (OMB Number 04–R2450) covering the State's highway safety improvement program during the previous July 1 through June 30 period. In its annual report, the State shall report on the progress made in implementing the hazard elimination program and the grade crossing improvement program, and shall evaluate the effectiveness of completed highway safety improvement projects in these programs.

(b) The preparation of the State's annual report may be financed with funds made available through 23 U.S.C. 402, 307(c), and, where applicable, 104(f).

Appendix – Parts of Title 23 U.S. Code

SUBCHAPTER A—PROCEDURES FOR STATE HIGHWAY SAFETY PROGRAMS

PART 1200—UNIFORM PROCEDURES FOR STATE HIGHWAY SAFETY PROGRAMS

Subpart A—General

Sec.
1200.1 Purpose.
1200.2 Applicability.
1200.3 Definitions.

Subpart B—The Highway Safety Plan

1200.10 Preparation and submission.
1200.11 Review and approval.
1200.12 Apportionment and obligation of Federal funds.
1200.13 Changes.

Subpart C—Implementation and Management of the Highway Safety Program

1200.20 General.
1200.21 Equipment.
1200.22 Vouchers and project agreements.
1200.23 Program income.
1200.24 Compliance.
1200.25 Appeals.

Subpart D—Closeout

1200.30 Expiration of the HSP.
1200.31 Extension of the HSP.
1200.32 Final voucher.
1200.33 Annual evaluation report.
1200.34 Disposition of unexpended balances.
1200.35 Post-grant adjustments.
1200.36 Continuing requirements.

AUTHORITY: 23 U.S.C. 402; delegations of authority at 49 CFR 1.48 and 1.50.

SOURCE: 58 FR 41033, Aug. 2, 1993, unless otherwise noted.

Subpart A—General

§ 1200.1 Purpose.

This part establishes the requirements governing submission and approval of State Highway Safety Plans and prescribes uniform procedures for the implementation and management of State highway safety programs.

§ 1200.2 Applicability.

The provisions of this part apply to States conducting highway safety programs in accordance with 23 U.S.C. 402, beginning with Highway Safety Plans for Fiscal Year 1994.

§ 1200.3 Definitions.

As used in this subchapter—

Annual evaluation report means the report submitted each year by each State which describes the accomplishments of its highway safety program for the preceding fiscal year.

Approving official means a Regional Administrator of the National Highway Traffic Safety Administration for issues concerning NHTSA funds or program areas and a Division Administrator of the Federal Highway Administration for issues concerning FHWA funds or program areas.

Carry-forward funds means those funds which a State has obligated but not expended in the fiscal year in which they were apportioned, that are being reprogrammed.

Contract authority means the statutory language which authorizes the agencies to enter into an obligation without the need for a prior appropriation or further action from Congress. When exercised, contract authority creates a binding obligation on the United States for which Congress must make subsequent appropriations in order to liquidate the obligations incurred pursuant to this authority.

Contractor means the recipient of a contract or subcontract under the HSP.

FHWA means the Federal Highway Administration.

Fiscal year means the Federal fiscal year, consisting of twelve months beginning each October 1 and ending the following September 30.

Governor means the Governor of any of the fifty States, Puerto Rico, the Virgin Islands, Guam, American Samoa, or the Commonwealth of the Northern Mariana Islands, the Mayor of the District of Columbia, or, for the application of this part to Indians as provided in 23 U.S.C. 402(i), the Secretary of the Interior.

Governor's Representative means the Governor's Representative for Highway

460

NHTSA and FHWA, DOT

Safety, the official appointed by the Governor to implement the State's highway safety program or, for the application of this part to Indians as provided in 23 U.S.C. 402(i), an official of the Bureau of Indian Affairs who is duly designated by the Secretary of the Interior to implement the Indian highway safety program.

Highway Safety Plan or *HSP* means the section 402 grant application document consisting of the plan submitted by a State describing the State's highway safety problems, identifying countermeasures, and detailing the projects the State plans to undertake which implement those countermeasures.

Highway safety program includes all of the projects planned or undertaken by a State or its subgrantees or contractors to address highway safety problems in the State.

Major equipment means tangible, nonexpendable, personal property having a useful life of more than one year and an acquisition cost of $5,000 or more per unit, or such other definition as a State may elect in accordance with the procedures in §1200.21(c)(1) of this part.

National Priority Program Area means a program area identified in §1205.3 of this chapter as eligible for Federal funding pursuant to 23 U.S.C. 402 because it encompasses a major highway safety problem which is of national concern and for which effective countermeasures have been identified.

NHTSA means the National Highway Traffic Safety Administration.

Non-major equipment means all tangible, personal property which does not meet the definition of major equipment, including supplies.

Program area means any area, including but not limited to a National Priority Program Area, which is eligible or approved for Federal funding pursuant to 23 U.S.C. 402.

Program income means gross income received by the State or any of its subgrantees or contractors which is directly or indirectly generated by a Federally-supported project during the project performance period.

Project means any of the activities proposed or implemented under an HSP to address discrete or localized high-

§ 1200.10

way safety problems falling within one or more program areas.

Project agreement means the written agreement between a State and a subgrantee or contractor under which the State agrees to provide section 402 funds in exchange for the subgrantee's or contractor's performance of one or more projects supporting the section 402 program.

Reprogramming means applying carry-forward funds to projects in a fiscal year subsequent to the one for which the funds were originally apportioned to the State.

Section 402 means section 402 of title 23 of the United States Code.

State means any of the fifty States of the United States, the District of Columbia, Puerto Rico, the Virgin Islands, Guam, American Samoa, the Commonwealth of the Northern Mariana Islands, or, for the application of this part to Indians as provided in 23 U.S.C. 402(i), the Secretary of the Interior.

Subgrantee means a recipient of an award of financial assistance by a State in the form of money, or equipment in lieu of money, under an HSP.

Subpart B—The Highway Safety Plan

§ 1200.10 Preparation and submission.

(a) *Time period covered by the HSP.* (1) Except as provided in paragraph (a)(2) of this section, a State shall submit an HSP for each fiscal year. The time period for which the State shall identify activities to address highway safety problems under each such HSP shall be one fiscal year, unless extended in accordance with the provisions of §1200.31 of this part.

(2) A State may elect to submit an HSP once every three fiscal years, provided advance notice is given to the approving officials. (States are encouraged to provide notice of at least 90 days.) The time period for which the State shall identify activities to address highway safety problems under each such HSP shall be three consecutive fiscal years, unless extended in accordance with the provisions of §1200.31 of this part. Obligation of Federal funds and authority to incur costs, however, shall be based on each fiscal

461

Appendix – Parts of Title 23 U.S. Code

§ 1200.10

year. A State submitting a three-year HSP shall, nevertheless, submit the trend data required by paragraph (b)(2) of this section on August 1 of each year.

(b) *Content of the HSP.* Each State's HSP shall contain the following elements: (1) *Certifications and assurances.* A statement containing certifications and assurances shall be signed by the Governor's Representative and shall satisfy the requirements of 49 CFR part 18 and other applicable law. A sample statement, which may from time to time be amended to reflect changes in applicable law, shall be made available by each approving official.

(2) *Problem identification summary.* The problem identification summary will highlight highway safety problems throughout the State and briefly describe the countermeasures the State will employ to address these problems. It shall be supported by statistical evidence of recent trends in fatal, injury, and property damage crashes.

(3) *Description of and justification for program areas to be funded.* A State may identify and seek funding for projects, and related equipment purchases, within any National Priority Program Area or any other program area. National Priority Program Areas are identified in § 1205.3 of this chapter. Other program areas may, from time to time, be identified by statute, by rule, or by a State. For each program area for which Federal funding is sought, the following procedures shall apply:

(i) For National Priority Program Areas, the funding procedures at § 1205.4 of this chapter.

(ii) For program areas identified by statute, the funding procedures prescribed by the statute and implementing regulations or, in the absence of prescribed procedures, the funding procedures at § 1205.4 of this chapter.

(iii) For program areas identified by a State and not falling within paragraphs (b)(3)(i) or (b)(3)(ii) of this section, the funding procedures at § 1205.5 of this chapter.

(4) *Discussion of planning and administration needs.* Planning and administration needs shall be discussed in sufficient detail to justify proposed expenditures. Proposed and actual expenditures shall comply with the Federal

23 CFR Ch. II (4-1-95 Edition)

contribution and State matching requirements of part 1252 of this chapter.

(5) *Description of training needs.* Training needed to support or further the objectives of the HSP should be described in adequate detail to justify proposed expenditures. Only training that supports or furthers the objectives of the HSP shall be eligible for funding.

(6) *Supporting financial documentation.* Financial documentation to be submitted with the HSP shall include the Highway Safety Program Cost Summary (HS Form 217) reflecting the State's proposed allocation of funds (including estimated carry-forward funds) among program areas and for planning and administration needs, completed in accordance with the form's written instructions, and such other financial documentation as may be required by law. A State electing to submit a three-year HSP shall include with the HSP, one completed HS Form 217 for each of the fiscal years covered by the three-year HSP.

(c) *Special funding conditions.* The planning and contents of an HSP shall reflect the following funding requirements:

(1) *Political subdivision participation.* Proposed expenditures under the HSP shall comply with the requirements for political subdivision participation contained in part 1250 of this chapter.

(2) *NHTSA project length.* NHTSA funding support under the section 402 program for a specific project under an HSP shall ordinarily not exceed three years. However, a funding extension beyond three years may be approved in writing by the NHTSA approving official on a year-by-year basis, provided the project has demonstrated great merit or the potential for significant long-range benefits and includes a cost assumption plan requiring, at a minimum, 35 percent non-Federal support for the fourth year and 50 percent non-Federal support for each year thereafter. A denial of a project funding extension shall be in writing by the NHTSA approving official and shall be subject to the appeal procedures of § 1200.25 of this part. The project length requirements are not applicable to planning and administration activities; program management (e.g., program area coordinators' or managers' over-

462

NHTSA and FHWA, DOT

sight of the continuing development, implementation and evaluation of 402 or related State/locally supported activities); mandatory (earmarked) programs; training projects which support activities within an identified program area; or other activities which are required by Federal statute.

(d) *Due date.* The completed HSP must be received by the approving officials no later than August 1 preceding the fiscal year to which it applies or, in the case of a 3-year HSP, no later than August 1 preceding the first year to which it applies. For a State operating under a 3-year HSP, the trend data identified in § 1200.10(a)(2) and (b)(2) must be received by the approving officials on a yearly basis on August 1, and any other HSP updates for the second and third years which the State elects to submit must be received by the approving officials no later than August 1 preceding the fiscal year for which the updates apply. The State shall furnish three copies of its HSP (or trend data or HSP update, as appropriate) to each of its NHTSA and FHWA approving officials. Failure to meet these deadlines may result in delayed approvals.

(Approved by the Office of Management and Budget under Control Number 2127-0003)

§ 1200.11 **Review and approval.**

(a) *Review.* (1) Each approving official shall verify that each HSP complies with the basic requirements of § 1200.10 (b) and (c) of this part. Where an HSP is found not in compliance, the approving official will advise the State to take such action as is necessary to bring the HSP into compliance.

(2) An HSP determined to satisfy the basic requirements of § 1200.10 (b) and (c) of this part shall be further reviewed to ensure that the State has proposed a highway safety program which justifies the commitment of Federal funds. Each approving official shall have the discretion to require further clarification or amendment of any portion of an HSP which does not adequately establish the existence of a bona fide highway safety problem, the selection of countermeasures and projects reasonably calculated to address the problem, and the efficient proposed use of Federal funds.

§ 1200.11

(3) Each approving official shall provide States with reasonable notice and opportunity to amend portions of HSPs which are found inadequate. Such notice and opportunity to amend shall facilitate the informal resolution of problems in the HSP, to the maximum extent practicable, before the time by which the approving official must render a written decision under paragraph (b)(2) of this section.11(b) *Approval/conditional approval/disapproval.* (1) If after reasonable notice and opportunity to amend pursuant to paragraph (a)(3) of this section, the approving official determines that a State has provided information in the HSP which is inadequate to justify a proposed use of Federal funds, or has failed to comply with other requirements of this part or applicable law, the approving official shall conditionally approve or disapprove the relevant portion(s) of the HSP, as appropriate. Otherwise, the approving official shall approve the HSP, except that in no case shall the approving official approve an HSP which is submitted without the statement required by § 1200.10(b)(1) of this part until receipt of such statement.

(2) Approval, conditional approval, or disapproval of the HSP, in whole or in part, shall be in writing, dated, and signed by the approving official(s), and shall be sent to the Governor, with a copy to the Governor's Representative, within 30 days after receipt of the HSP by the agency, unless extended by mutual agreement of the approving official(s) and the Governor's Representative.

(3) For any portion of the HSP which is conditionally approved or disapproved, a detailed explanation of conditions or reasons for disapproval shall be provided in writing by the approving official to the Governor's Representative. Conditional approval may include a requirement for project by project approval of federally funded activities.

(4) All approvals and conditional approvals sent to the Governor's Representative shall state the total Federal dollar amount of the program approved or conditionally approved and shall contain the following statement:

By this letter, (STATE)'s _____ fiscal year 19___ Highway Safety Plan, as

463

§ 1200.12

submitted on (DATE) _____, is hereby approved, subject to any conditions or limitations set forth below. This approval does not constitute an obligation of Federal funds for the fiscal year identified above or an authorization to incur costs against those funds. The obligation of section 402 program funds against the approved HSP will be effected in writing by the NHTSA/FHWA Administrator, as appropriate, at the commencement of the fiscal year identified above. However, Federal funds reprogrammed from a prior-year HSP will be available for immediate use by the State under the approved HSP on October 1. Reimbursement will be contingent upon the submission of an updated HS Form 217, consistent with the requirements of 23 CFR 1200.12(d), within 30 days after either the beginning of the fiscal year identified above or the date of this letter, whichever is later.

§ 1200.12 Apportionment and obligation of Federal funds.

(a) Except as provided in paragraph (b) of this section, on October 1 of each fiscal year covered by an HSP, the NHTSA/FHWA Administrator, as appropriate, shall, in writing, distribute funds available for obligation under section 402 to the States and provide a statement of any conditions or limitations imposed by law on the use of the funds.

(b) In the event that authorizations exist but no applicable appropriation act has been enacted by October 1 of a fiscal year covered by an HSP, the NHTSA/FHWA Administrator, as appropriate, shall, in writing, distribute a part of the funds authorized under section 402 contract authority to ensure program continuity and shall provide a statement of any conditions or limitations imposed by law on the use of the funds. Upon appropriation of section 402 funds, the NHTSA/FHWA Administrator, as appropriate, shall, in writing, promptly adjust the apportionment, in accordance with law.

(c) The funds distributed under paragraph (a) or (b) of this section shall be available for expenditure by the states to satisfy the Federal share of expenses under an approved HSP, and shall constitute a contractual obligation of the Federal Government, subject to any conditions or limitations identified in the distributing document or in the written explanation of conditions required under § 1200.11(b)(3) of this part, and not exceeding the total dollar amount of the approved or conditionally approved program identified in § 1200.11(b)(4) of this part.

(d)(1) Notwithstanding the provisions of paragraph (c) of this section, reimbursement of State expenses shall be contingent upon the submission of an updated HS Form 217, within 30 days after either the beginning of the fiscal year or the date of the written approval required under § 1200.11(b) of this part, whichever is later. A State submitting a three-year HSP shall, nevertheless, submit the updated HS Form 217 on a yearly basis.

(2) The updated HS Form 217 required under paragraph (d)(1) of this section shall reflect the State's allocation of section 402 funds made available for expenditure during the fiscal year, including known carry-forward funds, except that—

(i) The total of the Federal funds reflected on the form shall not exceed the total Federal dollar amount of the approved program identified in § 1200.11(b)(4) of this part; and

(ii) None of the federally funded amounts identified under any of the program areas on the form shall exceed the federally funded amounts identified under the same program areas on the HS Form 217 submitted under § 1200.10(b)(6) of this part.

(3) An updated HS Form 217 not meeting the requirements of paragraph (d)(2)(i) or (d)(2)(ii) of this section shall be accompanied by such information as is necessary to explain the deviation and shall require approval, in writing, by the approving officials.

§ 1200.13 Changes.

(a) *Changes requiring prior approval.* Each State shall obtain the written approval of the approving official prior to implementing or allowing subgrantees or contractors to implement any of the following changes:

(1) Any revision which would result in the need for additional Federal funding beyond that which is already obligated or approved for reprogramming under the current HSP;

(2) Any extension of the period during which costs may be incurred under an HSP, as provided in § 1200.31 of this part;

NHTSA and FHWA, DOT

(3) Any extension of the length of a NHTSA project beyond three years, as provided in § 1200.10(c)(2) of this part;

(4) Any movement of funds into or out of a program area which, either singly or netted with all past movements of funds, exceeds ten percent of the Federal funding for the program area (e.g., 8% in+3% out+7% in=12% net fund movement), provided that—

(i) For the duration of the HSP, the ten percent threshold for each program area shall be based on the total federally funded program amounts identified on the updated HS Form 217 required under § 1200.12(d) of this part, without regard to amounts contained on any later submitted or approved HS Form 217; and

(ii) Once the ten percent threshold is exceeded with respect to a given program area, all fund movements made thereafter with respect to that program area require prior approval; or

(5) Any change in the scope or objectives of a project (regardless of whether there is an associated fund movement requiring prior approval).

(b) *Approval procedures.* (1) States shall request prior approval for changes by submitting a written request to the approving official, accompanied by HS Form 217 and such other information as is necessary to explain the proposed change.

(2) The approving official shall indicate approval in writing to the Governor's representative, normally within 10 working days after receipt of the request.

(3) Subject to the appeal provisions of § 1200.25 of this part, the approving official may disapprove a change in writing, with a brief explanation of the reasons therefor. Any such disapproval shall normally be made within 10 working days after receipt of the request.

(c) *Procedures for changes not requiring prior approval.* States shall provide documentary evidence of changes not requiring prior approval to the approving official by submitting an amended HS Form 217 and such other information as is necessary to explain the change.

§ 1200.21

Subpart C—Implementation and Management of the Highway Safety Program

§ 1200.20 General.

Except as otherwise provided in this subpart and subject to the provisions herein, the requirements of 49 CFR part 18 and applicable cost principles govern the implementation and management of State highway safety programs carried out under 23 U.S.C. 402. Cost principles include those referenced in 49 CFR 18.22 and those set forth in applicable Department of Transportation, NHTSA, or FHWA Orders.

§ 1200.21 Equipment.

(a) *All equipment.* (1) *Title.* Except as provided in paragraphs (a)(3) and (b)(1) of this section, title to equipment acquired under the HSP will vest upon acquisition in the State or its subgrantee, as appropriate.

(2) *Use.* All equipment shall be used for the originally authorized grant purposes for as long as needed for those purposes, as determined by the approving officials, and neither the State nor any of its subgrantees or contractors shall encumber the title or interest while such need exists.

(3) *Right to transfer title.* The NHTSA or the FHWA may reserve the right to transfer title to equipment acquired under the HSP to the Federal Government or to a third party when such third party is otherwise eligible under existing statutes. Any such transfer shall be subject to the following requirements:

(i) The property shall be identified in the grant or otherwise made known to the State in writing;

(ii) The NHTSA or the FHWA, as applicable, shall issue disposition instructions within 120 calendar days after the end of the project for which the property was acquired, in the absence of which the State shall follow the applicable procedures in 49 CFR part 18.

(b) *Federally-owned equipment.* In the event a State or its subgrantee is provided federally-owned equipment:

465

§ 1200.22

(1) Title shall remain vested in the Federal Government;
(2) Management shall be in accordance with Federal rules and procedures, and an annual inventory listing shall be submitted;
(3) The State or its subgrantee shall request disposition instructions from NHTSA or FHWA, as appropriate, when the item is no longer needed in the program.

(c) *Major equipment.* (1) *Choice of definition.* A State may elect to use its own definition of major equipment, provided such definition would at least include all items captured by the definition appearing in § 1200.3 of this part. Such election shall be made in writing by the Governor's Representative to the approving official, in the absence of which the definition in § 1200.3 of this part shall apply.

(2) *Management and disposition.* Subject to the requirements of paragraphs (a)(2), (b)(2), and (b)(3) of this section, States and their subgrantees and contractors shall manage and dispose of major equipment acquired under the HSP in accordance with State laws and procedures.

(d) *Non-major equipment.* Subject to the requirements of paragraphs (a)(2), (b)(2), and (b)(3) of this section and except as otherwise provided in 49 CFR 18.33, States and their subgrantees and contractors shall manage and dispose of non-major equipment acquired under the HSP in accordance with State laws and procedures.

§ 1200.22 Vouchers and project agreements.

Each State shall submit to the approving official vouchers for total expenses incurred, regardless of whether the State receives advance payments or is reimbursed for expenditures under the HSP. Copies of the project agreement(s) and supporting documentation for the vouchers, and any amendments thereto, shall be made available for review by the approving official upon request.

(a) *Content of vouchers.* At a minimum, each voucher shall provide the following information for expenses claimed in each program area:
(1) Program Area/Project Number;
(2) Federal funds obligated;
(3) Amount of Federal funds allocated to local benefit (provided mid-year (by March 31) and with the final voucher);
(4) Cumulative Total Cost to Date;
(5) Cumulative Federal Funds Expended;
(6) Previous Amount Claimed;
(7) Amount Claimed this Period;
(8) Special matching rate (i.e., sliding scale rate) authorized under 23 U.S.C. 120(a), if used, determined in accordance with the applicable NHTSA Order.

(b) *Submission requirements.* At a minimum, vouchers shall be submitted to the approving official on a quarterly basis, no later than 15 working days after the end of each quarter, except that where a state receives funds by electronic transfer at an annualized rate of one million dollars or more, vouchers shall be submitted on a monthly basis, no later than 15 working days after the end of each month. Failure to meet these deadlines may result in delayed reimbursement.

§ 1200.23 Program income.

(a) *Inclusions.* Program income includes income from fees for services performed, from the use or rental of real or personal property acquired with grant funds, from the sale of commodities or items fabricated under the grant agreement, and from payments of principal and interest on loans made with grant funds.

(b) *Exclusions.* Program income does not include interest on grant funds, rebates, credits, discounts, refunds, taxes, special assessments, levies, fines, proceeds from the sale of real property or equipment, income from royalties and license fees for copyrighted material, patents, and inventions, or interest on any of these.

(c) *Use of program income.* (1) *Addition.* Program income shall ordinarily be added to the funds committed to the HSP. Such program income shall be used to further the objectives of the project under which it was generated.

(2) *Cost sharing or matching.* Program income may be used to meet cost sharing or matching requirements only upon written approval of the approving official. Such use shall not increase the commitment of Federal funds.

NHTSA and FHWA, DOT § 1200.33

§ 1200.24 Compliance.

Where a State is found to be in noncompliance with the terms of the HSP or applicable law, the approving official may apply the special conditions for high-risk grantees or the enforcement procedures of 49 CFR part 18, as appropriate and in accordance with their terms.

§ 1200.25 Appeals.

Review of any written decision by an approving official under this part, including a denial of a funding extension under § 1200.10(c)(2) of this part, a disapproval or conditional approval of any part of an HSP under § 1200.11(b) of this part, a disapproval of a request for a change under § 1200.13(b)(3) of this part, and a decision to impose special conditions or restrictions or to seek remedies under § 1200.24 of this part, may be obtained by submitting a written appeal of such decision, signed by the Governor's Representative, to the approving official. Such appeal shall be forwarded promptly to the NHTSA Associate Administrator for Regional Operations or the FHWA Regional Administrator with jurisdiction over the specific division, as appropriate. The decision of the NHTSA Associate Administrator or the FHWA Regional Administrator shall be final and shall be transmitted to the Governor's Representative through the cognizant approving official.

Subpart D—Closeout

§ 1200.30 Expiration of the HSP.

Unless extended in accordance with the provisions of § 1200.31 of this part, a one-year HSP shall expire on the last day of the fiscal year to which it pertains and a three-year HSP shall expire on the last day of the third fiscal year to which it pertains. The State and its subgrantees and contractors may not incur costs past the expiration date.

§ 1200.31 Extension of the HSP.

Upon written request by the State, specifying the reasons therefor, the approving official may extend the expiration date for some portion of an HSP by a maximum of 90 days. The approval of any such request for extension shall be in writing, shall specify the new expiration date, and shall be signed by the approving official. If an extension is granted, the State and its subgrantees and contractors may continue to incur costs under the HSP until the new expiration date, and the due dates for other submissions covered by this subpart shall be based upon the new expiration date. However, in no case shall any extension be deemed to authorize the obligation of additional Federal funds beyond those already obligated to the State by the Federal Government, nor shall any extension be deemed to extend the due date for submission of the annual evaluation report. Only one extension shall be allowed for each HSP.

§ 1200.32 Final voucher.

Each State shall submit a final voucher which satisfies the requirements of § 1200.22(a) of this part within 90 days after the expiration of each fiscal year, unless extended in accordance with the provisions of § 1200.31 of this part. The final voucher constitutes the final financial reconciliation for each one-year HSP or for each year of a three-year HSP.

§ 1200.33 Annual evaluation report.

Within 90 days after the expiration of the fiscal year, each State, whether operating under a one-year or a three-year HSP, shall submit to the approving official an annual evaluation report describing the accomplishments of the highway safety program under the HSP for that fiscal year. The report shall include the following information:

(a) *Statewide overview.* A three to five page overview of statewide accomplishments in highway safety, regardless of funding source;

(b) *Report by program area.* For each funded program area, a description of individual projects conducted thereunder, detailing costs and accomplishments (and status if a project has not been completed), identifying progress toward self-sustainment and contributions of independent groups, and including an accounting for any program income earned or used under the project;

(c) *Legislative and administrative accomplishments.* A discussion of signifi-

§ 1200.34

cant legislative and administrative accomplishments which promoted the goals of highway safety; and

(d) *Status of remedial actions.* An evaluation of the progress the State is making in correcting deficiencies identified through program and financial management reviews conducted by the approving officials or by the State.

§ 1200.34 Disposition of unexpended balances.

Any funds which remain unexpended after reconciliation of the final voucher shall be carried forward, credited to the State's highway safety account for the new fiscal year, and made immediately available for reprogramming under a new HSP or under the next year of a continuing three-year HSP, subject to the approval requirements of § 1200.11(b) of this part. Carry-forward funds must be identified by the program area from which they are removed when they are reprogrammed from the previous fiscal year. Once so identified, such funds are available for use without regard to the program area from which they were carried forward, unless specially earmarked by the Congress.

§ 1200.35 Post-grant adjustments.

The closeout of an HSP does not affect the ability of NHTSA or FHWA to disallow costs and recover funds on the basis of a later audit or other review or the State's obligation to return any funds due as a result of later refunds, corrections, or other transactions.

§ 1200.36 Continuing requirements.

The following provisions shall have continuing applicability, notwithstanding the closeout of an HSP:

(a) The requirement to use all equipment for the originally authorized grant purposes for as long as needed for those purposes, as provided in § 1200.21(a)(2) of this part;

(b) The management and disposition requirements for equipment, as provided in § 1200.21 of this part;

(c) The audit requirements and records retention and access requirements of 49 CFR part 18.

SUBCHAPTER B—GUIDELINES

PART 1204—UNIFORM GUIDELINES FOR STATE HIGHWAY SAFETY PROGRAMS

AUTHORITY: 23 U.S.C. 402; delegations of authority at 49 CFR 1.48 and 1.50.

Subpart A—(Reserved)

Subpart B—Guidelines

§ 1204.4 Highway Safety Program Guidelines.

The Uniform Guidelines for State Highway Safety Programs are set forth in this subpart.

HIGHWAY SAFETY PROGRAM GUIDELINE NUMBERS AND TITLES

No.
1 Periodic motor vehicle inspection.
2 Motor vehicle registration.
3 Motorcycle safety.
4 Driver education.
5 Drive licensing.
6 Codes and laws.
7 Traffic courts.
8 Alcohol in relation to highway safety.
9 Identification and surveillance of accident locations.
10 Traffic records.
11 Emergency medical services.
12 Highway design, construction and maintenance.
13 Traffic engineering services.
14 Pedestrian safety.
15 Police traffic services.
16 Debris hazard control and cleanup.
17 Pupil transportation safety.
18 Accident investigation and reporting.

HIGHWAY SAFETY PROGRAM GUIDELINE NO. 1

PERIODIC MOTOR VEHICLE INSPECTION

Each State should have a program for periodic inspection of all registered vehicles or other experimental, pilot, or demonstration program approved by the Secretary, to reduce the number of vehicles with existing or potential conditions which cause or contribute to

468

NHTSA and FHWA, DOT

accidents or increase the severity of accidents which do occur, and should require the owner to correct such conditions.

I. A model program would provide, at a minimum, that:

A. Every vehicle registered in the State is inspected either at the time of initial registration and at least annually thereafter, or at such other time as may be designated under an experimental, pilot or demonstration program approved by the Secretary.

B. The inspection is performed by competent personnel specifically trained to perform their duties and certified by the State.

C. The inspection covers systems, subsystems, and components having substantial relation to safe vehicle performance.

D. The inspection procedures equal or exceed criteria issued or endorsed by the National Highway Traffic Safety Administration.

E. Each inspection station maintains records in a form specified by the State, which include at least the following information:
1. Class of vehicle.
2. Date of inspection.
3. Make of vehicle.
4. Model year.
5. Vehicle identification number.
6. Defects by category.
7. Identification of inspector.
8. Mileage or odometer reading.

F. The State publishes summaries of records of all inspection stations at least annually, including tabulations by make and model of vehicle.

II. The program should be periodically evaluated by the State and the National Highway Traffic Safety Administration should be provided with an evaluation summary.

HIGHWAY SAFETY PROGRAM GUIDELINE NO. 2

MOTOR VEHICLE REGISTRATION

Each State should have a motor vehicle registration program.

I. A model registration program would be such that every vehicle operated on public highways is registered and the following information is readily available for each vehicle:

A. Make.

§ 1204.4, Guide 3

B. Model year.
C. Identification number (rather than motor number).
D. Type of body.
E. License plate number.
F. Name of current owner.
G. Current address of owner.
H. Registered gross laden weight of every commercial vehicle.

II. Each program should have a records system that provides at least the following services.

A. Rapid entry of new data into the records or data system.

B. Controls to eliminate unnecessary or unreasonable delay in obtaining data.

C. Rapid audio or visual response upon receipt at the records station of any priority request for status of vehicle possession authorization.

D. Data available for statistical compilation as needed by authorized sources.

E. Identification and ownership of vehicle sought for enforcement or other operation needs.

III. This program should be periodically evaluated by the State, and the National Highway Traffic Safety Administration should be provided with an evaluation summary.

HIGHWAY SAFETY PROGRAM GUIDELINE NO. 3

MOTORCYCLE SAFETY

For the purposes of this guideline a motorcycle is defined as any motordriven vehicle having a seat or saddle for the use of the rider and designed to travel on not more than three wheels in contact with the ground, but excluding tractors and vehicles on which the operator and passengers ride within an enclosed cab.

Each State should have a motorcycle safety program to insure that only persons physically and mentally qualified will be licensed to operate a motorcycle; that protective safety equipment for drivers and passengers will be worn; and that the motorcycle meets guidelines for safety equipment.

I. The program should provide as a minimum that:

A. Each person who operates a motorcycle:

469

Appendix – Parts of Title 23 U.S. Code

§ 1204.4, Guide 4

1. Passes an examination or reexamination designed especially for motorcycle operation.
2. Holds a license issued specifically for motorcycle use or a regular license endorsed for each purpose.

B. Each motorcycle operator wears an approved safety helmet and eye protection when he is operating his vehicle on streets and highways.

C. Each motorcycle passenger wears an approved safety helmet, and is provided with a seat and footrest.

D. Each motorcycle is equipped with a rear-view mirror.

E. Each motorcycle is inspected at the time it is initially registered and at least annually thereafter, or in accordance with the State's inspection requirements.

II. The program should be periodically evaluated by the State for its effectiveness in terms of reductions in accidents and their end results, and the National Highway Traffic Safety Administration should be provided with an evaluation summary.

HIGHWAY SAFETY PROGRAM GUIDELINE NO. 4

DRIVER EDUCATION

Each State, in cooperation with its political subdivisions, should have a driver education and training program. This program should provide at least that:

I. There is a driver education program available to all youths of licensing age which:

A. Is taught by instructors certified by the State as qualified for these purposes.

B. Provides each student with practice driving and instruction in at least the following:

1. Basic and advanced driving techniques including techniques for handling emergencies.
2. Rules of the road, and other State laws and local motor vehicle laws and ordinances.
3. Critical vehicle systems and subsystems requiring preventive maintenance.
4. The vehicle, highway and community features:
   a. That aid the driver in avoiding crashes.

23 CFR Ch. II (4-1-95 Edition)

b. That protect him and his passengers in crashes.

c. That maximize the salvage of the injured.

5. Signs, signals, and highway markings and highway design features which require understanding for safe operation of motor vehicles.
6. Differences in characteristics of urban and rural driving including safe use of modern expressways.
7. Pedestrian safety.

C. Encourages students participating in the program to enroll in first aid training.

II. There is a State research and development program including adequate research, development and procurement of practice driving facilities, simulators, and other similar teaching aids for both school and other driver training use.

III. There is a program for adult driver training and retraining.

IV. Commercial driving schools are licensed and commercial driving instructors are certified in accordance with specific criteria adopted by the State.

V. The program should be periodically evaluated by the State, and the National Highway Traffic Safety Administration should be provided with an evaluation summary.

HIGHWAY SAFETY PROGRAM GUIDELINE NO. 5

DRIVER LICENSING

Each State should have a driver licensing program: (a) To insure that only persons physically and mentally qualified will be licensed to operate a vehicle on the highways of the State, and (b) to prevent needlessly removing the opportunity of the citizen to drive. A model program would provide, as a minimum, that:

I. Each driver holds only one license, which identifies the type(s) of vehicle(s) he is authorized to drive.

II. Each driver submits acceptable proof of date and place of birth in applying for his original license.

III. Each driver:

A. Passes an initial examination demonstrating his:

1. Ability to operate the class(es) of vehicle(s) for which he is licensed.

NHTSA and FHWA, DOT　　　　　　　　　　　　§ 1204.4, Guide 7

2. Ability to read and comprehend traffic signs and symbols.
3. Knowledge of laws relating to traffic (rules of the road) safe driving procedures, vehicle and highway safety features, emergency situations that arise in the operation of an automobile, and other driver responsibilities.
4. Visual acuity, which must meet or exceed State guidelines.
B. Is reexamined at an interval not to exceed 4 years, for at least visual acuity and knowledge of rules of the road.
IV. A record on each driver should be maintained which includes positive identification, current address, and driving history. In addition, the record system should provide the following services:
A. Rapid entry of new data into the system.
B. Controls to eliminate unnecessary or unreasonable delay in obtaining data which is required for the system.
C. Rapid audio or visual response upon receipt at the records station of any priority request for status of driver license validity.
D. Ready availability of data for statistical compilation as needed by authorized sources.
E. Ready identification of drivers sought for enforcement or other operational needs.
V. Each license should be issued for a specific term, and should be renewed to remain valid. At time of issuance or renewal each driver's record should be checked.
VI. There should be a driver improvement program to identify problem drivers for record review and other appropriate actions designed to reduce the frequency of their involvement in traffic accidents or violations.
VII. There should be:
A. A system providing for medical evaluation of persons whom the driver licensing agency has reason to believe have mental or physical conditions which might impair their driving ability.
B. A procedure which will keep the driver license agency informed of all licensed drivers who are currently applying for or receiving any type of tax,

welfare or other benefits or exemptions for the blind or nearly blind.
C. A medical advisory board or equivalent allied health professional unit composed of qualified personnel to advise the driver license agency on medical criteria and vision guidelines.
VIII. The program should be periodically evaluated by the State, and the National Highway Traffic Safety Administration should be provided with an evaluation summary. The evaluation should attempt to ascertain the extent to which driving without a license occurs.

HIGHWAY SAFETY PROGRAM GUIDELINE NO. 6

CODES AND LAWS

Each State should develop and implement a program to achieve uniformity of traffic codes and laws throughout the State. The program should provide at least that:
I. There is a plan to achieve uniform rules of the road in all of its jurisdictions.
II. There is a plan to make the State's unified rules of the road consistent with similar unified plans of other States. Toward this end, each State should undertake and maintain continuing comparisons of all State and local laws, statutes and ordinances with the comparable provisions of the Rules of the Road section of the Uniform Vehicle Code.

HIGHWAY SAFETY PROGRAM GUIDELINE NO. 7

TRAFFIC COURTS

Each State in cooperation with its political subdivisions should have a program to assure that all traffic courts in it complement and support local and statewide traffic safety objectives. The program should provide at least that:
I. All convictions for moving traffic violations should be reported to the State traffic records system.
II. Program Recommendations.
In addition the State should take appropriate steps to meet the following recommended conditions:

471

§ 1204.4, Guide 8

A. All individuals charged with moving hazardous traffic violations are required to appear in court.

B. Traffic courts are financially independent of any fee system, fines, costs, or other revenue such as posting or forfeiture of bail or other collateral resulting from processing violations of motor-vehicle laws.

C. Operating procedures, assignment of judges, staff and quarters insure reasonable availability of court services for alleged traffic offenders.

D. There is a uniform accounting system regarding traffic violation notices, collection of fines, fees and costs.

E. There are uniform rules governing court procedures in traffic cases.

F. There are current manuals and guides for administration, court procedures, and accounting.

HIGHWAY SAFETY PROGRAM GUIDELINE NO. 8

ALCOHOL IN RELATION TO HIGHWAY SAFETY

Each State, in cooperation with its political subdivisions, should develop and implement a program to achieve a reduction in those traffic accidents arising in whole or in part from persons driving under the influence of alcohol. The program should provide at least that:

I. There is a specification by the State of the following with respect to alcohol related offenses:

A. Chemical test procedures for determining blood-alcohol concentrations.

B. (1) The blood-alcohol concentrations, not higher than .10 percent by weight, which define the terms "intoxicated" or "under the influence of alcohol," and

(2) A provision making it either unlawful, or presumptive evidence of illegality, if the blood-alcohol concentration of a driver equals or exceeds the limit so established.

II. Any person placed under arrest for operating a motor vehicle while intoxicated or under the influence of alcohol is deemed to have given his consent to a chemical test of his blood, breath, or urine for the purpose of determining the alcohol content of his blood.

23 CFR Ch. II (4-1-95 Edition)

III. To the extent practicable, there are quantitative tests for alcohol:

A. On the bodies of all drivers and adult pedestrians who die within 4 hours of a traffic accident.

B. On all surviving drivers in accidents fatal to others.

IV. There are appropriate procedures established by the State for specifying:

A. The qualifications of personnel who administer chemical tests used to determine blood, breath, and other body alcohol concentrations.

B. The methods and related details of specimen selection, collection, handling, and analysis.

C. The reporting and tabulation of the results.

V. The program should be periodically evaluated by the State, and the National Highway Traffic Safety Administration should be provided with an evaluation summary.

HIGHWAY SAFETY PROGRAM GUIDELINE NO. 9

IDENTIFICATION AND SURVEILLANCE OF ACCIDENT LOCATIONS

Each State, in cooperation with county and other local governments, should have a program for identifying accident locations and for maintaining surveillance of those locations having high accident rates or losses.

I. A model program would provide, as a minimum, that:

A. There is a procedure for accurate identification of accident locations on all roads and streets.

1. To identify accident experience and losses on any specific sections of the road and street system.

2. To produce an inventory of:

a. High accident locations.

b. Locations where accidents are increasing sharply.

c. Design and operating features with which high accident frequencies or severities are associated.

3. To take appropriate measures for reducing accidents.

4. To evaluate the effectiveness of safety improvements on any specific section of the road and street system.

B. There is a systematically organized program:

NHTSA and FHWA, DOT

1. To maintain continuing surveillance of the roadway network for potentially high accident locations.
2. To develop methods for their correction.

II. The program should be periodically evaluated by the State and the Federal Highway Administration should be provided with an evaluation summary.

Highway Safety Program Guideline No. 10

Traffic Records

Each State, in cooperation with its political subdivisions, should maintain a Statewide traffic records system.

A model program would provide, as a minimum that:

I. Information on vehicles and system capabilities should include (conforms to Motor Vehicle Registration guideline):
  A. Make.
  B. Model year.
  C. Identification number (rather than motor number).
  D. Type of body.
  E. License plate number.
  F. Name and current owner.
  G. Current address of owner.
  H. Registered gross laden weight of every commercial vehicle.
  I. Rapid entry of new data into the records or data system.
  J. Controls to eliminate unnecessary or unreasonable delay in obtaining data.
  K. Rapid audio or visual response upon receipt at the records station of any priority request for status of vehicle possession authorization.
  L. Data available for statistical compilation as needed by authorized sources.
  M. Identification and ownership of vehicles sought for enforcement or other operational needs.

II. Information on drivers and system capabilities should include (conforms to Driver Licensing guideline):
  A. Positive identification.
  B. Current address.
  C. Driving history.
  D. Rapid entry of new data into the system.

§ 1204.4, Guide 11

  E. Controls to eliminate unnecessary or unreasonable delay in obtaining data which is required for the system.
  F. Rapid audio or visual response upon receipt at the records station of any priority request for status of driver license validity.
  G. Ready availability of data for statistical compilation as needed by authorized sources.
  H. Ready identification of drivers sought for enforcement or other operational needs.

III. Information on types of accidents should include:
  A. Identification of location in space and time.
  B. Identification of drivers and vehicles involved.
  C. Type of accident.
  D. Description of injury and property damage.
  E. Description of environmental conditions.
  F. Causes and contributing factors, including the absence of or failure to use available safety equipment.

IV. There should be methods to develop summary listings, cross tabulations, trend analyses and other statistical treatments of all appropriate combinations and aggregations of data items in the basic minimum data record of drivers and accident and accident experience by specified groups.

V. All traffic records relating to accidents collected hereunder should be open to the public in a manner which does not identify individuals.

VI. The program should be periodically evaluated by the State and the National Highway Traffic Safety Administration should be provided with an evaluation summary.

Highway Safety Program Guideline No. 11

Emergency Medical Services

Each State, in cooperation with its local political subdivisions, should have a program to ensure that persons involved in highway accidents receive prompt emergency medical care under the range of emergency conditions encountered. The program should provide, as a minimum, that:

I. There are training, licensing, and related requirements (as appropriate)

473

§ 1204.4, Guide 12

for ambulance and rescue vehicle operators, attendants, drivers, and dispatchers.

II. There are requirements for types and number of emergency vehicles including supplies and equipment to be carried.

III. There are requirements for the operation and coordination of ambulances and other emergency care systems.

IV. There are first aid training programs and refresher courses for emergency service personnel, and the general public is encouraged to take first aid courses.

V. There are criteria for the use of two-way communications.

VI. There are procedures for summoning and dispatching aid.

VII. There is an up-to-date, comprehensive plan for emergency medical services, including:
  A. Facilities and equipment.
  B. Definition of areas of responsibility.
  C. Agreements for mutual support.
  D. Communications systems.

VIII. This program should be periodically evaluated by the State and the National Highway Traffic Safety Administration should be provided with an evaluation summary.

HIGHWAY SAFETY PROGRAM GUIDELINE NO. 12

HIGHWAY DESIGN, CONSTRUCTION AND MAINTENANCE

Every State in cooperation with county and local governments should have a program of highway design, construction, and maintenance to improve highway safety. Guidelines applicable to specific programs are those issued or endorsed by the Federal Highway Administrator.

I. The program should provide, as a minimum that:

A. There are design guidelines relating to safety features such as sight distance, horizontal and vertical curvature, spacing of decision points, width of lanes, etc., for all new construction or reconstruction, at least on expressways, major streets and highways, and through streets and highways.

23 CFR Ch. II (4-1-95 Edition)

B. Street systems are designated to provide a safe traffic environment for pedestrians and motorists when subdivisions and residential areas are developed or redeveloped.

C. Roadway lighting is provided or upgraded on a priority basis at the following locations:
1. Expressways and other major arteries in urbanized areas.
2. Junctions of major highways in rural areas.
3. Locations or sections of streets and highways having high ratios of night-to-day motor vehicle and/or pedestrian accidents.
4. Tunnels and long underpasses.

D. There are guidelines for pavement design and construction with specific provisions for high skid resistance qualities.

E. There is a program for resurfacing or other surface treatment with emphasis on correction of locations or sections of streets and highways with low skid resistance and high or potentially high accident rates susceptible to reduction by providing improved surfaces.

F. There is guidance, warning and regulation of traffic approaching and traveling over construction or repair sites and detours.

G. There is a systematic identification and tabulation of all rail-highway grade crossings and a program for the elimination of hazards and dangerous crossings.

H. Roadways and the roadsides are maintained consistent with the design guidelines which are followed in construction, to provide safe and efficient movement of traffic.

I. Hazards within the highway right-of-way are identified and corrected.

J. There are highway design and construction features wherever possible for accident prevention and survivability including at least the following:
1. Roadsides clear of obstacles, with clear distance being determined on the basis of traffic volumes, prevailing speeds, and the nature of development along the street or highway.
2. Supports for traffic control devices and lighting that are designed to yield or break away under impact wherever appropriate.

NHTSA and FHWA, DOT

3. Protective devices that afford maximum protection to the occupants of vehicles wherever fixed objects cannot reasonably be removed or designed to yield.
4. Bridge railings and parapets which are designed to minimize severity of impact, to retain the vehicle, to redirect the vehicle so that it will move parallel to the roadway, and to minimize danger to traffic below.
5. Guardrails, and other design features which protect people from out-of-control vehicles at locations of special hazard such as playgrounds, schoolyards and commercial areas.

K. There is a post-crash program which includes at least the following:
1. Signs at freeway interchanges directing motorists to hospitals having emergency care capabilities.
2. Maintenance personnel trained in procedures for summoning aid, protecting others from hazards at accident sites, and removing debris.
3. Provisions for access and egress for emergency vehicles to freeway sections where this would significantly reduce travel time without reducing the safety benefits of access control.

II. This program should be periodically evaluated by the State for its effectiveness in terms of reductions in accidents and their end results, and the Federal Highway Administration should be provided with an evaluation summary.

HIGHWAY SAFETY PROGRAM GUIDELINE No. 13

TRAFFIC ENGINEERING SERVICES

Each State, in cooperation with its political subdivisions, and each Federal department or agency which controls highways open to public travel or supervises traffic operations, should have a program for applying traffic engineering measures and techniques, including the use of traffic control devices, to reduce the number and severity of traffic accidents.

I. The program as a minimum should consist of:
A. A comprehensive manpower development plan to provide the necessary traffic engineering capability, including:

§ 1204.4, Guide 13

1. Provisions for supplying traffic engineering assistance to those jurisdictions unable to justify a full-time traffic engineering staff.
2. Provisions for upgrading the skills of practicing traffic engineers, and providing basic instruction in traffic engineering techniques to subprofessionals and technicians.

B. Utilization of traffic engineering principles and expertise in the planning, design, construction, and maintenance of the public roadways, and in the application of traffic control devices.

C. A traffic control devices plan including:
1. An inventory of all traffic control devices.
2. Periodic review of existing traffic control devices, including a systematic upgrading of substandard devices to conform with guidelines issued or endorsed by the Federal Highway Administrator.
3. A maintenance schedule adequate to insure proper operation and timely repair of control devices, including daytime and nighttime inspections.
4. Where appropriate, the application and evaluation of new ideas and concepts in applying control devices and in the modification of existing devices to improve their effectiveness through controlled experimentation.

D. An implementation schedule to utilize traffic engineering manpower to:
1. Review road projects during the planning, design, and construction stages to detect and correct features that may lead to operational safety difficulties.
2. Install safety-related improvements as a part of routine maintenance and/or repair activities.
3. Correct conditions noted during routine operational surveillance of the roadway system to rapidly adjust for the changes in traffic and road characteristics as a means of reducing accident frequency or severity.
4. Conduct traffic engineering analyses of all high accident locations and development of corrective measures.
5. Analyze potentially hazardous locations, such as sharp curves, steep grades, and railroad grade crossings

475

§ 1204.4, Guide 14

and develop appropriate countermeasures.

6. Identify traffic control needs and determine short and long range requirements.

7. Evaluate the effectiveness of specific traffic control measures in reducing the frequency and severity of traffic accidents.

8. Conduct traffic engineering studies to establish traffic regulations such as fixed or variable speed limits.

II. This Program should be periodically evaluated by the State, or appropriate Federal department or agency where applicable, and the Federal Highway Administration should be provided with an evaluation summary.

HIGHWAY SAFETY PROGRAM GUIDELINE NO. 14

PEDESTRIAN SAFETY

Every State in cooperation with its political subdivisions should develop and implement a program to insure the safety of pedestrians of all ages. A model program would provide, as a minimum that:

I. There should be a continuing statewide inventory of pedestrian-motor vehicle accidents, identifying specifically:

A. The locations and times of all such accidents.

B. The age of all of the pedestrians injured or killed.

C. Where feasible, to determine whether the exterior features of the vehicle produced or aggravated an injury.

D. The color and shade of clothing worn by pedestrians when injured or killed, and the visibility conditions which prevailed at the time.

E. The extent to which alcohol is present in the blood of fatally injured pedestrians 16 years of age and older.

F. Where possible, to determine, the extent to which pedestrians involved in accidents have physical or mental disabilities.

II. There should be established Statewide operational procedures for improving the protection of pedestrians through reduction of potential conflicts with vehicles:

A. By application of traffic engineering practices including pedestrian signals, signs, markings, parking regulations, and other pedestrian and vehicle traffic control devices.

B. By land-use planning in new and redevelopment areas for safe pedestrian movement.

C. By provision of pedestrian bridges, barriers, sidewalks and other means of physically separating pedestrian and vehicle pathways.

D. By provision of environmental illumination at high pedestrian volume and/or potentially hazardous pedestrian crossings.

III. There should be established a Statewide program for familiarizing drivers with the pedestrian problem and with ways to avoid pedestrian collisions.

A. The program content should include emphasis on:

(1) Behavior characteristics of the three types of pedestrians most commonly involved in accidents with vehicles: (i) Children; (ii) persons under the influence of alcohol; (iii) the elderly;

(2) Accident avoidance techniques that take into account the hazardous conditions, and behavior characteristics displayed by each of the three high risk pedestrian groups listed in subparagraph (1).

B. Emphasis on this program content should be included in:

(1) All driver education and training courses;

(2) Driver improvement courses; and

(3) Driver license examinations.

IV. There should be statewide programs for training and educating all members of the public as to safe pedestrian behavior on or near the streets and highways.

A. For children, youths and adults enrolled in schools, beginning at the earliest possible age.

B. For the general population via the public media.

V. There should be a statewide program for the protection of children walking to and from school, entering and leaving school buses, and in neighborhood play.

VI. There should be a statewide program for establishment and enforcement of traffic regulations designed to achieve orderly pedestrian and vehicle movement and to reduce vehicle-pedestrian conflicts.

NHTSA and FHWA, DOT

VII. This program should be periodically evaluated by the States, and the National Highway Traffic Safety Administration and the Federal Highway Administration should be provided with an evaluation summary.

HIGHWAY SAFETY PROGRAM GUIDELINE NO. 15

POLICE TRAFFIC SERVICES

Every State in cooperation with its political subdivisions should have a program to insure efficient and effective police services utilizing traffic patrols: To enforce traffic laws; to prevent accidents; to aid the injured; to document the particulars of individual accidents; to supervise accident cleanup and to restore safe and orderly traffic movement.

I. The program should provide as a minimum that there are:

A. Uniform training procedures in all aspects for police supervision of vehicular and pedestrian traffic related to highway safety, including use of appropriate instructional materials and techniques for recruit, advanced, in-service, and special course training.

B. Periodic in-service training courses for uniformed and other police department employees assigned to traffic duties dealing with:

(1) Administration and management of police, vehicular and pedestrian traffic services.

(2) Analysis, interpretation and use of traffic records data.

(3) Insurance of prompt reliable postaccident response, including skilled aid to the injured.

(4) Accomplishing postaccident responsibilities.

C. Procedures for the selective assignment of trained police personnel to supervise vehicular and pedestrian traffic duties including enforcement patrols in hazardous or congested areas based on: time and location of

(1) Traffic volume.
(2) Accident experience.
(3) Traffic violation frequency.
(4) Emergency and service needs.

D. Procedures for investigating, recording and reporting accidents pertaining to:

§ 1204.4, Guide 16

(1) The human, vehicular, and highway causative factors in individual accidents.

(2) The human, vehicular, and highway causative factors of injuries and deaths, including failure to use safety belts.

(3) The efficiency of the postaccident response.

E. Procedures for recognizing and reporting, to the appropriate agencies hazardous highway defects and conditions, including:

(1) Condition of drivers;
(2) Operational condition of motor vehicles;
(3) Defective signs, signals, controls, construction and maintenance deficiencies.

a. Data listed in (3) above should be readily available to the public.

F. Appropriate agreements by the State and its political subdivisions regarding primary responsibility and authority for police traffic supervision, and cooperative responsibilities where concurrent jurisdictional boundaries and problems exist, and appropriate participation of each law enforcement agency in the comprehensive highway safety program of the State and its political subdivisions.

II. The programs should be periodically evaluated by the State, and the National Highway Traffic Safety Administration should be provided with an evaluation summary.

III. Nothing in this guideline should preclude the use of personnel other than police officers in carrying out the minimum requirements in accordance with laws and policies established by State and/or local governments.

HIGHWAY SAFETY PROGRAM GUIDELINE NO. 16

DEBRIS HAZARD CONTROL AND CLEANUP

Each State in cooperation with its political subdivisions should have a program which provides for rapid, orderly, and safe removal from the roadway of wreckage, spillage, and debris resulting from motor vehicle accidents, and for otherwise reducing the likelihood of secondary and chain-reaction collisions, and conditions hazardous to the public health and safety.

Appendix – Parts of Title 23 U.S. Code

§ 1204.4, Guide 17

I. The program should provide as a minimum that:

A. Operational procedures are established and implemented for:

(1) Enabling rescue and salvage equipment personnel to get to the scene of accidents rapidly and to operate effectively on arrival:
   a. On heavily traveled freeways and other limited access roads;
   b. In other types of locations where wreckage or spillage of hazardous materials on or adjacent to highways endangers the public health and safety;

(2) Extricating trapped persons from wreckage with reasonable care—both to avoid injury or aggravating existing injuries;

(3) Warning approaching drivers and detouring them with reasonable care past hazardous wreckage or spillage;

(4) Safe handling of spillage or potential spillage of materials that are:
   a. Radioactive
   b. Flammable
   c. Poisonous
   d. Explosive
   e. Otherwise hazardous.

(5) Removing wreckage or spillage from roadways or otherwise causing the resumption of safe, orderly traffic flow.

B. Adequate numbers of rescue and salvage personnel are properly trained and retrained in the latest accident cleanup techniques.

C. A communications system is provided, adequately equipped and manned, to provide coordinated effort in incident detection, and the notification, dispatch, and response of appropriate services.

II. The program should be periodically evaluated by the State, and the National Highway Traffic Safety Administration should be provided with an evaluation summary.

HIGHWAY SAFETY PROGRAM GUIDELINE NO. 17

PUPIL TRANSPORTATION SAFETY

I. *Scope.* This guideline establishes minimum recommendations for a State highway safety program for pupil transportation safety including the identification, operation, and maintenance of buses used for carrying students; training of passengers, pedestrians, and bicycle riders; and administration.

II. *Purpose.* The purpose of this guideline is to minimize, to the greatest extent possible, the danger of death or injury to school children while they are traveling to and from school and school-related events.

III. *Definitions.*

Bus is a motor vehicle designed for carrying more than 10 persons (including the driver).

Federal Motor Carrier Safety Regulations (FMCSR) are the regulations of the Federal Highway Administration (FHWA) for commercial motor vehicles in interstate commerce, including buses with a gross vehicle weight rating (GVWR) greater than 10,000 pounds or designed to carry 16 or more persons (including the driver), other than buses used to transport school children from home to school and from school to home. (The FMCSR are set forth in 49 CFR parts 383–399.)

School-chartered bus is a "bus" that is operated under a short-term contract with State or school authorities who have acquired the exclusive use of the vehicle at a fixed charge to provide transportation for a group of students to a special school-related event.

School bus is a "bus" that is used for purposes that include carrying students to and from school or related events on a regular basis, but does not include a transit bus or a school-chartered bus.

IV. *Pupil Transportation Safety Program Administration and Operations.—Recommendation.* Each State, in cooperation with its school districts and other political subdivisions, should have a comprehensive pupil transportation safety program to ensure that school buses and school-chartered buses are operated and maintained so as to achieve the highest possible level of safety.

A. *Administration.* 1. There should be a single State agency having primary administrative responsibility for pupil transportation, and employing at least one full-time professional to carry out these responsibilities.

2. The responsible State agency should develop an operating system for collecting and reporting information needed to improve the safety of operat-

NHTSA and FHWA, DOT

ing school buses and school-chartered buses. This includes the collection and evaluation of uniform crash data consistent with the criteria set forth in Highway Safety Program Guidelines No. 10, "Traffic Records" and No. 18, "Accident Investigation and Reporting."

B. *Identification and equipment of school buses.* Each State should establish procedures to meet the following recommendations for identification and equipment of school buses.

1. All school buses should:

a. Be identified with the words "School Bus" printed in letters not less than eight inches high, located between the warning signal lamps as high as possible without impairing visibility of the lettering from both front and rear, and have no other lettering on the front or rear of the vehicle, except as required by Federal Motor Vehicle Safety Standards (FMVSS), 49 CFR part 571.

b. Be painted National School Bus Glossy Yellow, in accordance with the colorimetric specification of National Institute of Standards and Technology (NIST) Federal Standard No. 595a, Color 13432, except that the hood should be either that color or lusterless black, matching NIST Federal Standard No. 595a, Color 37038.

c. Have bumpers of glossy black, matching NIST Federal Standard No. 595a, Color 17038, unless, for increased visibility, they are covered with a reflective material.

d. Be equipped with safety equipment for use in an emergency, including a charged fire extinguisher, that is properly mounted near the driver's seat, with signs indicating the location of such equipment.

e. Be equipped with device(s) demonstrated to enhance the safe operation of school vehicles, such as a stop signal arm.

f. Be equipped with a system of signal lamps that conforms to the school bus requirements of FMVSS No. 108, 49 CFR 571.108.

g. Have a system of mirrors that conforms to the school bus requirements of FMVSS No. 111, 49 CFR 571.111.

h. Comply with all FMVSS applicable to school buses at the time of their manufacture.

§ 1204.4, Guide 17

2. Any school bus meeting the identification recommendations of sections 1.a–h above that is permanently converted for use wholly for purposes other than transporting children to and from school or school-related events should be painted a color other than National School Bus Glossy Yellow, and should have the stop arms and school bus signal lamps described by sections 1.e & f removed.

3. School buses, while being operated on a public highway and transporting primarily passengers other than school children, should have the words "School Bus" covered, removed, or otherwise concealed, and the stop arm and signal lamps described by sections 1.e & f should not be operated.

4. School-chartered buses should comply with all applicable FMCSR and FMVSS.

C. *Operations.* Each State should establish procedures to meet the following recommendations for operating school buses and school-chartered buses:

1. *Personnel.* a. Each State should develop a plan for selecting, training, and supervising persons whose primary duties involve transporting school children in order to ensure that such persons will attain a high degree of competence in, and knowledge of, their duties.

b. Every person who drives a school bus or school-chartered bus occupied by school children should, as a minimum:

(1) Have a valid State driver's license to operate such a vehicle. All drivers who operate a vehicle designed to carry 16 or more persons (including the driver) are required by FHWA's Commercial Driver's License Standards by April 1, 1992 (49 CFR part 383) to have a valid commercial driver's license;

(2) Meet all physical, mental, moral and other requirements established by the State agency having primary responsibility for pupil transportation, including requirements for drug and/or alcohol misuse or abuse; and

(3) Be qualified as a driver under the Federal Motor Carrier Safety Regulations of the FHWA, 49 CFR part 391, if the driver or the driver's employer is subject to those regulations.

479

§ 1204.4, Guide 17

2. *Vehicles.* a. Each State should enact legislation that provides for uniform procedures regarding school buses stopping on public highways for loading and discharge of children. Public information campaigns should be conducted on a regular basis to ensure that the driving public fully understands the implications of school bus warning signals and requirements to stop for school buses that are loading or discharging school children.

b. Each State should develop plans for minimizing highway use hazards to school bus and school-chartered bus occupants, other highway users, pedestrians, bicycle riders and property. They should include, but not be limited to:

(1) Careful planning and annual review of routes for safety hazards;

(2) Planning routes to ensure maximum use of school buses and school-chartered buses, and to ensure that passengers are not standing while these vehicles are in operation;

(3) Providing loading and unloading zones off the main traveled part of highways, whenever it is practical to do so;

(4) Establishing restricted loading and unloading areas for school buses and school-chartered buses at or near schools;

(5) Ensuring that school bus operators, when stopping on a highway to take on or discharge children, adhere to State regulations for loading and discharging including the use of signal lamps as specified in section B.1.f. of this guideline;

(6) Prohibiting, by legislation or regulation, operation of any school bus unless it meets the equipment and identification recommendations of this guideline; and

(7) Replacing, consistent with the economic realities which typically face school districts, those school buses which are not manufactured to meet the April 1, 1977 FMVSS for school buses, with those manufactured to meet the stricter school bus standards, and not chartering any pre-1977 school buses.

(8) Informing potential buyers of pre-1977 school buses that these buses may not meet current standards for newly manufactured buses and of the need for continued maintenance of these buses and adequate safety instruction.

c. Use of amber signal lamps to indicate that a school bus is preparing to stop to load or unload children is at the option of the State. Use of red warning signal lamps as specified in section B.1.f. of this guideline for any purpose or at any time other than when the school bus is stopped to load or discharge passengers should be prohibited.

d. When school buses are equipped with stop arms, such devices should be operated only in conjunction with red warning signal lamps, when vehicles are stopped.

e. Seating. (1) Standing while school buses and school-chartered buses are in motion should not be permitted. Routing and seating plans should be coordinated so as to eliminate passengers standing when a school bus or school-chartered bus is in motion.

(2) Seating should be provided that will permit each occupant to sit in a seat intended by the vehicle's manufacturer to provide accommodation for a person at least as large as a 5th percentile adult female, as defined in 49 CFR 571.208. Due to the variation in sizes of children of different ages, States and school districts should exercise judgment in deciding how many students are actually transported in a school bus or school-chartered bus.

(3) There should be no auxiliary seating accommodations such as temporary or folding jump seats in school buses.

(4) Drivers of school buses and school-chartered buses should be required to wear occupant restraints whenever the vehicle is in motion.

(5) Passengers in school buses and school-chartered buses with a gross vehicle weight rating (GVWR) of 10,000 pounds or less should be required to wear occupant restraints (where provided) whenever the vehicle is in motion. Occupant restraints should comply with the requirements of FMVSS Nos. 208, 209 and 210, as they apply to multipurpose vehicles.

f. Emergency exit access. Baggage and other items transported in the passenger compartment should be stored and secured so that the aisles are kept clear and the door(s) and emergency exit(s) remain unobstructed at all

NHTSA and FHWA, DOT

times. When school buses are equipped with interior luggage racks, the racks should be capable of retaining their contents in a crash or sudden driving maneuver.

D. *Vehicle maintenance.* Each State should establish procedures to meet the following recommendations for maintaining buses used to carry school children:

1. School buses should be maintained in safe operating condition through a systematic preventive maintenance program.

2. All school buses should be inspected at least semiannually. In addition, school buses and school-chartered buses subject to the Federal Motor Carrier Safety Regulations of FHWA should be inspected and maintained in accordance with those regulations (49 CFR Parts 393 and 396).

3. School bus drivers should be required to perform daily pre-trip inspections of their vehicles, and the safety equipment thereon (especially fire extinguishers), and to report promptly and in writing any problems discovered that may affect the safety of the vehicle's operation or result in its mechanical breakdown. Pre-trip inspection and condition reports for school buses and school-chartered buses subject to the Federal Motor Carrier Safety Regulations of FHWA should be performed in accordance with those regulations (49 CFR 392.7, 392.8, and 396).

E. *Other Aspects of Pupil Transportation Safety.* 1. At least once during each school semester, each pupil transported from home to school in a school bus should be instructed in safe riding practices, proper loading and unloading techniques, proper street crossing to and from school bus stops and should participate in supervised emergency evacuation drills, which are timed. Prior to each departure, each pupil transported on an activity or field trip in a school bus or school-chartered bus should be instructed in safe riding practices and on the location and operation of emergency exits.

2. Parents and school officials should work together to select and designate the safest pedestrian and bicycle routes for the use of school children.

3. All school children should be instructed in safe transportation prac-

§ 1204.4, Guide 18

tices for walking to and from school. For those children who routinely walk to school, training should include preselected routes and the importance of adhering to those routes.

4. Children riding bicycles to and from school should receive bicycle safety education, wear bicycle safety helmets, and not deviate from preselected routes.

5. Local school officials and law enforcement personnel should work together to establish crossing guard programs.

6. Local school officials should investigate programs which incorporate the practice of escorting students across streets and highways when they leave school buses. These programs may include the use of school safety patrols or adult monitors.

7. Local school officials should establish passenger vehicle loading and unloading points at schools that are separate from the school bus loading zones.

V. *Program evaluation.* The pupil transportation safety program should be evaluated at least annually by the State agency having primary administrative responsibility for pupil transportation.

HIGHWAY SAFETY PROGRAM GUIDELINE NO. 18

ACCIDENT INVESTIGATION AND REPORTING

I. *Scope.* This guideline establishes the requirement that each State should have a highway safety program for accident investigation and reporting.

II. *Purpose.* The purpose of this guideline is to establish a uniform, comprehensive motor vehicle traffic accident investigation program for gathering information—who, what, when, where, why, and how—on motor vehicle traffic accidents and associated deaths, injuries, and property damage; and entering the information into the traffic records system for use in planning, evaluating, and furthering highway safety program goals.

III. *Definitions.* For the purpose of this guideline the following definitions apply:

Accident—an unintended event resulting in injury or damage, involving one or more motor vehicles on a high-

Appendix – Parts of Title 23 U.S. Code

§ 1204.4, Guide 18

way that is publicly maintained and open to the public for vehicular travel.

Highway—the entire width between the boundary lines of every way publicly maintained when any part thereof is open to the use of the public for purposes of vehicular travel.

Motor vehicle—any vehicle driven or drawn by mechanical power manufactured primarily for use on the public streets, roads, and highways, except any vehicle operated exclusively on a rail or rails.

IV. *Requirements.* Each State, in cooperation with its political subdivisions, should have an accident investigation program. A model program would be structured as follows:

A. *Administration.* 1. There should be a State agency having primary responsibility for administration and supervision of storing and processing accident information, and providing information needed by user agencies.

2. There should be employed at all levels of government adequate numbers of personnel, properly trained and qualified, to conduct accident investigations and process the resulting information.

3. Nothing in this guideline should preclude the use of personnel other than police officers, in carrying out the requirements of this guideline in accordance with laws and policies established by State and/or local governments.

4. Procedures should be established to assure coordination, cooperation, and exchange of information among local, State, and Federal agencies having responsibility for the investigation of accidents and subsequent processing of resulting data.

5. Each State should establish procedures for entering accident information into the statewide traffic records system established pursuant to Highway Safety Program Guideline No. 10, Traffic Records, and for assuring uniformity and compatibility of this data with the requirements of the system, including as a minimum:

a. Use of uniform definitions and classifications acceptable to the National Highway Traffic Safety Administration and identified in the Highway Safety Program Manual.

23 CFR Ch. II (4-1-95 Edition)

b. A guideline format for input of data into the statewide traffic records system.

c. Entry into the statewide traffic records system of information gathered and submitted to the responsible State agency.

B. *Accident reporting.* Each State should establish procedures which require the reporting of accidents to the responsible State agency within a reasonable time after occurrence.

C. *Owner and driver reports.* 1. In accidents involving only property damage, where the vehicle can be normally and safely driven away from the scene, the drivers or owners of vehicles involved should be required to submit a written report consistent with State reporting requirements, to the responsible State agency. A vehicle should be considered capable of being normally and safely driven if it does not require towing and can be operated under its own power, in its customary manner, without further damage or hazard to itself, other traffic elements, or the roadway. Each report so submitted should include, as a minimum, the following information relating to the accident:

a. Location.
b. Time.
c. Identification of driver(s).
d. Identification of pedestrian(s), passenger(s), or pedal-cyclist(s).
e. Identification of vehicle(s).
f. Direction of travel of each unit.
g. Other property involved.
h. Environmental conditions existing at the time of the accident.
i. A narrative description of the events and circumstances leading up to the time of impact, and immediately after impact.

2. In all other accidents, the drivers or owners of motor vehicles involved should be required to immediately notify the police of the jurisdiction in which the accident occurred. This includes, but is not limited to accidents involving: (1) Fatal or nonfatal personal injury, or (2) damage to the extent that any motor vehicle involved cannot be driven under its own power, in its customary manner, without further damage or hazard to itself, other traffic elements, or the roadway, and therefore requires towing.

482

NHTSA and FHWA, DOT § 1205.1

D. *Accident investigation.* Each State should establish a plan for accident investigation and reporting which should meet the following criteria:

1. Police investigation should be conducted of all accidents as identified in section IV.C.2. of this guideline 18. Information gathered should be consistent with the police mission of detecting and apprehending law violators, and should include, as a minimum, the following:

a. Violation(s), if any occurred, cited by section and subsection, numbers and titles of the State code, that (1) contributed to the accident where the investigating officer has reason to believe that violations were committed regardless of whether the officer has sufficient evidence to prove the violation(s); and (2) for which the driver was arrested or cited.

b. Information necessary to prove each of the elements of the offense(s) for which the driver was arrested or cited.

c. Information, collected in accordance with the program established under Highway Safety Program Guideline No. 15, Police Traffic Services, section I-D, relating to human, vehicular, and highway factors causing individual accidents, injuries, and deaths, including failure to use safety belts.

2. Accident investigation teams should be established, representing different interest areas, such as police; traffic; highway and automotive engineering; medical, behavioral, and social sciences. Data gathered by each member of the investigation team should be consistent with the mission of the member's agency, and should be for the purpose of determining probable causes of accidents, injuries, and deaths. These teams should conduct investigations of an appropriate sampling of accidents in which there were one or more of the following conditions:

a. Locations that have a similarity of design, traffic engineering characteristics, or environmental conditions, and that have a significantly large or disproportionate number of accidents.

b. Motor vehicles or motor vehicle parts that are involved in a significantly large or disproportionate number of accidents or injury-producing accidents.

c. Drivers, pedestrians, and vehicle occupants of a particular age, sex, or other grouping, who are involved in a significantly large or disproportionate number of motor vehicle traffic accidents or injuries.

d. Accidents in which causation or the resulting injuries and property damage are not readily explainable in terms of conditions or circumstances that prevailed.

e. Other factors that concern State and national emphasis programs.

V. *Evaluation.* The program should be evaluated at least annually by the State. Substance of the evaluation report should be guided by Chapter V of the Highway Safety Program Manual. The National Highway Traffic Safety Administration should be provided with a copy of the evaluation report.

[33 FR 16337, Nov. 7, 1968. Redesignated at 38 FR 10811, May 2, 1973]

EDITORIAL NOTE: For FEDERAL REGISTER citations affecting § 1204.4, see the List of CFR Sections Affected in the Finding Aids section of this volume.

PART 1205—HIGHWAY SAFETY PROGRAMS; DETERMINATIONS OF EFFECTIVENESS

Sec.
1205.1 Scope.
1205.2 Purpose.
1205.3 Identification of National Priority Program Areas.
1205.4 Funding procedures for National Priority Program Areas.
1205.5 Funding procedures for other program areas.

AUTHORITY: 23 U.S.C. 402; delegations of authority at 49 CFR 1.48 and 1.50.

SOURCE: 47 FR 15120, Apr. 8, 1982, unless otherwise noted.

§ 1205.1 Scope.

This part identifies those highway safety programs that are eligible for Federal funding under the State and Community Highway Safety Grant Program (23 U.S.C. 402) and specifies the Federal funding requirements for those programs.

483

Appendix – Parts of Title 23 U.S. Code

§ 1205.2

§ 1205.2 Purpose.

The purpose of this part is to establish national highway safety priorities and establish program areas within which highway safety programs developed by the states would be eligible to receive Federal funding.

§ 1205.3 Identification of National Priority Program Areas.

(a) Under statutory provisions administered by NHTSA, the following NHTSA-administered highway safety program areas have been identified as encompassing a major highway safety problem which is of national concern, and for which effective countermeasures have been identified. Programs developed in such areas are eligible for Federal funding, pursuant to guidelines issued by the National Highway Traffic Safety Administration and the review procedure set forth in § 1205.4:

(1) Alcohol and Other Drug Countermeasures
(2) Police Traffic Services
(3) Occupant Protection
(4) Traffic Records
(5) Emergency Medical Services
(6) Motorcycle Safety

(b) Under statutory provisions administered by FHWA, the following FHWA-administered highway safety program area has been identified as encompassing a major highway safety problem which is of national concern, and for which effective countermeasures have been identified. The program developed in this area is eligible for Federal funding, pursuant to provisions of 23 U.S.C. 402(g), guidelines issued by the Federal Highway Administration and the review procedures set forth in § 1205.4: *Roadway Safety.*

(c) Under statutory provisions jointly administered by NHTSA and FHWA, the following highway safety program areas, jointly administered by NHTSA and FHWA, have been identified as encompassing a major highway safety problem which is of national concern, and for which effective countermeasures have been identified. Programs developed in such areas are eligible for Federal funding, pursuant to guidelines issued by NHTSA and FHWA and the review procedures set forth in § 1205.4:

(1) Pedestrian and Bicycle Safety
(2) Speed Control

[47 FR 15120, Apr. 8, 1982, as amended at 53 FR 11270, Apr. 6, 1988; 56 FR 50255, Oct. 4, 1991; 59 FR 64127, Dec. 13, 1994]

§ 1205.4 Funding procedures for National Priority Program Areas.

A State planning to use funds under 23 U.S.C. 402 to support a program that is within a National Highway Safety Priority Program Area shall be subject to the following procedures:

(a) The State shall describe each highway safety problem within such Priority Area and any countermeasure proposed to decrease or stabilize the problem, and provide recent statistical trend data concerning injury, fatal, and property damage crashes to support the problem and countermeasure identifications.

(b) The State shall list the specific projects proposed to implement such countermeasures and the criteria for project selection.

(c) The NHTSA and/or the FHWA, as applicable, shall review the information provided under paragraphs (a) and (b) of this section in accordance with the procedures of § 1200.11 of this chapter.

[47 FR 15120, Apr. 8, 1982, as amended at 58 FR 41038, Aug. 2, 1993]

§ 1205.5 Funding procedures for other program areas.

If a State intends to use funds under 23 U.S.C. 402 to support a project that is not within a National Highway Safety Priority Program Area, the State shall describe the project in its annual Highway Safety Plan, and shall, at its option, select one or both of the following procedures:

(a) *Formal decisionmaking.* Under this procedure, the State shall first develop and submit as part of its annual Highway Safety Plan or by a separate submission a formal administrative decisionmaking process for identifying highway safety problems and corresponding countermeasures. Upon approval of the Plan and adoption by the State of the process involved, a State may thereafter certify in subsequent Plan submissions that it has developed each proposed project in accordance with the described process. NHTSA or

NHTSA and FHWA, DOT § 1206.1

FHWA shall on such subsequent submissions consider the findings and determinations made by the State pursuant to such process to be determinative and shall review proposed projects only pursuant to the limited review criteria applicable to the projects subject to § 1205.4. NHTSA and/or FHWA, as applicable, shall review and approve proposed State administrative processes pursuant to the following general criteria:

(1) Use of State data on traffic accidents to determine the magnitude and severity of the highway safety problems by geographic area and target group.

(2) Determination of related system deficiencies and driver behavior deficiencies that can be stabilized or remedied by countermeasure approaches.

(3) Development of countermeasures to remedy the problems. Priorities should be assigned based on the following considerations:

(i) Estimates of the impact on accidents and injuries;
(ii) Cost effectiveness;
(iii) Past program and project results;
(iv) Innovative approaches;
(v) Comprehensiveness of programs;
(vi) Catalytic and leverage effects; and
(vii) Prospects for activities to be self-supporting or continued with State/local resources after Federal funds are discontinued.

(4) Development of projects from the countermeasure approaches that ensure consultation with affected groups and participation by the public. This shall be accomplished by conducting public meetings to identify traffic safety problems and to recommend alternate countermeasure solutions.

(5) Development of administrative and impact evaluations for the projects, as appropriate.

(b) *Problem identification.* Under this procedure, the State shall submit information on individual proposed projects. NHTSA or FHWA, as applicable, shall approve each project if it addresses the identified problem in a manner reasonably calculated to decrease or stabilize the problems. The State shall submit, at a minimum, the following information:

(1) The State and local data on traffic accidents used to determine the magnitude and severity of the particular highway safety problem by geographic area and target group.

(2) The impact each project is estimated to have on traffic accidents and injuries.

(3) Estimates of the resources necessary to carry out planned activities and projects.

(4) The relation of each project to a comprehensive, balanced program.

(5) The improvements in program operational efficiency and/or cost effectiveness which are expected as a result of the implementation of each project.

(6) The commitment of State and/or local resources to each project.

(7) The prospects for activities to be self-supporting or continued with State/local resources after Federal funds are discontinued.

(8) The criteria to be used to conduct administrative and impact evaluations of products, as appropriate.

PART 1206—RULES OF PROCEDURE FOR INVOKING SANCTIONS UNDER THE HIGHWAY SAFETY ACT OF 1966

Sec.
1206.1 Scope.
1206.2 Purpose.
1206.3 Definitions.
1206.4 Sanctions.
1206.5 Commencement of proceedings.
1206.6 Contents of notice of proposed recommended determination.
1206.7 Hearing officers.
1206.8 Prehearing conference.
1206.9 Consent determination.
1206.10 Hearing.
1206.11 Recommended determination.
1206.12 Final determination.

AUTHORITY: 23 U.S.C. 116, 315, 402.

SOURCE: 39 FR 19206, May 31, 1974, unless otherwise noted.

§ 1206.1 Scope.

This part establishes procedures governing determinations to invoke the sanctions applicable to any State that does not comply with the highway safety program requirements in the Highway Safety Act of 1966, as amended (23 U.S.C. 402), and highway safety program standards issued thereunder.

485

§ 1206.2

§ 1206.2 Purpose.

The purpose of this part is to prescribe procedures for determining whether and the extent to which the 23 U.S.C. 402 sanctions should be invoked, and to ensure a full airing of views on the issues relevant to such determinations by affording the affected State and all other interested persons an opportunity to participate in a public hearing.

§ 1206.3 Definitions.

As used in this part:

(a) *Administrators* means the Administrators of the Federal Highway Administration and the National Highway Traffic Safety Administration.

(b) *Affected State* means the State with respect to which a proposed recommended determination has been made pursuant to this part.

(c) *Highway safety program* means a State program consisting of both (1) a Comprehensive Highway Safety Plan, a multi-year plan of the State and its political subdivisions for implementing the highway safety program standards and (2) an Annual Highway Safety Work Program, a program detailing the activities and proposed expenditures of the State and its political subdivisions for implementing selected components of the State's Comprehensive Highway Safety Plan during a single year.

(d) *Highway safety program standards* means the standards in § 1204.4 of this chapter, issued under 23 U.S.C. 402, for State highway safety programs.

(e) *Implementing* includes, but is not limited to, complying with conditions upon which the Secretary has granted approval for a State's highway safety program.

(f) *Secretary* means the Secretary of Transportation.

§ 1206.4 Sanctions.

(a)(1) Except as provided in paragraph (a)(2) of this section, the Secretary withholds all of a State's Federal highway safety funds and 10 percent of its Federal-aid highway funds that would otherwise have been apportioned to the State under 23 U.S.C. 402 and 104, respectively, and withholds approval of further highway safety programs in the State, if he finally determines pursuant to this part that a State:

(i) Does not have a highway safety program approved by him in accordance with 23 U.S.C. 402; or

(ii) Is not implementing such an approved program.

(2) The Secretary may suspend, for such periods as he deems necessary, the application to a State of the provision in 23 U.S.C. 402 regarding the withholding of funds apportioned under 23 U.S.C. 104, if he determines that such suspension is in the public interest.

(3) When the Secretary withholds funds and program approval from a State pursuant to paragraph (a)(1) of this section, he continues doing so until the State has an approved program or is implementing an approved program to his satisfaction, as appropriate.

(b)(1) The Governor of each State shall implement, or cause to be implemented, a highway safety program approved by the Secretary pursuant to 23 U.S.C. 402. If the Governor lacks legal authority to implement such program or part thereof, he shall enter into formal agreements for its implementation with appropriate officials of the appropriate State agency, political subdivision, or other governmental instrumentality. However, the existence of such agreement does not relieve the Governor of his responsibility for implementation of the entire highway safety program.

(2) If the Secretary finally determines pursuant to this part that a State has failed to comply with the Federal laws or regulations with respect to its highway safety program, or has failed to implement a highway safety program approved by him pursuant to 23 U.S.C. 402, he may withhold payment to the State on account of (i) the State's current annual highway safety work program, with respect to obligations incurred after the date of the Secretary's determination or a date established in such determination, whichever is later, and (ii) the State's subsequent annual highway safety work programs, and may also withhold approval of further highway safety programs of the State, until the State complies or takes remedial action to

486

NHTSA and FHWA, DOT

the satisfaction of the Secretary, as appropriate.

§ 1206.5 Commencement of proceedings.

(a) The Administrators initiate the proceedings pursuant to this part by making a proposed recommended determination to invoke the sanctions specified in § 1206.4

(b) The Administrators send the Governor of the affected State by certified mail and publish in the FEDERAL REGISTER a notice of the proposed recommended determination.

§ 1206.6 Contents of notice of proposed recommended determination.

The notice of proposed recommended determination includes:

(a) A statement of the reasons for the proposed action, including the specific highway safety program deficiencies upon which the proposed recommended determination is based; and

(b) The time, date, and place for a hearing at which the affected State and any interested person may present evidence and oral and written views, or both, concerning the specified deficiencies. Hearings are held in Washington, DC, at the headquarters of the Department of Transportation.

§ 1206.7 Hearing officers.

(a) A three-member hearing board is established consisting of an Office of the Secretary official appointed by the Secretary, a Federal Highway Administration official appointed by the Administrator of that Administration, and a National Highway Traffic Safety Administration official appointed by the Administrator of that Administration. The official from the Office of the Secretary serves as presiding officer. The appointment of the hearing board is announced in a notice published in the FEDERAL REGISTER by the presiding officer. A copy of the notice is sent to the Governor of the affected State by certified mail.

(b) The presiding officer has power to take any action and to make all necessary rules and regulations to govern the conduct of the hearing. His powers include the following:

(1) Changing the date and time of the hearing upon reasonable notice to the affected State and other hearing participants and publication of such change in the FEDERAL REGISTER;

(2) Continuing the hearing in whole or in part;

(3) Regulating the course of the hearing and the conduct of the participants and counsel therein;

(4) Examining witnesses; and

(5) Taking any other action authorized by this part.

§ 1206.8 Prehearing conference.

(a) At any time before the hearing begins, the presiding officer, on his own initiative or at the written request of the Governor of the affected State, may, after publication of a notice in the FEDERAL REGISTER giving notice to interested persons, convene a public prehearing conference to consider the following:

(1) Simplification and clarification of the issues;

(2) Stipulations as to the facts, and contents and authenticity of documents;

(3) Disclosure of the names and addresses of witnesses and provision of documents intended to be offered in evidence; or

(4) Any other matter that will tend to simplify the issues or expedite the proceedings.

§ 1206.9 Consent determination.

At any time prior to the commencement of the hearing, the affected State and the Administrators may execute an appropriate agreement for disposing of the matter, on which the proposed recommended determination is based, by mutual consent for the consideration of the Secretary. The agreement is submitted to the Secretary, who may:

(a) Accept it and publish a notice in the FEDERAL REGISTER setting forth its terms;

(b) Reject it and direct that the proceedings in the matter continue; or

(c) Take such other action as he deems appropriate.

§ 1206.10 Hearing.

(a) Hearings held pursuant to this part are informal, legislative type hearings and are open to the public.

487

§ 1206.11

(b) The affected State and any interested person participating in a hearing conducted pursuant to this part may, except as specified by the presiding officer pursuant to § 1206.7(b):

(1) Appear by counsel or other authorized representative;

(2) Present evidence orally or by documents; and

(3) Present oral or written argument.

(c) The hearing is stenographically transcribed verbatim and reported by an official reporter designated by the presiding officer.

(d) As soon as practicable after the presiding officer receives the official transcript of the hearing, exhibits, and other documents filed at the hearing, he forwards them to the Administrators.

§ 1206.11 Recommended determination.

As soon as practicable, the Administrators review the materials forwarded to them by the presiding officer, any prehearing conference notices, and the evidence of the Administrations regarding the affected State's program deficiencies cited in the proposed recommended determination. On the basis of the review, they issue a recommended determination and submit it and the material they reviewed to the Secretary.

§ 1206.12 Final determination.

(a) As soon as practicable, the Secretary reviews the recommended determination and the material forwarded him by the Administrators. On the basis of the review, the Secretary may adopt or reject the recommended determination, in whole or in part. The Secretary then issues a final determination which includes:

(1) His decision, and reasons therefor, on the question of whether and to what extent the sanctions provided in § 1206.4 will be invoked; and

(2) A specification of the funds and program approval, if any, to be withheld.

(b) A copy of the final determination is sent by certified mail to the Governor of the affected State and to each other hearing participant, and is published in the FEDERAL REGISTER.

23 CFR Ch. II (4-1-95 Edition)

PART 1208—NATIONAL MINIMUM DRINKING AGE

Sec.
1208.1 Scope.
1208.2 Purpose.
1208.3 Definitions.
1208.4 Adoption of National Minimum Drinking Age.
1208.5 Period of availability of withheld funds.
1208.6 Apportionment of withheld funds after compliance.
1208.7 Period of availability of subsequently apportioned funds.
1208.8 Effect of noncompliance.
1208.9 Procedures affecting States in noncompliance.

AUTHORITY: 23 U.S.C. 158; delegation of authority at 49 CFR 1.48 and 1.50.

SOURCE: 51 FR 10380, Mar. 26, 1986, unless otherwise noted.

§ 1208.1 Scope.

This part prescribes the requirements necessary to implement 23 U.S.C. 158, which establishes the National Minimum Drinking Age.

§ 1208.2 Purpose.

The purpose of this part is to clarify the provisions which a State must have incorporated into its laws in order to prevent the withholding of Federal-aid highway funds for noncompliance with the National Minimum Drinking Age.

§ 1208.3 Definitions.

As used in this part:

Alcoholic beverage means beer, distilled spirits and wine containing one-half of one percent or more of alcohol by volume. Beer includes, but is not limited to, ale, lager, porter, stout, sake, and other similar fermented beverages brewed or produced from malt, wholly or in part or from any substitute therefor. Distilled spirits include alcohol, ethanol or spirits or wine in any form, including all dilutions and mixtures thereof from whatever process produced.

Public possession means the possession of any alcoholic beverage for any reason, including consumption on any street or highway or in any public place or in any place open to the public (including a club which is *de facto* open to the public). The term does not apply to the possession of alcohol for an es-

NHTSA and FHWA, DOT

tablished religious purpose; when accompanied by a parent, spouse or legal guardian age 21 or older; for medical purposes when prescribed or administered by a licensed physician, pharmacist, dentist, nurse, hospital or medical institution; in private clubs or establishments; or to the sale, handling, transport, or service in dispensing of any alcoholic beverage pursuant to lawful employment of a person under the age of twenty-one years by a duly licensed manufacturer, wholesaler, or retailer of alcoholic beverages.

Purchase means to acquire by the payment of money or other consideration.

§ 1208.4 Adoption of National Minimum Drinking Age.

(a) The Secretary shall withhold five percent of the amount required to be apportioned to any State under each of §§ 104(b)(1), 104(b)(2), 104(b)(5) and 104(b)(6) of title 23 U.S.C. on the first day of the fiscal year succeeding the first fiscal year beginning after September 30, 1985, in which the purchase or public possession in such State of any alcoholic beverage by a person who is less than twenty-one years of age is lawful.

(b) The Secretary shall withhold ten percent of the amount required to be apportioned to any State under each of §§ 104(b)(1), 104(b)(2), 104(b)(5) and 104(b)(6) of title 23 U.S.C. on the first day of each fiscal year after the second fiscal year beginning after September 30, 1985, in which the purchase or public possession in such State of any alcoholic beverage by a person who is less than twenty-one years of age is lawful.

(c) A State that has in effect a law which permits the purchase and public possession in the State of any alcoholic beverage by a person who is less than 21 years of age, but 18 years or older on the day preceding the effective date of the State law and at that time could lawfully purchase or publicly possess any alcoholic beverage in the State, will be deemed to be in compliance with paragraphs (a) and (b) of this section in each fiscal year in which the law is in effect, provided:

(1) The law must be in effect before the later of (i) October 1, 1986 or (ii) the tenth day following the last day of the first session the legislature of the State convenes after April 7, 1986; and

(2) The State law otherwise makes unlawful the purchase and public possession in the State of any alcoholic beverage by a person who is less than 21 years of age.

[51 FR 10380, Mar. 26, 1986, as amended at 53 FR 31321, Aug. 18, 1988]

§ 1208.5 Period of availability of withheld funds.

(a) Funds withheld under § 1208.4 from apportionment to any State on or before September 30, 1988 will remain available for apportionment as follows:

(1) If the funds would have been apportioned under 23 U.S.C. 104(b)(5)(A) but for this section, the funds will remain available until the end of the fiscal year for which the funds are authorized to be appropriated.

(2) If the funds would have been apportioned under 23 U.S.C. 104(b)(5)(B) but for this section, the funds will remain available until the end of the second fiscal year following the fiscal year for which the funds are authorized to be appropriated.

(3) If the funds would have been apportioned under 23 U.S.C. 104(b)(1), 104(b)(2) or 104(b)(6) but for this section, the funds will remain available until the end of the third fiscal year following the fiscal year for which the funds are authorized to be appropriated.

(b) Funds withheld under § 1208.4 from apportionment to any State after September 30, 1988 will not be available for apportionment to the State.

[53 FR 31322, Aug. 18, 1988]

§ 1208.6 Apportionment of withheld funds after compliance.

Funds withheld under § 1208.4 from apportionment, which remain available for apportionment under § 1208.5(a), will be apportioned to any State that makes effective a law prohibiting the purchase or public possession in the State of any alcoholic beverage by a person who is less than 21 years of age before the last day of the period of availability as defined in § 1208.5(a). The funds will be apportioned to the State on the day following the effective date of the law.

[53 FR 31322, Aug. 18, 1988]

§ 1208.7

§ 1208.7 Period of availability of subsequently apportioned funds.

(a) Funds apportioned pursuant to § 1208.6 will remain available for expenditure as follows:

(1) Funds apportioned under 23 U.S.C. 104(b)(5)(A) will remain available until the end of the fiscal year succeeding the fiscal year in which the funds are apportioned.

(2) Funds apportioned under 23 U.S.C. 104(b)(1), 104(b)(2), 104(b)(5)(B), or 104(b)(6) will remain available until the end of the third fiscal year succeeding the fiscal year in which the funds are apportioned.

(b) Sums apportioned to a State pursuant to § 1208.6 and not obligated at the end of the periods defined in § 1208.7(a), shall lapse or, in the case of funds apportioned under 23 U.S.C. 104(b)(5), shall lapse and be made available by the Secretary for projects in accordance with 23 U.S.C. 118(b).

[53 FR 31322, Aug. 18, 1988]

§ 1208.8 Effect of noncompliance.

If a State has not made effective a law prohibiting the purchase and public possession in the State of any alcoholic beverage by a person who is less than 21 years of age at the end of the period for which funds withheld under § 1208.4 from apportionment are available for apportionment to a State under § 1208.5, then such funds shall lapse or, in the case of funds withheld from apportionment under 23 U.S.C. 104(b)(5), shall lapse and be made available by the Secretary for projects in accordance with 23 U.S.C. 118(b).

[53 FR 31322, Aug. 18, 1988]

§ 1208.9 Procedures affecting States in noncompliance.

(a) Every fiscal year, each State determined to be in noncompliance with the National Minimum Drinking Age, based on NHTSA's and FHWA's preliminary review of its statutes for compliance or non-compliance, will be advised of the funds expected to be withheld under § 1208.4 from apportionment, as part of the advance notice of apportionments required under 23 U.S.C. 104(e), normally not later than ninety days prior to final apportionment.

(b) If NHTSA and FHWA determine that the State is in noncompliance with the National Minimum Drinking Age based on their preliminary review, the State may, within 30 days of its receipt of the advance notice of apportionments, submit documentation showing why it is in compliance. Documentation shall be submitted to the National Highway Traffic Safety Administration, 400 Seventh Street SW, Washington, DC 20590.

(c) Every fiscal year, each State determined to be in noncompliance with the National Minimum Drinking Age, based on NHTSA's and FHWA's final determination of compliance or noncompliance, will receive notice of the funds being withheld under § 1208.4 from apportionment, as part of the certification of apportionments required under 23 U.S.C. 104(e), which normally occurs on October 1 of each fiscal year.

[53 FR 31322, Aug. 18, 1988]

PART 1212—DRUG OFFENDER'S DRIVER'S LICENSE SUSPENSION

Sec.
1212.1 Scope.
1212.2 Purpose.
1212.3 Definitions.
1212.4 Adoption of drug offender's driver's license suspension.
1212.5 Certification requirements.
1212.6 Period of availability of withheld funds.
1212.7 Apportionment of withheld funds after compliance.
1212.8 Period of availability of subsequently apportioned funds.
1212.9 Effect of noncompliance.
1212.10 Procedures affecting states in noncompliance.

AUTHORITY: Pub. L. 101-516; Pub. L. 102-143; delegation of authority at 49 CFR 1.48 and 1.50.

SOURCE: 57 FR 35999, Aug. 12, 1992, unless otherwise noted.

§ 1212.1 Scope.

This part prescribes the requirements necessary to implement 23 U.S.C. § 159, which encourages States to enact and enforce drug offender's driver's license suspensions.

§ 1212.2 Purpose.

The purpose of this part is to specify the steps that States must take in

NHTSA and FHWA, DOT

§ 1212.5

order to avoid the withholding of Federal-aid highway funds for noncompliance with 23 U.S.C. 159.

§ 1212.3 Definitions.

As used in this part:
(a) *Convicted* includes adjudicated under juvenile proceedings.
(b) *Driver's license* means a license issued by a State to any individual that authorizes the individual to operate a motor vehicle on highways.
(c) *Drug offense* means:
(1) The possession, distribution, manufacture, cultivation, sale, transfer, or the attempt or conspiracy to possess, distribute, manufacture, cultivate, sell, or transfer any substance the possession of which is prohibited under the Controlled Substances Act, or
(2) The operation of a motor vehicle under the influence of such a substance.
(d) *Substance the possession of which is prohibited under the Controlled Substances Act* or *substance* means a controlled or counterfeit chemical, as those terms are defined in subsections 102 (6) and (7) of the Comprehensive Drug Abuse Prevention and Control Act of 1970 (21 U.S.C. 802 (6) and (7) and listed in 21 CFR 1308.11–.15.

[57 FR 35999, Aug. 12, 1992; 58 FR 62415, Nov. 26, 1993; 59 FR 39256, Aug. 2, 1994]

§ 1212.4 Adoption of drug offender's driver's license suspension.

(a) The Secretary shall withhold five percent of the amount required to be apportioned to any State under each of sections 104(b)(1), 104(b)(3), and 104(b)(5) of title 23 of the United States Code on the first day of fiscal years 1994 and 1995 if the States does not meet the requirements of this section on that date.
(b) The Secretary shall withhold ten percent of the amount required to be apportioned to any State under each of sections 104(b)(1), 104(b)(3), and 104(b)(5) of title 23 of the United States Code on the first day of fiscal year 1996 and any subsequent fiscal year if the State does not meet the requirements of this section on that date.
(c) A State meets the requirements of this section if:
(1) The State has enacted and is enforcing a law that requires in all circumstances, or requires in the absence of compelling circumstances warranting an exception:
(i) The revocation, or suspension for at least 6 months, of the driver's license of any individual who is convicted, after the enactment of such law, of
(A) Any violation of the Controlled Substances Act, or
(B) Any drug offense, and
(ii) A delay in the issuance or reinstatement of a driver's license to such an individual for at least 6 months after the individual otherwise would have been eligible to have a driver's license issued or reinstated if the individual does not have a driver's license, or the driver's license of the individual is suspended, at the time the individual is so convicted, or
(2) The Governor of the State:
(i) Submits to the Secretary no earlier than the adjournment sine die of the first regularly scheduled session of the State's legislature which begins after November 5, 1990, a written certification stating that he or she is opposed to the enactment or enforcement in the State of a law described in paragraph (c)(1) of this section relating to the revocation, suspension, issuance, or reinstatement of driver's licenses to convicted drug offenders; and
(ii) Submits to the Secretary a written certification that the legislature (including both Houses where applicable) has adopted a resolution expressing its opposition to a law described in paragraph (c)(1) of this section.
(d) A State that makes exceptions for compelling circumstances must do so in accordance with a State law, regulation, binding policy directive or Statewide published guidelines establishing the conditions for making such exceptions and in exceptional circumstances specific to the offender.

§ 1212.5 Certification requirements.

(a) Each State shall certify to the Secretary of Transportation by April 1, 1993 and by January 1 of each subsequent year that it meets the requirements of 23 U.S.C. 159 and this regulation.
(b) If the State believes it meets the requirements of 23 U.S.C. 159 and this regulation on the basis that it has en-

491

§ 1212.6

acted and is enforcing a law that suspends or revokes the driver's license of drug offenders, the certification shall contain:

(1) A statement by the Governor of the State that the State has enacted and is enforcing a Drug Offender's Driver's License Suspension law that conforms to 23 U.S.C. 159(a)(3)(A). The certifying statement may be worded as follows: I, (Name of Governor), Governor of the (State or Commonwealth) of _____, do hereby certify that the (State or Commonwealth) of _____, has enacted is enforcing a Drug Offender's Driver's License Suspension law that conforms to section 23 U.S.C. 159(a)(3)(A).

(2) Until a State has been determined to be in compliance with the requirements of 23 U.S.C. 159 and this regulation, the certification shall include also:

(i) A copy of the State law, regulation, or binding policy directive implementing or interpreting such law or regulation relating to the suspension, revocation, issuance or reinstatement or driver's licenses of drug offenders, and

(ii) A statement describing the steps the State is taking to enforce its law with regard to within State convictions, out-of-State convictions, Federal convictions and juvenile adjudications. The statement shall demonstrate that, upon receiving notification that a State driver has been convicted of a within State, out-of-State or Federal conviction or juvenile adjudication, the State is revoking, suspending or delaying the issuance of that drug offender's driver's license; and that, when the State convicts an individual of a drug offense, it is notifying the appropriate State office or, if the offender is a non-resident driver, the appropriate office in the driver's home State. If the State is not yet making these notifications, the State may satisfy this element by submitting a plan describing the steps it is taking to establish notification procedures.

(c) If the State believes it meets the requirements of 23 U.S.C. 159(a)(3)(B) on the basis that it opposes a law that requires the suspension, revocation or delay in issuance or reinstatement of the driver's license of drug offenders that conforms to 23 U.S.C. 159(a)(3)(A), the certification shall contain:

(1) A statement by the Governor of the State that he or she is opposed to the enactment or enforcement of a law that conforms to 23 U.S.C. 159(a)(3)(A) and that the State legislature has adopted a resolution expressing its opposition to such a law. The certifying statement may be worded as follows: I, (Name of Governor), Governor of the (State or Commonwealth) of _____, do hereby certify that I am opposed to the enactment or enforcement of a law that conforms to 23 U.S.C. 159(a)(3)(A) and that the legislature of the (State or Commonwealth) of _____, has adopted a resolution expressing its opposition to such a law.

(2) Until a State has been determined to be in compliance with the requirements of 23 U.S.C. 159(a)(3)(B) and this regulation, the certification shall include a copy of the resolution.

(d) The Governor each year shall submit the original and four copies of the certification to the local FHWA Division Administrator. The FHWA Division Administrator shall retain the original and forward two copies each to the Regional Administrator of NHTSA and FHWA. The Regional Administrators shall each retain one copy and forward one copy of the submission, with any pertinent comments, to their respective Washington Headquarters, attention of the Chief Counsel.

(e) Any changes to the original certification or supplemental information necessitated by the review of the certifications as they are forwarded, State legislative changes or changes in State enforcement activity (including failure to make progress in a plan previously submitted) shall be submitted in the same manner as the original.

§ 1212.6 Period of availability of withheld funds.

(a) Funds withheld under § 1212.4 from apportionment to any State on or before September 30, 1995, will remain available for apportionment as follows:

(1) If the funds would have been apportioned under 23 U.S.C. 104(b)(5)(A) but for this section, the funds will remain available until the end of the fiscal year for which the funds are authorized to be appropriated.

NHTSA and FHWA, DOT § 1212.10

(2) If the funds would have been apportioned under 23 U.S.C. 104(b)(5)(B) but for this section, the funds will remain available until the end of the second fiscal year following the fiscal year for which the funds are authorized to be appropriated.

(3) If the funds would have been apportioned under 23 U.S.C. 104(b)(1) or 104(b)(3) but for this section, the funds will remain available until the end of the third fiscal year following the fiscal year for which the funds are authorized to be appropriated.

(b) Funds withheld under § 1212.4 from apportionment to any State after September 30, 1995 will not be available for apportionment to the State.

§ 1212.7 Apportionment of withheld funds after compliance.

Funds withheld under § 1212.4 from apportionment, which remain available for apportionment under § 1212.6(a), will be made available to any State that conforms to the requirements of § 1212.4 before the last day of the period of availability as defined in § 1212.6(a).

[57 FR 35999, Aug. 12, 1992, as amended at 59 FR 39256, Aug. 2, 1994]

§ 1212.8 Period of availability of subsequently apportioned funds.

(a) Funds apportioned pursuant to § 1212.7 will remain available for expenditure as follows:

(1) Funds originally apportioned under 23 U.S.C. 104(b)(5)(A) will remain available until the end of the fiscal year succeeding the fiscal year in which the funds are apportioned.

(2) Funds originally apportioned under 23 U.S.C. 104(b)(1), 104(b)(2), 104(b)(5)(B), or 104(b)(6) will remain available until the end of the third fiscal year succeeding the fiscal year in which the funds are apportioned.

(b) Sums apportioned to a State pursuant to § 1212.7 and not obligated at the end of the periods defined in § 1212.8(a), shall lapse or, in the case of funds apportioned under 23 U.S.C. 104(b)(5), shall lapse and be made available by the Secretary for projects in accordance with 23 U.S.C. 118(b).

§ 1212.9 Effect of noncompliance.

If a State has not met the requirements of 23 U.S.C. 159(a)(3) at the end of the period for which funds withheld under § 1212.4 are available for apportionment to a State under § 1212.6, then such funds shall lapse or, in the case of funds withheld from apportionment under 23 U.S.C. 104(b)(5), shall lapse and be made available by the Secretary for projects in accordance with 23 U.S.C. 118(b).

§ 1212.10 Procedures affecting states in noncompliance.

(a) Each fiscal year, each State determined to be in noncompliance with 23 U.S.C. 159, based on NHTSA's and FHWA's preliminary review of its statutes, will be advised of the funds expected to be withheld under § 1212.4 from apportionment, as part of the advance notice of apportionments required under 23 U.S.C. 104(e), normally not later than ninety days prior to final apportionment.

(b) If NHTSA and FHWA determine that the State is not in compliance with 23 U.S.C. 159 based on the agencies' preliminary review, the State may, within 30 days of its receipt of the advance notice of apportionments, submit documentation showing why it is in compliance. Documentation shall be submitted to the National Highway Traffic Safety Administration, 400 Seventh Street, SW., Washington, DC 20590.

(c) Each fiscal year, each State determined not to be in compliance with 23 U.S.C. 159(a)(3), based on NHTSA's and FHWA's final determination, will receive notice of the funds being withheld under § 1212.4 from apportionment, as part of the certification of apportionments required under 23 U.S.C. 104(e), which normally occurs on October 1 of each fiscal year.

PART 1215—USE OF SAFETY BELTS AND MOTORCYCLE HELMETS—COMPLIANCE AND TRANSFER-OF-FUNDS PROCEDURES

Sec.
1215.1 Scope.
1215.2 Purpose.
1215.3 Definitions.
1215.4 Compliance criteria.
1215.5 Exemptions.
1215.6 Review and notification of compliance status.

§ 1215.1

1215.7 Transfer of funds.
1215.8 Use of transferred funds.

AUTHORITY: 23 U.S.C. 153; delegation of authority at 49 CFR 1.50.

SOURCE: 58 FR 44759, Aug. 25, 1993, unless otherwise noted.

§ 1215.1 Scope.

This part establishes criteria, in accordance with 23 U.S.C. 153, for determining compliance with the requirement that States not having safety belt and motorcycle helmet use laws be subject to a transfer of Federal-aid highway apportionments under 23 U.S.C. 104 (b)(1), (b)(2), and (b)(3) to the highway safety program apportionment under 23 U.S.C. 402.

§ 1215.2 Purpose.

This part clarifies the provisions which a State must incorporate into its laws to prevent the transfer of a portion of its Federal-aid highway funds to the section 402 highway safety program apportionment, describes notification and transfer procedures, and establishes parameters for the use of transferred funds.

§ 1215.3 Definitions.

As used in this part:

FHWA means the Federal Highway Administration.

Motor vehicle means any vehicle driven or drawn by mechanical power manufactured primarily for use on public highways, except any vehicle operated exclusively on a rail or rails.

Motorcycle means a motor vehicle which is designed to travel on not more than 3 wheels in contact with the surface.

NHTSA means the National Highway Traffic Safety Administration.

Passenger vehicle means a motor vehicle which is designed for transporting 10 individuals or less, including the driver, except that such term does not include a vehicle which is constructed on a truck chassis, a motorcycle, a trailer, or any motor vehicle which is not required on the date of the enactment of this section under a Federal motor vehicle safety standard to be equipped with a belt system.

Safety belt means, with respect to open-body passenger vehicles, including convertibles, an occupant restraint system consisting of a lap belt or a lap belt and a detachable shoulder belt; and with respect to other passenger vehicles, an occupant restraint system consisting of integrated lap shoulder belts.

§ 1215.4 Compliance criteria.

(a) In order to avoid the transfer specified in § 1215.7, a State must have a law which makes unlawful throughout the State the operation of a motorcycle if any individual on the motorcycle is not wearing a motorcycle helmet.

(b) In order to avoid the transfer specified in § 1215.7, a State must have a law which makes unlawful throughout the State the operation of a passenger vehicle whenever an individual in the front seat of the vehicle (other than a child who is secured in a child restraint system) does not have a safety belt properly fastened about the individual's body.

(c) A State that enacts the laws specified in paragraphs (a) and (b) of this section will be determined to comply with 23 U.S.C. 153, provided that any exemptions are consistent with § 1215.5.

§ 1215.5 Exemptions.

(a) The following provisions shall be deemed to comply with 23 U.S.C. 153:

(1) Safety belt laws exempting persons with medical excuses, persons in emergency vehicles, persons in the custody of police, persons in public and livery conveyances, persons in parade vehicles, persons in positions not equipped with safety belts, and postal, utility and other commercial drivers who make frequent stops in the course of their business.

(2) Motorcycle helmet laws exempting riders in enclosed cabs.

(b) The following provisions shall be deemed not to comply with 23 U.S.C. 153:

(1) Safety belt laws exempting vehicles equipped with air bags.

(2) Motorcycle laws of less than universal application (e.g. laws applying only to minors or novice motorcycle operators) or whose enforcement is by any means other than primary enforcement.

(c) An exemption not identified in paragraph (a) of this section shall be

NHTSA and FHWA, DOT

deemed to comply with 23 U.S.C. 153 only if NHTSA and FHWA determine that it is consistent with the intent of § 1215.4 (a) or (b), as applicable, and applies to situations in which the risk to occupants is very low or in which there are exigent justifications.

§ 1215.6 Review and notification of compliance status.

(a) Review of each State's laws and notification of compliance status for fiscal year 1994 shall occur in accordance with the following procedures:

(1) NHTSA and FHWA will review appropriate State laws and notify States by certified mail of their initial assessment of compliance with 23 U.S.C. 153 by September 30, 1993.

(2) If NHTSA and FHWA initially find that a State complies with 23 U.S.C. 153, the notice shall so inform the State. Otherwise, the notice shall state the reasons for the non-compliance and shall inform the State that it may, within 30 calendar days after its receipt of the notice, submit documentation showing why it is in compliance to the Associate Administrator for Regional Operations, NHTSA, 400 Seventh Street SW., Washington, DC 20950. For each State initially found in non-compliance, NHTSA and FHWA will provide a final determination of compliance or non-compliance with 23 U.S.C. 153 by January 31, 1994.

(b) Review of each State's laws and notification of compliance status for fiscal year 1995 and beyond shall occur in accordance with the following procedures:

(1) NHTSA and FHWA will review appropriate State laws for compliance with 23 U.S.C. 153. States initially found to be in non-compliance will be notified of such funding and of funds expected to be transferred under § 1215.7 through the advance notice of apportionments required under 23 U.S.C. 104(e), normally not later than ninety days prior to final apportionment.

(2) A State notified of non-compliance under paragraph (b)(1) of this section may, within 30 days after its receipt of the advance notice of apportionments, submit documentation showing why it is in compliance to the Associate Administrator for Regional Operations, NHTSA, 400 Seventh Street SW., Washington, DC 20950.

(3) Every fiscal year, each State determined to be in non-compliance with 23 U.S.C. 153 will receive notice of the funds being transferred under § 1215.7 through the certification of apportionments required under 23 U.S.C. 104(e), normally on October 1.

§ 1215.7 Transfer of funds.

(a) If, at any time in fiscal year 1994, a State does not have in effect the laws described in § 1215.4, the Secretary shall transfer 1½ percent of the funds apportioned to the State for fiscal year 1995 under 23 U.S.C. 104 (b)(1), (b)(2) and (b)(3) to the apportionment of the State under 23 U.S.C. 402.

(b) If, at any time in a fiscal year beginning after September 30, 1994, a State does not have in effect the laws described in § 1215.4, the Secretary shall transfer 3 percent of the funds apportioned to the State for the succeeding fiscal year under 23 U.S.C. 104 (b)(1), (b)(2), and (b)(3) to the apportionment of the State under 23 U.S.C. 402.

(c) Any obligation limitation existing on the transferred construction funds prior to transfer will apply, proportionately, to those funds after transfer.

§ 1215.8 Use of transferred funds.

(a) Any funds transferred under § 1215.7 may be used for approved projects in any section 402 program area.

(b) Any funds transferred under § 1215.7 shall not be subject to Federal earmarking of any amounts or percentages for specific program activities.

(c) The Federal share of the cost of any project carried out under section 402 with the transferred funds shall be 100 percent.

(d) In the event of a transfer of funds under § 1215.7, the 40 percent political subdivision participation in State highway safety programs and the 10 percent limitation on the Federal contribution for Planning and Administration activities carried out under section 402 shall be based upon the sum of the funds transferred and amounts otherwise available for expenditure under section 402.

§ 1230.1

PART 1230—HIGHWAY SAFETY PROGRAM STANDARDS—APPLICABILITY TO FEDERALLY ADMINISTERED AREAS

Sec.
1230.1 Scope.
1230.2 Purpose.
1230.3 Applicability.
1230.4 Requirements.

AUTHORITY: 23 U.S.C. 402, 38 FR 12147, 49 CFR 1.48.

SOURCE: 38 FR 18665, July 13, 1973, unless otherwise noted.

§ 1230.1 Scope.

This part establishes requirements for implementation by Federal Departments or agencies of highway safety program standards set out in this chapter, in federally administered areas where a Federal department or agency controls highways open to public travel or supervises traffic operations.

§ 1230.2 Purpose.

The purpose of this part is to ensure that uniform standards established to regulate highway safety activities apply uniformly throughout the United States to those highways and activities administered by a Federal agency.

§ 1230.3 Applicability.

Pursuant to 23 U.S.C. 402, the highway safety program standards set forth in this chapter are applicable to Federal departments and agencies that control highways open to public travel within federally administered areas or supervise traffic operations on such highways, to the extent that they engage in activities covered by the highway safety program standards set out in this chapter.

§ 1230.4 Requirements.

(a) Each department or agency shall implement the highway safety program standards, to the extent that they are relevant to the activities of the department or agency. Implementation activities shall include but not be limited to:

(1) In cooperation with the FHWA and NHTSA, review of the department's or agency's activities to determine which are covered by the highway safety program standards.

(2) Review of the current status of those activities with regard to the relevant requirements of the standards.

(3) Development, submission to the FHWA and NHTSA, and periodic updating and implementation of a multiyear Comprehensive Plan for highway safety in accordance with the highway safety program standards.

(b) Each department or agency shall submit annually to the Secretary of Transportation a comprehensive report on the administration of its highway safety program for the preceding calendar year. The report shall be suitable for inclusion in the report to the President for transmittal to the Congress as required by section 202(a) of the Highway Safety Act of 1966 (Pub. L. 89–564), and shall include but not be limited to:

(1) Thorough statistical data on fatal, injury and property damage accidents which occurred within its federally administered area.

(2) The scope of observance of applicable Federal standards.

(3) The effectiveness of its highway safety programs.

PART 1235—UNIFORM SYSTEM FOR PARKING FOR PERSONS WITH DISABILITIES

Sec.
1235.1 Purpose.
1235.2 Definitions.
1235.3 Special license plates.
1235.4 Removable windshield placards.
1235.5 Temporary removable windshield placards.
1235.6 Parking.
1235.7 Parking space design, construction and designation.
1235.8 Reciprocity.

APPENDIX A TO PART 1235—SAMPLE REMOVABLE WINDSHIELD PLACARD
APPENDIX B TO PART 1235—SAMPLE TEMPORARY REMOVABLE WINDSHIELD PLACARD

AUTHORITY: Pub. L. 100–641, 102 Stat. 3335 (1988); 23 U.S.C. 101(a), 104, 105, 109(d), 114(a), 135, 217, 307, 315, and 402(a); 23 CFR 1.32 and 1204.4; and 49 CFR 1.48(b).

SOURCE: 56 FR 10329, Mar. 11, 1991, unless otherwise noted.

§ 1235.1 Purpose.

The purpose of this part is to provide guidelines to States for the establish-

NHTSA and FHWA, DOT

ment of a uniform system for handicapped parking for persons with disabilities to enhance access and the safety of persons with disabilities which limit or impair the ability to walk.

§ 1235.2 Definitions.

Terms used in this part are defined as follows:

(a) *International Symbol of Access* means the symbol adopted by Rehabilitation International in 1969 at its Eleventh World Congress on Rehabilitation of the Disabled.

(b) *Persons with disabilities which limit or impair the ability to walk* means persons who, as determined by a licensed physician:

(1) Cannot walk two hundred feet without stopping to rest; or

(2) Cannot walk without the use of, or assistance from, a brace, cane, crutch, another person, prosthetic device, wheelchair, or other assistive device; or

(3) Are restricted by lung disease to such an extent that the person's forced (respiratory) expiratory volume for one second, when measured by spirometry, is less than one liter, or the arterial oxygen tension is less than sixty mm/hg on room air at rest; or

(4) Use portable oxygen; or

(5) Have a cardiac condition to the extent that the person's functional limitations are classified in severity as Class III or Class IV according to standards set by the American Heart Association; or

(6) Are severely limited in their ability to walk due to an arthritic, neurological, or orthopedic condition.

(c) *Special license plate* means a license plate that displays the International Symbol of Access:

(1) In a color that contrasts to the background, and

(2) In the same size as the letters and/or numbers on the plate.

(d) *Removable windshield placard* means a two-sided, hanger-style placard which includes on each side:

(1) The International Symbol of Access, which is at least three inches in height, centered on the placard, and is white on a blue shield;

(2) An identification number;

(3) A date of expiration; and

§ 1235.4

(4) The seal or other identification of the issuing authority.

(e) *Temporary removable windshield placard* means a two-sided, hanger-style placard which includes on each side:

(1) The International Symbol of Access, which is at least three inches in height, centered on the placard, and is white on a red shield;

(2) An identification number;

(3) A date of expiration; and

(4) The seal or other identification of the issuing authority.

§ 1235.3 Special license plates.

(a) Upon application of a person with a disability which limits or impairs the ability to walk, each State shall issue special license plates for the vehicle which is registered in the applicant's name. The initial application shall be accompanied by the certification of a licensed physician that the applicant meets the § 1235.2(b) definition of persons with disabilities which limit or impair the ability to walk. The issuance of a special license plate shall not preclude the issuance of a removable windshield placard.

(b) Upon application of an organization, each State shall issue special license plates for the vehicle registered in the applicant's name if the vehicle is primarily used to transport persons with disabilities which limit or impair the ability to walk. The application shall include a certification by the applicant, under criteria to be determined by the State, that the vehicle is primarily used to transport persons with disabilities which limit or impair the ability to walk.

(c) The fee for the issuance of a special license plate shall not exceed the fee charged for a similar license plate for the same class vehicle.

§ 1235.4 Removable windshield placards.

(a) The State system shall provide for the issuance and periodic renewal of a removable windshield placard, upon the application of a person with a disability which limits or impairs the ability to walk. The State system shall require that the issuing authority shall, upon request, issue one addi-

497

§ 1235.5

tional placard to applicants who do not have special license plates.

(b) The initial application shall be accompanied by the certification of a licensed physician that the applicant meets the § 1235.2(b) definition of persons with disabilities which limit or impair the ability to walk.

(c) The State system shall require that the removable windshield placard is displayed in such a manner that it may be viewed from the front and rear of the vehicle by hanging it from the front windshield rearview mirror of a vehicle utilizing a parking space reserved for persons with disabilities. When there is no rearview mirror, the placard shall be displayed on the dashboard.

§ 1235.5 Temporary removable windshield placards.

(a) The State system shall provide for the issuance of a temporary removable windshield placard, upon the application of a person with a disability which limits or impairs the ability to walk. The State system shall require that the issuing authority issue, upon request, one additional temporary removable windshield placard to applicants.

(b) The State system shall require that the application shall be accompanied by the certification of a licensed physician that the applicant meets the § 1235.2(b) definition of persons with disabilities which limit or impair the ability to walk. The certification shall also include the period of time that the physician determines the applicant will have the disability, not to exceed six months.

(c) The State system shall require that the temporary removable windshield placard is displayed in such a manner that it may be viewed from the front and rear of the vehicle by hanging it from the front windshield rearview mirror of a vehicle utilizing a parking space reserved for persons with disabilities. When there is no rearview mirror, the placard shall be displayed on the dashboard.

(d) The State system shall require that the temporary removable windshield placard shall be valid for a period of time for which the physician has determined that the applicant will have the disability, not to exceed six months from the date of issuance.

§ 1235.6 Parking.

Special license plates, removable windshield placards, or temporary removable windshield placards displaying the International Symbol of Access shall be the only recognized means of identifying vehicles permitted to utilize parking spaces reserved for persons with disabilities which limit or impair the ability to walk.

§ 1235.7 Parking space design, construction, and designation.

(a) Each State shall establish design, construction, and designation standards for parking spaces reserved for persons with disabilities, under criteria to be determined by the State. These standards shall:

(1) Ensure that parking spaces are accessible to, and usable by, persons with disabilities which limit or impair the ability to walk;

(2) Ensure the safety of persons with disabilities which limit or impair the ability to walk who use these spaces and their accompanying accessible routes; and

(3) Ensure uniform sign standards which comply with those prescribed by the "Manual on Uniform Traffic Control Devices for Streets and Highways" (23 CFR part 655, subpart F) to designate parking spaces reserved for persons with disabilities which limit or impair the ability to walk.

(b) The design, construction, and alteration of parking spaces reserved for persons with disabilities for which Federal funds participate must meet the Uniform Federal Accessibility Standards.

§ 1235.8 Reciprocity.

The State system shall recognize removable windshield placards, temporary removable windshield placards and special license plates which have been issued by issuing authorities of other States and countries, for the purpose of identifying vehicles permitted to utilize parking spaces reserved for persons with disabilities which limit or impair the ability to walk.

NHTSA and FHWA, DOT   **Pt. 1235, App. A**

APPENDIX A TO PART 1235—SAMPLE REMOVABLE WINDSHIELD PLACARD

123456

EXPIRES:

12-12-89

Name or Seal of
Issuing Authority

COLORS
SYMBOL & LEGEND — WHITE
BACKGROUND — BLUE

499

Pt. 1235, App. B   23 CFR Ch. II (4-1-95 Edition)

APPENDIX B TO PART 1235—SAMPLE TEMPORARY REMOVABLE WINDSHIELD PLACARD

123456

EXPIRES:

12-12-89

`Name or Seal of Issuing Authority

COLORS
SYMBOL & LEGEND—WHITE
BACKGROUND—RED

SUBCHAPTER C—GENERAL PROVISIONS

PART 1250—POLITICAL SUBDIVISION PARTICIPATION IN STATE HIGHWAY SAFETY PROGRAMS

Sec.
1250.1 Scope.
1250.2 Purpose.
1250.3 Policy.
1250.4 Determining local share.
1250.5 Waivers.

AUTHORITY: 23 U.S.C. 315, 402(b); and delegations of authority at 49 CFR 1.48 and 1.50.

SOURCE: 41 FR 23948, June 14, 1976, unless otherwise noted.

§ 1250.1 Scope.

This part establishes guidelines for the States to assure their meeting the requirements for 40 percent political subdivision participation in State highway safety programs under 23 U.S.C. 402 (b)(1)(C).

§ 1250.2 Purpose.

The purpose of this part is to provide guidelines to determine whether a State is in compliance with the requirement that at least 40 percent of all Federal funds apportioned under 23 U.S.C. 402 will be expended by political subdivisions of such State.

§ 1250.3 Policy.

To assure that the provisions of 23 U.S.C. 402(b)(1)(C) are complied with, the NHTSA and FHWA field offices will:

(a) Prior to approving the State's Annual Work Program (AWP), review the AWP and each of the subelement plans which make up the AWP. The NHTSA Regional Administrator will review the 14½ safety standard areas for which NHTSA is responsible and the FHWA Division Administrator will review the 3½ safety standard areas for which FHWA is responsible. The narrative description for each subelement plan should contain sufficient information to identify the funds to be expended by, or for the benefit of the political subdivisions.

(b) Withhold approval of a State's AWP, as provided in Highway Safety Program Manual volume 103, chapter III, paragraph 3c, where the program does not provide at least 40 percent of Federal funds for planned local program expenditures.

(c) During the management review of the State's operations, determine if the political subdivisions had an active voice in the initiation, development and implementation of the programs for which such sums were expended.

§ 1250.4 Determining local share.

(a) In determining whether a State meets the requirement that at least 40 percent of Federal 402 funds be expended by political subdivisions, FHWA and NHTSA will apply the 40 percent requirement sequentially to each fiscal year's apportionments, treating all apportionments made from a single fiscal year's authorizations as a single entity for this purpose. Therefore, at least 40 percent of each State's apportionments from each year's authorizations must be used in the highway safety programs of its political subdivisions prior to the period when funds would normally lapse. The 40 percent requirement is applicable to the State's total federally funded safety program irrespective of Standard designation or Agency responsibility.

(b) When Federal funds apportioned under 23 U.S.C. 402 are expended by a political subdivision, such expenditures are clearly part of the local share. Local safety project related expenditures and associated indirect costs, which are reimbursable to the grantee local governments, are classifiable as the local share of Federal funds. Illustrations of such expenditures are the cost incurred by a local government in planning and administration of project related safety activities, driver education activities, traffic court programs, traffic records system improvements, upgrading emergency medical services, pedestrian safety activities, improved traffic enforcement, alcohol countermeasures, highway debris removal programs, pupil transportation programs, accident investigation, surveillance of high accident locations, and traffic engineering services.

§ 1250.5

(c) When Federal funds apportioned under 23 U.S.C. 402 are expended by the State or a State agency for the benefit of a political subdivision, such funds may be considered as part of the local share, provided that the political subdivision benefitted has had an active voice in the initiation, development, and implementation of the programs for which such funds are expended. In no case may the State arbitrarily ascribe State agency expenditures as "benefitting local government." Where political subdivisions have had an active voice in the initiation, development, and implementation of a particular program, and a political subdivision which has not had such active voice agrees in advance of implementation to accept the benefits of the program, the Federal share of the cost of such benefits may be credited toward meeting the 40 percent local participation requirement. Where no political subdivisions have had an active voice in the initiation, development, and implementation of a particular program, but a political subdivision requests the benefits of the program as part of the local government's highway safety program, the Federal share of the cost of such benefits may be credited toward meeting the 40 percent local participation requirement. Evidence of consent and acceptance of the work, goods or services on behalf of the local government must be established and maintained on file by the State, until all funds authorized for a specific year are expended and audits completed.

(d) State agency expenditures which are generally not classified as local are within such standard areas as vehicle inspection, vehicle registration and driver licensing. However, where these Standards provide funding for services such as: driver improvement tasks administered by traffic courts, or where they furnish computer support for local government requests for traffic record searches, these expenditures are classifiable as benefitting local programs.

§ 1250.5 Waivers.

While the 40 percent requirement may be waived in whole or in part by the Secretary or his delegate, it is expected that each State program will generate political subdivision participation to the extent required by the Act so that requests for waivers will be minimized. Where a waiver is requested, however, it will be documented at least by a conclusive showing of the absence of legal authority over highway safety activities at the political subdivision levels of the State and will recommend the appropriate percentage participation to be applied in lieu of the 40 percent.

PART 1251—STATE HIGHWAY SAFETY AGENCY

Sec.
1251.1 Purpose.
1251.2 Policy.
1251.3 Authority.
1251.4 Functions.

AUTHORITY: 23 U.S.C. 402; 23 U.S.C. 315; 49 CFR 1.48 and 1.50.

SOURCE: 45 FR 59145, Sept. 8, 1980, unless otherwise noted.

§ 1251.1 Purpose.

The purpose of this part is to prescribe the minimum authority and functions of the State Highway Safety Agency established in each State by the Governor under the authority of the Highway Safety Act (23 U.S.C. 402).

§ 1251.2 Policy.

In order for a State to receive funds under the Highway Safety Act, the Governor shall exercise his or her responsibilities through a State Highway Safety Agency that has "adequate powers and is suitably equipped and organized to carry out the program to the satisfaction of the Secretary." 23 U.S.C. 402(b)(1)(A). Accordingly, it is the policy of this part that approval of a State's Highway Safety Plan will depend upon the State's compliance with §§ 1251.3 and 1251.4 of this part.

§ 1251.3 Authority.

Each State Highway Safety Agency shall be authorized to:

(a) Develop and implement a process for obtaining information about the highway safety programs administered by other State and local agencies.

(b) Periodically review and comment to the Governor on the effectiveness of highway safety plans and activities in the State regardless of funding source.

NHTSA and FHWA, DOT

(c) Provide or facilitate the provision of technical assistance to other State agencies and political subdivisions to develop highway safety programs.

(d) Provide financial and technical assistance to other State agencies and political subdivisions in carrying out highway safety programs.

§ 1251.4 Functions.

Each State Highway Safety Agency shall:

(a) Develop and prepare the Highway Safety Plan prescribed by volume 102 of the Highway Safety Program Manual (23 CFR 1204.4, Supplement B), based on evaluation of highway accidents and safety problems within the State.

(b) Establish priorities for highway safety programs funded under 23 U.S.C. 402 within the State.

(c) Provide information and assistance to prospective aid recipients on program benefits, procedures for participation, and development of plans.

(d) Encourage and assist local units of government to improve their highway safety planning and administration efforts.

(e) Review the implementation of State and local highway safety plans and programs, regardless of funding source, and evaluate the implementation of those plans and programs funded under 23 U.S.C. 402.

(f) Monitor the progress of activities and the expenditure of section 402 funds contained in the State's approved Highway Safety Plan.

(g) Assure that independent audits are made of the financial operations of the State Highway Safety Agency and of the use of section 402 funds by any subrecipient.

(h) Coordinate the State Highway Safety Agency's Highway Safety Plan with other federally and non-federally supported programs relating to or affecting highway safety.

(i) Assess program performance through analysis of data relevant to highway safety planning.

§ 1252.2

PART 1252—STATE MATCHING OF PLANNING AND ADMINISTRATION COSTS

Sec.
1252.1 Purpose.
1252.2 Definitions.
1252.3 Applicability.
1252.4 Policy.
1252.5 Procedures.
1252.6 Responsibilities.

AUTHORITY: 23 U.S.C. 402 and 315; 49 CFR 1.48(b) and 1.50.

SOURCE: 45 FR 47145, July 14, 1980, unless otherwise noted.

§ 1252.1 Purpose.

This part establishes the National Highway Traffic Safety Administration (NHTSA) and the Federal Highway Administration (FHWA) policy on planning and administration (P&A) costs for State highway safety agencies. It defines planning and administration costs, describes the expenditures that may be used to satisfy the State matching requirement, prescribes how the requirement will be met, and when States will have to comply with the requirement.

§ 1252.2 Definitions.

(a) *Fiscal year* means the twelve months beginning each October 1, and ending the following September 30.

(b) *Direct costs* are those costs which can be identified specifically with a particular planning and administration or program activity. The salary of a data analyst on the State highway safety agency staff is an example of a direct cost attributable to P&A. The salary of an emergency medical technician course instructor is an example of direct cost attributable to a program activity.

(c) *Indirect costs* are those costs (1) incurred for a common or joint purpose benefiting more than one program activity and (2) not readily assignable to the program activity specifically benefited. For example, centralized support services such as personnel, procurement, and budgeting would be indirect costs.

503

§ 1252.3

(d) *Planning and administration (P&A) costs* are those direct and indirect costs that are attributable to the overall development and management of the Highway Safety Plan. Such costs could include salaries, related personnel benefits, travel expenses, and rental costs.

(e) *Program management costs* are those costs attributable to a program area (e.g., salary of an emergency medical services coordinator, the impact evaluation of an activity, or the travel expenses of a local traffic engineer).

(f) *State highway safety agency* is the agency directly responsible for coordinating the State's highway safety program authorized by 23 U.S.C. 402.

§ 1252.3 Applicability.

The provisions of this part apply to obligations incurred after November 6, 1978, for planning and administration costs under 23 U.S.C. 402.

§ 1252.4 Policy.

Federal participation in P&A activities shall not exceed 50 percent of the total cost of such activities, or the applicable sliding scale rate in accordance with 23 U.S.C. 120. The Federal contribution for P&A activities shall not exceed 10 percent of the total funds the State receives under 23 U.S.C. 402. In accordance with 23 U.S.C. 120(i), the Federal share payable for projects in the Virgin Islands, Guam, American Samoa and the Commonwealth of the Northern Mariana Islands shall be 100 percent. The Indian State, as defined by 23 U.S.C. 402 (d) and (i), is exempt from the provisions of this part. NHTSA funds shall be used only to finance P&A activities attributable to NHTSA programs and FHWA funds shall be used only to finance P&A costs attributable to FHWA programs.

[47 FR 15121, Apr. 8, 1982]

§ 1252.5 Procedures.

(a) P&A tasks and related costs shall be described in the P&A module of the State's Highway Safety Plan. The State's matching share shall be determined on the basis of the total P&A costs in the module. Federal participation shall not exceed 50 percent (or the applicable sliding scale) of the total P&A costs. A State shall not use NHTSA funds to pay more than 50 percent of the P&A costs attributable to NHTSA programs nor use FHWA funds to pay more than 50 percent of the P&A costs attributable to FHWA programs. In addition, the Federal contribution for P&A activities shall not exceed 10 percent of the total funds in the State received under 23 U.S.C. 402.

(b) FHWA and NHTSA funds may be used to pay for the Federal share of P&A costs up to the amounts determined by multiplying the Federal share by the ratio between the P&A costs attributable to FHWA programs and the P&A costs attributable to NHTSA programs. For example: A State's total P&A costs are $40,000. The State's share is 50 percent or $20,000. To pay the remaining $20,000, the State first ascertains the amount spent out of the total costs for each agency's programs, then applies the ratio between these two amounts to the $20,000. If $36,000 of the total costs are spent for NHTSA programs and $4,000 for FHWA programs, the ratio would be 9/1 and the corresponding allocation of the Federal share would be $18,000 to NHTSA and $2,000 to FHWA.

(c) A State at its option may allocate salary and related costs of State highway safety agency employees to one of the following:

(1) The administration and planning functions in the P&A module;

(2) The program management functions in one or more Program modules; or

(3) A combination of administration and planning functions in the P&A module and the program management functions in one or more program modules.

(d) If an employee is principally performing administration and planning functions under a P&A module, the total salary and related costs may be allocated to the P&A module. If the employee is principally performing program management functions under one or more program modules, the total salary and related costs may be charged directly to the appropriate module(s). If an employee is spending time on a combination of administration and planning functions and program management functions, the total salary and related costs may be

NHTSA and FHWA, DOT

charged to the appropriate module(s) based on the actual time worked under each module. If the State highway safety agency elects to allocate costs based on acutal time spent on an activity, the State highway safety agency must keep accurate time records showing the work activities for each employee. The State's record keeping system must be approved by the appropriate FHWA and NHTSA officials.

(e) Those tasks and related costs contained in the P&A module, not defined as P&A costs under § 1252.2(d) of this part, are not subject to the planning and administration cost matching requirement.

[45 FR 47145, July 14, 1980, as amended at 47 FR 15121, Apr. 8, 1982]

§ 1252.6 Responsibilities.

During the Highway Safety Plan approval process, the responsible FHWA and NHTSA officials shall approve a P&A module only if the projected State expenditure is at least 25 percent (or the appropriate sliding scale rate) of the total P&A costs identified in the module. If a State elects to prorate P&A and program management costs, the appropriate NHTSA and FHWA officials must approve the method that the State highway safety agency will use to record the time spent on these activities. During the process of reimbursement, the responsible FHWA and NHTSA officials shall assure that Federal reimbursement for P&A costs at no time exceeds 75 percent (of the applicable sliding scale rate) of the costs accumulated at the time of reimbursement.

PART 1260—CERTIFICATION OF SPEED LIMIT ENFORCEMENT

Sec.
1260.1 Purpose.
1260.3 Objective.
1260.5 Definitions.
1260.7 Adoption of national maximum speed limits.
1260.9 Formulation of a plan for monitoring speeds.
1260.11 Guidelines and evaluations of operations.
1260.13 Certification requirement.
1260.15 Certification content.
1260.17 Certification and statistical submittal.

§ 1260.5

1260.19 Effect of failure to certify or to meet compliance standards.
1260.21 Penalty reduction and notification of noncompliance.
Appendix to Part 1260—Speed Monitoring Program Procedural Manual

AUTHORITY: 23 U.S.C. 118, 141, 154, 315 and delegations of authority at 49 CFR 1.48 and 1.50.

SOURCE: 58 FR. 54820, Oct. 22, 1993, unless otherwise noted.

§ 1260.1 Purpose.

This part prescribes requirements for administering a program for monitoring speeds on public highways in order to provide reliable data to be included in a State's annual certification of speed limit enforcement.

§ 1260.3 Objective.

To establish a valid statistical method of measuring a sample of vehicle speeds on a sample of highways to estimate the percentage of vehicles exceeding the speed limit on highways posted at 55 mph and on highways posted at 65 mph with sufficient accuracy to support a determination of compliance by a State's motoring public with the National Maximum Speed Limits in accordance with 23 U.S.C. 154; to prescribe the compliance reporting requirements for the States; and to specify fund transfer provisions for noncompliance with the National Maximum Speed Limits.

§ 1260.5 Definitions.

As used in this part:

(a) *FHWA* means the Federal Highway Administration.

(b) *Fiscal year* means the Federal fiscal year, consisting of twelve months beginning each October 1 and ending the following September 30.

(c) *Freeway* means a divided arterial highway for through traffic with full control of access and grade separated intersections.

(d) *Governor* means the Governor of any of the fifty States, Puerto Rico or the Mayor of the District of Columbia.

(e) *Highway* means all streets, roads or parkways under the jurisdiction of State, including its political subdivisions, open for use by the general public, and including toll facilities.

505

About the Author

Dr. John C. Glennon, President of John C. Glennon, Chartered, Overland Park, Kansas is a world-renown roadway safety consultant and expert. Over his 33-year career, he has led and participated in research projects dealing with most of the major topics in this book. He has published extensively in professional papers, research reports, training manuals, and books. He has also given presentations at many major roadway safety meetings over the last 25 years. As an expert witness, retained by either the plaintiff or the defendent, he has participated in several hundred roadway defect cases covering every topic in this book.

As a professional engineer, Dr. Glennon has a broad understanding of both the tort liability for roadway defects and the role of the expert witness as an independent and unbiased advisor to judges and juries. His expert testimony encompasses roadway safety, roadway design, traffic engineering, accident reconstruction, and the human factors of driver behavior.

Index

A AASHTO Maintenance Manual 25, 375, 395, 409
Abandoned Crossings 250-251, 260
Abandoned Railroad Right-of-Way 251
Abrasive Application 401
Accident Investigation 26, 409
Accident Rates 173-175, 273, 280, 295, 341, 354
Accident Reconstruction Aspects 24, 142, 156, 187, 266, 293
Accident Statistics 2, 183, 225
Accumulation of Packed Snow 399
Active Advance Warning Signs 255
Adhesion 160-164, 181
ADT 37
Advance Warning Arrow Displays 317
Advance Warning Signs 23, 228, 234, 238, 253, 255, 261-262, 264, 283, 287, 317, 320, 338, 370
Advanced Pavement Markings Required 239
Advisory Speed Plates 138, 253, 283, 286, 289, 294, 333
Aggregate Polishing 382-383
Aggregate Surfaced Shoulders 388
Alignment Series of Warning Signs 284
American National Standard Practice 256, 271
American Public Works Association 22, 25, 375, 409
American Railway Engineering Association 257, 271
Angled Breakaway Slip Base 54
Angled Too Severely 112
Appearance of Abandonment 235
Appendix 2, 411
Application of Arrowboards 323
Application of Intersection Control Beacons 354

Approach Zone 228
Aquaplaning 190
Arrangement of Lenses 346
Arrow Displays 317-318
Asphalt Flushing 160, 171-172
Asphalt Pavement Maintenance 378
ASTM Method E-274 170, 189
Authoritative Sources of Information 21
Automatic Gates 225, 238, 241-242, 249, 254, 259, 262
Automobile Dynamics 222
Automobile Tire Hydroplaning 190
Automobiles Tires Rolling 191

B Backslope 40, 42, 74, 93
Backward Splice Overlap 110
Ball-Bank Indicator 283-284, 286
Barrier End Treatment 77
BCT 75-77
Bicycle Accommodations 411
Blading 375-376, 378, 387-388
Blatnik 27, 65
Blocked-out Thrie-beam Guardrails 90
Blocked-out Thrie-beam Median Barriers 91
Blocked-out W-beam Guardrails 73, 75
Blocked-out W-beam Median Barriers 89, 90
Blowups 385
Box Beam Guardrails 90,91
Box Beam Median Barriers 90
Braking Distance 129, 272
Breakaway Sign Slip Base 51
Breakaway Signs 23, 50
Breakaway Wooden Sign Posts 54
Breakway Transformer Base. 47
Bridge Abutments 30-31, 35, 63, 74
Bridge Decks 406, 409
Bridge Rail Defects 102, 113

513

Bridge Rail Maintenance Checklist 392
Bridge Rail Retrofit 86
Bridge Rails 31, 43, 63, 67-69, 81-85, 113-114, 335, 337, 339, 390-392
Build-up of Turf Shoulders 186
Bumps 282, 292, 375, 381, 403, 409
Bureau of Public Roads 21, 27-28, 144

C Cantilevered Flashing Light Signals 241
Cantilevered Signs 49
Centerlines 319-320, 334-336, 372
Channelization 301
Channelizing Devices 300, 302, 306, 315-317, 319, 321
Channelizing Lines 335, 337, 372
Characteristics of Roadside Encroachments 32
Chevron Alignment Sign 287
Clear Zone 38-40, 43-45, 50, 56, 58-63, 67, 69, 74, 83, 95, 274, 321, 389-390
Clear Zone Standards 38-39, 56
Clear Zone Width 40, 59, 274
Coefficients of Friction 185, 245
Collision Zone 230
Combination Curves 286
Commuter Train 272
Compaction 179, 183, 378, 382
Comparative Negligence 18, 135
Compression Crash Cushion Physics 96
Compression Crash Cushions 95, 393
Concrete Bridge Rail 115
Concrete Median Barrier 73, 90, 92, 141
Concrete Pavement Maintenance 383
Concrete Safety Shape 70, 73, 82, 86
Congress 159, 190, 205, 208, 222, 225
Congressional Blatnik Committee 27
Conservation of Energy 268
Consolidate Crossings 260
Conspicuity 258, 260
Construction Zones 16, 203, 211-212, 216, 218-219, 222, 299, 315, 319-320, 323
Continuous Roadway Lighting 46

Coordination of Edge Profiles 274
Corrugations 376-379
Cracks 379, 381-383, 385-386
Crash Cushion Defects 102, 118
Crash Cushion Maintenance Checklist 393
Crash Cushions 23, 27, 43, 48, 58, 67, 93, 95, 97, 101, 121, 392-393
Crashworthy 16, 50, 63, 67-68, 74, 86, 92-93, 102, 106, 303, 306, 391, 393
Criterion Press 7, 9-10, 14, 39, 41, 131, 146, 148, 195, 218, 229, 245, 248, 271, 285-286, 301, 353
Critical Stopping Zone 228, 230
Cross Slope Rate 182
Crossbuck Signs 237, 249
Crossing Angle 234-235, 246, 256, 261
Crossing Consolidation 249, 271
Crosswalk Closure 305
Crosswalk Markings 338
Crush Damage Method 267
Culverts 31, 35, 43, 61, 186, 389-390, 406
Curbs 30-31, 35, 40, 53, 56, 79, 85, 95, 101, 110, 303, 387
Curve Delineation 280, 295
Curve Warning Signs 294, 331, 359, 368

D Damaged Traffic Signs 368
Debris 186, 299, 321, 382, 386, 390-393
Delineation 16, 53, 203, 280, 288-289, 291-292, 295, 315, 317, 374
Delineator Spacing 288
Delineators 287-288, 294, 317, 332, 335, 340, 394-395
Depth of Compacted Wheel Tracks 178
Design Immunity 20, 64, 121, 125, 142, 289, 293, 362
Design Speed 40, 79, 81, 130, 132, 135, 139, 278-279, 282-283, 302, 321
Design Standards 21-22, 127, 142-143, 257, 278, 294, 411
Detour Signs 314, 321
Develop Beam Strength 15, 104-105
Directional Sign 358
Dirty Pavements 160, 182
Distracting 230, 254

Index

DO NOT ENTER Signs 334
DO NOT STOP ON TRACKS Signs 261-262
DOUBLE LINES 335
Downgrades 132, 186
Downstream End 79, 117
Downstream Tapers 300
Dragging 378
Drainage Facilities 40, 43, 375, 390
Driver 1-2, 4-14, 16, 25, 38, 40, 48, 59, 64, 78, 121, 126-129, 133, 137-138, 141-143, 145-149, 151, 153-154, 156, 160, 166-168, 174, 179-180, 185, 188-189, 194, 196, 198, 200, 202-203, 205, 219-222, 225, 227-232, 234-235, 237, 241, 243-244, 246-247, 250, 252-253, 255-259, 261-262, 264-266, 269-270, 273, 278, 282, 287-290, 292, 294, 296-297, 302, 306, 313, 317, 322, 326-327, 330-331, 333-335, 337-338, 340, 357, 361-362, 364, 366, 369, 373-374, 386, 390, 395, 403-404, 407-408, 511
Driver Brake Reaction Time 25
Driver Expectancy 6, 10, 12-14, 25, 289, 296, 374, 403
Driver Information Needs 229, 327
Driver Needs 14, 231, 244, 297, 327, 390
Driver Perception-Reaction Time 6-7, 10, 205, 330
Driver Performance 4, 6, 11-12, 133, 189
Driveways 154, 256-257
Driving Task 5, 12, 14, 222, 327
Drums 211, 300, 315, 317
Dry Pavement 170, 175, 395, 399
Dry Snowfall 399
Dust 160, 182, 376, 378, 394, 403
Dynamic Hydroplaning 179-181, 183, 185, 187-188, 191

E Earth Shoulders 387-388
Edge Drops 15, 24, 64, 193-194, 196-198, 200, 203-209, 212-216, 218-221, 290, 315, 321, 382, 386-387, 389, 402, 406
Edge Height 199, 205-206, 216

Edge of Travel Lanes 78
Effective Sight Distance 243, 262
Effectiveness of Alternative Skid Reduction Measures 190
Embankments 27, 30-31, 40, 69, 122, 125-126, 150-151, 153, 391
Emergency Stopping Distance 267
End of Road Markers 369, 371
End Treatments 16, 74-75, 92-93, 106, 123
Energite 97, 100, 118, 120
Energite Sand Barrel 120
Energite Sand Barrels 100
Energite System 97, 100
Energy Absorption Systems 76-77, 97-100
Engineering News-Record 144
Equivalent Barrier Speed 268-269
Estimating Train Speed 272
Excedance Distribution of Encroachment Angles 33
Excedance Distribution of Lateral Displacements 33

F Faded Pavement Markings 372
False Cues 291
Federal Aid Highway Program Manual 323
Federal Highway Administration Instructional Memorandum 29
Federal Highway Safety Act 225
Federal Highway Trust Fund 159
Federal Railroad Administration 249, 271
Federal Tort Claims Act 19
Federal-Aid Highway Program Manual 271
Fitch Inertial System 97, 100, 118
Fixed Object Fatalities 31, 43
Fixed Objects 27, 30, 32, 38, 40, 56, 58-60, 67-68, 78, 92, 97, 100, 281
Flaggers 302-303, 307
Flagging 299-300, 302-303, 306, 321
Flare Rate 80-81
Flared Ends 93
Flashing Light Signals 239, 241-242, 253

Flashing Signals 23, 225, 230, 238, 241-243, 249, 252, 254-255, 259-262, 348, 358
Flexible Barriers 87, 392
Flexible Guardrails 70-71
Flushing 160, 171-172, 183, 381, 383, 386
Four Hour Volume Warrant 345
Friction Properties of Pavements 190, 295
Frictional Requirements of Traffic 164
Frost Heave 379
Full Scale Tire 189

G Geometric Design Criteria 190
Grade Intersections 333-334
Gravel Shoulders 405
Gravel-Bed Attenuator 97, 100
GREAT 12-13, 31, 56, 93-94, 97, 99-100, 125, 135, 151, 231, 261, 320, 357, 364, 387, 396, 398, 400
GREEN ARROW 348, 352
GREEN ARROW Left Turn 348
GREEN ARROW Right Turn 348
GREEN ARROW Straight Through 348
GREEN BALL 348-349, 352, 354
Guardrail Angled Too Severely 112
Guardrail BCT End Treatment 76
Guardrail Defects 102
Guardrail End Treatments 74-75, 123
Guardrail Energy Absorbing Terminal 97
Guardrail ET-2000 Extruder End Treatment 78
Guardrail Flare Rate 80
Guardrail Installation 107, 122
Guardrail Maintenance Checklist 391
Guardrail Mounted Too High 104
Guardrail Mounted Too Low 103
Guardrail Outside Shy Line 81
Guardrail Terminal Layout 80
Guardrail Too Short 105
Guardrail Turned-down End Treatment 75
Guardrails 15-16, 27, 31, 35, 43-45, 58, 67-71, 73, 75, 79, 81, 83, 85, 102-104, 106, 112, 122, 335, 390-392
Guide Signs-Expressways 359

H Hand Signal Devices 304
Handbook of Highway Safety Design 23, 25, 123
Handbook of Traffic Control Practices 295
Hazard Identification Beacons 255, 341
Headwalls 43-46, 63, 339
Height of Signal Faces 350
Hex-Foam Sandwich System 97
HI-Dro 97-98, 102, 119, 393
HI-Dro Cell Cluster 97-98
HI-Dro Sandwich 97-98
HI-Dro Sandwich Crash Cushion 98
HI-Dro Sandwich System 97
High Exit Angle 194
High-Speed Passing Maneuvers 189
Highway Bridges 122
Highway Construction 297, 323
Highway Construction Zones 323
Highway Curve Design 189, 295
Highway Design 23, 25-27, 64-66, 157, 295, 374
Highway Design Consistency 25
Highway Guardrail 122
Highway Maintenance 22, 25, 375, 409
Highway Maintenance Manual 25, 375, 409
Highway Research Board 21, 65, 122, 189, 191, 295
Highway Research Board Proceedings 65, 189, 295
Highway Research Record 65, 144, 189-190, 295
Highway Safety Consciousness 65
Highway Safety Funds 411
Highway Safety Handbook 123, 409
Hillcrests 16, 129, 131, 135, 137-140, 336, 362, 403, 406
Horizontal Curves 126, 181, 189
Human Factors 12, 25, 511
HVOSM 203
Hydraulics 44, 411
Hydrodynamics of Tire Hydroplaning 190
Hydroplaning Cases 184
Hydroplaning Conditions 161, 293

Index

Hydroplaning Phenomena 190
Hydroplaning Sections 159, 186, 407
Hysteresis 162-164

I Ice Control 396, 398
Icy Bridges 406
Illuminating Society of America 271
Illumination 237, 255-256, 329, 395
Illusive Traffic Signs 364
Impact Forces 53, 71, 73, 79, 86, 92-93, 95, 103, 118
Impaling Elements 102
Incident Management Operations 297, 323
Inconsistent 12, 14, 61, 283, 299, 359
Inertial Crash Cushion Physics 96
Inflation Pressure 178, 180, 184, 209
Information Handling 5, 327
Institute of Mechanical Engineers 190
Institute of Traffic Engineers 21, 26
Institute of Transportation Engineers 21, 25, 144, 157, 374
Intersection Control Beacon 354
Intersection Sight Distance 145-146, 149-151, 154, 156, 389
Intersection Sight Requirements 157
Isolated Traffic Signals 372

J Joint Distress 383

L Lack of Advance Warnings 264
Lack of Continuity 355, 358, 361
Lack of Preview 234
Lamps 253, 258, 395-396
Lane Width 204-207, 282, 399
Lane-closure 16
Large Signs 63
Large Trees 30, 35, 56, 61, 390
Large Trucks 73, 82, 92, 100, 129, 133, 157, 159, 164, 179, 183, 234, 243, 261-262, 382, 386
Lateral Clearance 211, 329
Lateral Extent of Hazard 79
Lateral Offset 78
Left-Turn Signal Displays 354
Legal Requirements of Drivers 227

Legal Responsibilities of Railroads 226
Legibility 264, 326, 328, 330, 394
Length of Curve 274, 280
Length of Need 78-79, 102, 104, 106-107
Length of Spiral 274
Leveling Wing 400
Light Poles 30, 45, 47, 61, 92
Lighting 40, 45-46, 48, 256, 258, 260, 271, 306, 319, 326, 328, 340, 375, 394-395
Locked-Wheel Skid Trailer 170
Longitudinal Lines 335
Low-Volume Rural Roads 374
Luminaires 396

M Maintain Adequate Crossing Width 260
Maintain Adequate Road Width 260
Maintain Advance Warning Devices 260
Maintain Crossbucks 260
Maintain Smooth Crossing Surfaces 260
Maintenance Defects 102, 112, 401-403
Maintenance Function 101, 394
Maintenance Guidelines 200, 222
Maintenance of Highway Guardrails 122
Maintenance of Pavement Edge Drops 389
Maintenance Zone Defect Cases 320
Maintenance Zones 24, 297, 300, 306, 312-313, 315, 321
Manual of Traffic Signal Design 374
Manual of Uniform Traffic Control Devices 355
Marking Removal 395
Markings 13, 22-23, 219, 225, 227-228, 231, 234-235, 237-239, 249, 253, 260, 264, 294, 300, 306, 315, 317, 319, 321, 325, 334-338, 340, 358-359, 362, 372, 375, 394-395
Maximum Lateral Acceleration 282
Maximum Safe Re-entry Angle 204
Maximum Superelevation Rates 279
Measurement of Pavement Skid Resistance 169
Mechanics of Hydroplaning 179

Median 7-9, 23, 27, 37, 58, 63, 65, 67-68, 73, 86-95, 97, 102, 118, 122, 141, 302, 307, 309, 321, 330, 333-334, 337, 343, 351, 390, 392-393
Median Barrier 27, 73, 87-94, 102, 118, 141, 393
Median Barrier Defects 102, 118
Median Barrier Maintenance Checklist 393
Merging Taper 300-301
Microtexture 160-161, 171-172, 178-179, 182-183, 185
Minimum Curve Radius 278-279
Minimum Dynamic Hydroplaning Speed 191
Minimum Pedestrian Volume 342
Minimum Radius 273, 279
Minimum Safe Re-entry Angle 205
Ministerial Acts 20
Missing Bolts 102, 110
Missing End of Road Markers 369
Mobile Operations 300, 303
Model Traffic Ordinances 271
Moderate Exit Angle 194
Moderate Return Angle 196
Motorcycles 208, 219, 292, 379, 383, 386, 405
Motorgrader 376
Mudjacking 381, 386
MUTCD Contradictions 357
MUTCD Definitions 325
MUTCD Revision 302
MUTCD Word Meanings 355

N Narrow Bridge 135, 314, 333
Narrow Median 102, 141, 393
Narrow Roadway 141, 290, 403, 405
National Academy of Forensic Engineers 272
National Academy of Sciences 21
National Highway Safety Act 2
National Highway Traffic Safety Administration 172, 222, 409
National Safety Council 25

National Transportation Safety Board 172, 189, 272
National Weather Service 172
Negative Superelevation 279, 292
Negligence 17-19, 24, 59, 67, 135, 149, 153, 266, 322, 355, 369-370, 373, 402, 407
No Crashworthy End Treatment 106
No-Passing Zone Markings 335-336
Non-Functional Roadside Elements 56
Notice 19, 21, 117, 139, 143, 156, 188-189, 221, 242, 294, 313, 323, 362, 366, 368, 373, 398, 407

O Object Markers 332, 335, 337-339, 359, 372, 394-395
ONE WAY Signs 334
One-Lane Two-Way Traffic Control 308
One-Way Pavement 335
One-way Plow 400
One-Way Traffic Control 300, 302, 319
Ordinances 151, 271, 326
Overloading 328
Overpass 31, 50
Oversteering 166, 221, 408

P Parking Space Markings 338
Passing Sight Distance 336, 374
Passive Rail-Highway Grade Crossings 260
Paved Roadway 13, 169, 383, 386, 405
Paved Shoulders 169, 198, 254, 387, 389
Pavement Cross Slope 186, 191
Pavement Design 190-191
Pavement Drainage 161, 181, 379
Pavement Edge Drop 13, 16, 193-195, 197-198, 200, 203-204, 206-209, 213, 215, 219, 221-222, 314, 321, 331, 365, 402
Pavement Edge Drop Traversal 194
Pavement Edge Lines 337, 340
Pavement Grooving 180, 190
Pavement Irregularities 220, 282, 379, 383
Pavement Markings 13, 219, 227-228, 234-235, 237-239, 249, 253, 264,

Index

294, 300, 317, 319, 321, 335-338, 340, 359, 372, 394-395
Pavement Texture 179-180
Pavement Water Depth 191
Pavement Wheel Ruts 182
Peak Hour Delay 344
Peak Hour Volume Warrant 346
Pedestrian Detours 305, 307
Pedestrian Safety 300, 303
Pedestrian Volume 342, 353
Perception-Reaction Time 6-7, 9-10, 127-128, 157, 205, 246-247, 330
Piers 31, 35, 43, 63, 69, 73-74, 92, 335, 338
Pipe Support 54
Placement of Guardrail 61, 78, 104
Plowing 391-393, 398-401, 406-407
Pocketing 85, 103
Polishing 160-161, 182, 382-383
Poor Flashing Signal Visibility 264
Poor Maintenance of Signs 262
Poor Visibility 132, 234, 362, 399
Positive Guidance 10, 13-14, 25, 225, 289, 297, 299, 303, 306, 317, 335, 360, 374
Post-Mounted Flashing Light Signal 240
Pothole Repair Regimen 381
Potholes 292, 378-379, 402-405
Primacy 6
Primary Traffic Barrier Defects 102
Produce Loss of Control 165
Profile 139, 142, 256
Progressive Movement 343
Proprietary Functions 19
Provide Adequate Sight Distance 260
Provide Relatively Flat Crossing Grades 260
Pullout 257
Pumping 386

R Radius of Curvature 126, 273-274, 279
Rail-Highway Crossing Accidents 266
Rail-Highway Grade Crossing Improvements 248, 260, 270

Rail-Highway Grade Crossing Maneuver 228
Rail-Highway Grade Crossing Sight Triangles 148, 245
Railroad Litigation 271
Railroad-Highway Grade Crossing Handbook 25, 242, 244, 271
Railroad-Highway Grade Crossing Surfaces 257, 271
Railway Engineering 257, 271
RED ARROW Left Turn 348
RED ARROW Right Turn 348
RED BALL 348, 354
RED MARKINGS 335
Reflectorization 322, 326, 329, 335, 373
Regulatory Sign 353, 359
Remove Devices 250
Remove Signs 260
Restricted Stopping Sight Distance 129
Results of Studies of Highway Grooving 190
Resurfacing Roadways 159, 216
Retaining Walls 35, 45, 95
Retrofitting Substandard Bridge Rails 85
Retroreflectivity of Roadway Signs 374
Reverse Curve Sign 283, 287
Reverse Turn Sign 283, 287
Reversible Plow 400
Rigid Barriers 77, 95, 102, 391-392
Rigid Bridge Rail 85, 95, 106, 117
Rigid Guardrail 70, 73
Rigid Traffic Signal Support 55
Road Closed 250, 307, 312, 314, 321, 333
Road Closure 306-307, 310, 312, 321, 328, 333, 359
Road Closure Signs 312, 328, 333
Road Surface Discontinuities 222
Road Width 260, 274
Roadside Barriers 58, 67-68, 82, 101
Roadside Clear Zone Standards 39
Roadside Design Guide 25, 38-39, 65, 69, 78, 122, 223
Roadside Drainage 389

Roadside Hazard 30-31, 34, 36, 43, 56, 59, 68, 79, 390
Roadside Hazard Model 34
Roadside Improvements 31, 123
Roadside Objects 2, 65, 296
Roadside Safety 23, 27, 32, 48, 58-59, 64-65, 390, 402
Roadside Safety Improvement Programs 65
Roadside Slopes 35, 40, 282
Roadway Consistency 10, 13
Roadway Construction 24, 126, 297, 299
Roadway Crossing 251, 261
Roadway Curve Accident Characteristics 280
Roadway Curve Design Standards 278
Roadway Delineation Practice Handbook 374
Roadway Delineators 340
Roadway Design 4, 10, 12, 15, 20-21, 23, 27, 126, 142, 166, 293-294, 362, 390, 511
Roadway Grade 181-182
Roadway Guardrail 23, 70
Roadway Lighting 40, 46, 48, 256, 271, 394-395
Roadway Lighting Handbook 256, 271
Roadway Maintenance 20, 22, 24, 375, 390, 395, 401-402
Roadway Surfaces 170, 181, 189, 376, 389, 400
Roadway Widening Projects 59
Rotary Plows 400
Rough Crossings 226, 234, 266
Rub Rail 90-91
Rules of Thumb 24, 135, 215, 243
Runout Length 79, 81
Rural Uncontrolled Intersection 152
Rutted Pavements 186, 402
Rutting 179, 181-183, 216, 376, 387

S Safe Re-entry Angle 204-206
Safe Side Friction Factors 189, 278
Safety Consideration In Median Design 65
Safety Criteria 295

Safety Effectiveness Evaluation 189
Safety Effectiveness of Highway Design Features 157, 295
Safety Effects of Cross-Section Design 65
Safety Improvements 65, 225, 248, 259-260, 271
Safety of Passively-Controlled Crossings 243
Safety Restoration During Snow Removal 409
Sag Vertical Curve 186
Sand Barrel Crash Cushion Shielding 94
Sand Barrel Crash Cushions 93, 393
Sand-filled Plastic Barrels 97, 100
Scaling 383
School Crossing 343, 353
School Crossing Warrant 353
Scrubbing Condition 201
Scrubbing Re-entry 196-197, 200, 202-204, 206-207, 221
Second International Skid Prevention Conference 161, 189
Semi-Rigid Guardrails 70-71
Sensitivity of Vehicle Dynamics 295
SENTRE 76-77
Severe Sight Restrictions 138, 154, 232
Severity Index 35, 37
Shallow Return Angle 194
Sharp Angle Crossing Approach 262
Sharp Angled Crossing 263
Sharp Crossing Intersection Angle 234
Sharp Curvature With Deficient Warning 288
Sharp Hillcrest 13, 289, 362-363, 403
Sharp Roadway Approach Angles 243
Sharp Roadway Curves 15-16, 133, 138, 140-141, 287, 292, 294, 362, 365, 368, 403, 406-407
Sheet-Flow Conditions 186
Shield Dangerous Body of Water 106
Shield Steep Side Slope 107
Shielding of Signal Faces 352
Shifting Taper 300-302
Shoulder Characteristics 222

Index

Shoulder Drop-offs 222, 409
Shoulder Maintenance 290, 387
Shoulder Slope 274
Shoulder Taper 300-301
Shoulder Width 274, 387
Shoving 379
Shrinkage 381-382
Shy Line Offset 79
Sight Distance Obstructions 15, 24, 149
Sight Distance Profiles 140
Sight Distance Requirements 127, 132, 144-147, 243, 271, 283
Sight Distance Standards 127, 129, 131, 146, 148, 151, 156, 289
Sight Restrictions 133, 138, 142, 153-154, 232, 251-252, 259, 261, 269
Sight Triangle Obstructions 261
Sight-Restricted Crossing 232
Sight-Restricted Roadway Curve 290
Sign 13, 16, 27, 31, 48-54, 61-62, 92, 95, 138-139, 147-148, 211, 237-240, 252-253, 255, 262, 283-284, 286-287, 289, 294, 297, 312-314, 317, 321, 325, 328-334, 338, 344, 350, 353-354, 357-360, 362-370, 391-392, 394
Sign Supports 48, 50, 53, 61, 92, 391-392
SIGNAL AHEAD Sign 331, 350
Signal Faces 332, 346-347, 350-352, 372
Signal Needs 352
Signal Out of Operation 352
Signal Visibility Defects 372
Signals 7, 16, 22-23, 53, 66, 154, 167, 225, 230-231, 235, 238-239, 241-243, 249, 252-255, 259-262, 265, 269, 302, 306-307, 319, 325, 329-330, 337, 341, 343-346, 348, 350, 352, 354, 358, 362, 372, 394-395
Signing 16, 22, 141, 215, 218, 258, 283, 285, 291-292, 303, 312, 314, 319, 321, 327, 359
Signing No-Passing 22
Signposts 30, 35
Simple Reaction Time 9-10

Single-Vehicle Collision Characteristics 30
Size of Signal Lenses 349
Skid Numbers 173-174, 176-177, 185, 275
Skid Resistance Measurements 168, 170, 189
Skid Resistance Requirements 173, 175, 189
Skid Resistant Pavement Design 191
Skid Testing 189
Skidding Characteristics of Automobile Tires 189
Skidding Problems 170, 189
Sleet 398-399
Slip-base 50, 77
Slippery Pavements 159, 185, 292, 382, 396, 402
Slotted Fuse Plate 50, 52
Slow-moving Vehicles 279
Small Signs 54, 63
Snag Elements 102
Snow 71, 112, 184, 279, 322, 340, 375, 391-394, 396, 398-402, 406-407, 409
Snow Plow Cloud 407
Snow Plowing 398-399, 406-407
Sod Shoulders 388
Soft Shoulders 406
Soft Spots 376, 378
SOLID LINES 335
Sovereign Immunity 17, 19
Spalling 385, 392
Speared 109
Speed Limit 6, 135-136, 156, 187-188, 258-259, 286, 288-289, 300, 303, 315, 322, 359, 370, 408
Stabilization 376-377
Standard of Care 18
Standard Specifications 21, 66, 122, 322, 378
State Highway Safety Programs 411
Statutes 17-18, 20, 289
Steep Approach Grade 236, 261

Steep Crossing Grade 265
Steep Side Slope 60, 79, 106-107
Steering Wheel Angle 201
STOP AHEAD Sign 138, 362
Stop Lines 337-338
STOP Signs 13, 16, 138, 140, 146-147, 150-151, 154, 167, 252-253, 259-261, 328-330, 337, 357, 360, 362, 364, 398
Stop-Controlled Intersections 16, 147, 153-154, 407
Stopped Roadway Vehicle 247
Stopping Sight Distance 25, 125, 127-130, 132, 134-137, 139, 141, 143-146, 168, 189, 230, 244, 246, 254, 274, 283, 289-291, 389, 406
Stopping Sight Distance Around Curve 274
Stopping Sight Distance Design Standards 127
Storm Watch 398
Straight-through GREEN ARROW 348
Straight-through RED ARROWS 348
Structural Supports 66
Studies of Braking Systems 222
Subdrains 378, 381, 386
Substandard 16, 85, 135, 156, 282, 288-290, 294, 320, 408
Superelevation Rate 167, 273-274, 279, 292
Superelevation Runoff Length 274
Supplemental Alignment Signs 287
Supplementary Curves Signs 287
Supreme Court 19
Surface Texture 5, 160, 191, 383
Surface Texture Distress 383
Surface Treated Shoulders 389
Symbol Markings 335, 337-338
Systems Warrant 344

T T-intersection 13, 250
Tall Wall Median Barrier 93
Tangent Roadway With False Visual Cues 291
Taper Length 300, 337

Tapered-Down Ends 93
Tapers 16, 300-301
Target Value 6, 16, 149, 315, 328, 352, 360
Technical Aspects of Intersection Sight Distance Defect Cases 149
Technical Aspects of Maintenance Defect Cases 402
Technical Aspects of Pavement Edge Drop Cases 219
Technical Aspects of Rail-Highway Grade Crossing Defect Cases 259
Technical Aspects of Roadside Hazard Cases 59
Technical Aspects of Roadway Curve Defect Cases 288
Technical Aspects of Slippery Pavement 184
Technical Aspects of Stopping Sight Distance Defects Cases 135
Technical Aspects of Traffic Barrier Defect Cases 101
Technical Aspects of Traffic Control Device Defect Cases 362
Temporary Traffic Control Zone 298, 322
Temporary Traffic Signals 302, 319
Tentative Service Requirements 122
Tentative Skid-Resistance Requirements 173
Terrain Effects 79, 118
Testimony 30, 185, 187, 266, 320, 511
Texturing 161, 190
Thrie-beam 70, 73-74, 77, 85-86, 90-91, 97, 392
Tire 130, 159-164, 170, 172, 178-181, 183-185, 188-191, 196-198, 200, 202-203, 209, 211, 274, 404-405
Tire Hydroplaning 190-191
Tire Marks Left 196-197
Tire Pavement Skid Resistance 274
Tire Traction 190
Tire Tread Rubber Reversion Hydroplaning 179
Tire Tread Wear 178

Index

Tire-Pavement Interface 159, 162, 164, 275, 278
Tolerability of Pavement Edge Drops 207
Tort Liability Overview 1
Tower Lighting 48
Traction Studies 190
Tractor Semi-Trailer Accidents 191
Traffic Accident Investigation 409
Traffic Barrier Maintenance 390
Traffic Barriers 23, 25, 34, 38, 40-41, 43, 45, 53, 56, 58, 63, 65, 67-69, 79, 82, 87, 101, 121-122, 321, 390, 402
Traffic Control Device Maintenance 394
Traffic Control Device Requirements 326
Traffic Control Devices Handbook2 325
Traffic Control Plan 297, 299, 320
Traffic Engineering 14, 23, 25, 295, 327, 374, 394, 511
Traffic Engineering Handbook 23, 25, 374
Traffic Markings 334-335, 362, 395
Traffic Safety 2, 144, 149, 157, 172, 214, 222, 295, 298, 320, 323, 325, 409
Traffic Sign Obscured 370
Traffic Signal Supports 53
Traffic Signal Warrants 341
Traffic Signals 16, 53, 66, 154, 239, 243, 249, 255, 262, 302, 307, 319, 329, 337, 341, 343, 345-346, 352, 362, 372, 394-395
Traffic Signals Near Grade Crossings 243
Traffic Signs 23, 327-328, 337, 362-364, 366, 368-369, 394
Trailers 171, 189, 219, 257
Train Accident Reconstruction 266, 271
Train Braking Performance 266
Train Detection 242, 255
Train Speed 231, 246-247, 259, 266-267, 269, 272
Training 2, 125, 306, 320, 390, 398, 511
Transitions 85, 95, 118, 273, 297, 391-393
Transportation Research Board 21, 25, 65-66, 68, 122-123, 144, 189, 205, 208, 222-223, 295, 323, 374, 409
Transportation Research Record 25, 143-144, 157, 271, 295, 323, 374

Transverse Pavement Markings 337
Traversals of Vertical Face Edge Drops 206
Tread 160-161, 172, 178-181, 183-184
Tree Hazard 61
Trees 30-31, 35, 40, 56, 61, 133, 151, 153, 246, 283, 362, 389-390
TREND 2, 27, 77, 159
Truck Driver 129, 133, 234-235
Truck Operations 129, 133, 179, 222
Truck Tire Hydroplaning 191
Trucks 4, 69, 73, 82, 92, 95, 100, 129, 133, 157, 159, 164, 179, 183, 185-186, 208, 216, 219, 226, 234-235, 242-243, 251-252, 254, 261-262, 271, 283, 290, 333, 339, 374, 378, 382, 386, 400-401, 405
Tubular Markers 315
Turn Prohibition Signs 239, 255, 262
Turn Signs 138, 283-284, 286, 289, 366
Two-way Left-Turn Lanes 337
Two-way Traffic Taper 300
Type of Shoulder Material 274
Type of Taper 300
Typical Applications 306, 311, 353

U U. S. Code 236, 325, 411
U. S. Congress 159, 205, 208, 225
U. S. Department of Transportation 21, 225, 259, 271
U-posts 53
Unanchored Approach Guardrail 106, 109, 113
Unanchored Guardrail Approach 117
Unclear Messages 366
Uncontrolled Intersections 16, 145-146, 150-151, 360
Uncrashworthy Bridge Rails 114
Underbody Plows 400
Undercarriage Drag 207, 213
Uneven Pavements 292
Unexpected Information 8
Unfamiliar Driver 138, 357, 361
Uniform Meanings of Longitudinal Pavement Markings 335

Uniform Procedures 411
Uniform Traffic Laws 271
Uniform Vehicle Code 227, 271
Uniformity 326-327, 360
Unpaved Roadways 376-378, 403
Unpaved Surfaces Maintenance 375
Unprotected Bridge Rail End 117
Unprotected Functional Objects 63
Unrepaired Damage 102, 118
Unshielded Large Sign Post 62
Unshielded Median Bridge Support 63
Unstabilized Shoulders 193-194
Untreated Guardrail End 106, 109
Upheaval 379
Utility Poles 30-31, 35, 40, 56, 61, 95, 254

V Vegetation Control 149, 390, 409
Vehicle Cornering 273
Vehicle Crash Testing of Highway Appurtenances 122
Vehicle Curve Transition Behavior 282
Vehicle Handling 222, 409
Vehicle Loss of Control 409
Vehicle Performance 128, 190
Vehicle Stability 222
Vehicle Tires 183, 194
Vehicle-Train Conflict Circumstances 230
Vertical Face Pavement Edge Drop 204
Vertical Panels 211, 315, 317
Vertical Signal Face 348
Viscous Hydroplaning 179
Visibility of Signal Faces 350-351

W W-Beam Guardrail 71-72, 77, 79, 102, 110, 409
Warning Lights 317
Warning Signs 13, 23, 138, 142, 179, 216, 227-228, 234, 238, 253, 255, 260-262, 264, 270, 283-284, 287, 289, 292, 294, 303, 306, 313-314, 317, 320-321, 328-333, 338, 357, 359-360, 363-366, 368-370, 406

Warrants 68-70, 87-88, 122, 253, 261, 328-330, 341, 343-344
Washboard 282, 292, 376-377, 402-403
Weak Bridge Rails 114
Weak-post W-beam Guardrail 71-72
Weak-post W-beam Median Barrier 89
Wet Pavements 160, 164-166, 172, 178-180, 182-183, 185, 274-275, 292
Wet Road Friction 190
Wet Weather Speed Limits 189
Wet-Pavement Accident Rates 173, 175
Wet-Pavement Safety Programs 189
Wheel Ruts 182, 186, 378, 382, 386, 403, 407
WHITE LINES 335-338
Wide Median 102, 333-334
Widening 59-60, 127, 385, 387
Winding Road Sign 283, 287
Windrows of Gravel 290, 402-403
Winter Maintenance Policies 396
Work Zones 23, 313
Worker Safety 300, 306
Worn Tires 160
Wrong Way Traffic Control 328, 333

Y Y Intersection 331
YELLOW ARROW 348, 354
YELLOW ARROW Left Turn 348
YELLOW ARROW Right Turn 348
YELLOW BALL 348-349, 353-354
Yellow Book 27, 30, 65
YELLOW LINES 335-337
Yellow Phase-Change Intervals 352
Yield Signs 145-146, 151, 156, 239, 252, 260, 302, 328, 330, 365

Other Quality Products Available From:

Lawyers & Judges Publishing Co.

Train Accident Reconstruction and FELA & Railroad Litigation, Second Edition #0969

by James R. Loumiet, BSME and William G. Jungbauer, Esq.

Train collisions usually result in serious injury or death to motorists and pedestrians, often resulting in litigation. Yet the railroad industry presents complexities not known to the average person, making it difficult for an attorney to prepare a successful case. Even experienced accident reconstructionists are usually unfamiliar with the operation of railroads and trains. Yet the ability to reconstruct train collisions is often vital to the successful conduct of railroad litigation.

This ground breaking reference text by Loumiet and Jungbauer is the first of its kind in the are of train accident reconstruction, FELA and Railroad litigation. It explains how the railroads operate, the design and operation of trains, how to calculate train speed and successfully conduct FELA and railroad litigation. This text is for both the novice and experienced litigator and reconstructionist. It contains information and formulas that can be found nowhere else. $6^{1}/_{8}$" x $9^{1}/_{4}$", 686 pages.

Aircraft Accident Reconstruction and Litigation #5155

by Dr. Barnes W. McCormick, Ph.D., P.E. and Myron P. Papadakis, Esq. (1996)

This book teaches you how to investigate accidents and discover evidence. Your ability to reconstruct aircraft accidents can make the difference between successful and failed aviation litigation. Yet even experienced accident reconstructionists are often unfamiliar with the myriad complexities of aircraft. Now, whether you're a reconstruction expert or an experienced litigator, you can access a wealth of information in this ground-breaking text. In fact, much of the data, formulas, illustrations and investigative tools available to translate wreckage reconstruction data into credible and probative evidence suitable for introduction in the courtroom are available here exclusively, saving you hours of tedious calculations and research. It's the most comprehensive listing of aviation state law and cases ever published. Written in a hands-on format to give you the vital tools needed to investigate, reconstruct, and litigate in the technical and sophisticated arena of aircraft accident reconstruction. 6" x 9", hardbound, 720 pages, also includes 84 page Addendum List of Experts.

Bicycle Accident Reconstruction and Litigation #0918

by James M. Green, Paul F. Hill and Douglas Hayduk (1996)

Your ability to reconstruct bicycle accidents can make the difference between successful and failed litigation. Yet even experienced accident reconstructionists are often unfamiliar with the myriad complexities of bicycles. Now, whether you're a novice or an experienced litigator, you can access a wealth of information in this ground-breaking text. In fact, much of the case research data and formulas are available here exclusively, saving you hours of tedious calculations and research. This book is divided into three sections: engineering analysis; bicycle metallurgy; and legal analysis. You'll have informative diagrams and drawings, plus a complete index, appendix and citations. 6" x 9", hardbound, 784 pages.

Motorcycle Accident Reconstruction: Understanding Motorcycles #5031
by Kenneth Obenski, P.E. (1994)

Riding a motorcycle has been estimated as five times more complex as driving an automobile...and reconstructing an accident can be equally difficult. This text, written by a professional engineer who has investigated over 170 motorcycle accidents—and has safely ridden motorcycles for more than 35 years—gives you the detailed information you need to reconstruct and litigate accidents. In clear, jargon-free language the text sets out the basics of motorcycle operation, and the unique interactions between riders, passengers, machines, and other traffic. You'll understand why motorcycles frequently leave less evidence at the scene of an accident, and how a trained reconstructionist can observe and translate even scant evidence.

Of special significance is the section that explains what motorcycle operators can and cannot do, as well as what they should and should not do. Also included is training information published by the State of California, which is a basic source of motorcycle safe operating instruction. Your single source for motorcycle safety standards because its supplemental appendix contains the Department of Transportation (D.O.T.) Motorcycle Safety Standards as well as the preambles and superseded standards. With this detailed history you'll understand D.O.T.'s philosophy in arriving at each standard. $6\frac{1}{8}$" x $9\frac{1}{4}$", hardbound, 296 pages.

Snowmobile Accident Reconstruction: A Technical and Legal Guide #5023
by Richard Hermance, ACTAR (1995)

The increased popularity of snowmobiling over the past few years has resulted in a rise in snowmobile-related accidents and injuries, and reconstruction of these types of accidents poses a unique challenge for you. This ground-breaking reference and law text is the first of its kind in the area of snowmobile accident reconstruction and litigation. You'll find detailed reference material covering the technical aspects of snowmobile safety, trail design and administration, maintenance, mapping, accident investigation and reconstruction, as well as snowmobile law.

The author is a nationally recognized and accredited accident reconstruction expert who has been involved in over 50 snowmobile accident investigations and reconstructions. An avid snowmobiler for the past 25 years, Hermance has ridden extensively in the northeastern and northwestern United States and Canada. He also designed and taught the New York State Snowmobile Accident Reconstruction course given to New York's State Troopers, Forest Park Police, Sheriff's Department, and local police. 6" x 9", hardbound, 700 pages.

Pedestrian Accident Reconstruction with Videotape #0977
by Jerry Eubanks (1994)

Pedestrian accidents involving a motor vehicle vary in both nature and scope from accidents involving only passenger cars. Intensified collision damage and injuries frequently result, often due to the inherent disadvantages a pedestrian has in an accident. Reconstruction of these types of collisions pose a unique challenge for accident investigators, attorneys, and insurance professionals. The new Pedestrian Accident Reconstruction book and video set discusses, demonstrates, and teaches you the technical fundamentals and mechanics of collisions between motor vehicles and pedestrians; providing a single source of state-of-the-art research dealing with these types of collisions. This valuable set contains chapters on: history and statistics of motor vehicle/pedestrian collisions; pedestrian walking speeds and relative motion; vehicle speed and pedestrian collision formulas; human factors and conspicuity; photography of a motor vehicle/pedestrian collision; case discovery and preparation; and a workbook section covering the videotape of pedestrian accident investigation and reconstruction. $6\frac{1}{8}$" x $9\frac{1}{4}$", 288 pages, hardbound, includes 60-minute VHS video cassette tape.

Boat Accident Reconstruction and Litigation #5082
by Roy Scott Hickman, P.E. (1996)

The clearest guide to understanding accidents on and in the water. Your capability to reconstruct boating accidents can make the difference between successful and failed litigation. Yet even experienced engineers and attorneys are often unaccustomed with the myriad complexities of boating accidents which involve concepts and methods not present in other types of accidents. Now, whether you're a beginner or experienced litigator, you can access a wealth of information in this 496 page ground-breaking text. In fact, much of the information at hand here will save you hours of research time hunting through numerous books and on-line services.

You'll start with the basics of how boats function and progress to the way boats are designed which is a key element in the reconstruction of most boat accidents. You'll be introduced to Fluid mechanics and the numerous formulae and methods normally used to analyze accidents. Of particular value is the extensive series of appendices which includes numerous Coast Guard Regulations and Rules controlling boating traffic and safety equipment. You can quickly access the most complete description of major boating accident case citations ever published. You'll be presented with chapters on safety equipment, transportation issues, boat descriptions and designs, and marina issues. All types of boats are covered including Personal Water Craft, Water-skiing and the sport of SCUBA diving. You'll also learn about the special problems of Carbon Monoxide Poisoning with boats. 6" x 9", hardbound, 496 pages.

These and many more useful products are available through our catalog.

For a FREE catalog, write or call:

Lawyers & Judges Publishing Company, Inc.

P.O. Box 30040 • Tucson, Arizona 85751-0040

(602) 323-1500 • (520) 323-1500 • FAX (800) 330-8795

e-mail address: lj_publ@azstarnet.com

Find us on the Internet at:

http://www.azstarnet.com/~lj_publ